Biotechnology of Microbial Enzymes

Biotechnology of Microbial Enzymes

Edited by Hazel Nygel

SYRAWOOD
PUBLISHING HOUSE

New York

Published by Syrawood Publishing House,
750 Third Avenue, 9th Floor,
New York, NY 10017, USA
www.syrawoodpublishinghouse.com

Biotechnology of Microbial Enzymes
Edited by Hazel Nygel

International Standard Book Number: 978-1-68286-645-0 (Hardback)

Cataloging-in-Publication Data

Biotechnology of microbial enzymes / edited by Hazel Nygel.
    p. cm.
Includes bibliographical references and index.
ISBN 978-1-68286-645-0
1. Biotechnology. 2. Microbial enzymes. I. Nygel, Hazel.
TP248.2 .B56 2019
660.6--dc23

TABLE OF CONTENTS

PREFACE

Enzymes are biological catalysts, which help in accelerating chemical reactions in the cells and thereby helps in sustaining life. Microorganisms are a primary source for deriving enzymes. Bacteria, fungi and yeast are some of the commonly used microorganisms used to extract enzymes for diverse industries such as pharmaceuticals, food, textiles, animal feed, etc. Microbes can be genetically modified which makes them a better source of enzymes. Such selected concepts that redefine this field of study have been presented in this book. The various advancements in microbial enzymes are glanced at and their applications as well as ramifications are looked at in detail. For all readers who are interested in this field, the case studies included in this book will serve as an excellent guide to develop a comprehensive understanding.

Significant researches are present in this book. Intensive efforts have been employed by authors to make this book an outstanding discourse. This book contains the enlightening chapters which have been written on the basis of significant researches done by the experts.

Finally, I would also like to thank all the members involved in this book for being a team and meeting all the deadlines for the submission of their respective works. I would also like to thank my friends and family for being supportive in my efforts.

Editor

Enzymatic process optimization for the in vitro production of isoprene from mevalonate

Tao Cheng[1,3†], Hui Liu[1†], Huibin Zou[1,2*], Ningning Chen[2], Mengxun Shi[2], Congxia Xie[3], Guang Zhao[1*] and Mo Xian[1*]

Abstract

Background: As an important bulk chemical for synthetic rubber, isoprene can be biosynthesized by robust microbes. But rational engineering and optimization are often demanded to make the in vivo process feasible due to the complexities of cellular metabolism. Alternative synthetic biochemistry strategies are in fast development to produce isoprene or isoprenoids in vitro.

Results: This study set up an in vitro enzyme synthetic chemistry process using 5 enzymes in the lower mevalonate pathway to produce isoprene from mevalonate. We found the level and ratio of individual enzymes would significantly affect the efficiency of the whole system. The optimized process using 10 balanced enzyme unites (5.0 μM of MVK, PMK, MVD; 10.0 μM of IDI, 80.0 μM of ISPS) could produce 6323.5 μmol/L/h (430 mg/L/h) isoprene in a 2 ml in vitro system. In a scale up process (50 ml) only using 1 balanced enzyme unit (0.5 μM of MVK, PMK, MVD; 1.0 μM of IDI, 8.0 μM of ISPS), the system could produce 302 mg/L isoprene in 40 h, which showed higher production rate and longer reaction phase with comparison of the in vivo control.

Conclusions: By optimizing the enzyme levels of lower MVA pathway, synthetic biochemistry methods could be set up for the enzymatic production of isoprene or isoprenoids from mevalonate.

Keywords: Mevalonate, Isoprene, Synthetic biochemistry, Isoprenoids, Bio-based chemicals, Enzymatic process

Background

Commodity chemicals are traditionally produced from petroleum via the energy-intensive chemical methods. Alternative process that does not rely on the petroleum resources and chemo-process is under fast development for chemical production, using the advanced tools of synthetic biology and metabolic engineering [1, 2]. Recently, isoprene and a variety of isoprenoids can be produced from renewable feedstock through engineered microbes [3–7]. The general strategy is engineering the whole isoprenoid biosynthesis pathways in the chassis strains [6], and the mevalonate (MVA) pathway other than the methylerythritol 4-phosphate (MEP) pathway is usually selected to construct robust strains with higher production titers [8].

However, the biosynthesis of isoprene and isoprenoids by the cell-based strategy faced one major bottleneck: the nutrient limitation and the accumulation of toxic intermediates or products may lead to cell growth inhibition and limited yields and titers; thus rational engineering and optimization are often demanded to make the process economically feasible [9–12]. Corresponding synthetic biochemistry tool has also been demonstrated to overcome the major bottleneck of the cell-based strategy, especially in rational optimization of synthetic multi-enzyme pathways [13]. The emergence of cell-free system has several advantages over the cell-base system: (1) process can be operated continuously; (2) can synthesize hazard products which are toxic towards living cell system; (3) enzyme can be quantitatively and biochemically adjusted to optimize flux through the metabolic pathways; (4) reduce the operation (like gas-stripping) and

*Correspondence: huibinzou@hotmail.com; zhaoguang@qibebt.ac.cn; xianmo@qibebt.ac.cn

†Tao Cheng and Hui Liu contributed equally to the work

[1] CAS Key Laboratory of Bio-based Materials, Qingdao Institute of Bioenergy and Bioprocess Technology, Chinese Academy of Sciences, No. 189 Songling Road, Laoshan District, Qingdao 266101, China

Full list of author information is available at the end of the article

purification requirements; (5) higher yields without competing cellular pathways. In case of isoprene biosynthesis, a recent cell-free approach for the conversion of the glycolysis intermediate phosphoenolpyruvate into isoprene at nearly 100% molar yield was promoted [14]. The in vitro system involves enzymes through the full MVA pathway. While challenges also exist for the synthetic biochemistry platform, as the cofactors of ATP, NADPH, and acetyl-CoA need to be balanced in the complex MVA pathway: NADPH and acetyl-CoA are involved in the upper MVA pathway (from pyruvate to MVA), ATP is involved in the lower MVA pathway (from MVA to isoprene). Thus additional enzymes are needed to balance the supply/consumption of these co-factors which increased the total enzymes of 12 in the complex in vitro system for bioisoprene [13, 14].

This study focused on implementing an enzymatic process only consisting of lower MVA pathway for the biosynthesis of isoprene directly from MVA. In the demonstrated system, the cofactors of NADPH and acetyl-CoA are not involved, which decreases the needs of additional enzymes to balance these factors. Moreover, the enzyme activities and quantities of the simplified system (only involves 5 enzymes) can be easily and precisely adjusted to optimize flux through the biochemical steps, which would improve the conversion efficiency comparing to the current in vitro platform for bioisoprene.

Results and discussion
Purification of the enzymes in lower MVA pathway

In this study, four biosynthetic enzymes of lower MVA pathway and isoprene synthase were individually prepared by recombinant E. coli strains (Table 1) and purified before in vitro production. Through these enzymes mevalonate can be converted to dimethylallyl pyrophosphate (DMAPP) by primary mevalonate kinase (MVK, EC 2.7.1.36) and secondary phosphorylation (phosphomevalonate kinase, PMK, EC 2.7.4.2), decarboxylation (diphosphomevalonaet decarboxylase, MVD, EC 4.1.1.33) and isomerization (isopentenyl diphosphate isomerase, IDI, EC 5.3.3.2), then the isoprene synthase (ISPS, EC 4.2.3.27) catalyzes the formation of isoprene from DMAPP. For each molecular of isoprene from MVA, 3 ATP is needed and NADPH is not demanded for the in vitro bioconversion, which is different with the in vivo system from pyruvate [14, 15]. In order to prevent the formation of inclusion body and ensure correct enzyme folding, each enzyme was purified and checked

Table 1 Bacterial strains and plasmids used in this study

Strain/plasmid/primer	Relevant genotype/property/sequence	Source/reference
Strains		
E. coli BL21(DE3)	F$^-$ompThsdS$_B$(r$_B^-$ m$_B^-$) gal dcm rne131 (DE3)	Invitrogen
MVK producer	BL21(DE3)/pET-ERG12	This study
MVD producer	BL21(DE3)/pET-ERG8	This study
PMK producer	BL21(DE3)/pET-ERG19	This study
IDI producer	BL21(DE3)/pET-IDI	This study
ISPS producer	BL21(DE3)/pET-ISPS	This study
Plasmids		
pYJM14	pTrcHis2B derivative carryinggenes ERG8, ERG12, ERG19 and IDI, Trc promoter, Ap^R	[16]
pET-ERG12	pET30a(+) derivative carryinggenes gene ERG12, T7 promoter, Kan^R	This study
pET-ERG8	pET30a(+) derivative carryinggenes gene ERG8, T7 promoter, Kan^R	This study
pET-ERG19	pET30a(+) derivative carryinggenes geneERG19, T7 promoter, Kan^R	This study
pET-IDI	pET30a(+) derivative carryinggenes gene IDI, T7 promoter, Kan^R	This study
pET-ISPS	pET30a(+) derivative carryinggenes gene ISPS, T7 promoter, Kan^R	This study
pACY-ISPS	pACYDuet-1 derivative carryinggenes gene ISPS, T7 promoter, Cm^R	This study
Primers		
ERG12_F	5'-CCCAAGCTTGGTCATTACCGTTCTTAACTTC-3'	
ERG12_R	5'-CCGCTCGAGTTATGAAGTCCATGGTAAAT-3'	
ERG8_F	5'-CCGGAATTCTCAGAGTTGAGAGCCTTCAG-3'	
ERG8_R	5'-CCGCTCGAGTTATTTATCAAGATAAGTTT-3'	
ERG19_F	5'-CGCGGATCCACCGTTTACACAGCATCCGT-3'	
ERG19_R	5'-CCGCTCGAGTTATTCCTTTGGTAGACCAG-3'	
IDI_F	5'-CGCGGATCCACTGCCGACAACAATAGTAT-3'	
IDI_R	5'- CCGCTCGAGTTATAGCATTCTATGAATTT-3'	

their molecular size by SDS-page before enzymatic assay experiments. We found that keeping the cultures at 20 °C after IPTG induction in the fermentation process helped to improve the yield of soluble recombinant enzymes. The purity and molecular size of the purified enzymes can be seen in Fig. 1.

In vitro production of isoprene from mevalonate

With the purified enzymes, we further tested whether the lower mevalonate pathway can be reconstituted in vitro by mixing of substrate and ATP with the purified enzymes. The addition of ATP is compulsory as three ATP are used in the phosphorylation and decarboxylation of mevalonate to IPP (Fig. 1). In the 2 ml in vitro system with purified enzymes (0.5 μM each), substrate (2.5 mM) and ATP (12 mM), we took samples and analyzed total isoprene at different time points and calculate the rate of reaction (total isoprene/L/h) at the different time points. The results showed that the maximum isoprene production (76.5 μmol/L/h, 5.2 mg/L/h) occurred at 4 h after the addition of ATP and mevalonate (Fig. 2a),

similar with the time course curve of in vitro isoprene production from PEP in the previous study [14].

We next tested the optimum mevalonate/ATP ratio in the 2 ml in vitro system. The optimum concentration of mevalonate (substrate) was analyzed by starting with a fixed amount of ATP (12 mM) and sequentially increasing amount of mevalonate (Fig. 2b). The results showed that the maximum isoprene production was achieved when 2.5 mM mevalonate was added in the system, and the isoprene production was decreased when the initial concentration of mevalonate was over 2.5 mM. We also tested the optimum concentration of ATP with fixed amount of mevalonate (2.5 mM). The results showed that supplementation of 10 mM ATP resulted in the maximum isoprene production (Fig. 2c). The production of isoprene was increasing with the increasing of ATP from 2 to 12 mM, it is suggested that the lower pathway of MVA was inhibited when the concentration of substrate of mevalonate was higher than 2.5 mM. The optimum ratio for substrate (mevalonate) and cofactors (ATP) is around 1:4, which is excess to the theoretical

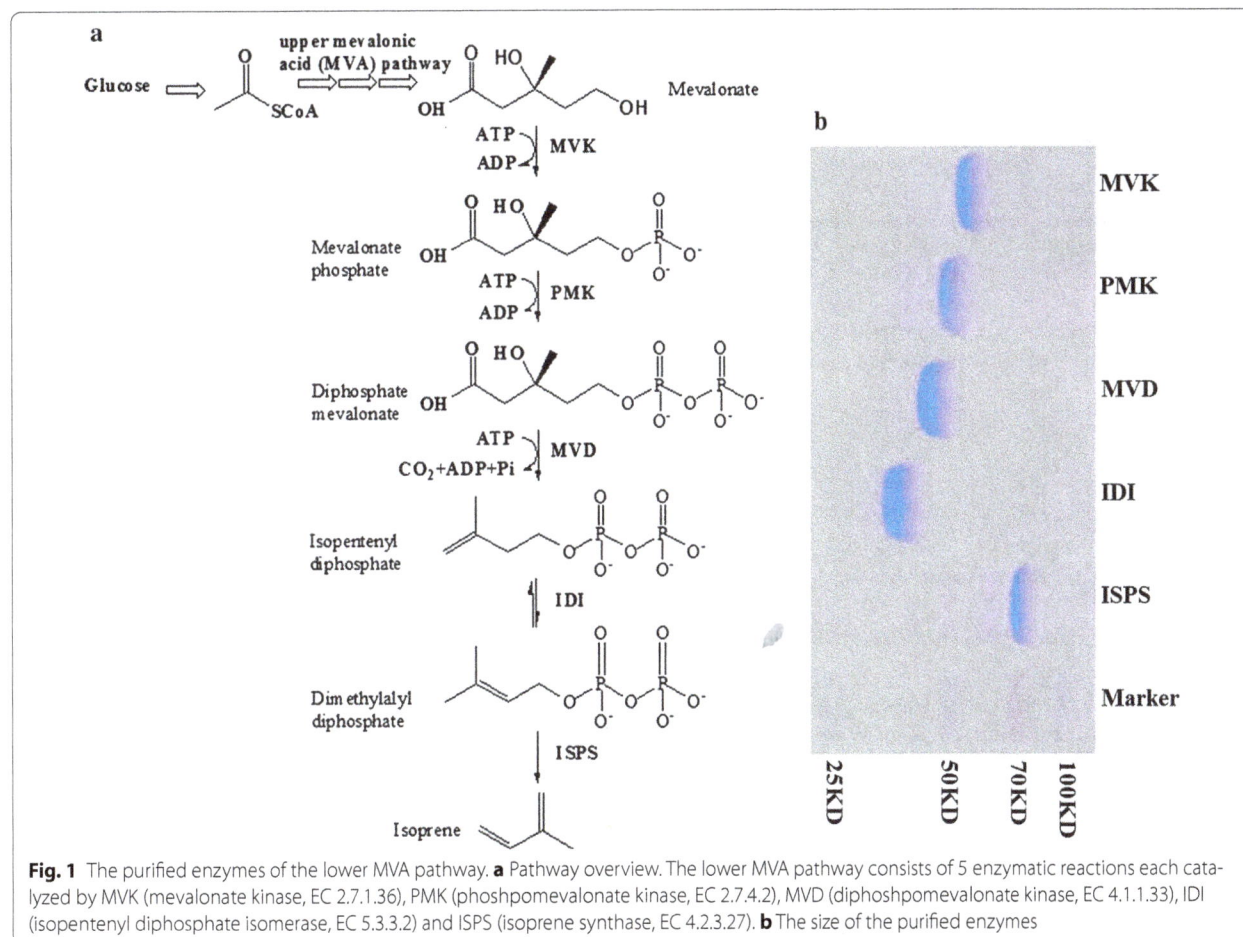

Fig. 1 The purified enzymes of the lower MVA pathway. **a** Pathway overview. The lower MVA pathway consists of 5 enzymatic reactions each catalyzed by MVK (mevalonate kinase, EC 2.7.1.36), PMK (phoshpomevalonate kinase, EC 2.7.4.2), MVD (diphoshpomevalonate kinase, EC 4.1.1.33), IDI (isopentenyl diphosphate isomerase, EC 5.3.3.2) and ISPS (isoprene synthase, EC 4.2.3.27). **b** The size of the purified enzymes

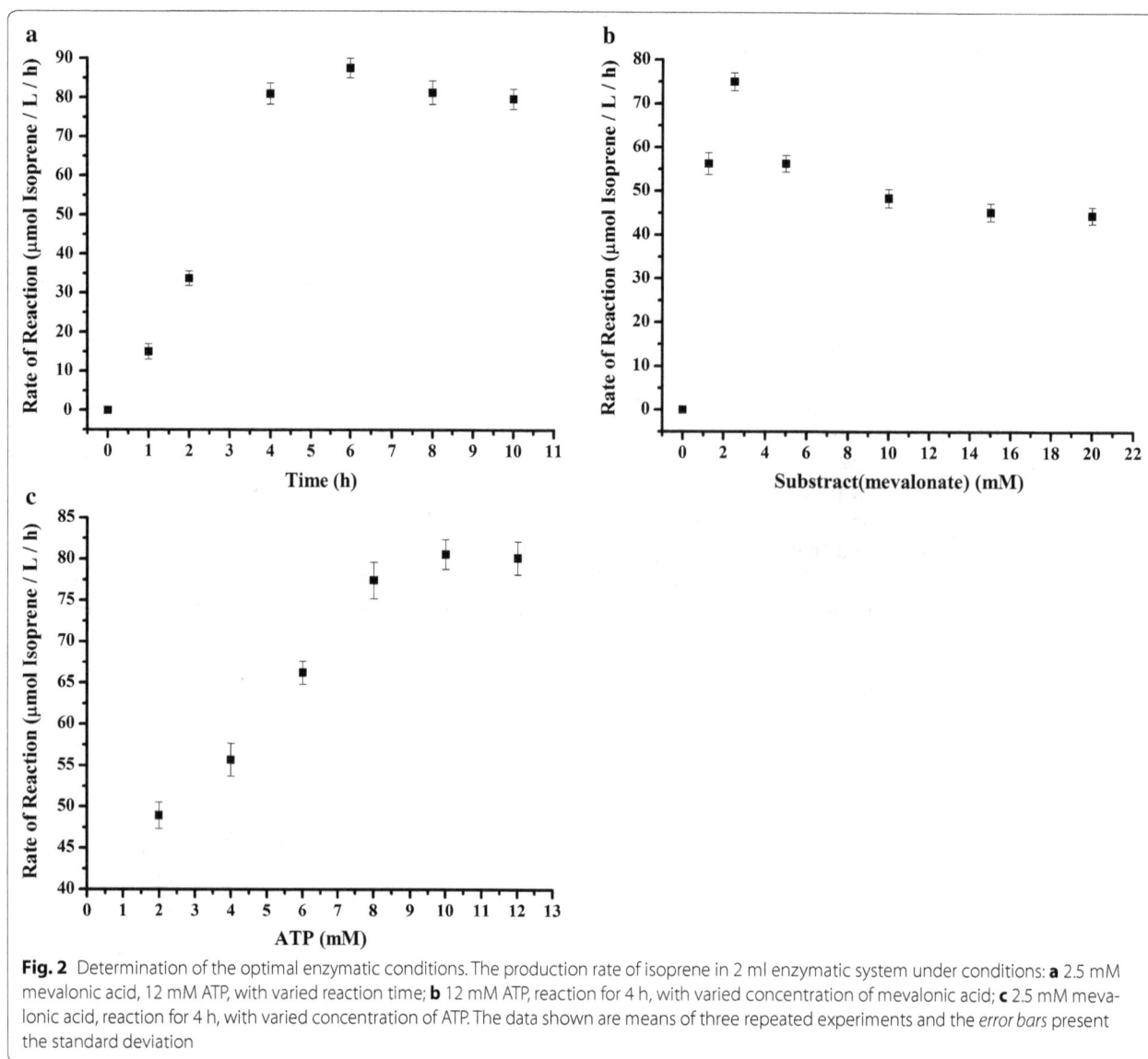

Fig. 2 Determination of the optimal enzymatic conditions. The production rate of isoprene in 2 ml enzymatic system under conditions: **a** 2.5 mM mevalonic acid, 12 mM ATP, with varied reaction time; **b** 12 mM ATP, reaction for 4 h, with varied concentration of mevalonic acid; **c** 2.5 mM mevalonic acid, reaction for 4 h, with varied concentration of ATP. The data shown are means of three repeated experiments and the *error bars* present the standard deviation

ATP consumption (1:3) in the lower mevalonate pathway (Fig. 1). The excess addition of ATP is similar with previous study when acetyl-CoA was utilized as substrate for the in vitro production of isoprene [14].

Quantitatively balancing enzyme levels to maximum the isoprene production

In the in vitro isoprene biosynthesis system, the enzyme quantity can be precisely adjusted to optimize the flux through the biochemical steps comparing with the isoprenoids biosynthesis using living cells, in which promoter induction is often highly cooperative and fine control is difficult [11]. To quantitatively construct balanced in vitro system, we firstly screened the bottleneck enzymes which significantly affect the isoprene production. The effects of enzyme levels towards isoprene production were tested by varying the levels between 0.02 and 5 μM with the constant levels of other four enzymes at 0.5 μM (Fig. 3a) in the 2 ml in vitro system. The results showed that the enzymes had different effects towards the isoprene production. PMK and MVD were not belonging to the bottleneck enzymes in the pathway, as their levels did not significantly influence the isoprene production. For the enzymes of MVK and IDI, their increasing levels generated a 30–90% increase in isoprene production: MVK reached the maximum isoprene production at the level of 0.5 μM while IDI reached the maximum isoprene production at 1.0 μM (Fig. 3a). The last enzyme of ISPS, which catalyzes the production of isoprene from DMAPP, belonged to the bottleneck

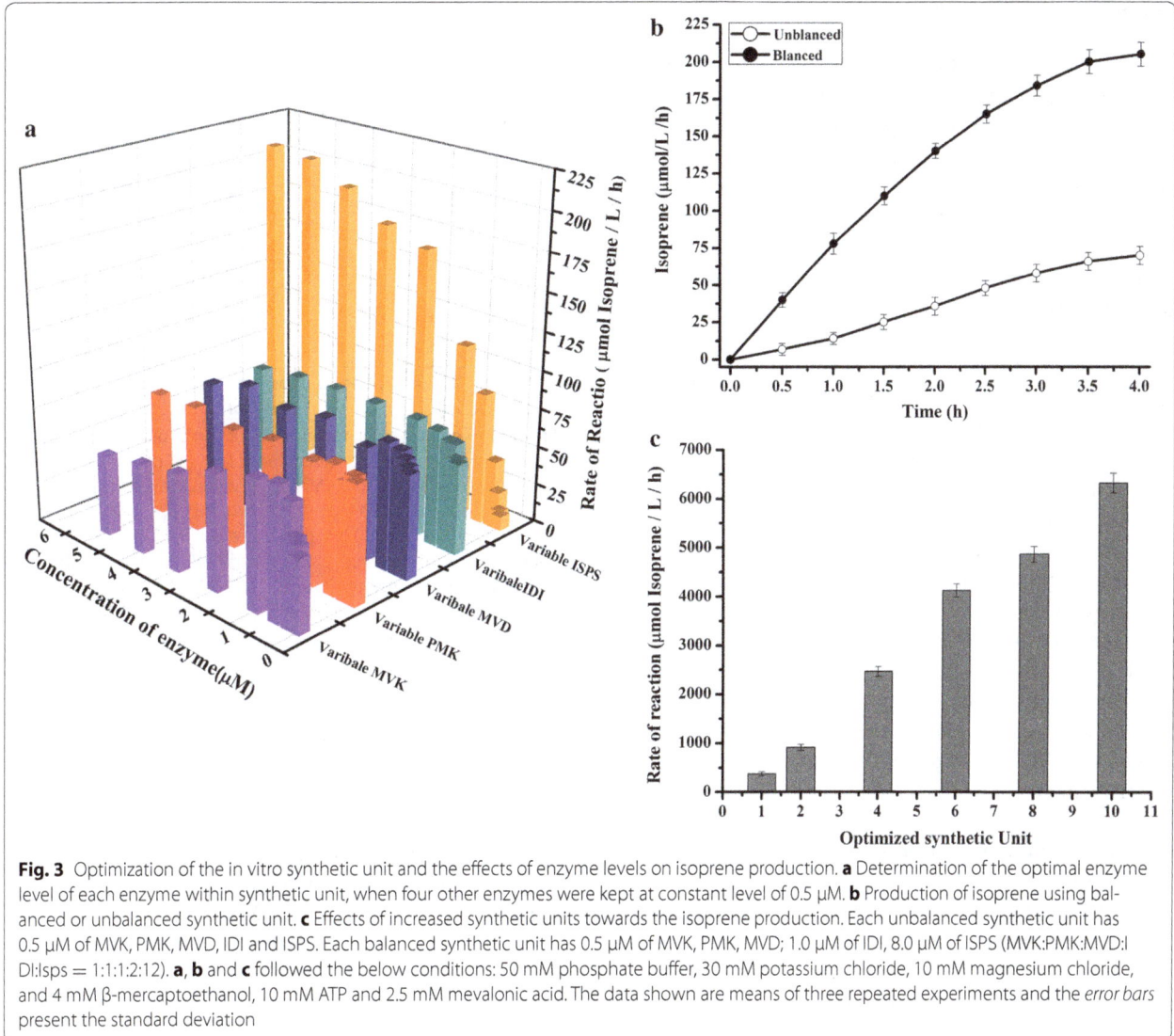

Fig. 3 Optimization of the in vitro synthetic unit and the effects of enzyme levels on isoprene production. **a** Determination of the optimal enzyme level of each enzyme within synthetic unit, when four other enzymes were kept at constant level of 0.5 μM. **b** Production of isoprene using balanced or unbalanced synthetic unit. **c** Effects of increased synthetic units towards the isoprene production. Each unbalanced synthetic unit has 0.5 μM of MVK, PMK, MVD, IDI and ISPS. Each balanced synthetic unit has 0.5 μM of MVK, PMK, MVD; 1.0 μM of IDI, 8.0 μM of ISPS (MVK:PMK:MVD:IDI:Isps = 1:1:1:2:16). **a**, **b** and **c** followed the below conditions: 50 mM phosphate buffer, 30 mM potassium chloride, 10 mM magnesium chloride, and 4 mM β-mercaptoethanol, 10 mM ATP and 2.5 mM mevalonic acid. The data shown are means of three repeated experiments and the *error bars* present the standard deviation

enzyme of this in vitro system, as its increasing levels from 0.02 to 5 μM gave a 30-fold increase in isoprene production. The results indicated that IDI and ISPS were two bottleneck enzymes in the in vitro production of isoprene from mevalonate, and their levels needed to be adjusted to balance the in vitro system. From the data of Fig. 1a, we could deduce that the optimized molarity ratio of MVK: PMK: MVD: IDI: ISPS was 1:1:1:2:16. One such balanced synthetic unit (0.5 μM of MVK, PMK, MVD; 1.0 μM of IDI, 8.0 μM of ISPS) could produce up to 382.4 μmol/L/h (26 mg/L/h) isoprene, which is nearly five folds of the isoprene production by the unbalanced synthetic unit (0.5 μM of each enzyme, Fig. 3b). Moreover, when multiple synthetic unit is supplemented in the 2 ml in vitro system, the production efficiency of isoprene could be significantly improved, 10 balanced synthetic units (5.0 μM of MVK, PMK, MVD; 10.0 μM of IDI, 80.0 μM of ISPS) would increase the isoprene production to 6323.5 μmol/L/h (430 mg/L/h) (Fig. 3c).

Our results demonstrated that balancing of enzyme levels could significantly increase the in vitro production of isoprene, which followed the results of in vivo studies that the balanced expression levels of heterologous enzymes is a key determinant in optimizing isoprenoid production [10, 11]. The in vivo studies aim to balance the heterologous pathways to reduce the growth inhibition effects towards the microbial hosts while the in vitro studies more focus on overcoming the limiting biochemical steps in heterologous flux towards the objective products. For example, we have noted that the enzymes levels of MVK did not apparently affect the in vitro production of isoprene (Fig. 3a) and they were kept at minimum level

(0.5 μM) in our balanced synthetic unit. While in the in vivo studies for isoprene [3, 16] and isoprenoids biosynthesis [11], the expression of MVK were adjusted at higher levels, in which MVK was believed to be a key enzyme and was expressed by stronger promoters than PMK and MVD. We hypothesized that the in vivo system has specific mechanism to adjust the intra- and extracellular levels of mevalonate (MVA), and higher level of MVK helps to convert the intracellular MVA before it is transported off the cell. The last enzyme ISPS were both found bottleneck enzyme in previous study [17] as well as in this study. From the data of this study, ISPS catalyzed the rate-limiting biochemical step in heterologous flux from DMAPP to the isoprene. The in vivo models also supported this conclusion, it was proved that lower expression level of key enzyme downstream the IPP/DMAPP will lead to the accumulation of C5 building blocks (IPP and DMAPP), and will inhibit normal cell growth [9].

Comparing with the previous in vitro experiment [14] which incorporated as much as 12 enzymes of upper and lower MVA pathways, this study demonstrated a simpler system which only involved 5 enzymes and their levels were precisely adjusted to optimize flux through the biochemical steps. After optimization, the isoprene production was significantly improved (from 214.5 to 6323.5 μmol/L/h) comparing with the previous in vitro experiment [14].

In vitro and in vivo production of isoprene from MVA: a comparison

We further set up a 50 ml in vitro system and compared its isoprene production with the in vivo flask fermentation (50 ml), from the same starting concentration of mevalonate substrate. The results showed that the in vitro model apparently had higher isoprene production rate than the in vivo model during the earlier 12 h of isoprene production (Fig. 4). The initial isoprene production rate is about 220.6 μmol/L/h (15.0 mg/l/h, total 160 mg/l isoprene within 12 h), which is similar with the recent study for the in vitro isoprene production system [14]. Moreover, isoprene production remained longer for the in vitro system: from 20 to 40 h, the production of isoprene was suspended in the in vivo system, but was consistent in the in vitro system. Until 40 h, the production of isoprene from mevalonate by the in vitro system reached to 4442.4 μmol/L (302.0 mg/l). From the comparison of the isoprene production by the in vitro and the in vivo models, we could deduce that the balanced in vitro enzyme system led to higher production rate and prolonged isoprene production. We estimated that the unbalanced intracellular enzymes levels, accumulated toxic intermediates, competitive consumption of co-factors may affect the isoprene production by the in vivo methods.

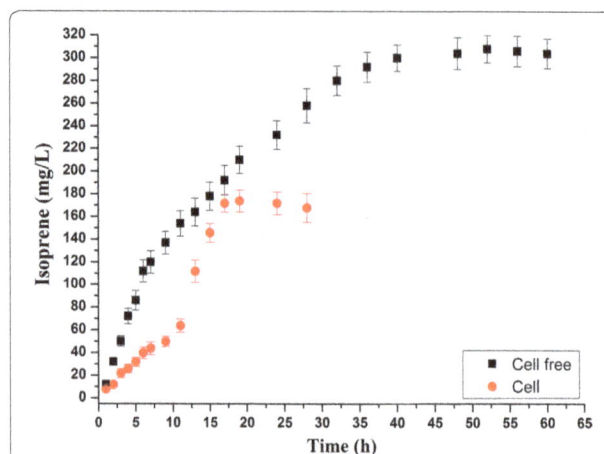

Fig. 4 The comparison of isoprene production by 50 ml in vitro and in vivo systems. The concentration of MVK (0.5 μM) was similar in both in vitro (*black*) and in vivo (*red*) systems. The cell free in vitro system has 1 synthetic unit (0.5 μM of MVK, PMK, MVD; 1.0 μM of IDI, 8.0 μM of ISPS) with 50 mM phosphate buffer, 30 mM potassium chloride, 10 mM magnesium chloride, and 4 mM β-mercaptoethanol. 50 ml culture of engineered *E. coli* (BL21(DE3)/pYJM14/pACY-ISPS, Table 1) which expressed the five enzymes of the lower MVA pathway was utilized as the in vivo control. Both systems was initiated by addition of 2.5 mM mevalonic acid and incubated at 30 °C in rotary shaker (180 rpm). Samples were taken to analyze their isoprene production at different time points. The data shown are means of three repeated experiments and the *error bars* present the standard deviation

Conclusion

In this study, isoprene was produced from mevalonate by an optimized enzymatic process using the five recombinant enzymes of MVK, PMK, MVD, IDI and ISPS from the lower MVA pathway. Balancing and increasing the enzyme levels could apparently enhance isoprene production, indicated that the production rate could be further increased by raising the concentrations of the mevalonate and enzymes, which is hard to achieved by the in vivo approaches. When the balanced enzyme units were increased to ten, the proposed process would produce 6323.5 μmol/L/h (430 mg/L/h in a 2 ml system) isoprene from mevalonate. Moreover, this study showed that the proposed process has the advantages of longer producing period of 40 h comparing with the controlled in vivo process. The improved production efficiency indicated that the proposed strategy is useful for the enzymatic production of isoprene or isoprenoids.

Methods
Strains and plasmids

Bacterial strains and plasmids used in this study were listed in Table 1. *Escherichia coli* strain DH5α and BL21 (DE3) and *Saccharomyces cerevisiae* used in this study were purchased from Invitrogen. The *E. coli* DH5α strain and *E. coli* BL21 (DE3) were used for plasmids preparation and for protein overexpression respectively. *S. cerevisiae* was used for gene for cloning. *E. coli* DH5α and *E. coli* BL21 (DE3) were cultured in Luria–Bertani (LB) broth in construction of strains and plasmids. *S. cerevisiae* was cultured in YPD medium. Antibiotics were added at final concentration of 50 µg/mL for kanamycin, 34 µg/mL for chloramphenicol and 100 µg/mL for ampicillin when necessary.

Protein preparation and purification

All PCRs were done using PrimerSTAR Max DNA polymerase (TAKARA, Dalian, China). Four genes of the enzymes MVK, PMK, MVD, IDI were amplified from *S. cerevisiae* genome (obtained from ATCC201508D) and cloned into the plasmid pET-30a(+). The plasmid pET-ERG12 was constructed by cloning the *ERG12* gene (for MVK) of *S. cerevisiae* into *Hind*III and *Xho*I sites of vector pET-30a(+) with primers ERG12_F and ERG12_R. The plasmid pET-ERG8 was constructed by cloning the *ERG8* gene (for PMK) from *S. cerevisiae* into *Eco*RI and *Xho*I sites of vector pET-30a(+) with primers ERG8_F and ERG8_R. The plasmid pET_ERG19 was constructed by cloning the *ERG19* gene (for MVD) from *S.cerevisiae* into *Bam*HI and *Xho*I sites of vector pET-30a(+) with primers ERG19_F and ERG19_R. The fourth gene for enzyme IDI was amplified by PCR using primers IDI_F and IDI_R and cloned into *Bam*HI and *Xho*I sites of vector pET-30a(+). The resulting plasmid was named pET-IDI. The isoprene synthase (ISPS) from poplar was synthesized after code optimization and digested with enzymes *Bam*HI and *Xho*I, then ligated into the pET-30a(+), the resulting plasmid was named pET-ISPS (Table 1).

Individual recombinant enzymes were extracted and purified from the strains of *E.coli* BL21(DE3) harboring the relevant plasmids. The cultures were incubated in 200 ml LB medium with 50 µg/ml Kanamycin at 37 °C until the OD600 reached 0.6–0.8. The cultures were added 0.2 mM IPTG for induction and were cooled to 20 °C for protein expression. After further growth at 20 °C for 12 h, cells were harvested by centrifugation at 8000*g* and suspended in 10 mL 50 mM phosphate buffer (pH7.4) containing 4 mM β-mercaptoethanol. The suspension was lysed by sonication (at 60% output for 3 s pulses with 3 s intervals between each cycle) for 40 min at 4 °C with tube jacketed in wet ice and centrifuged at 18,000*g* for 10 min at 4 °C. The supernate filtered by 0.22 µm PALL filter was added into the Nickel Column, which was washed by 10 ml water and 10 ml binding buffer in order to ensure that the column was equilibrated, and then the protein containing 6 His-tag was able to specifically bind to nickel column. Unbound protein was washed out with 10 ml washing buffer 1, then washing Buffer 2 was used to wash any nonspecific binding protein. Elution buffer was added to wash the specific protein with collection of 10 ml fractions. The column was then re-equilibrated with buffer. Protein concentrations were measured with BCA protein assay kit using a spectrophotometer. Recombinant enzymes were stored at −80 °C after flash freezing in liquid nitrogen.

In vitro reaction system

The reaction system was performed as previously described [18] to ensure the correct concentration of individually enzyme. For the demonstrated assay, a variety levels of each enzyme component and ATP were added to the reaction buffer which contain 50 mM potassium phosphate, 30 mM potassium chloride, 10 mM magnesium chloride, and 4 mM β-mercaptoethanol. The reaction was initiated by addition of 2.5 mM mevalonic acid which was made by the saponification of mevalonolactone with KOH at 1.05:1 (vol:vol) KOH:mevlonolactone for 30 min at 37 °C, and then incubated at 30 °C for 4 h. To confirm isoprene had accumulated in the 2 ml reaction system, 0.2 ml of the headspace gas of sealed 10 ml vial were analyzed by gas chromatography using Agilent 7890B GC (Agilent, American) equipped with a flame ionization detector and a Agilent HP-INNWOX column, designed to detect short-chain hydrocarbons. Amounts of isoprene produced in the recombinant system were calculated by comparison with an isoprene standard (Aladdin, China).

Gas chromatography (GC) analysis of isoprene

1 ml of off-gas samples from the headspace of the fermentor were analyzed as described earlier [16] using a GC (Agilent 7890A, America) equipped with a flame ionization detector (FID) and a HP-INNOWAX column (30 m × 320 µm × 8 µm). N_2 was used as carrier gas with a linear velocity of 1 ml/min. The product was characterized by direct comparison with standard isoprene (TCI-EP, Tokyo, Japan). The peak area was converted to isoprene concentration by comparing with a standard curve plotted with a set of known concentration of isoprene. Then isoprene accumulation was measured every 30 min by GC.

The comparison of production of isoprene in vitro and in vivo

To compare isoprene production in shake flasks (in vivo) and in vitro, the engineered strain *E. coli* BL21(DE3)/pYJM14/pACY-ISPS (Table 1) were cultured in 500 ml sealed glass flasks containing 50 ml of M9 medium supplemented with 10 g/L glucose and 0.5 g/L yeast extract, 34 µg/mL chloramphenicol, and 100 µg/mL Ampicillin. IPTG was added to the medium when the cell density

OD_{600} reached to 0.6 and the cultures were performed at 30 °C in rotary shaker (180 rpm). For isoprene production, the substrate 2.5 mM mevalonate was added after the cell was induced and cultivated at 30 °C for 4 and 16 h respectively. During the cultivation, samples of both the flask headspace and the culture were taken at multiple times points.

To set up similar enzyme levels in the in vitro system comparing to the in vivo control. The intracellular MVK concentration of the in vivo strain was firstly quantified using Quantity One Software (Biorad) described previously [19]. When OD_{600} reached 0.6, intracellular MVK concentration was about 0.4 μM. According to the optimized molarity ration for the in vitro system, 0.4 μM of MVK, PMK, MVD, 0.8 μM of IDI, 6.4 μM of ISPS was added in the 50 ml in vitro reaction system. 10 mM ATP, 50 mM potassium phosphate, 30 mM potassium chloride, 10 mM magnesium chloride, and 4 mM β-mercaptoethanol were also added in the reaction system. The reaction was initiated by addition of 2.5 mM mevalonic acid which was made by the saponification of mevalonolactone with KOH at 1.05:1 (vol:vol) KOH:mevlonolactone for 30 min at 37 °C, and then incubated at 30 °C in rotary shaker (180 rpm). During the cultivation, samples of both the flask headspace and the culture were taken at multiple times points.

Abbreviations

MVA pathway: mevalonate pathway; MEP pathway: methylerythritol 4-phosphate pathway; MVK: mevalonate kinase; PMK: phosphomevalonate kinase; MVD: diphosphomevalonaet decarboxylase; IDI: isopentenyl diphosphate isomerase; ISPS: isoprene synthase; IPTG: isopropyl-β-d-thiogalactoside; GC: gas chromatography.

Authors' contributions

TC and HL conceived of the study, participated in its design, carried out the process control studies and drafted the manuscript. MS and NC participated in the coordination of this study, contributed to the data analysis and the process control studies. GZ and CX participated in its design and helped to draft the manuscript. MX and HZ conceived of the study, and participated in its design and coordination and helped to draft the manuscript. All authors read and approved the final manuscript.

Author details

[1] CAS Key Laboratory of Bio-based Materials, Qingdao Institute of Bioenergy and Bioprocess Technology, Chinese Academy of Sciences, No. 189 Songling Road, Laoshan District, Qingdao 266101, China. [2] College of Chemical Engineering, Qingdao University of Science and Technology, Qingdao 266042, China. [3] State Key Laboratory Base of Eco-Chemical Engineering, College of Chemistry and Molecular Engineering, Qingdao University of Science and Technology, Qingdao 266042, China.

Acknowledgements

Not applicable.

Competing interests

The authors declare that they have no competing interests.

Funding

The present study was supported by Shandong Province Natural Science Foundation (ZR2015BM011), Technology Development Project of Shandong Province (2016GSF121013), National Natural Science Foundation (21106170, 21376129, 21572242, 31670493).

References

1. Rabinovitch-Deere CA, Oliver JWK, Rodriguez GM, Atsumi S. Synthetic biology and metabolic engineering approaches to produce biofuels. Chem Rev. 2013;113(7):4611–32.
2. Keasling JD. Synthetic biology and the development of tools for metabolic engineering. Metab Eng. 2012;14(3):189–95.
3. Whited GM, Feher FJ, Benko DA. Development of a gas-phase bioprocess for isoprene-monomer production using metabolic pathway engineering. Ind Biotechnol. 2010;6(3):152–63.
4. Chang MCY, Keasling JD. Production of isoprenoid pharmaceuticals by engineered microbes. Nat Chem Biol. 2006;2(12):674–81.
5. George KW, Chen A, Jain A, Batth TS, Baidoo EEK, Wang G, Adams PD, Petzold CJ, Keasling JD, Lee TS. Correlation analysis of targeted proteins and metabolites to assess and engineer microbial isopentenol production. Biotechnol Bioeng. 2014;111(8):1648–58.
6. Gronenberg LS, Marcheschi RJ, Liao JC. Next generation biofuel engineering in prokaryotes. Curr Opin Chem Biol. 2013;17(3):462–71.
7. Zhang H, Liu Q, Cao Y, Feng X, Zheng Y, Zou H, Liu H, Yang J, Mo X. Microbial production of sabinene—a new terpene-based precursor of advanced biofuel. Microb Cell Fact. 2014;13(1):452–7.
8. Immethun CM, Hoynes-O'Connor AG, Balassy A, Moon TS. Microbial production of isoprenoids enabled by synthetic biology. Front Microbiol. 2013;4:75.
9. Martin VJ, Pitera DJ, Withers ST, Newman JD, Keasling JD. Engineering a mevalonate pathway in Escherichia coli for production of terpenoids. Nat Biotechnol. 2003;21(7):796–802.
10. Pitera DJ, Paddon CJ, Newman JD, Keasling JD. Balancing a heterologous mevalonate pathway for improved isoprenoid production in Escherichia coli. Metab Eng. 2007;9(2):193–207.
11. Ajikumar PK, Xiao W-H, Tyo KEJ, Wang Y, Simeon F, Leonard E, Mucha O, Phon TH, Pfeifer B, Stephanopoulos G. Isoprenoid pathway optimization for taxol precursor overproduction in Escherichia coli. Science. 2010;330(6000):70–4.
12. Dueber JE, Wu GC, Malmirchegini GR, Moon TS, Petzold CJ, Ullal AV, Prather KL, Keasling JD. Synthetic protein scaffolds provide modular control over metabolic flux. Nat Biotechnol. 2009;27(8):753–9.
13. Hodgman CE, Jewett MC. Cell-free synthetic biology: thinking outside the cell. Metab Eng. 2012;14(3):261–9.
14. Korman TP, Sahachartsiri B, Li D, Vinokur JM, Eisenberg D, Bowie JU. A synthetic biochemistry system for the in vitro production of isoprene from glycolysis intermediates. Protein Sci. 2014;23(5):576–85.
15. Opgenorth PH, Korman TP, Bowie JU. A synthetic biochemistry molecular purge valve module that maintains redox balance. Nat Commun. 2014;17(5):4113.
16. Yang J, Xian M, Su S, Zhao G, Nie Q, Jiang X, Zheng Y, Liu W. Enhancing production of bio-isoprene using hybrid MVA pathway and isoprene synthase in E. coli. PLoS ONE. 2012;7(4):e33509.
17. Ilmén M, Oja M, Huuskonen A, Lee S, Ruohonen L, Jung S. Identification of novel isoprene synthases through genome mining and expression in Escherichia coli. Metab Eng. 2015;31:153–62.
18. Yu X, Liu T, Zhu F, Khosla C. In vitro reconstitution and steady-state analysis of the fatty acid synthase from Escherichia coli. Proc Natl Acad Sci USA. 2015;108(46):18643–8.
19. Jin K, Peel AL, Mao XO, Xie L, Cottrell BA, Henshall DC, Greenberg DA. Increased hippocampal neurogenesis in Alzheimer's disease. Proc Natl Acad Sci USA. 2004;101:343–7.

2

Engineering microbial cell factories for the production of plant natural products: from design principles to industrial-scale production

Xiaonan Liu[1,2†], Wentao Ding[1†] and Huifeng Jiang[1*]

Abstract

Plant natural products (PNPs) are widely used as pharmaceuticals, nutraceuticals, seasonings, pigments, etc., with a huge commercial value on the global market. However, most of these PNPs are still being extracted from plants. A resource-conserving and environment-friendly synthesis route for PNPs that utilizes microbial cell factories has attracted increasing attention since the 1940s. However, at the present only a handful of PNPs are being produced by microbial cell factories at an industrial scale, and there are still many challenges in their large-scale application. One of the challenges is that most biosynthetic pathways of PNPs are still unknown, which largely limits the number of candidate PNPs for heterologous microbial production. Another challenge is that the metabolic fluxes toward the target products in microbial hosts are often hindered by poor precursor supply, low catalytic activity of enzymes and obstructed product transport. Consequently, despite intensive studies on the metabolic engineering of microbial hosts, the fermentation costs of most heterologously produced PNPs are still too high for industrial-scale production. In this paper, we review several aspects of PNP production in microbial cell factories, including important design principles and recent progress in pathway mining and metabolic engineering. In addition, implemented cases of industrial-scale production of PNPs in microbial cell factories are also highlighted.

Keywords: Plant natural products, Synthetic biology, Microbial cell factories, Metabolic engineering, Industrial production

Background

Thousands of plant natural products (PNPs) can be utilized as drugs, cosmetics, dyes, seasonings, nutraceuticals, and industrial chemicals, all of which play important roles in human life [1]. Unfortunately, in recent years over-exploitation has endangered more than 15,000 medicinal plant species in their natural habitats, which resulted in a precarious and unsustainable supply of the invaluable natural products [2, 3]. At the same time, seasonal, climatic or other environmental variations can also threaten the supply of these natural products [4]. The yield of natural products from plants also cannot satisfy the demands of the growing market. Due to the structural complexity of most natural products, total chemical synthesis approaches are often inefficient and accompanied by large amounts of waste and heavy pollution [5]. Therefore, engineering microbial cell factories to produce these high-value PNPs has become a promising solution to protect endangered plants and prevent pollution from chemical synthesis [4, 6]. Current developments in biological "omics" techniques and synthetic biology provide excellent tools to remove obstacles in the construction of microbial cell factories to produce PNPs [5, 7, 8]. Over the past few decades, there has been great progress in multiple aspects of heterologous synthesis of PNPs, including the identification of biosynthetic pathways, the

*Correspondence: jiang_hf@tib.cas.cn
†Xiaonan Liu and Wentao Ding contributed equally to this work
[1] Key Laboratory of Systems Microbial Biotechnology, Tianjin Institute of Industrial Biotechnology, Chinese Academy of Sciences, Tianjin, China
Full list of author information is available at the end of the article

construction of microbial cell factories, and the development of industrial production [9]. Engineered microbial cell factories have been successfully applied to produce PNPs from renewable carbon sources at an industrial scale. Further progress in the field of microbial cell factories will not only potentially save endangered plants, but also profoundly change the traditional ways of PNP production.

In this review, we summarize the recent progress of microbial cell factory engineering, identification of novel biosynthetic pathways, and the production of PNPs on an industrial scale. Firstly, two tentative strategies based on biological "omics" and synthetic biology technologies for gene mining from the biosynthetic pathways of PNPs are introduced, followed by a discussion of three pivotal steps in the optimization of microbial cell factories, including the improvement of precursor supply, enzyme activity and product transport. Finally, several milestone examples of industrial PNP production in microbial cell factories are showcased (Fig. 1).

Mining of PNP biosynthetic pathways

The first step in the construction of a PNP-producing microbial cell factory is to identify the original biosynthetic pathway. The rapid development of molecular biology, "omics" technologies and bioinformatics, has enabled great breakthroughs in the identification of PNP-biosynthetic pathways. In this section, recent approaches in gene exploration for the production of natural compounds are reviewed.

Sequencing-guided pathway exploration

Traditionally, reverse transcription-polymerase chain reaction (RT-PCR), rapid amplification of cDNA ends (RACE) [10–12], RNA interference (RNAi) [13, 14], virus-induced gene silencing (VIGS) [15] and isotopic tracer methods [16] had been used to uncover novel genes in the biosynthetic pathways of PNPs. However, these technologies are usually time-consuming, laborious, and consequently expensive. With the rapid development of sequencing technology, "omics" analysis and high-throughput screening technologies, thousands of functional genes related to PNP-biosynthetic pathways have been identified in recent years [17]. For example, the assembly of the *Salvia miltiorrhiza* transcriptome provided a valuable resource for the investigation of the complete biosynthetic pathway of tanshinone [18]. High-density genetic linkage mapping using recombinant inbred lines (RILs) had been used to decipher the genetic networks underlying flavonoid biosynthesis in *Brassica napus* [19]. Genes related to cucurbitacins biosynthesis were characterized by genome-wide association analysis based on the genomic variation map of

115 diverse cucumber lines [20, 21]. Whole-genome sequencing of *Siraitia grosvenorii*, combined with transcriptomic (RNA-Seq) and bioinformatic analyses made a great contribution to illuminate the biosynthetic pathway of mogroside V [22]. Glycosyltransferases of ginsenosides were also successfully cloned based on the ESTs (Expressed Sequence Tags) and cDNA database of *Panax ginseng* [23, 24]. Six enzymes acquired by RNA-Seq from mayapple were assembled into the complete biosynthetic pathway of the etoposide aglycone [25]. Therefore, "omics" technologies have greatly accelerated the pace of identification of novel genes from the biosynthetic pathways of PNPs.

Recombined artificial biosynthetic pathways

Currently, various online databases providing enormous amounts of genomic data also facilitate the identification of novel genes. For example, the 1000 plants project (1KP) proposed to collect the transcriptomes of 1000 plant species [26]. The Medicinal Plant Genomics Resource provides transcriptome and metabolome resources from medicinal plants [27]. Phytozome is a comparative platform for green plant genomics, providing a view of the evolutionary history of each plant's genes, as well as access to the sequences and functional annotations [28]. Moreover, based on detailed information about genes, enzymes, reactions and their regulation, comprehensive metabolic networks of PNPs have been composed based on reactions mined from different species using bioinformatics databases such as KEGG (Kyoto Encyclopedia of Genes and Genomes) [29]. From these databases, it is therefore possible to "dig up" genes from different species to construct artificial biosynthetic pathways of PNPs for which the natural pathway is unknown. For instance, the biosynthesis of opioids in yeast is considered to be one of the most remarkable landmarks of synthetic biology, representing a highly sophisticated feat of engineering a very complex metabolic pathway in microbe [30–32]. So far, 21 enzymes were successfully reconstructed in yeast for the heterologous biosynthesis of opioids by combining the building blocks derived from multiple species, such as *Papaver somniferum, Papaver bracteatum, Coptis japonica, Eschscholzia californica, Rattus norvegicus, Pseudomonas putida*, and yeast [31]. Additionally, raspberry ketone [33], salidroside [34], gastrodin [35], and salvianic acid A (SAA) [36] were also successfully synthesized in microorganisms using recombined artificial pathways. Therefore, it has become feasible to construct the biosynthetic pathways of PNPs without genetic information from the original plant. Unprecedented microbial synthesis of PNPs via recombined artificial pathways provides a new perspective in the construction of biosynthetic pathways for even the most complex PNPs. With

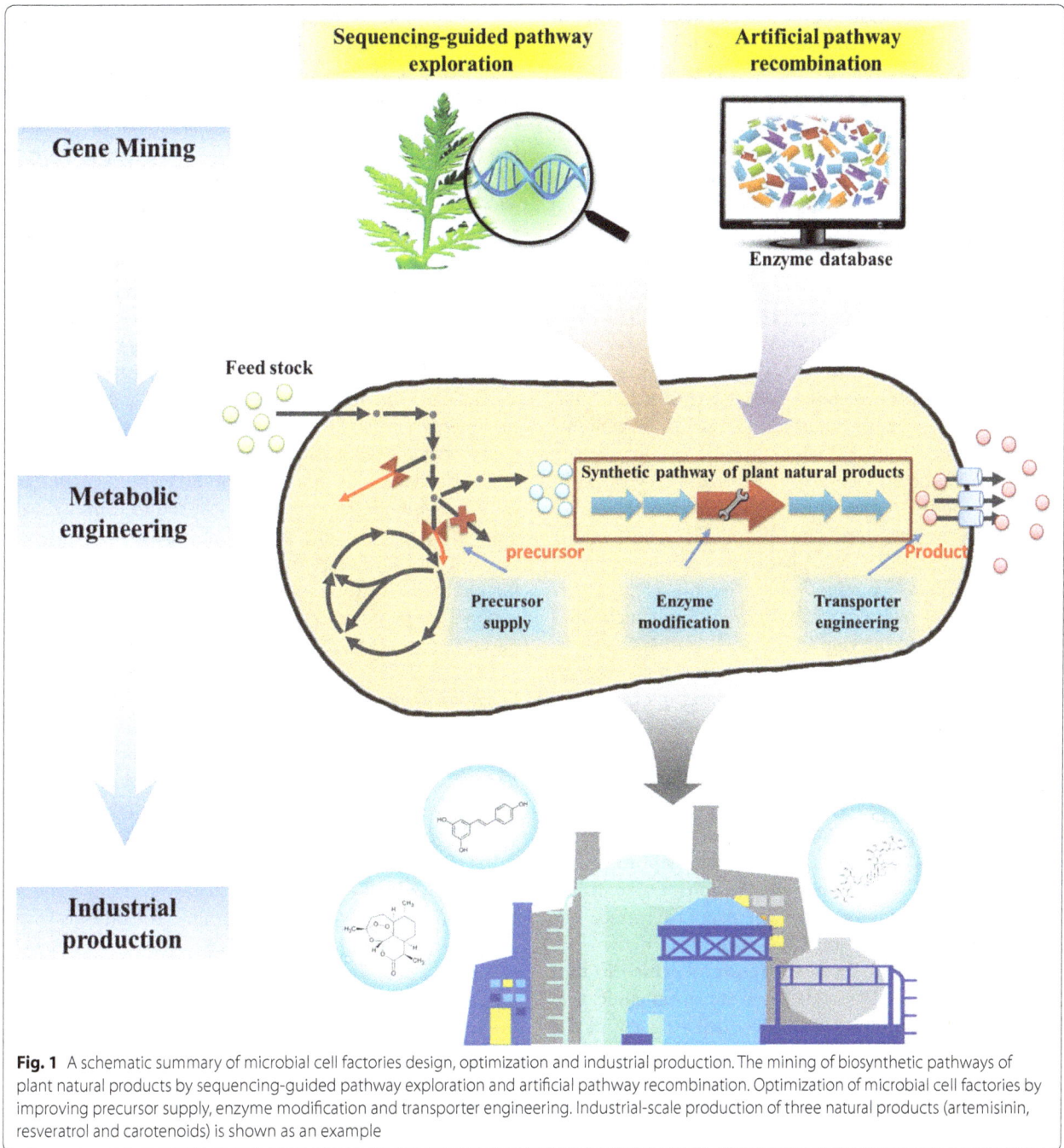

Fig. 1 A schematic summary of microbial cell factories design, optimization and industrial production. The mining of biosynthetic pathways of plant natural products by sequencing-guided pathway exploration and artificial pathway recombination. Optimization of microbial cell factories by improving precursor supply, enzyme modification and transporter engineering. Industrial-scale production of three natural products (artemisinin, resveratrol and carotenoids) is shown as an example

increasing information on functional genes and enzymes, recombined artificial pathways will shed light on ways to produce new PNPs in cell factories.

Optimization of microbial cell factories

Although many biosynthetic pathways of PNPs have been identified, producing PNPs in microbial cell factories is still a tough challenge for a number of reasons. First, most PNPs are secondary metabolites, which receive

relatively low levels of carbon metabolic flux in comparison with primary metabolites. Thus, it is necessary to rewire the metabolic fluxes toward their precursors in order to increase the production of PNPs. Second, due to the complexity of the molecular structures of most high-value PNPs, multiple genes are involved in the biosynthesis of these compounds. Consequently, improving the catalytic activities of so many genes in the synthetic pathway is another bottleneck. Last but not least, when the microbial host is producing PNPs, the accumulation of precursors or products in vivo can cause feedback inhibition or toxicity. In order to rescue cells from metabolite accumulation and improve the production of heterologous PNPs, transporter engineering strategies have been implemented to secrete the target metabolites outside of the cells. In this section, recent progresses in metabolic network rewiring, as well as engineering of enzymes and PNP transporters are summarized.

Improving precursor supply

The limited supply of precursors is one of the main challenges for heterologous PNP synthesis in microbial hosts. One typical case is the heterologous production of terpenoids. The carbon skeletons of all terpene molecules are composed of the five-carbon precursors isopentenyl pyrophosphate (IPP) and dimethylallyl pyrophosphate (DMAPP), which are synthesized through the MEP pathway (2-C-methyl-D-erythritol 4-phosphate pathway) or MVA pathway (mevalonate pathway) [37]. However, the flux toward the MEP or MVA pathway is considered to be limited in terpenoid production in microbial hosts. Overexpression of a truncated *HMG1* (*tHMG1*), which catalyzes the rate-limiting step in the MVA pathway, can be used to avoid accumulating toxic amounts of β-hydroxy-β-methylglutaryl-CoA (HMG-CoA), and increase the supply of IPP or DMAPP in *Saccharomyces cerevisiae* [38]. Overexpression of all genes from the MVA pathway in *S. cerevisiae* using the GAL promoter also increased the metabolic flux [39]. Moreover, decreasing or eliminating the fluxes going toward competitive pathways is also a conventional strategy for pathway optimization. For example, to improve the production of artemisinic acid, an inducible promoter was used to reduce the expression of the *ERG9* gene and resulted in a decrease in the metabolic flux from IPP to ergosterol, which in turn pushed the conversion of IPP toward artemisinic acid [40]. Furthermore, a sequential control strategy of the precursor farnesyl pyrophosphate (FPP) has been developed in *S. cerevisiae*, which improved the production of β-carotenoids [41]. This offers a practical and cost-efficient approach to improve the biosynthetic production of natural compounds.

Flavonoids are widespread PNPs with high pharmaceutical value, but the production of flavonoids in recombinant microbes is low. Tyrosine or phenylalanine and malonyl-CoA are the main precursors in flavonoid biosynthesis. Introducing a tyrosine insensitive 3-deoxy-D-arabinose-heptulosonate-7-phosphate synthase mutant (encoded by *ARO4*^{G226S}), knocking-out *ARO3*, and phenylpyruvate decarboxylase genes (*PDC1*, *PDC5* and *PDC6*), together with overexpression of chalcone synthase (CHS) and tyrosine ammonia lyase (TAL)

resulted in a 40-fold increase of extracellular naringenin titer in glucose-grown shake-flask cultures [42, 43]. In *Escherichia coli*, the acetyl-CoA carboxylase complex (*accABCD*), biotin ligase (*BirA*) from *Photorhabdus luminescens*, and enzymes in the acetate assimilation pathway [acetate kinase A (*ackA*), phosphate acetyltransferase (*pta*) and acetyl-CoA synthase (*acs*)] were overexpressed, which largely increased flavonoid production [44]. Additionally, overexpression of β-ketoacyl-ACP synthase II (*FabF*) was able to increase cellular malonyl-CoA levels and pinocembrin production [45]. In recent years, a number of groups have developed strategies for dynamic regulation, which usually depend on appropriate biosensors. These strategies allow the rebalancing of fluxes according to changing conditions in the cell or the fermentation medium [46]. For example, a hybrid *cis*- and *trans*- regulatory promoter from *Bacillus subtilis*, which responds to a broad concentration range of malonyl-CoA during metabolic processes, was introduced into *E. coli* and led to the dynamic control of malonyl-CoA-associated fluxes [47]. Therefore, various metabolic engineering strategies for the regulation of precursor supply have enabled improvements in yields and titers of a variety of natural products produced in microorganisms.

Enzyme modification

The production of metabolites is often impeded by insufficient catalytic activity of enzymes. Enzyme engineering thus is a key solution for the improvement of metabolic fluxes towards PNPs. Two main strategies have been used in enzyme engineering for higher catalytic activities. One is based on directed evolution, utilizing methods such as error-prone PCR (Polymerase Chain Reaction), random mutations and site-specific mutations. The other is based on semi-rational or rational design.

Directed evolution is a widely used strategy for altering the catalytic characteristics of enzymes. For example, the directed evolution of lycopene cyclase (CrtYB) was employed to inactivate the lycopene cyclase function but retain the phytoene synthase function for improving lycopene production in engineered *S. cerevisiae*. The catalytic activity of geranylgeranyl diphosphate synthase (CrtE) was also improved by directed evolution in order to enhance the synthesis of geranylgeranyl pyrophosphate (GGPP), and reached a production of 1.61 g/L lycopene in the engineered diploid strain by fed-batch fermentation [48]. Bai et al. reported that the catalytic properties of a glycosyltransferase UGT73B6 toward phenolic alcohol were improved through directed evolution, and the resulting strain produced much higher yield of gastrodin, reaching 545 mg/L in 48 h [35]. Through in vivo evolution of stilbene synthase, pinosylvin production was increased up to 23-fold when cerulenin was added [49].

Despite the heavy workload and low efficiency compared to rational design methods, directed evolution is still a simple and accessible method for enzyme modification, especially for those proteins whose structure is unknown.

Rational design methods are usually based on the knowledge of protein structure and catalytic mechanism. Keasling's team engineered P450BM3 (a substrate-promiscuous P450 enzyme) from *Bacillus megaterium* via a ROSETTA-based energy minimization method, which enabled the P450BM3 mutants to conduct selective oxidation of amorphadiene, and produced artemisinic-11S,12-epoxide at titers greater than 250 mg/L in *E. coli* [50]. Ajikumar's team modified the N-terminus of CYP725A4 and achieved the highest titer of oxygenated taxanes so far (570 ± 45 mg/L) in *E. coli* [51]. Liu et al. reported that the catalytic activity of isopentenyl phosphate kinase (IPK) was improved for about seven-fold by rationally analyzing the coevolution of IPK protein sequences, so that the recombinant *E. coli* strain produced 97% more β-carotenoids than the starting strain [52]. Morita et al. demonstrated the synthesis of several unnatural polyketide-alkaloid scaffolds by exploiting a type III Polyketide synthase (PKS) using precursor-directed and structure-based approaches. The catalytic versatility of the type III PKS provides an excellent platform for further development of novel biocatalysts [53]. Computational, rational, or directed evolution engineering strategies can tailor a promiscuous enzyme for greater catalytic activity, thermostability or substrate specificity, and further increase the conversion efficiency from precursor to product.

Transporter engineering

In microbes, excessive accumulation of plant secondary metabolites is usually toxic to the host, which can hamper cell growth and decrease PNP productivity. Thus, engineering transporters from both microorganisms and plants can improve the production of PNPs in microbial hosts [54, 55].

For example, overexpression of an efflux transporter from *Alcanivorax borkumensis* increased limonene production about 1.5-fold in engineered *E. coli* [56]. The native ABC transporter (ATP-Binding Cassette transporter) SNQ2 was overexpressed in the engineered *S. cerevisiae* that produced resveratrol, and the yield of resveratrol was increased from 48 to 61 mg/L after 48 h of fermentation [57]. Tripartite efflux pumps (pleiotropic resistant pumps) were constructed by combining TMDs (transmembrane domains) and NBDs (nucleotide binding domains) from endogenous transporters (AcrAB-TolC and MdtEF-TolC from *E. coli*) and heterologous transporters (MexAB-OprM from *Pseudomonas aeruginosa*), to improve isoprenoid production in *E. coli* [58].

The resulting chimeric transporter TolC-TolC-AcrB improved the specific yield of amorphadiene by 118% and kaurene by 104% [58]. Moreover, the co-overexpression of multiple transporters (*tolC* combined with *macAB*, *emrAB* and *emrKY*) enhanced the titer of amorphadiene more than threefold [59]. Recently, ginsenoside efflux pumps from *Panax ginseng* were identified as PDR (pleiotropic drug resistance homolog) transporter subfamily, which is used for the export of ginsenosides from microbial cell factories engineered for ginsenoside production [60, 61]. In addition to pumping out the product, transporters are also involved in substrate uptake in microbial hosts. Leonard et al. introduced *Rhizobium trifolii* MatB and MatC (encoding malonate synthetase and malonate carrier protein, respectively) into a recombinant *E. coli* strain, which introduced malonate uptake mechanism to increase synthesis of malonyl-CoA in flavonoid biosynthesis pathway [62].

With the broad and deep development of pathway identification and microbial metabolic engineering strategies, the accumulation and transportation of substrates, intermediates and products at the cellular/subcellular level has become more significant in cell factories, and has attracted growing attention. Accumulated knowledge about the function, structure and mechanism of transporters will facilitate the rearrangement of mass transport for the construction of more sophisticated cell machines in cell factories through transporter engineering.

Engineered microbial cell factories for industrial-scale application

Several microbial cell factories have been improved through decades of efforts, reaching a level of productivity which is acceptable for industrial applications. The production of artemisinic acid, resveratrol and lycopene has increased by tens or hundreds of times, and has reached or is close to an industrial scale. Here, we summarize some successful industrialized cases of PNP production by transgenic microbial hosts to emphasize the viability and prospects of the commercial application of microbial cell factories.

Artemisinin

Artemisinin is acknowledged as an effective pharmaceutical compound for the treatment of malaria [63], a serious disease that in 2015 alone affected 214 million people (African 88%, South-East Asia Region 10%), causing 438,000 deaths [64]. The demand for artemisinin is exponentially increasing every year because of the increased incidence of drug-resistant malaria throughout the world [65]. However, the concentration of artemisinin in the plant *Artemisia annua* is very low (0.01–1.1%), and improvement of the yield of artemisinin through plant breeding or total organic synthesis remains a challenge. Nevertheless, the synthesis or semi-synthesis of artemisinin using recombinant microorganisms is a promising solution. For more than 10 years, efforts have been made to improve the microbial production of precursors of artemisinin, and remarkable achievements have been made [39, 40, 66–69]. By engineering the genotype and carbon flux of *S. cerevisiae*, the yield of artemisinic acid reached 0.65 g/L in flask fermentation, starting from an initial 0.11 g/L [39]. When the recombinant strain was cultured in a well-controlled fermenter, and when the extractive solvent isopropyl myristate (IPM) was added, 25 g/L artemisinic acid was produced, which was about 35-fold higher than in the flask. Therefore, the production of artemisinic acid has been improved from 0.1 to 25 g/L in engineered *S. cerevisiae* through in vivo carbon flux rewiring and optimization of fermentation conditions [66], achieving the semi-synthesis of artemisinin (combined with one-step photochemical catalysis) at an industrial scale. Amyris Inc. has been engaged in pushing the semi-synthesis of artemisinin into commercial production.

Resveratrol

Resveratrol is a polyphenolic compound found in several plant species, such as bush berries, peanuts, cranberries, and grapes. Resveratrol has been proved to decrease the risk of heart disease, diabetes and cancer (reviewed in [70]). It is widely used in medicine, as well as the health and cosmetic industries [71]. It is one of the fastest growing nutritional supplements in the flavonoid market [3]. According to a Frost & Sullivan report, in 2012 the global supply market value of resveratrol was about $50 million [72]. Due to the complexity and contamination problems in chemical synthesis of resveratrol, currently it is mostly extracted from plants [73]. Since the concentration of resveratrol in plants is extremely low, its production is limited and unsustainable. Recently, efforts have been focused on engineering microorganisms to synthesize resveratrol through fermentation [73]. Beekwilder et al. first introduced 4CL$_2$ (4-coumaroyl-CoA ligase) from *Nicotiana tabacum* and STS (stilbene synthase) from *Vitis vinifera* into *E. coli*, and produced 16 mg/L resveratrol from 4-coumaric acid [74]. Many studies have also focused on metabolic engineering of *E. coli* cell factories towards resveratrol production [75–77]. Leam et al. reported an engineered *E. coli* with high resveratrol yield, which was constructed by using a stilbene synthase modification and enhanced intracellular malonyl-CoA supply, resulting in a final resveratrol titer of 2.3 g/L [78]. The budding yeast *S. cerevisiae* has also been engineered as a host for resveratrol synthesis. Durhuus et al. reported

a recombinant *S. cerevisiae* strain producing about 5 g/L resveratrol [79], which is the highest resveratrol titer from microbial cell factories until now. The Evolva company has successfully accomplished industrial production of resveratrol in a yeast cell factory.

Carotenoids

Carotenoids are a class of tetraterpenoids containing 40 carbons, with many important members such as lycopene, α-carotene, β-carotene, canthaxanthin, zeaxanthin, astaxanthin and lutein, which are used in various industries [80]. Especially, lycopene and astaxanthin are two of the most potent antioxidants among the dietary carotenoids, and may help lower the risk of chronic diseases including cancer and heart diseases [81–83]. The global market of carotenoids was $1.5 billion in 2014, and is expected to reach nearly $1.8 billion in 2019, with a compound annual growth rate (CAGR) of 3.9% [84]. In recent years, producing carotenoids through microbial fermentation has attracted intense attention [85]. Alper et al. reported that by using systematic (model-based) and combinatorial (transposon-based) methods to identify gene knockout targets, they obtained a maximum lycopene yield of 18 mg/g DCW (Dry Cell Weight), which represents an 8.5-fold increase over the recombinant *E. coli* K12 without the knockouts [86]. Cho et al. reported that the production of lycopene could reach as high as 1754 mg/L (38.1 mg/g DCW) in *E. coli* transformed with the *mvaK1*, *mvaD*, *mvaK2*, *mvaE*, *mvaS* and *idi* genes [87]. Although the engineered model microorganisms like *E. coli* and *S. cerevisiae* have not yet been successfully applied at the industrial scale, some microorganisms that endogenously produce carotenoids can be made applicable for industrial production through screening or mutation. Marcos et al. reported that a selected *Blakeslea trispora* strain produced 3.6 g/L lycopene in a dichloromethane-extracted fermentation [88]. Recently, lycopene and astaxanthin have been successfully produced commercially through fermentation of *Blakeslea trispora* [89] and *Phaffia rhodozyma* [90], respectively.

Challenges in the fermentation process

Although developments in "omics" technology and synthetic biology have accelerated the pace of construction of microbial cell factories, it is still difficult to drive cell factories on an industrial scale. Only a few PNPs can be produced commercially so far. The essential problem in industrialization of microbial cell factories is how to cut down the production costs. In addition to the summarized strategies for optimizing microbial production in this context, there are still challenges in the field of fermentation process engineering. Two challenges we are facing right now are the engineering of microbial hosts toward the utilization of low-cost feedstocks, and improving microbial host resistance toward robust fermentation conditions.

Glucose is the most widely used material for the production of PNPs, including flavonoids, terpenoids, alkaloids and so on. However, glucose is not favorable in large-scale production because of the high price of raw materials and potential threats to food security. Therefore, engineering microbial hosts to use low-cost, non-food materials is extremely pertinent. Price et al. developed an engineering strategy to manipulate supramolecular enzyme assemblies which dramatically enhanced the carbon flux from methanol to the key intermediate fructose-6-phosphate in the microbial metabolic network, which provides a platform for biological conversion of methanol to higher value-added chemicals [91]. In 2016, Antonovsky et al. reported a non-native carbon fixation cycle that can synthesize sugars and other major biomass components from CO_2 in *E. coli*, where all of the pathway intermediates and products are solely synthesized using CO_2 as an inorganic carbon source and pyruvate as an energy source [92]. Although the engineering of microbial hosts to utilize low-cost materials is still not efficient, it will no doubt bring revolutionary progress to the entire cell factory industry, and therefore deserves further attention.

Since the 1930s, microorganisms have been used for industrial fermentation. *E. coli* and *S. cerevisiae* are two widely used hosts, as models for prokaryotic and eukaryotic expression systems, respectively. Recently, other microorganisms such as *Corynebacterium glutamicum* [93], *Streptomyces* spp. [94], *Yarrowia lipolytica* [80] and *Pichia pastoris* [95] have also been used as potential hosts for the production of PNPs. However, to achieve sufficiently high yields, the fermentation process of these microbial hosts requires precise control of factors such as temperature, pH, aeration, stirring, carbon source, nutritional supplements and antibiotics, resulting in high operating costs. Thus, engineering microbial hosts to be more competitive under robust fermentation conditions is another way to achieve cost saving. Shaw et al. engineered *E. coli*, *S. cerevisiae* and *Y. lipolytica* fermentations by supplying essential growth nutrients in the form of xenobiotic or ecologically rare chemicals, omitting the need for sterilization or antibiotics [96]. Yue et al. reported that a recombinant halophilic *Halomonas campaniensis* LS21 can produce the bioplastic PHB (polyhydroxybutyrate) in an energy-saving (non-sterilization), seawater-based, long-lasting and continuous open process, which largely saved fermentation costs [97]. This method provides microbial competitive advantages with minimal external risks, given that the engineered microbial hosts possess improved fitness within the customized fermentation environments.

Conclusions

This review summarizes recent progress in the identification of novel biosynthetic pathways, engineering of microbial cell factories and industrial fermentation for the production of plant natural products (PNPs). Taking advantage of biological "omics" technologies and synthetic biology, more and more PNPs are being synthesized using microbial cell factories. Nevertheless, a huge number of PNPs remains to be investigated and overcoming current challenges will certainly open the doors for the valorization of a multitude of natural compounds supply through synthetic biology.

Authors' contributions

XL, WD and HJ conceived and designed the manuscript. XL and WD performed the bibliographic research and drafted the manuscript. HJ outlined the structure and reviewed the manuscript. All authors read and approved the final manuscript.

Author details

[1] Key Laboratory of Systems Microbial Biotechnology, Tianjin Institute of Industrial Biotechnology, Chinese Academy of Sciences, Tianjin, China. [2] University of Chinese Academy of Sciences, Beijing, China.

Acknowledgements

We thank Dr. Dan Li and Dr. Lucija Tomljenovic from University of British Columbia, and Prof. Jing Cai from Marco University for revising the manuscript. We are also very grateful to the editor and two reviewers for their valuable comments on the earlier version of this paper.

Competing interests

The authors declare that they have no competing interests.

Funding

This work was supported by the 973 Program (2015CB755704), the Hundred Talent Program of the Chinese Academy of Sciences and the National Natural Science Foundation of China (31470215) to HJ, as well as by Tianjin Research Program of Application Foundation and Advanced Technology (15JCY-BJC24200), and the National Natural Science Foundation of China (31501041) to WD.

References

1. Facchini PJ, Bohlmann J, Covello PS, De LV, Mahadevan R, Page JE, Ro DK, Sensen CW, Storms R, Martin VJ. Synthetic biosystems for the production of high-value plant metabolites. Trends Biotechnol. 2012;30:127–31.
2. Brower V. Back to nature: extinction of medicinal plants threatens drug discovery. Cancerspectrum Knowl Environ. 2008;100:838–9.
3. Wang Y, Chen S, Yu O. Metabolic engineering of flavonoids in plants and microorganisms. Appl Microbiol Biotechnol. 2011;91:949.
4. Li JWH, Vederas JC. Drug discovery and natural products: end of an era or an endless frontier? Science. 2009;325:161–5.
5. Du J, Shao ZY, Zhao HM. Engineering microbial factories for synthesis of value-added products. J Ind Microbiol Biotechnol. 2011;38:873–90.

6.   Huang B, Guo J, Yi B, Yu X, Sun L, Chen W. Heterologous production of secondary metabolites as pharmaceuticals in *Saccharomyces cerevisiae*. Biotechnol Lett. 2008;30:1121–37.

7.   Chemler JA, Koffas MA. Metabolic engineering for plant natural product biosynthesis in microbes. Curr Opin Biotechnol. 2008;19:597–605.

8.   Keasling JD. Synthetic biology and the development of tools for metabolic engineering. Metab Eng. 2012;14:189–95.

9.   Suzuki S, Koeduka T, Sugiyama A, Yazaki K, Umezawa T. Microbial production of plant specialized metabolites. Plant Biotechnol. 2014;31:465–82.

10. Berim A, Gang DR. The roles of a flavone-6-hydroxylase and 7-*O*-demethylation in the flavone biosynthetic network of sweet basil. J Biol Chem. 2013;288:1795–805.

11. Miyahara T, Hamada A, Okamoto M, Hirose Y, Sakaguchi K, Hatano S, Ozeki Y. Identification of flavonoid 3′-hydroxylase in the yellow flower of *Delphinium zalil*. J Plant Physiol. 2016;202:92–6.

12. Xiong S, Tian N, Long J, Chen Y, Qin Y, Feng J, Xiao W, Liu S. Molecular cloning and characterization of a flavanone 3-Hydroxylase gene from *Artemisia annua* L. Plant Physiol Biochem. 2016;105:29–36.

13. Gosch C, Halbwirth H, Stich K. Phloridzin: biosynthesis, distribution and physiological relevance in plants. Phytochemistry. 2010;71:838–43.

14. Guo J, Ma X, Cai Y, Ma Y, Zhan Z, Zhou YJ, Liu W, Guan M, Yang J, Cui G, et al. Cytochrome P450 promiscuity leads to a bifurcating biosynthetic pathway for tanshinones. New Phytol. 2016;210:525–34.

15. Hileman LC, Drea S, Martino G, Litt A, Irish VF. Virus-induced gene silencing is an effective tool for assaying gene function in the basal eudicot species *Papaver somniferum* (opium poppy). Plant J. 2005;44:334–41.

16. Di P, Zhang L, Chen J, Tan H, Xiao Y, Dong X, Zhou X, Chen W. [13]C tracer reveals phenolic acids biosynthesis in hairy root cultures of *Salvia miltiorrhiza*. ACS Chem Biol. 2013;8:1537.

17. Medema MH, Osbourn A. Computational genomic identification and functional reconstitution of plant natural product biosynthetic pathways. Nat Prod Rep. 2016;33:951–62.

18. Xu Z, Peters RJ, Weirather J, Luo H, Liao B, Zhang X, Zhu Y, Ji A, Zhang B, Hu S, et al. Full-length transcriptome sequences and splice variants obtained by a combination of sequencing platforms applied to different root tissues of *Salvia miltiorrhiza* and tanshinone biosynthesis. Plant J. 2015;82:951–61.

19. Qu C, Zhao H, Fu F, Zhang K, Yuan J, Liu L, Wang R, Xu X, Lu K, Li JN. Molecular mapping and QTL for expression profiles of flavonoid genes in *Brassica napus*. Front Plant Sci. 2016;7:1691.

20. Shang Y, Ma Y, Zhou Y, Zhang H, Duan L, Chen H, Zeng J, Zhou Q, Wang S, Gu W, et al. Biosynthesis, regulation, and domestication of bitterness in cucumber. Science. 2014;346:1084–8.

21. Zhou Y, Ma Y, Zeng J, Duan L, Xue X, Wang H, Lin T, Liu Z, Zeng K, Zhong Y, et al. Convergence and divergence of bitterness biosynthesis and regulation in Cucurbitaceae. Nat Plants. 2016;2:16183.

22. Itkin M, Davidovich-Rikanati R, Cohen S, Portnoy V, Doron-Faigenboim A, Oren E, Freilich S, Tzuri G, Baranes N, Shen S, et al. The biosynthetic pathway of the nonsugar, high-intensity sweetener mogroside V from *Siraitia grosvenorii*. Proc Natl Acad Sci USA. 2016;113:E7619–28.

23. Yan X, Fan Y, Wei W, Wang P, Liu Q, Wei Y, Zhang L, Zhao G, Yue J, Zhou Z. Production of bioactive ginsenoside compound K in metabolically engineered yeast. Cell Res. 2014;24:770–3.

24. Wang P, Wei Y, Fan Y, Liu Q, Wei W, Yang C, Zhang L, Zhao G, Yue J, Yan X, Zhou Z. Production of bioactive ginsenosides Rh2 and Rg3 by metabolically engineered yeasts. Metab Eng. 2015;29:97–105.

25. Lau W, Sattely ES. Six enzymes from mayapple that complete the biosynthetic pathway to the etoposide aglycone. Science. 2015;349:1224–8.

26. Matasci N, Hung LH, Yan Z, Carpenter EJ, Wickett NJ, Mirarab S, Nguyen N, Warnow T, Ayyampalayam S, Barker M. Data access for the 1,000 Plants (1KP) project. GigaScience. 2014;3:17.

27. Chen S, Xiang L, Guo X, Li Q. An introduction to the medicinal plant genome project. Front Med. 2011;5:178–84.

28. Goodstein DM, Shu S, Howson R, Neupane R, Hayes RD, Fazo J, Mitros T, Dirks W, Hellsten U, Putnam N. Phytozome: a comparative platform for green plant genomics. Nucleic Acids Res. 2012;40:D1178–86.

29. Kanehisa MGS. KEGG: kyoto encyclopedia of genes and genomes. Nucleic Acids Res. 2000;28:27–30.

30. Tan GY, Deng Z, Liu T. Recent advances in the elucidation of enzymatic function in natural product biosynthesis. F1000Res. 2015. doi:10.12688/f1000research.7187.2.

31. Galanie S, Thodey K, Trenchard IJ, Interrante MF, Smolke CD. Complete biosynthesis of opioids in yeast. Science. 2015;349:1095–100.

32. Nakagawa A, Matsumura E, Koyanagi T, Katayama T, Kawano N, Yoshimatsu K, Yamamoto K, Kumagai H, Sato F, Minami H. Total biosynthesis of opiates by stepwise fermentation using engineered *Escherichia coli*. Nat Commun. 2016;7:10390.

33. Lee D, Lloyd ND, Pretorius IS, Borneman AR. Heterologous production of raspberry ketone in the wine yeast *Saccharomyces cerevisiae* via pathway engineering and synthetic enzyme fusion. Microb Cell Fact. 2016;15:49.

34. Bai Y, Bi H, Zhuang Y, Liu C, Cai T, Liu X, Zhang X, Liu T, Ma Y. Production of salidroside in metabolically engineered *Escherichia coli*. Sci Rep. 2014;4:6640.

35. Bai Y, Yin H, Bi H, Zhuang Y, Liu T, Ma Y. De novo biosynthesis of Gastrodin in *Escherichia coli*. Metab Eng. 2016;35:138–47.

36. Luo Y, Li BZ, Liu D, Zhang L, Chen Y, Jia B, Zeng BX, Zhao H, Yuan YJ. Engineered biosynthesis of natural products in heterologous hosts. Chem Soc Rev. 2015;44:5265–90.

37. Tholl D. Biotechnology of isoprenoid. Adv Biochem Eng Biotechnol. 2015;148:63–106.

38. Dai Z, Liu Y, Zhang X, Shi M, Wang B, Wang D, Huang L, Zhang X. Metabolic engineering of *Saccharomyces cerevisiae* for production of ginsenosides. Metab Eng. 2013;20:146–56.

39. Ro DK, Paradise EM, Ouellet M, Fisher KJ, Newman KL, Ndungu JM, Ho KA, Eachus RA, Ham TS, Kirby J, et al. Production of the antimalarial drug precursor artemisinic acid in engineered yeast. Nature. 2006;440:940–3.

40. Westfall PJ, Pitera DJ, Lenihan JR, Eng D, Woolard FX, Regentin R, Horning T, Tsuruta H, Melis DJ, Owens A, et al. Production of amorphadiene in yeast, and its conversion to dihydroartemisinic acid, precursor to the antimalarial agent artemisinin. Proc Natl Acad Sci USA. 2012;109:E111–8.

41. Xie W, Ye L, Lv X, Xu H, Yu H. Sequential control of biosynthetic pathways for balanced utilization of metabolic intermediates in *Saccharomyces cerevisiae*. Metab Eng. 2015;28:8–18.

42. Koopman F, Beekwilder J, Crimi B, Van HA, Hall RD, Bosch D, van Maris AJ, Pronk JT, Daran JM. De novo production of the flavonoid naringenin in engineered *Saccharomyces cerevisiae*. Microb Cell Fact. 2012;11:402–7.

43. Jiang H, Wood KV, Morgan JA. Metabolic engineering of the phenylpropanoid pathway in *Saccharomyces cerevisiae*. Appl Environ Microbiol. 2005;71:2962–9.

44. Leonard E, Lim KH, Saw PN, Koffas MA. Engineering central metabolic pathways for high-level flavonoid production in *Escherichia coli*. Appl Environ Microbiol. 2007;73:3877–86.

45. Cao W, Ma W, Zhang B, Wang X, Chen K, Li Y, Ouyang P. Improved pinocembrin production in *Escherichia coli* by engineering fatty acid synthesis. J Ind Microbiol Biotechnol. 2016;43:557–66.

46. Brockman IM, Prather KL. Dynamic metabolic engineering: new strategies for developing responsive cell factories. Biotechnol J. 2015;10:1360–9.

47. Xu P, Li L, Zhang F, Stephanopoulos G, Koffas M. Improving fatty acids production by engineering dynamic pathway regulation and metabolic control. Proc Natl Acad Sci USA. 2014;111:11299–304.

48. Xie W, Lv X, Ye L, Zhou P, Yu H. Construction of lycopene-overproducing Saccharomyces cerevisiae by combining directed evolution and metabolic engineering. Metab Eng. 2015;30:69–78.

49. van Summeren-Wesenhagen PV, Marienhagen J. Metabolic engineering of *Escherichia coli* for the synthesis of the plant polyphenol pinosylvin. Appl Environ Microbiol. 2015;81:840–9.

50. Dietrich JA, Yoshikuni Y, Fisher KJ, Woolard FX, Ockey D, Mcphee DJ, Renninger NS, Chang MC, Baker D, Keasling JD. A novel semi-biosynthetic route for artemisinin production using engineered substrate-promiscuous P450(BM3). ACS Chem Biol. 2009;4:261–7.

51. Biggs BW, Lim CG, Sagliani K, Shankar S, Stephanopoulos G, De Mey M, Ajikumar PK. Overcoming heterologous protein interdependency to optimize P450-mediated Taxol precursor synthesis in *Escherichia coli*. Proc Natl Acad Sci USA. 2016;113:3209–14.

52. Liu Y, Yan Z, Lu X, Xiao D, Jiang H. Improving the catalytic activity of isopentenyl phosphate kinase through protein coevolution analysis. Sci Rep. 2016;6:24117.

53. Morita H, Yamashita M, Shi SP, Wakimoto T, Kondo S, Kato R, Sugio S, Kohno T, Abe I. Synthesis of unnatural alkaloid scaffolds by exploiting plant polyketide synthase. Proc Natl Acad Sci USA. 2011;108:13504–9.

54. Nicolaou SA, Gaida SM, Papoutsakis ET. A comparative view of metabolite and substrate stress and tolerance in microbial bioprocessing: from biofuels and chemicals, to biocatalysis and bioremediation. Metab Eng. 2010;12:307–31.

55. Lv H, Li J, Wu Y, Garyali S, Wang Y. Transporter and its engineering for secondary metabolites. Appl Microbiol Biotechnol. 2016;100:1–12.

56. Dunlop MJ, Dossani ZY, Szmidt HL, Chu HC, Lee TS, Keasling JD, Hadi MZ, Mukhopadhyay A. Engineering microbial biofuel tolerance and export using efflux pumps. Mol Syst Biol. 2011;7:361–73.

57. Katz M, Durhuus T, Smits HP, Förster J. Production of metabolites US 20130209613 A1. US; 2013.

58. Wang JF, Xiong ZQ, Li SY, Wang Y. Enhancing isoprenoid production through systematically assembling and modulating efflux pumps in Escherichia coli. Appl Microbiol Biotechnol. 2013;97:8057–67.

59. Zhang C, Chen X, Stephanopoulos G, Too HP. Efflux transporter engineering markedly improves amorphadiene production in Escherichia coli. Biotechnol Bioeng. 2016;113:1755–63.

60. Zhang R, Huang J, Zhu J, Xie X, Tang Q, Chen X, Luo J, Luo Z. Isolation and characterization of a novel PDR-type ABC transporter gene PgPDR3 from Panax ginseng C.A. Meyer induced by methyl jasmonate. Mol Biol Rep. 2013;40:6195–204.

61. Cao H, Nuruzzaman M, Xiu H, Huang J, Wu K, Chen X, Li J, Wang L, Jeong JH, Park SJ. Transcriptome analysis of methyl jasmonate-elicited panax ginseng adventitious roots to discover putative ginsenoside biosynthesis and transport genes. Int J Mol Sci. 2015;16:3035–57.

62. Leonard E, Yan Y, Fowler ZL, Li Z, Lim CG, Lim KH, Koffas MA. Strain improvement of recombinant Escherichia coli for efficient production of plant flavonoids. Mol Pharm. 2008;5:257–65.

63. Barbacka K, Baer-Dubowska W. Searching for artemisinin production improvement in plants and microorganisms. Curr Pharm Biotechnol. 2011;12:1743–51.

64. WHO. World Malaria Report. 2015. http://www.who.int/malaria/publications/world-malaria-report-2015/report/en/.

65. Abdin MZ, Alam P. Genetic engineering of artemisinin biosynthesis: prospects to improve its production. Acta Physiol Plant. 2015;37:1–12.

66. Paddon CJ, Westfall PJ, Pitera DJ, Benjamin K, Fisher K, McPhee D, Leavell MD, Tai A, Main A, Eng D, et al. High-level semi-synthetic production of the potent antimalarial artemisinin. Nature. 2013;496:528–32.

67. Lenihan JR, Tsuruta H, Diola D, Renninger NS, Regentin R. Developing an industrial artemisinic acid fermentation process to support the cost-effective production of antimalarial artemisinin-based combination therapies. Biotechnol Prog. 2008;24:1026–32.

68. Baadhe RR, Mekala NK, Parcha SR, Prameela Devi Y. Combination of ERG9 repression and enzyme fusion technology for improved production of amorphadiene in Saccharomyces cerevisiae. J Anal Methods Chem. 2013;2013:140469.

69. Martin VJ, Pitera DJ, Withers ST, Newman JD, Keasling JD. Engineering a mevalonate pathway in Escherichia coli for production of terpenoids. Nat Biotechnol. 2003;21:796–802.

70. Chun-Fu WU, Yang JY, Wang F, Wang XX. Resveratrol: botanical origin, pharmacological activity and applications. Chin J Nat Med. 2013;11:1–15.

71. Dgw A, Wyl B, Gtc A. School AM, University NT, Tong N: a simple method for the isolation and purification of resveratrol from Polygonum cuspidatum. J Pharm Anal. 2013;3:241–7.

72. Ge S, Yin T, Xu B, Gao S, Hu M. Curcumin affects phase II disposition of resveratrol through inhibiting efflux transporters MRP2 and BCRP. Pharm Res. 2016;33:590–602.

73. Mei YZ, Liu RX, Wang DP, Wang X, Dai CC. Biocatalysis and biotransformation of resveratrol in microorganisms. Biotech Lett. 2015;37:9–18.

74. Beekwilder J, Wolswinkel R, Jonker H, Hall R, de Vos CH, Bovy A. Production of resveratrol in recombinant microorganisms. Appl Environ Microbiol. 2006;72:5670–2.

75. Watts KT, Lee PC, Schmidt-Dannert C. Biosynthesis of plant-specific stilbene polyketides in metabolically engineered Escherichia coli. BMC Biotechnol. 2006;6:1–12.

76. Katsuyama Y, Funa N, Miyahisa I, Horinouchi S. Synthesis of unnatural flavonoids and stilbenes by exploiting the plant biosynthetic pathway in Escherichia coli. Chem Biol. 2007;14:613–21.

77. Wu J, Liu P, Fan Y, Han B, Du G, Zhou J, Chen J. Multivariate modular metabolic engineering of Escherichia coli to produce resveratrol from l-tyrosine. J Biotechnol. 2013;167:404–11.

78. Lim CG, Fowler ZL, Hueller T, Schaffer S, Koffas MA. High-yield resveratrol production in engineered Escherichia coli. Appl Environ Microbiol. 2011;77:3451–60.

79. Durhuus T, Förster J, Katz M, Smits HP. Production of metabolites. WO; 2011.

80. Ye VM, Bhatia SK. Pathway engineering strategies for production of beneficial carotenoids in microbial hosts. Biotech Lett. 2012;34:1405–14.

81. Omoni AO, Aluko RE. The anti-carcinogenic and anti-atherogenic effects of lycopene: a review. Trends Food Sci Technol. 2005;16:344–50.

82. Muller L, Caris-Veyrat C, Lowe G, Bohm V. Lycopene and its antioxidant role in the prevention of cardiovascular diseases-a critical review. Crit Rev Food Sci Nutr. 2016;56:1868–79.

83. Maritim AC, Sanders RA, Iii JBW. Diabetes, oxidative stress, and antioxidants: a review. J Biochem Mol Toxicol. 2003;17:24–38.

84. Research B. The global market for carotenoids FOD025E. 2015. http://www.bccresearch.com/market-research/food-and-beverage/carotenoids-global-market-report-fod025e.html.

85. Matagómez LC, Montañez JC, Méndezzavala A, Aguilar CN. Biotechnological production of carotenoids by yeasts: an overview. Microb Cell Fact. 2014;13:1–11.

86. Alper H, Miyaoku K, Stephanopoulos G. Construction of lycopene-overproducing E. coli strains by combining systematic and combinatorial gene knockout targets. Nat Biotechnol. 2005;23:612–6.

87. Cho NR, Park MS, Lee DH, Chung HS, Kim JK. Method of producing lycopene using recombinant Esherichia coli. US; 2014.

88. Marcos Rodríguez AT, Estrella De CA, Costa Perez J, Oliver Ruiz MA, Fraile Yecora N, De La FMJL, Rodríguez Saiz M, Diez Garcia B, Peiro Cezon E, Muñoz Ruiz A. Method of producing lycopene through the fermentation of selected strains of Blackeslea trispora, formulations and uses of the lycopene thus obtained; 2010.

89. Mehta BJ, Obraztsova IN, Cerdáolmedo E. Mutants and intersexual heterokaryons of Blakeslea trispora for production of beta-carotene and lycopene. Appl Environ Microbiol. 2003;69:4043–8.

90. Schmidt I, Schewe H, Gassel S, Jin C, Buckingham J, Hümbelin M, Sandmann G, Schrader J. Biotechnological production of astaxanthin with Phaffia rhodozyma/Xanthophyllomyces dendrorhous. Appl Microbiol Biotechnol. 2011;89:555–71.

91. Price JV, Chen L, Whitaker WB, Papoutsakis E, Chen W. Scaffoldless engineered enzyme assembly for enhanced methanol utilization. Proc Natl Acad Sci USA. 2016;113:12691–6. doi:10.1073/pnas.1601797113.

92. Antonovsky N, Gleizer S, Noor E, Zohar Y, Herz E, Barenholz U, Zelcbuch L, Amram S, Wides A, Tepper N, et al. Sugar synthesis from CO$_2$ in Escherichia coli. Cell. 2016;166:115–25.

93. Kallscheuer N, Vogt M, Stenzel A, Gatgens J, Bott M, Marienhagen J. Construction of a Corynebacterium glutamicum platform strain for the production of stilbenes and (2S)-flavanones. Metab Eng. 2016;38:47–55.

94. Park SR. Enhanced flavonoid production in Streptomyces venezuelae via metabolic engineering. J Microbiol Biotechnol. 2011;21:1143–6.

95. Araya-Garay JM, Feijoo-Siota L, Rosa-dos-Santos F, Veiga-Crespo P, Villa TG. Construction of new Pichia pastoris X-33 strains for production of lycopene and beta-carotene. Appl Microbiol Biotechnol. 2012;93:2483–92.

96. Shaw AJ, Lam FH, Hamilton M, Consiglio A, Macewen K, Brevnova EE, Greenhagen E, Latouf WG, South CR, Van DH. Metabolic engineering of microbial competitive advantage for industrial fermentation processes. Science. 2016;353:583–6.

97. Yue H, Ling C, Yang T, Chen X, Chen Y, Deng H, Wu Q, Chen J, Chen GQ. A seawater-based open and continuous process for polyhydroxyalkanoates production by recombinant Halomonas campaniensis LS21 grown in mixed substrates. Biotechnol Biofuels. 2014;7:1–12.

3

Purification and characterization of a novel cold adapted fungal glucoamylase

Mario Carrasco, Jennifer Alcaíno, Víctor Cifuentes and Marcelo Baeza*

Abstract

Background: Amylases are used in various industrial processes and a key requirement for the efficiency of these processes is the use of enzymes with high catalytic activity at ambient temperature. Unfortunately, most amylases isolated from bacteria and filamentous fungi have optimal activity above 45 °C and low pH. For example, the most commonly used industrial glucoamylases, a type of amylase that degrades starch to glucose, are produced by *Aspergillus* strains displaying optimal activities at 45–60 °C. Thus, isolating new amylases with optimal activity at ambient temperature is essential for improving industrial processes. In this report, a glucoamylase secreted by the cold-adapted yeast *Tetracladium* sp. was isolated and biochemically characterized.

Results: The effects of physicochemical parameters on enzyme activity were analyzed, and pH and temperature were found to be key factors modulating the glucoamylase activity. The optimal conditions for enzyme activity were 30 °C and pH 6.0, and the K_m and k_{cat} using soluble starch as substrate were 4.5 g/L and 45 min^{-1}, respectively. Possible amylase or glucoamylase encoding genes were identified, and their transcript levels using glucose or soluble starch as the sole carbon source were analyzed. Transcription levels were highest in medium supplemented with soluble starch for the potential glucoamylase encoding gene. Comparison of the structural model of the identified *Tetracladium* sp. glucoamylase with the solved structure of the *Hypocrea jecorina* glucoamylase revealed unique structural features that may explain the thermal lability of the glucoamylase from *Tetracladium* sp.

Conclusion: The glucoamylase secreted by *Tetracladium* sp. is a novel cold-adapted enzyme and its properties should render this enzyme suitable for use in industrial processes that require cold-active amylases, such as biofuel production.

Keywords: Fungal amylase, Cold-adapted amylase, *Tetracladium sp.*, Antarctic fungi

Background

A large proportion of the earth's biosphere is constantly below 5 °C and these cold environments are inhabited by cold-adapted microorganisms among other forms of life. Cold-adapted yeasts have attracted the attention of scientists because these yeast species have evolved to adapt to cold climates, and thus have significant potential for applications in diverse fields of industry [1–5]. A very well studied feature of cold-adapted yeasts is the presence of hydrolytic enzymes, which are secreted to aid the uptake of nutrients available in their surrounding environment. These cold-active enzymes have many applications in processes requiring high activity at low or mild temperatures [6–8]. Examples are the cold-active amylases, lipases, proteases, cellulases, pectinases and esterases, which are applied in food, wine, textile and detergent industries [4, 9]. Amylases hydrolyze α-glucosidic bonds in starch and according to their catalytic mechanism they are classified into three main groups: (i) α-amylase, which disrupts α-1,4-glycosidic linkages; (ii) β-amylase, which catalyzes the hydrolysis of the second α-1,4 glycosidic bonds from the non-reducing end of starch; and (iii) glucoamylase (an α-glucosidase), which acts on both, α-1,4 and α-1,6 glycosidic bonds from the non-reducing end of the starch molecule [10–13]. Amylases are used in several industrial processes, including the production of high-fructose corn syrup, as

---

*Correspondence: mbaeza@u.uchile.cl
Departamento de Ciencias Ecológicas, Facultad de Ciencias, Universidad de Chile, Las Palmeras 342, Casilla 653, Santiago, Chile

additives in detergent formulations, in wool treatment and to obtain fermentable sugars from starch-rich wastes that are used as a substrate for biofuels production [13, 14]. The efficient microbial production of biofuels from raw starch wastes requires the complete degradation of starch, which is currently accomplished by the addition of α-amylase and glucoamylase during the fermentative process to release glucose as the primary end product [15, 16]. The majority of glucoamylases present in bacteria and fungi have optimal activity above 45 °C and at low pH [16–18]. The glucoamylases used in industrial processes, mainly derived from *Aspergillus* strains, display the highest activity at temperatures between 45 and 60 °C [16]. Currently, there is strong interest in finding amylases with better performance at lower temperatures than commercially available amylases, because these enzymes would circumvent the requirement of heating during the reaction process thereby minimizing costs [19]. Several fungi isolated from soil samples from King George Island in the sub-Antarctic region grew on soluble starch as the sole carbon source and displayed extracellular amylase activity. The highest amylase activity was found in samples obtained from the yeast *Tetracladium* sp., and in preliminary characterizations the molecular weight of the enzyme was found to be ~80 kDa [20, 21].

In this report, an amylase from the cold-adapted yeast *Tetracladium* sp. was purified and biochemically characterized. In addition, the amylase encoding gene was identified and its expression was analyzed through RNA-seq when using soluble starch or glucose as the sole carbon source. A model of the enzyme was constructed, revealing several features that are characteristic in cold-adapted enzymes. The optimal conditions for enzyme activity, thermal stability and kinetic parameters were determined. The obtained results suggest that the characterized glucoamylase secreted by *Tetracladium* sp. is a novel cold-adapted enzyme that may be useful in processes where cold-active amylases are required, such as biofuel production.

Results

Enzyme purification and characterization

Tetracladium sp. extracellular protein samples were obtained by precipitation with ammonium sulfate at 80% saturation of cell-free supernatants of cultures grown using starch as the sole carbon source. Protein separation was attempted using ion-exchange or gel filtration chromatography, obtaining a suitable protein separation only with the last method (Additional file 1). The amylase activity and the protein profile of each fraction were determined. Two main peaks centered at fractions 42 (peak 1) and 72 (peak 2) were observed, but amylase activity was only detected in peak 1 (Additional file 1A).

A single protein band of 84 kDa was observed by SDS-PAGE analysis in fractions displaying amylase activity (Additional file 1A). This protein is glycosylated (Additional file 1B, C) and has a relative molecular weight (rMW) of 80 kDa under non-reducing conditions, as determined by gel filtration chromatography.

Characterization of enzymatic activity

To evaluate the specificity of the amylase secreted by *Tetracladium* sp., enzymatic activity assays were performed using either ethylidene-pNP-G7 (E-pNP-G7) or 4-nitrophenyl α-D-glucopyranoside (4-NPGP), which are α-amylase and α-glucosidase substrates, respectively. As shown in Fig. 1a, the enzyme was able to use 4-NPGP as a substrate, but not E-pNP-G7, indicating that the *Tetracladium* sp. enzyme is an α-glucosidase. A characteristic of α-glucosidases is the release of glucose as the end product of starch hydrolysis. To test the possible release of glucose, assays were performed using 5 g/L soluble starch as the substrate and glucose production was quantified during the reaction using the DNS method. As shown in Fig. 1b, the amylase from *Tetracladium* sp. releases glucose from soluble starch, showing that this amylase is an α-glucosidase. Two slopes could be distinguished along the reaction curve; the enzyme had a rapid reaction rate until 10 min reaching 2.5 g/L of glucose and then a slower reaction rate after 10 min (Fig. 1b).

A two-level Plackett–Burman design was applied to determine the influence of temperature, pH, Ca^{2+}, Mg^{2+} and soluble starch concentration on enzyme activity. The reaction was followed by measuring the release of glucose in each trial (eight in total) and the reaction velocities were calculated from the slopes of each curve. The effect of each variable was calculated, and it was found that temperature and pH were the two principal factors that influenced strongly the α-glucosidase activity, whereas the ranges tested of soluble starch concentration, and Ca^{2+} and Mg^{2+} levels, affected the enzyme activity to a lesser extent (Additional file 2).

The optimal pH and temperature for α-glucosidase activity on soluble starch were determined using a central composite design of two levels, and it was found that the highest α-glucosidase activity was reached at pH 6.0 and 30 °C (Fig. 2). The enzyme activity and stability were evaluated at temperatures from 4 to 60 °C using soluble starch as the substrate. The highest activity was observed between 30 and 45 °C. At lower and higher temperatures, the enzyme activity decreased significantly, with a more pronounced activity decrease at temperatures above 45 °C (Fig. 3a). At 22 and 4 °C, the enzyme retained 55 and 25% of its maximal activity, respectively.

The thermal stability of the enzyme was evaluated by incubating the protein sample for 1 h at different

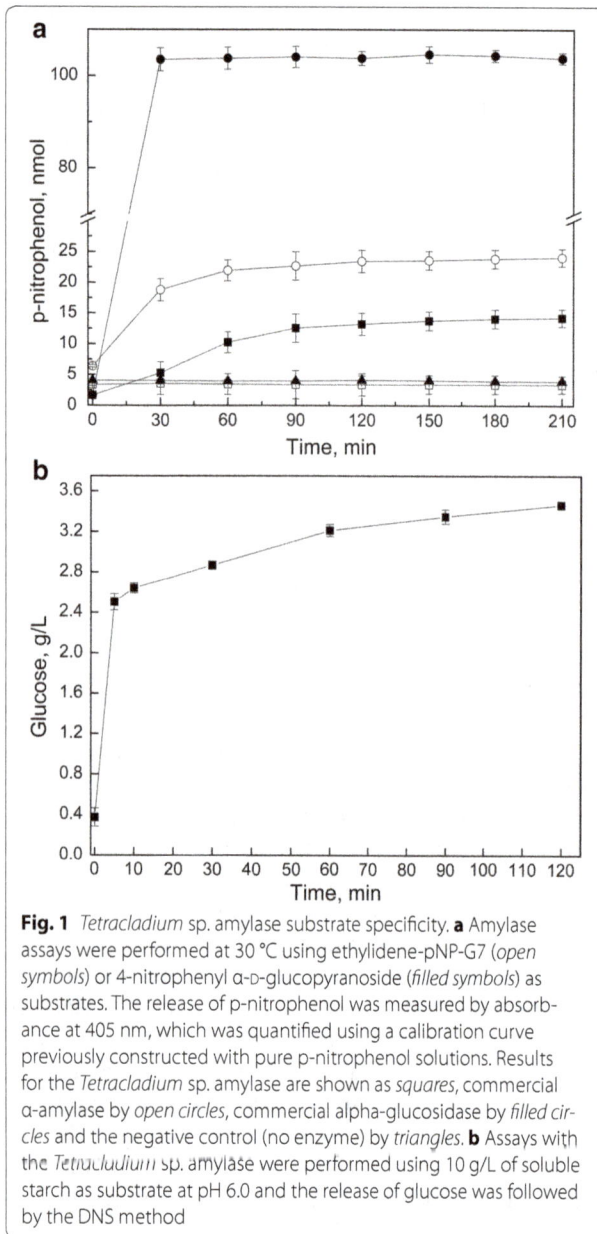

Fig. 1 *Tetracladium* sp. amylase substrate specificity. **a** Amylase assays were performed at 30 °C using ethylidene-pNP-G7 (*open symbols*) or 4-nitrophenyl α-D-glucopyranoside (*filled symbols*) as substrates. The release of p-nitrophenol was measured by absorbance at 405 nm, which was quantified using a calibration curve previously constructed with pure p-nitrophenol solutions. Results for the *Tetracladium* sp. amylase are shown as *squares*, commercial α-amylase by *open circles*, commercial alpha-glucosidase by *filled circles* and the negative control (no enzyme) by *triangles*. **b** Assays with the *Tetracladium* sp. amylase were performed using 10 g/L of soluble starch as substrate at pH 6.0 and the release of glucose was followed by the DNS method

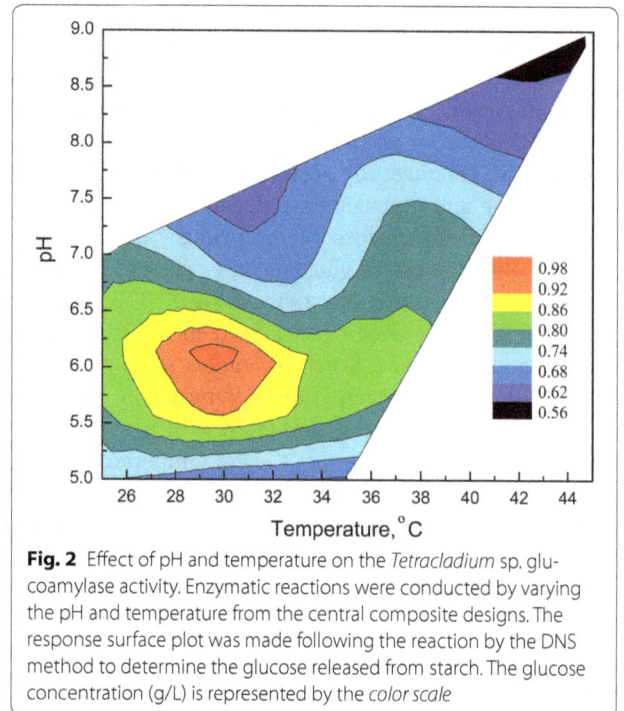

Fig. 2 Effect of pH and temperature on the *Tetracladium* sp. glucoamylase activity. Enzymatic reactions were conducted by varying the pH and temperature from the central composite designs. The response surface plot was made following the reaction by the DNS method to determine the glucose released from starch. The glucose concentration (g/L) is represented by the *color scale*

temperatures, and subsequently the enzyme activity was assayed under optimal conditions. This test revealed that the enzyme was stable at temperatures from 4 to 37 °C (Fig. 3a). However, the stability decreased at temperatures above 40 °C, maintaining only 10% of its optimal activity after incubation for 1 h at 50 °C. Figure 3b shows the enzymatic activity after 2–8 h of incubation at different temperatures. The enzyme maintained its activity after incubation at 22 and 30 °C even after 6–8 h of incubation. When incubated at 40 °C, the enzyme activity displayed a linear decrease as a function of incubation period with 70 and 40% of its optimal activity after 2 and 8 h incubation,

respectively. The activity loss was very pronounced when the protein sample was incubated at 50 °C for different periods, maintaining only 10% of the maximal activity after only 2 h of incubation at this temperature. Steady-state experiments were used to determine the kinetics of *Tetracladium* sp. α-glucosidase using soluble starch as substrate. Reaction velocities at different soluble starch concentrations and the double-reciprocal plot are shown in Fig. 4. The enzyme displayed Michaelis–Menten kinetics, and the calculated K_m, k_{cat} and k_{cat}/K_m values were 4.5 g/L, 45 min^{-1} and 10 g/L min, respectively.

Identification and characterization of the α-glucosidase encoding gene

The available genomic and transcriptomic data of *Tetracladium* sp. taken from our research efforts were used for the identification and characterization of the gene encoding the α-glucosidase. Bioinformatic tools and the NCBI databases were used to predict and annotate the genes and Open Reading Frames (ORFs) of *Tetracladium* sp. Two putative encoding α-amylase genes and one glucoamylase encoding gene were identified. The expression patterns of these putative genes were evaluated by analysis of the transcriptomes obtained from *Tetracladium* sp. when grown in medium supplemented either with glucose or soluble starch as the sole carbon source. The Fragments Per Kilobase Of Exon Per Million Fragments Mapped (FPKM) values were calculated for the three putative genes and are presented in Table 1. Higher

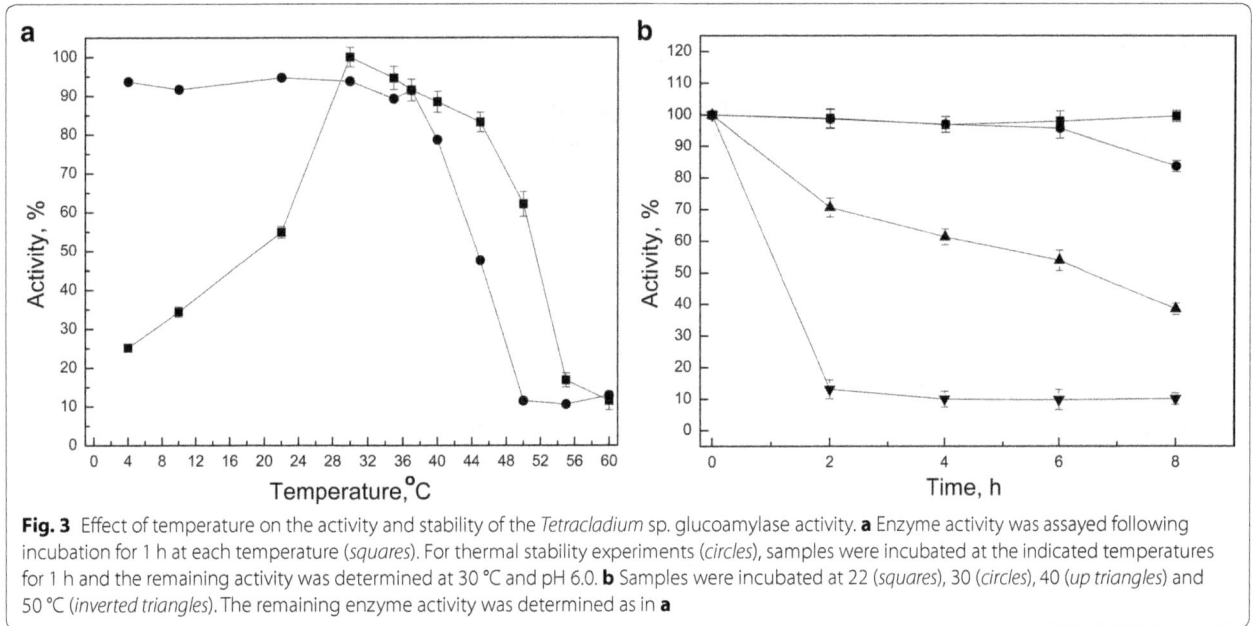

Fig. 3 Effect of temperature on the activity and stability of the *Tetracladium* sp. glucoamylase activity. **a** Enzyme activity was assayed following incubation for 1 h at each temperature (*squares*). For thermal stability experiments (*circles*), samples were incubated at the indicated temperatures for 1 h and the remaining activity was determined at 30 °C and pH 6.0. **b** Samples were incubated at 22 (*squares*), 30 (*circles*), 40 (*up triangles*) and 50 °C (*inverted triangles*). The remaining enzyme activity was determined as in **a**

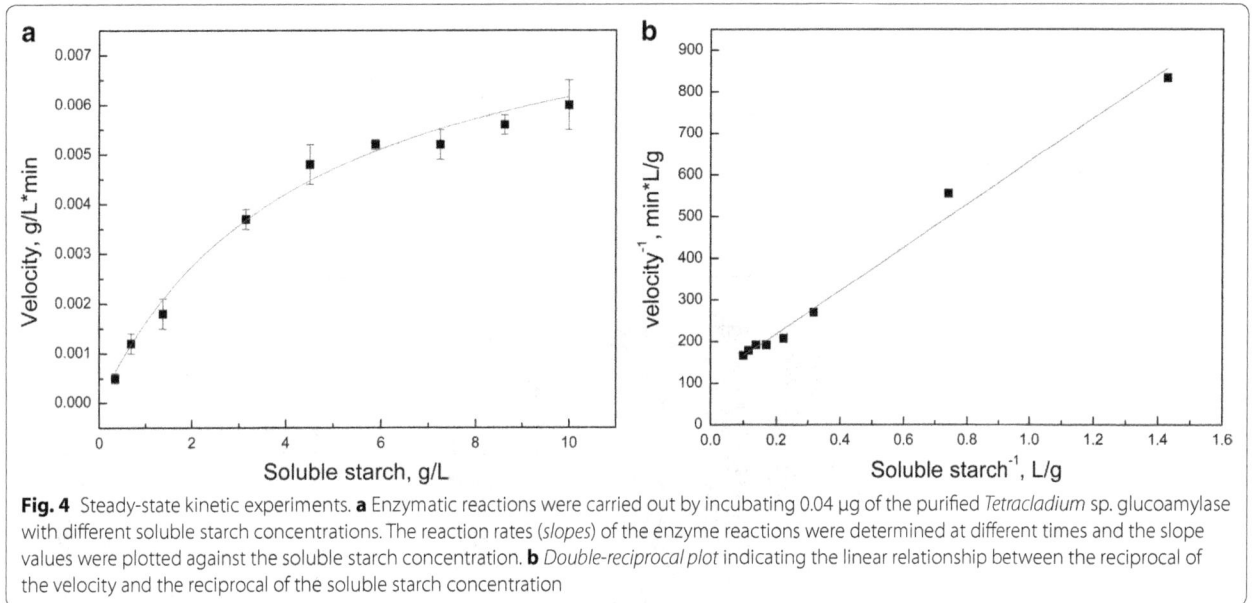

Fig. 4 Steady-state kinetic experiments. **a** Enzymatic reactions were carried out by incubating 0.04 μg of the purified *Tetracladium* sp. glucoamylase with different soluble starch concentrations. The reaction rates (*slopes*) of the enzyme reactions were determined at different times and the slope values were plotted against the soluble starch concentration. **b** *Double-reciprocal plot* indicating the linear relationship between the reciprocal of the velocity and the reciprocal of the soluble starch concentration

Table 1 Expression of putative amylase genes of *Tetracladium* sp. in medium supplemented with soluble starch or glucose

Putative gene	Amylase type	FPMK[a]	
		Glucose	Soluble starch
1.g909.t1	Glucoamylase	1545	1722
0.g2254.t1	Alpha amylase	122	184
2.g884.t1	Alpha amylase	76	7

[a] Fragments Per Kilobase Of Exon Per Million Fragments Mapped

expression levels were found for the putative glucoamylase encoding gene (1.g909.t1), with expression levels similar when either carbon source was used. The two putative α-amylase genes showed expression levels that were approximately tenfold lower than the putative glucoamylase gene. The translated sequences of these three putative genes were compared with peptide mass fingerprint results obtained previously [20]. None of the peptides mapped to putative α-amylase genes, whereas 20 peptides matched the translated sequence of the 1.g909.

t1 gene (eight with 100% identity). Twelve of the 20 peptides gave BLAST hits to a fungal glucoamylase precursor deposited in the NCBI protein database. The 1.g909.t1 gene (2.1 kb, sequence in Additional file 3) includes four exons (Fig. 5) with a 1965 nt ORF, which encodes for a protein that is 655 amino acids in length (henceforth called AmyT1). The promoter region was predicted (Fig. 5), which yields a large 5′ UTR region.

Phylogenetic analysis based on the encoding nucleotide sequence revealed that the *Tetracladium* sp. AmyT1 groups with sequences deposited as hypothetical Coding Sequences (CDS) from *Sclerotinia sclerotiorum* and *Botrytis cinerea*, and glycosyl hydrolase CDS from *Phialocephala* and *Trichoderma* species. When the analysis was performed using the deduced protein sequence of AmyT1, the amylase grouped with glucoamylases and hypothetical proteins from *Verticilium* and *Pestalotiopsis* species (Additional file 4). Alignment of the amino acid sequence of AmyT1 and glucoamylases from other fungi showed the presence of conserved residues that are implicated in catalysis (Y47, W51, W120, E179, R309, Y315 and E404) and starch binding (W525, T527, K560, V570, W572 and N577) (Fig. 6). Protein domain analysis predicted that AmyT1 has a 35 amino acid N-terminal export signal, an N-terminal catalytic domain of 407 amino acids, a C-terminal starch binding domain of 108 amino acids and a serine/threonine linker of 47 amino acids between the two domains (Fig. 6b).

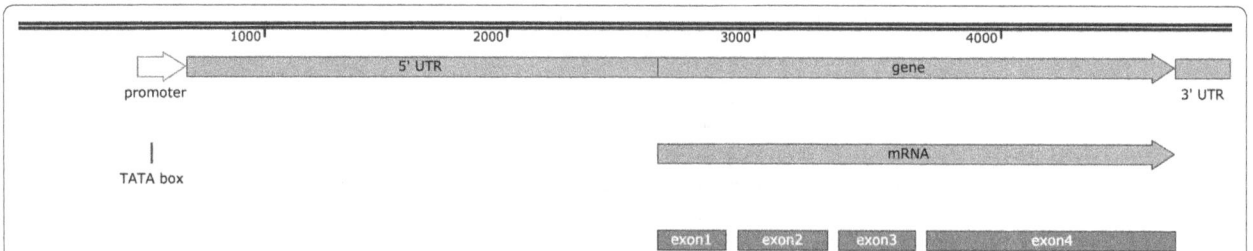

Fig. 5 *Tetracladium* sp. glucoamylase gene structure. The mature mRNA has an ORF of 1965 nt that encodes a protein 655 amino acids in length. The identified promoter (200 bp) is shown, which has a TATA-box located at the −40 position from the start site of transcription. The mRNA has a 5′ UTR and a 3′ UTR region of 1928 and 196 bp, respectively. The predicted transcription factors binding sites (SWI6, SAP1 and GCN4) are indicated

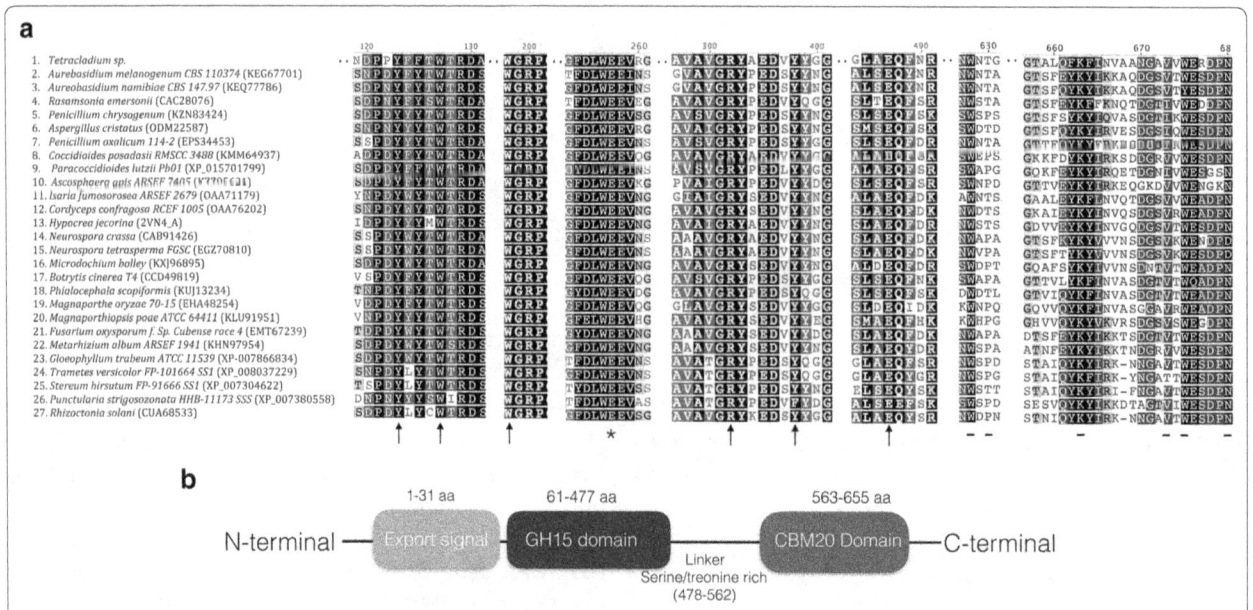

Fig. 6 Fungal glucoamylases alignment. **a** Amino acid sequence alignment was performed using fungal glucoamylases deposited in the NCBI database and access numbers are indicated in *parenthesis*. Amino acids involved in substrate binding in the active site, catalysis and in starch binding in the CBM20 domain are indicated by *black arrows*, *asterisks* and the *—symbol*, respectively. **b** Schematic representation of the predicted glucoamylase structure. The export peptide, GH15 domain, CBM20 and linker sequence, are shown

Discussion

According to the biochemical and molecular data presented herein, the glucoamylase Amy1T produced by *Tetracladium* sp. is a novel cold-adapted amylase. The predicted molecular weight based on the ORF is 66 kDa, which is lower than the determined rMW value of 80 kDa observed through SDS-PAGE. This difference is probably due to post-translational glycosylation, which is in accord with previous studies showing that microbial amylases undergo this post-translational modification [22–26]. The optimal pH for activity of the glucoamylase from *Tetracladium* sp. was 6.0, which is similar to the optimum pH value of other microbial glucoamylases [16]. However, its optimal temperature for activity was 30 °C, which is lower than other glucoamylases (i.e., between 40 and 70 °C). Furthermore, the K_m of the *Tetracladium* sp. glucoamylase towards soluble starch was 4.5 g/L, whereas reported K_m values for microbial glucoamylases are <1.0 g/L [16]. Generally, cold-adapted or cold-active enzymes have higher K_m values than their mesophilic or thermophilic counterparts, which is in accordance with our results.

At the structural level, cold-adapted enzymes are usually more flexible than their thermostable counterparts [27]. The structures of glucoamylases of other fungi such as *Aspergillus awamorii* and *Hypocrea jecorina* (now *Trichoderma reesei*) have been reported [28, 29]. The structure of an enzyme produced by *H. jecorina*, HjGa, was the first crystal structure to include both the catalytic and starch binding domains. This protein shares 69% identity with Amy1T from *Tetracladium* sp., having a 90% coverage. Amy1T was modeled using the crystallographic structure of the glucoamylase from *H. jecorina* (2VN7.1.A) as the template. The superimposition of both structures (Fig. 7a) shows high similarity for both the starch binding domain (SBD) and the catalytic domain (CD). However, differences between the structures were observed mainly for the variable loop of HjGa, which undergoes a large conformational change in the presence of the substrate [29], which is not observed in the Amy1T model. The linker region in the AmyT1 model is 10 residues longer than HjGa and has a lower proline (14.9 vs. 24.3%) and tyrosine (2.1 vs. 5.4%) content. Proline residues have a rigid side chain that confers low flexibility to proteins, whereas tyrosine residues have an aromatic ring that can stabilize protein structures through hydrophobic interactions. These amino acid content differences may explain the more flexible linker of Amy1T when compared with that of HjGa. Generally, cold-adapted enzymes have larger active sites than their mesophilic counterparts [30]; however, no significant differences in the distances between the side chains of the amino acids

involved in catalysis between the Amy1T model (Fig. 7b) and HjGa (Fig. 7c) were observed.

As mentioned above, the optimal temperature for activity of glucoamylase Amy1T was 30 °C. Therefore, this enzyme would be considered as a cold-adapted enzyme, but not as a cold-active enzyme. This is supported by the observed thermal stability of the enzyme, which was highly stable at temperatures between 4 and 37 °C, but activity was rapidly lost at temperatures over 40 °C. The stability and optimal temperature of activity of HjGa are between 45 and 65 °C. Numerous interactions between the SBD and CD are important for enzyme activity and stability of HjGa [29], including electrostatic and hydrogen bonding (T589/R27, H560, E652, A80, E157, H600 and D93) and hydrophobic interactions (F76, V594, I604 and V650). The same amino acid types are not present in Amy1T, which may explain the lower thermal stability of Amy1T when compared with that of HjGa.

The activities of fungal glucoamylases are generally affected by calcium [25, 31], with only a few enzymes showing no dependency on this cation for activity [32]. The activity of the glucoamylase from *Tetracladium* sp. showed no calcium dependency. From an industrial application perspective, this is a desirable characteristic because the addition of calcium salts to any process would increase costs and/or give rise to possible secondary effects. In saccharification processes, the pH must be adjusted from 6–6.5 to 4–4.5 prior to the addition of glucoamylase, because most glucoamylases currently used have low activity at pH 6.0–6.5 [33]. Therefore, the amylase from *Tetracladium* sp., that has optimal activity at pH 6.0, represents a good alternative enzyme for these processes. Furthermore, the thermal properties of Amy1T should facilitate process operations at lower temperatures. This feature should save energy and facilitates a simpler enzyme inactivation step by mild heating, and such mild heat denaturation also avoids interfering with downstream steps of the industrial process.

Conclusions

The glucoamylase secreted by *Tetracladium* sp. is a novel cold-adapted enzyme with optimal activity at pH 6.0 and 30 °C, and has no dependency on Ca^{2+} for its hydrolytic activity. Protein modeling analysis predicted that Amy1T is more flexible than thermostable counterparts, which could explain, at least in part, its higher activity at lower temperatures. The properties of the glucoamylase described in this work should render this enzyme suitable for use in industrial processes that require high starch degrading activity at mild temperatures, such as biofuel production.

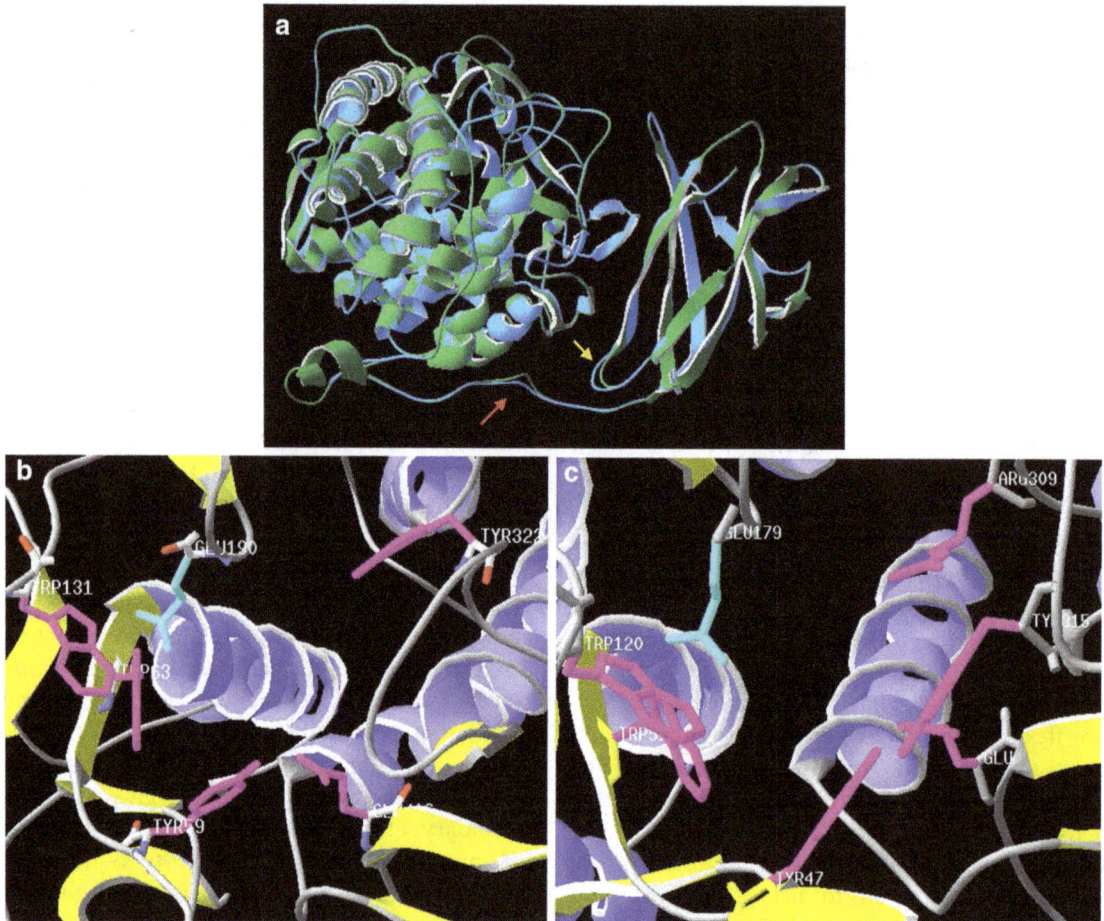

Fig. 7 Structural comparison between glucoamylases from *H. jecorina* (HjGa) and *Tetracladium* sp. (AmyT1). **a** Superimposition of the predicted glucoamylase model of AmyT1 (*green*) and the structure of the glucoamylase from HjGa (*blue*). The linker region (*red arrow*) and variable loop (*yellow arrow*) are indicated. Residues involved in substrate recognition and in the catalytic site are shown in *magenta* and *cyan*, respectively, for AmyT1 (**b**) and HjGa (**c**). The calculated distances in Amy1T between residues Y47 and R309, W51 and Y315, W120 and Y315, and E190 and Y315 are 8.5, 9.6, 8.7 and 10.2 Å, respectively. For HjGa, the calculated distances between the equivalent residues Y47 and R309, W51 and Y315, W120 and Y315, and E190 and Y315 were 8.5, 8.6, 8.3 and 10.1 Å, respectively

Methods

Strains and growth conditions

Tetracladium sp. was grown in YM medium (yeast extract 0.3%, malt extract 0.3%, peptone 0.5%, pH 7) supplemented with 1% glucose (YM-G) or soluble starch 1% (YM-S). For cultures of 300–500 mL, a decimal volume inoculum at $OD_{600} = 12$ was used, incubated at 22 °C with 150 rpm orbital agitation. Semisolid media were prepared by the addition of agar at 1.5%, before autoclaving at 121 °C for 20 min. *Tetracladium* sp. is conserved at the Genetic Laboratory Yeast Collection, Faculty of Sciences, Universidad de Chile.

Extraction and fractioning of extracellular proteins

300 mL cultures of *Tetracladium* sp. at the late exponential phase of growth ($OD_{600} = 12$) were centrifuged at 7000g for 10 min at 4 °C, and the supernatants were filtered through a sterile 0.45-µm pore size polyvinylidene fluoride membrane (Millipore, Billerica, MA, USA). Ammonium sulfate was added to the cell-free supernatants to a final concentration of 80% saturation, incubated on ice for 2 h, and centrifuged at 10,000g for 15 min at 4 °C. The pellet was suspended in 2 mL of potassium phosphate buffer (20 mM, pH 7.0 and 150 mM NaCl). For fractioning, ammonium sulfate was added to the

supernatants at concentrations from 20 to 80% and proteins were obtained in each fractioning step, as described above. The samples were desalted using a HiTrap desalting column (GE, Schenectady, New York, USA). The protein content of samples was quantified using a BCA assay kit (Thermo Scientific, IL, USA), according to manufacturer's instructions.

Amylase purification and glycosylation determination

The total proteins obtained by precipitation with 80% ammonium sulfate were dialyzed through a HiTrap desalting column against a 20 mM sodium phosphate buffer (pH 7.0, 150 mM NaCl). Aliquots of 500 µL protein samples were loaded onto a Superdex 75 10/300 GL column, equilibrated with 20 mM sodium phosphate buffer and a flow rate of 0.2 mL/min attached to an Akta Prime purification system (General Electrics, New York, USA). Fractions of 0.2 mL were collected and analyzed for amylase activity (see below) and protein content by SDS-PAGE. Fractions with amylase activity were pooled and concentrated at $1000g$ and 4 °C using Amicon filters with a 3 kDa molecular weight cut-off.

The rMW of the amylase was determined by comparison against the protein marker bands (PageRuler Plus Prestained Protein Ladder, Thermo Scientific, IL, USA). The calibration curve for the determination of the amylase molecular mass was prepared using a commercial protein standard kit (Gel filtration standard, Bio-Rad, CA, USA).

Glycosylation of the purified enzyme was analyzed by SDS-PAGE stained with the Pierce Glycoprotein Staining Kit (Thermo Scientific, IL, USA).

Amylase activity determination

The amylase activity in protein extracts was measured as the liberation of reducing sugars from soluble starch by the dinitrosalicylic acid (DNS) method [34]. Briefly, a mixture of 50 µL soluble starch solution at 10 g/L (Sigma-Aldrich Corporation, St Louis, USA) and 50 µL of the protein sample were incubated for 1 h. Then, 100 µLof the DNS (1.6% NaOH, 30% sodium potassium tartrate and 1% 3,5-dinitrosalicylic acid) solution was added, the mixture was incubated for a further 10 min at 100 °C and then for 5 min on ice. The absorbance of the aliquots at 540 nm was measured in 96 well microplates using an Epoch 2 microplate reader (Biotek Instruments Inc., Winooski, VT, USA). The values were normalized by the amount of protein present in each sample. Glucose release from starch was determined using a glucose determination kit (Megazyme, IL, USA).

The specificity of the purified amylase was determined using the chromogenic substrates ethylidene-pNP-G7 (Abnova, Taipei, Taiwan) and 4-Nitrophenyl

α-D-glucopyranoside (Sigma-Aldrich Corporation, St Louis, USA) following the supplier instructions. The α-amylase from Abnova and the α-glucosidase from Megazyme were used as positive controls for each specific substrate (glucose/fructose assay kit). The reactions were incubated at 30 °C for different times (0, 5, 10 and 15 min), and at each point the absorbance at 405 nm was measured.

The effect of different factors on the activity of the enzyme was evaluated in a Plackett–Burman design experiment. The parameters assayed were temperature (30, 50 °C), pH (5, 7), concentration of soluble starch (1, 10 g/L), calcium chloride (0, 10 mM) and magnesium chloride (0, 10 mM). Eight different reactions were carried out at different incubation times (between 0 and120 min) and 50 µL samples were taken and assayed by the DNS method, as described above.

To determine the optimum pH and temperature of the reaction catalyzed by the amylase. The reactions were conducted at different temperatures (25, 30, 32.5, 35, 37.5, 40, 42.5 and 45 °C) and pH values (5, 6, 6.5, 7, 7.5, 8, 8.5 and 9). The soluble starch concentration and the incubation time were 10 g/L and 30 min, respectively. Then, 100 µL of the reaction samples were assayed by the DNS method. The concentration of the released reducing sugars at each condition was plotted in a response surface plot.

To determine the kinetic parameters (K_m, k_{cat} and V_{max}), the enzyme concentration was varied such that it gave different reactions rates at different substrate concentrations. The reactions were performed using various soluble starch concentrations (4 to 10 g/L) and different enzyme concentrations (0.8–24.5 µg/mL). In the kinetic assays, the reactions were carried out at the determined optimum pH and temperature (6 and 30 °C) and 0.04 µg glucoamylase. Fifty microliter samples were taken at different time points (0, 30, 60, 90, 120, 150, 180 and 220 min) and assayed by the DNS method. The DNS values obtained in each condition were plotted against the reaction time for each substrate concentration. The slope of the linear phase of the reaction was determined to give the reaction rate. Subsequently, the kinetic parameters were determined using a double-reciprocal plot.

The activity of the glucoamylase at different temperatures was evaluated by incubating 50 µL glucoamylase solutions (6.1 µg/mL) at temperatures from 4 to 60 °C at pH 6 for 1 h. The amylase activity was then determined by the DNS method. The thermal stability of the enzyme was evaluated by incubating enzyme samples at temperatures from 4 to 60 °C for 1 h, prior to the determination of enzyme activity at the optimal conditions. Additionally, kinetic stability was determined by incubating samples at 22–50 °C for 0–8 h using the same procedure.

Alignment, modeling and bioinformatics analysis

Amino acid sequence alignments were made using the Geneious program v10. The glucoamylases sequences chosen for the amino acid sequence comparison had a minimum of 50% similarity and 50% coverage. The promoter prediction was performed using the FindM tool available in the single search analysis server (http://ccg.vital-it.ch/ssa/findm.php). The glucoamylase model was constructed using the Swiss-model platform [35]. The 2VN4_A PDB entry was chosen for modeling, which is the crystal structure of *Hypocrea jecorina* glucoamylase (HjGa), which has 90% coverage and 69% identity with AmyT1. Distance calculations and models were created using the spdb viewer [36].

Additional files

Additional file 1. Purification and characterization of the amylase enzyme. A, protein samples were loaded onto a Superdex 75 10/300 GL column equilibrated with 20 mM sodium phosphate buffer pH 7.0 and 150 mM NaCl, with a flow rate of 0.2 ml/min. The absorbance values at 280 nm (continuous line) and amylase activity (dotted line) were measured. The SDS PAGE analysis shows the active fractions (lanes 38 to 45). M, protein marker; E, sample from 80% ammonium sulfate precipitation. The identified amylase enzyme is indicated by the arrow. B and C show SDS PAGE gels stained with Coomassie blue reagent or a glycoprotein staining kit, respectively. Lanes 1 and 4, amylase from *Tetracladium* sp.; lanes 2 and 5, horse peroxidase glycoprotein (positive control); and lanes 3 and 6 trypsin inhibitor soybean protein (negative control).

Additional file 2. Influence of different parameters on AmyT1 activity.

Additional file 3. Sequence of ORF encoding for AmyT1 from *Tetracladium* sp. exons are indicated in grey.

Additional file 4. Molecular Phylogenetic analysis by Maximum Likelihood method. A, Based on the encoding nucleotide sequences; B, Based on the translated sequences. All positions with less than 95% site coverage were eliminated, that is, fewer than 5% alignment gaps, missing data, and ambiguous bases were allowed at any position. Evolutionary analyses were conducted using MEGA7. Box, Sequences from *Tetracladium* sp.

Authors' contributions

MB conceived the study; MC performed the experiments; MC and MB analyzed the results; MC, MB, JA and VC wrote the manuscript. All authors read and approved the final manuscript.

Acknowledgements

Not applicable.

Competing interests

The authors declare that they have no competing interests.

Funding

This study was financially supported through FONDECYT Grant 1130333.

References

1. Buzzini P, Margesin R. Cold-adapted yeasts: a lesson from the cold and a challenge for the XXI century. In: Cold-adapted Yeasts. Berlin: Springer; 2014. p. 3–22.
2. Margesin R, Miteva V. Diversity and ecology of psychrophilic microorganisms. Res Microbiol. 2011;162:346–61.
3. D'Amico S, Claverie P, Collins T, Georlette D, Gratia E, Hoyoux A, Meuwis MA, Feller G, Gerday C. Molecular basis of cold adaptation. Philos Trans R Soc Lond B Biol Sci. 2002;357:917–25.
4. Białkowska A, Turkiewicz M. Miscellaneous cold-active yeast enzymes of industrial importance. In: Cold-adapted yeasts: biodiversity, adaptation strategies and biotechnological significance. Berlin: Springer; 2014 p. 377–395.
5. Alcaíno J, Cifuentes V, Baeza M. Physiological adaptations of yeasts living in cold environments and their potential applications. World J Microbiol Biotechnol. 2015;31:1467–73.
6. Cavicchioli R, Charlton T, Ertan H, Mohd Omar S, Siddiqui KS, Williams TJ. Biotechnological uses of enzymes from psychrophiles. Microb Biotechnol. 2011;4:449–60.
7. Feller G. Psychrophilic enzymes: from folding to function and biotechnology. Scientifica (Cairo). 2013;2013:512840.
8. Gerday C. Fundamentals of cold-active enzymes. In: Buzzini P, Margesin R, editors. Cold-adapted yeasts: biodiversity, adaptation strategies and biotechnological significance. Berlin: Springer; 2014. p. 325–50.
9. Sarmiento F, Peralta R, Blamey JM. Cold and hot extremozymes: industrial relevance and current trends. Front Bioeng Biotechnol. 2015;3:148.
10. Janeček Š, Svensson B, MacGregor EA. α-Amylase: an enzyme specificity found in various families of glycoside hydrolases. Cell Mol Life Sci. 2014;71:1149–70.
11. Janeček Š, Ševčík J. The evolution of starch-binding domain. FEBS Lett. 1999;456:119–25.
12. Gupta R, Gigras P, Mohapatra H, Goswami VK, Chauhan B. Microbial α-amylases: a biotechnological perspective. Process Biochem. 2003;38:1599–616.
13. Gurung N, Ray S, Bose S, Rai V. A broader view: microbial enzymes and their relevance in industries, medicine, and beyond. Biomed Res Int. 2013;2013:329121.
14. Uthumporn U, Shariffa YN, Karim AA. Hydrolysis of native and heat-treated starches at sub-gelatinization temperature using granular starch hydrolyzing enzyme. Appl Biochem Biotechnol. 2012;166:1167–82.
15. van Zyl WH, Bloom M, Viktor MJ. Engineering yeasts for raw starch conversion. Appl Microbiol Biotechnol. 2012;95:1377–88.
16. Kumar P, Satyanarayana T. Microbial glucoamylases: characteristics and applications. Crit Rev Biotechnol. 2009;29:225–55.
17. Michelin M, Ruller R, Ward RJ, Moraes LA, Jorge JA, Terenzi HF, Polizeli ML. Purification and biochemical characterization of a thermostable extracellular glucoamylase produced by the thermotolerant fungus *Paecilomyces variotii*. J Ind Microbiol Biotechnol. 2008;35:17 25.
18. Aquino ACMM, Jorge JA, Terenzi HF, Polizeli MLTM. Thermostable glucose-tolerant glucoamylase produced by the thermophilic fungus *Scytalidium thermophilum*. Folia Microbiol. 2001;46:11–6.
19. Yingling B, Li C, Honglin W, Xiwen Y, Zongcheng Y. Multi-objective optimization of bioethanol production during cold enzyme starch hydrolysis in very high gravity cassava mash. Bioresour Technol. 2011;102:8077–84.
20. Carrasco M, Villarreal P, Barahona S, Alcaíno J, Cifuentes V, Baeza M. Screening and characterization of amylase and cellulase activities in psychrotolerant yeasts. BMC Microbiol. 2016;16:21.
21. Carrasco M, Rozas JM, Barahona S, Alcaíno J, Cifuentes V, Baeza M. Diversity and extracellular enzymatic activities of yeasts isolated from King George Island, the sub-Antarctic region. BMC Microbiol. 2012;12:251.
22. de Barros MC, do Nascimento Silva R, Ramada MH, Galdino AS, de Moraes LM, Torres FA, Ulhoa CJ. The influence of N-glycosylation on biochemical properties of Amy1, an alpha-amylase from the yeast *Cryptococcus flavus*. Carbohydr Res. 2009;344:1682–6.

23. Eriksen SH, Jensen B, Olsen J. Effect of N-linked glycosylation on secretion, activity, and stability of α-amylase from *Aspergillus oryzae*. Curr Microbiol. 1998;37:117–22.

24. Prieto JA, Bort BR, Martínez J, Randez-Gil F, Sanz P, Buesa C. Purification and characterization of a new α-amylase of intermediate thermal stability from the yeast *Lipomyces kononenkoae*. Biochem Cell Biol. 1995;73:41–9.

25. Vihinen M, Mantsiila P. Microbial amylolytic enzyme. Crit Rev Biochem Mol Biol. 1989;24:329–418.

26. Wanderley KJ, Torres FAG, Moraes LÄMP, Ulhoa CJ. Biochemical characterization of alpha-amylase from the yeast *Cryptococcus flavus*. FEMS Microbiol Lett. 2004;231:165–9.

27. Siddiqui KS, Cavicchioli R. Cold-adapted enzymes. Annu Rev Biochem. 2006;75:403–33.

28. Marín-Navarro J, Polaina J. Glucoamylases: structural and biotechnological aspects. Appl Microbiol Biotechnol. 2011;89:1267–73.

29. Bott R, Saldajeno M, Cuevas W, Ward D, Scheffers M, Aehle W, Karkehabadi S, Sandgren M, Hansson H. Three-dimensional structure of an intact glycoside hydrolase family 15 glucoamylase from *Hypocrea jecorina*. Biochemistry. 2008;47:5746–54.

30. Tsigos I, Mavromatis K, Tzanodaskalaki M, Pozidis C, Kokkinidis M, Bouriotis V. Engineering the properties of a cold active enzyme through rational redesign of the active site. Eur J Biochem. 2001;268:5074–80.

31. Benassi VM, Pasin TM, Facchini FD, Jorge JA, Polizeli MLTM. A novel glucoamylase activated by manganese and calcium produced in submerged fermentation by *Aspergillus phoenicis*. J Basic Microbiol. 2014;54:333–9.

32. Soni SK, Kaur A, Gupta JK. A solid state fermentation based bacterial α-amylase and fungal glucoamylase system and its suitability for the hydrolysis of wheat starch. Process Biochem. 2003;39:185–92.

33. Xu QS, Yan YS, Feng JX. Efficient hydrolysis of raw starch and ethanol fermentation: a novel raw starch-digesting glucoamylase from *Penicillium oxalicum*. Biotechnol Biofuels. 2016;9:216.

34. Miller GL. Use of DNS reagent for the measurement of reducing sugar. Anal Chem. 1959;31:426–8.

35. Biasini M, Bienert S, Waterhouse A, Arnold K, Studer G, Schmidt T, Kiefer F, Gallo Cassarino T, Bertoni M, Bordoli L, Schwede T. SWISS-MODEL: modelling protein tertiary and quaternary structure using evolutionary information. Nucleic Acids Res. 2014;42:252–8.

36. Guex N, Peitsch MC. SWISS-MODEL and the Swiss-Pdb Viewer: an environment for comparative protein modeling. Electrophoresis. 1997;18:2714–23.

4

Identification of nocamycin biosynthetic gene cluster from *Saccharothrix syringae* NRRL B-16468 and generation of new nocamycin derivatives by manipulating gene cluster

Xuhua Mo[1*], Chunrong Shi[1], Chun Gui[2], Yanjiao Zhang[1], Jianhua Ju[2] and Qingji Wang[1*]

Abstract

Background: Nocamycins I and II, produced by the rare actinomycete *Saccharothrix syringae*, belong to the tetramic acid family natural products. Nocamycins show potent antimicrobial activity and they hold great potential for antibacterial agent design. However, up to now, little is known about the exact biosynthetic mechanism of nocamycin.

Results: In this report, we identified the gene cluster responsible for nocamycin biosynthesis from *S. syringae* and generated new nocamycin derivatives by manipulating its gene cluster. The biosynthetic gene cluster for nocamycin contains a 61 kb DNA locus, consisting of 21 open reading frames (ORFs). Five type I polyketide synthases (NcmAI, NcmAII, NcmAIII, NcmAIV, NcmAV) and a non-ribosomal peptide synthetase (NcmB) are proposed to be involved in synthesis of the backbone structure, a Dieckmann cyclase NcmC catalyze the releasing of linear chain and the formation of tetramic acid moiety, five enzymes (NcmEDGOP) are related to post-tailoring steps, and five enzymes (NcmNJKIM) function as regulators. Targeted inactivation of *ncmB* led to nocamycin production being completely abolished, which demonstrates that this gene cluster is involved in nocamycin biosynthesis. To generate new nocamycin derivatives, the gene *ncmG*, encoding for a cytochrome P450 oxidase, was inactivated. Two new nocamycin derivatives nocamycin III and nocamycin IV were isolated from the *ncmG* deletion mutant strain and their structures were elucidated by spectroscopic data analyses. Based on bioinformatics analysis and new derivatives isolated from gene inactivation mutant strains, a biosynthetic pathway of nocamycins was proposed.

Conclusion: These findings provide the basis for further understanding of nocamycin biosynthetic mechanism, and set the stage to rationally engineer new nocamycin derivatives via combinatorial biosynthesis strategy.

Keywords: Nocamycins, Biosynthetic gene cluster, Cytochrome P450 oxidase, Post-tailoring modification, *Saccharothrix syringae*

Background

Nocamycins I and II (Fig. 1), isolated from the broth of *Saccharothrix syringae* NRRL B-16468 by Russian scientists in 1977, belong to the tetramic acid (2, 4-pyrrolidinedione) family natural products [1–3]. The original structural assignment of nocamycin I was incorrect and it was revised by a Japanese research group [4].

The Japanese research group reported two compounds Bu-2313A and Bu-2313B from the strain *Microtetraspora caesia* ATCC 31295 nearly at the same time [5]. Further structural elucidation showed that Bu-2313B was virtually identical to nocamycin I [4]. Beyond the common tetramic acid structure, a tricyclic ketal structure is another interesting motif in nocamycins. In terms of structural viewpoint, streptolydigin, tirandamycins and tirandalydigin are closely related to nocamycins (Fig. 1). Among these compounds, nocamycins I and II are unique because they have a fused oxolane ring system

*Correspondence: xhmo2013@163.com; qingjiwang2016@hotmail.com
[1] Shandong Key Laboratory of Applied Mycology, School of Life Sciences, Qingdao Agricultural University, Qingdao 266109, China
Full list of author information is available at the end of the article

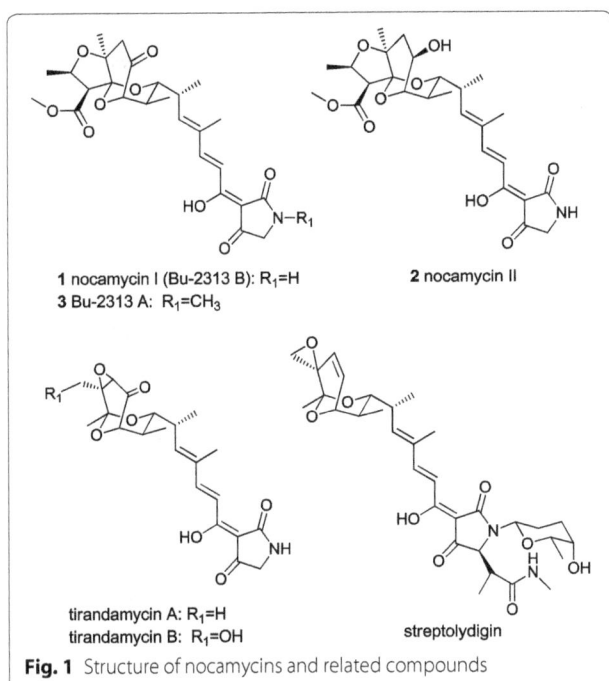

Fig. 1 Structure of nocamycins and related compounds

other than an oxirane (spiro or fused) ring in streptolydigin, tirandamycin and tirandalydigin.

Nocamycin I (Bu-2313B) displays broad antimicrobial activity toward a panel of Gram-positive and Gram-negative anaerobic bacteria as well as some aerobic bacteria. Inhibitions of anaerobic bacteria *Bacteroides fragilis*, *Clostridium* sp., *Fusobacterium* sp., *Sphaerophorus* sp. by nocamycins are particularly potent, and the minimum inhibitory concentrations (MICs) are in the range of 0.1–0.4 µg/mL [5–7]. Further in vivo experiments conducted in mice showed that nocamycin I is effective in protecting mice against *B. fragilis* A20928-1 and *Clostridium perfingens* A9635 when administered by both oral and subcutaneous routes [5]. In addition, nocamycins show antitumor effects [1]. Up to now, the exact antibacterial mold of action of nocamycins has not been investigated. The closely related compounds tirandamycin and streptolydigin are validated to be inhibitors of bacterial RNA polymerase (RNAP), thus nocamycins are probably to be inhibitors of RNAP. In recent years, the molecular evidences for the structural basis of the RNAP interaction mechanism of this class of natural products have been disclosed by co-crystal complexes of streptolydigin with RNAPs from *Escherichia coli* and *Thermus thermophilus* [8, 9]. The key affinities of both bicyclic ketal and tetramic acid structures with RNAPs have been observed from the co-crystal complexes, indicating the substitution or modification in these two structural motifs is critical for the biological activity [8, 9]. Meanwhile, results of antibacterial activities of streptolydigin, tirandamycin and

their congeners also demonstrated that the two featured motifs are closely related to the activity of this family of natural products [10–12].

The intriguing structure, action mold and biological activity of this small class of natural products attract more and more attentions from biochemists. So far, the gene clusters responsible for tirandamycin and streptolydigin biosynthesis have been identified from three different *Streptomyces* species by Sherman, Salas and Ju group, respectively [10, 13, 14]. Both tirandamycins and streptolydigin are assembled by hybrid iterative type I polyketide synthase (PKS) and non-ribosomal peptide synthase (NRPS). The functions of a number of genes related to post-tailoring, regulator and resistance involved in tirandamycin and streptolydigin biosynthetic pathway have been fully elucidated [12, 13, 15–20]. Some streptolydigin derivatives were generated by using combinatorial biosynthesis method [10]. In streptolydigin and tirandamycins biosynthetic pathway, a uniform strategy is employed to catalyze the formation of tetramic acid moiety [21]. The mechanism of bicyclic ketal structure formation remains unclear since no related gene candidates have been discovered in the two gene clusters. To fully understand the biosynthetic pathway of nocamycins, provide insights into the formation of bicyclic ketal structure and generate diversified nocamycin derivatives, we started to identify the nocamycin biosynthetic gene cluster from *S. syringae* NRRL B-16468. Here, we report the identification of nocamycin biosynthetic gene cluster and new nocamycin derivatives generated by manipulating the gene cluster.

Methods

Bacterial strains, plasmids, medium and culture conditions
The bacteria and plasmids used in this study are listed in Table 1. *S. syringae* was maintained on ISP4 agar medium. The medium used for fermentation of *S. syringae* and its mutant strains consists of 1% soybean, 3% glycerol, 0.5% mycose, 0.2% NaCl and 0.2% $CaCO_3$. All cultures for *S. syringae* were incubated at 28 °C. For *E. coli*, Luria–Bertani (LB) liquid or agar media were used with appropriate antibiotics at a final concentration of: 100 µg/mL ampicillin (Amp), 50 µg/mL apramycin (Apr), 50 µg/mL kanamycin (Kan), 25 µg/mL chloramphenicol (Cml) and 50 µg/mL trimethoprim (TMP).

DNA sequencing, assembly and analysis
After growing in TSB medium for 48–72 h, the genomic DNA of *S. syringae* NRRL B-16468 was extracted according to standard protocols [26]. Then, the genomic DNA was shotgun sequenced and annotated by Shanghai South Gene Technology Co. Ltd. (Shanghai, China). The gene cluster responsible for secondary metabolite

Table 1 Bacteria and plasmids used in this study

Strains or plasmids	Description	Reference or source
Strains		
E. coli LE392	Host strain of cosmid vector SuperCos I	Stratagene
E. coli DH5*a*	Host strain for general clone	Stratagene
E. coli ET12567/pUZ8002	Host strain for conjugation	[22]
E. coli BW25113	Host strain for PCR-targeting	[23]
S. syringae	Nocamycin-producing strain	NRRL
S. syringae MoS1001	*ncmB deletion mutant strains originated from Saccharothrix syringae*	This study
S. syringae MoS1002	*ncmL deletion mutant strains originated from Saccharothrix syringae*	This study
S. syringae MoS1003	*ncmG deletion mutant strains originated from Saccharothrix syringae*	This study
Plasmids		
SuperCosl	Amp^r, Kan^r, cosmid vector	Stratagene
pIJ790	Cml^r, including λ-RED (*gam, bet, exo*) for PCR-targeting	[24]
pIJ773	Apr^r, source of *acc(3)IV* and *oriT* fragment	[24]
pUZ8002	Kan^r, including *tra* for conjugation	[25]
p5-C-9	Amp^r, Kan^r, harboring *ncmL*gene	This study
p2-H-12	Amp^r, Kan^r, harboring *ncmG* gene	This study

biosynthesis was analyzed by antiSMASH online analysis tool (http://antismash.secondarymetabolites.org/). DNA and corresponding protein sequences in nocamycin gene cluster were analyzed by ORF finder program (http://www.ncbi.nlm.nih.gov/gorf/gorf.html), Frameplot 2.3.2 program (http://www.nih.go.jp/~jun/cgi-bin/frameplot.pl), and BLAST program (http://blast.ncbi.nlm.nih.gov/).

Construction and screening of *S. syringae* genomic library
Genomic library of *S. syringae* NRRL B-16468 was constructed using SuperCos1 Vector Kit according to manufacturer's instruction (Stratagene). The library was packaged using phage extracts and transduced into the *E. coli* LE392. About 2600 resulting transductants were picked up and transferred to twenty-seven 96-well microtiter plates containing 150 μL LB medium supplemented with Kan (50 μg/mL). After overnight incubation at 37 °C, 30 μL *E. coli* broth in every microtiter pore was absorbed and mixed every 12 clones in a horizontal line and every 8 clones in a vertical line for each 96-well plate. Glycerol was added to the remaining broth of the clones (20% final concentration) for permanent stock. The DNA of mixed clones was extracted as templates for PCR screening.

The primer pairs targeted the cytochrome P450 oxidase gene (NcmG-SF and NcmG-SR), Dieckmann cyclase gene (NcmC-SF and NcmC-SR) and DH domain at NcmAII gene (DH-SF and DH-SR) were designed and they were used as PCR primers to screen *S. syringae* NRRL B-16468 genomic library (Table 2). The positive clones were selected from the genomic library. The cosmids were extracted and further analyzed by

terminal-sequencing. The PCR reaction (20 mL volume) contained 2 μL 10 × PCR buffer, 1.6 μL dNTPs (2.5 mM), 0.4 μL forward primer (10 μM), 0.4 μL reverse primer (10 μM), 1 μL dimethylsulfoxide (DMSO), 1 μL DNA template, 0.1 μL rTaq (5 U/μL), and 13.5 μL ddH$_2$O. The following PCR program was used: 94 °C, 4 min, 30 cycles of 94 °C, 45 s, 59 °C, 45 s, 72 °C, 1 min, and a final extension cycle at 72 °C, 10 min. Eventually, two cosmids p5-C-9 and p2-H-12 were chosen for further gene-inactivation experiments.

Generation of *S. syringae* mutant strains
λ-RED recombination technology was employed to inactivate the target gene *ncmB*, *ncmL* and *ncmG* according to literature previously reported [14]. The primer pairs used for PCR-targeting are listed in Table 2. The fragment *oriT/acc(3)IV* cassette was used to replace partial gene region of *ncmB* or *ncmL* in p5-C-9 to generate cosmid pMoS1001 (*ΔncmB*) or pMoS1002 (*ΔncmL*). For *ncmG*, partial gene region was replaced by fragment *oriT/acc(3)IV* cassette in cosmid p2-H-12 and plasmid pMoS1003(*ΔncmG*) were generated. After verified by PCR and restriction enzyme digestion analysis, the correct mutated cosmids were introduced into *E. coli* ET12567/pUZ8002 and conjugated with wild type *S. syringae* spores. The wild type *S. syringae* spores were germinated in LB medium for 4–5 h at 30 °C, 200 rpm. The *E. coli* ET12567/pUZ8002 containing each mutated cosmid was grown in LB medium supplemented with Kan (50 μg/mL), Amp (100 μg/mL), Cml (25 μg/mL) and Apr (50 μg/mL) to OD$_{600}$ = 0.6–0.8. Then the cells were harvested, washed twice with LB medium, mixed

Table 2 Primer pairs used in this study

Primers sequences (5′–3′)	
NcmG-delF	CTGCTGGGCGCCGACGTGCCGCGCACCACGCCGCGGGTG ATTCCGGGGATCCGTCGACC
NcmG-delR	CAGGTCCGCCGCCGGTACCGCGAGCCGCAGGGTCGGGAA TGTAGGCTGGAGCTGCTTC
NcmB-delF	CTGGCCTGCGCCGAACCGCCCGCGCCCGCCTCGCCCGTC ATTCCGGGGATCCGTCGACC
NcmB-delR	GCCCCGGTGCCCCGCGGGCAGGCGGGCGCCGGGGCCCGC TGTAGGCTGGAGCTGCTTC
NcmL-delF	CGCAGCCTGGAGGTGTTCGACGACCTGGGCGTCGTCGAC ATTCCGGGGATCCGTCGACC
NcmL-delR	GAACCCGAAGAGCGTGAAGTGCGGGCCGCGCTGCGCGTC TGTAGGCTGGAGCTGCTTC
NcmG-tF	GAGGTCCGGCAGGTGCTGTC
NcmG-tR	GACGACCTTGGCGGTGTGCC
NcmB-tF	CGGGAGTACTGGCGGCAGC
NcmB-tR	GGTCCAGCAGGTCGGCCAGCA
NcmL-tF	CTGATCATCGACAAGGACTC
NcmL-tR	GGACGAGCACCAGCGCGTCC
DH-SF	GCTCGGTGTTCCTGGACCTGGC
DH-SR	GCAGTTCGAAGCCGCTCCACAG
NcmG-SF	GTCCACCGCGACGCCATAC
NcmG-SR	CGGCCAGGTAGTCCTTGAGCC
NcmC-SF	GGGCGGTGCTCGGGTTCT
NcmC-SR	GCAGGTCGGCGTGGTGGA

with germinated wild type spores and plated on ISP4 medium. The plates were incubated at 30 °C for 24 h. Then, each plate was covered by 800 µL sterile water supplemented with 30 µL TMP (50 mg/mL) and 30 µL Apr (50 mg/mL). The plates were continued incubated at 30 °C for 7–10 days until exconjugants appeared. Double cross-over mutants were first selected by the phenotype of Kan sensitive (Kan^S) and Apr resistant (Apr^R), and the genotype of the mutants were further confirmed by PCR. Finally, the mutant strains *S. syringae* MoS-1001 (Δ*ncmB*), *S. syringae* MoS-1002 (Δ*ncmL*) and *S. syringae* MoS-1003 (Δ*ncmG*) were obtained.

Fermentation and analysis of *S. syringae* and mutant strains

Saccharothrix syringae wild type and mutant strains were inoculated in 250 mL flasks with 50 mL medium and incubated on a rotary shaker at 28 °C, 200 rpm. After 7 days fermentation, each of the 50 mL culture was added with 100 mL ethyl acetate and then vigorously mixed for 30 min. The ethyl acetate phase was evaporated to dryness to yield a residue. The residue was dissolved in 1 mL methanol and centrifuged, then, the supernatant was subjected to HPLC analysis. Analytical HPLC was performed on Agilent 1260 HPLC system (Agilent

Technologies Inc., USA) equipped with a binary pump and a diode array detector using a Phenomenex Prodigy ODS column (150 × 4.60 mm, 5 µ) with UV detection at 355 nm. The mobile phase comprises solvent A and B. Solvent A consists of 15% CH_3CN in water supplemented with 0.1% trifluoroacetic acid (TFA). Solvent B consists of 85% CH_3CN in water supplemented with 0.1% TFA. Samples were eluted with a linear gradient from 5 to 90% solvent B in 20 min, followed by 90–100% solvent B for 5 min, then eluted with 100% solvent B for 3 min, at a flow rate of 1 mL/min and UV detection at 355 nm.

Isolation of new produced nocamycin derivatives from Δ*ncmG* mutant strain

Two-step fermentation was used to culture Δ*ncmG* mutant strain. 250 mL flask containing 50 mL medium was used as seed culture and 500 mL flask containing 100 mL medium was used as fermentation medium. Appropriate spores were inoculated to seed culture and grown at 28 °C, 200 rpm for 3 days. Then, 5 mL seed medium was inoculated to 100 mL fermentation medium and continued 7 days culture. 15 L liquid medium was used in total. After incubation, the culture broth was collected and centrifuged. The supernatant was extracted by ethyl acetate for three times and the mycelium was extracted by acetone for three times. Then, the entire organic phases were evaporated to dryness to yield crude extract. The crude extract was dissolved in a mixture of CH_3OH: $CHCl_3$ (1:1) and mixed with appropriate amount of silica gel (100–200 mesh, Qingdao Marine Chemical Corporation, China). The sample was applied on normal phase silica gel column chromatography and eluted with $CHCl_3$-CH_3OH (100:0–50:50) to give 10 fractions. All the fractions were analyzed by HPLC. Fraction 4 and 5 contained the major targeted compound nocamycin III and fractions 7 and 8 contained the major targeted compound nocamycin IV. The fractions 4–5 and fractions 7–8 were further purified on reverse phase C-18 silica gel (YMC, Japan) by using medium-pressure liquid chromatography (MPLC, Agela corporation, China) eluted by a linear gradient from 20 to 90% CH_3CN in water, respectively. The sub-fractions contained targeted compounds were further purified by Sephadex LH-20 (GE healthcare, Sweden) gel filtration chromatography to afford the purified nocamycin III and nocamycin IV.

Spectroscopy analysis of new produced nocamycin derivatives

^1H and ^{13}C NMR spectra of nocamycin III and nocamycin IV were recorded at 25 °C on Bruker AV 500 instruments. HR–ESI–MS spectra data were acquired on a Waters micro MS Q-Tof spectrometer.

Results

Sequencing and identification of nocamycin gene cluster

Saccharothrix syringae NRRL B-16468 genome was shotgun sequenced by Hiseq4000 technologies and the sequence reads were assembled into 10.8 Mb nucleotides. Then, *S. syringae* NRRL B-16468 genome data was analyzed by using online antiSMASH tool [27]. AntiSMASH analysis results demonstrated that a hybrid PKS-NRPS gene cluster designated as *Ncm* seems to be the candidate responsible for nocamycin biosynthesis since it shows high similarity to tirandamycin biosynthetic gene cluster. In the *Ncm* gene cluster, some deduced gene products such as NcmC, NcmE, NcmF and NcmB show high similarity to TrdC, TrdE, TrdF and TrdB originated from tirandamycin biosynthetic pathway, respectively [14]. Thus, we assumed that this cluster is probably involved in nocamycin biosynthesis. We then screened *S. syringae* genomic library by using PCR method with the primer pairs targeted at *ncmG*, *ncmC* and dehydratase (DH) domain at module 4. In total of eight positive cosmids were obtained. The eight cosmids were end-sequenced and two cosmids p2-H-12 and p5-H-9 were used for further gene inactivation experiments. To verify our hypothesis, a gene *ncmB* encoding a NRPS was inactivated to afford the strain *S. syringae* MoS-1001 (Additional file 1: Figure S1). HPLC analysis of the extract of *S. syringae* MoS-1001 fermentation broth revealed that *S. syringae* MoS-1001 failed to produce nocamycin I and II (Fig. 2I) completely, indicating *ncmB*'s involvement in nocamycin biosynthesis. This result also demonstrated this PKS-NRPS gene cluster is responsible for nocamycin biosynthesis. On basis of bioinformatics analysis, about 61 kb DNA locus consisted of 21 open reading frames (ORFs) whose deduced products are likely to be involved in nocamycin biosynthesis (Fig. 3; Table 3). Corresponding homologues and deduced function of each *ncm* gene are listed in Table 3. The sequence data of nocamycin biosynthesis in this study have been deposited in Genbank under accession number KY287782.

Linear chain assembly and releasing

Hybrid PKS-NRPS are employed to construct the backbone structure of nocamycin. Five type I PKS genes *ncmAI*, *ncmAII*, *ncmAIII*, *ncmAIV* and *ncmAV* transcribed in the same direction were identified in the gene cluster (Fig. 3). The deduced products of the five PKS genes were constituted by four, one, one, one and two modules respectively to assemble the polyketide backbone (Fig. 4). Each PKS module minimally contains ketosynthase (KS), acyltransferase (AT) and acyl carrier protein (ACP) domains. The conserved motifs from PKS modules in nocamycin gene cluster are listed in Additional file 1: Table S1. Except for loading module, M2 and

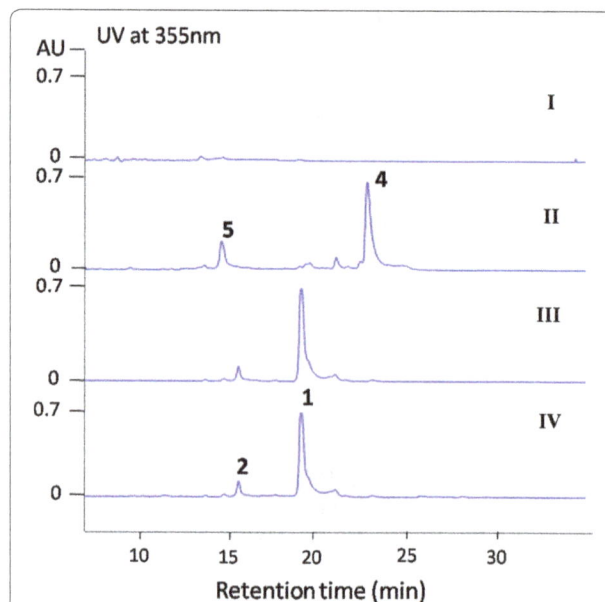

Fig. 2 HPLC analysis of *S. syringae* and its mutant strain. I *ncmB* deletion mutant strain *S. syringae* MoS1001; II *ncmG* deletion mutant strain *S. syringae* MoS1003; III *ncmL* deletion mutant strain *S. syringae* MoS1002; IV *S. syringae* wild type strain. 1, 2, 4, 5 represent for nocamycin I, II, III and IV, respectively

M8, each module possess a ketoreductase (KR) domain with conserved active motif. KR domain in module M5 is the only KR domain contains the characteristic of A-type KRs, and all the other KR domains display the conserved motif characteristic for the B-type KRs [28]. A characteristic KS^Q domain of loading module indicated that a malonoyl–CoA might be used to provide acetate as starter unit, and this phenomenon was observed in tirandamycin and streptolydigin gene clusters [10, 14]. As shown in Table 4 and Fig. 4, the AT domains in extension modules M3, M7 and M8 display conserved active motif specific for malonate-CoA incorporation [29, 30], whereas AT domains in extension modules M1, M2, M4, M5 and M6 show conserved active motif specific for methylmalonate-CoA incorporation [29, 30], which is in good agreement with the polyketide carbon skeleton. There are three DH domains with conserved active motif HXXXGXXXXP distributed in module M4, M6 and M7 [31].

NcmB, a NRPS, shows most similarity to TrdD (56% identity/66% similarity) from *Streptomyces* sp. SCSIO1666 involved in tirandamycin biosynthetic pathway [14]. Three domains condensation (C), adenylation (A), and peptidyl carrier protein (PCP) are found in NcmD. The amino acid binding pocket DILQLGVI located in A domain is predicted to activate glycine, which is accord to nocamycin structure.

NcmC shows most similarity to TrdC (45% identity/58% similarity) from *Streptomyces* sp. SCSIO1666

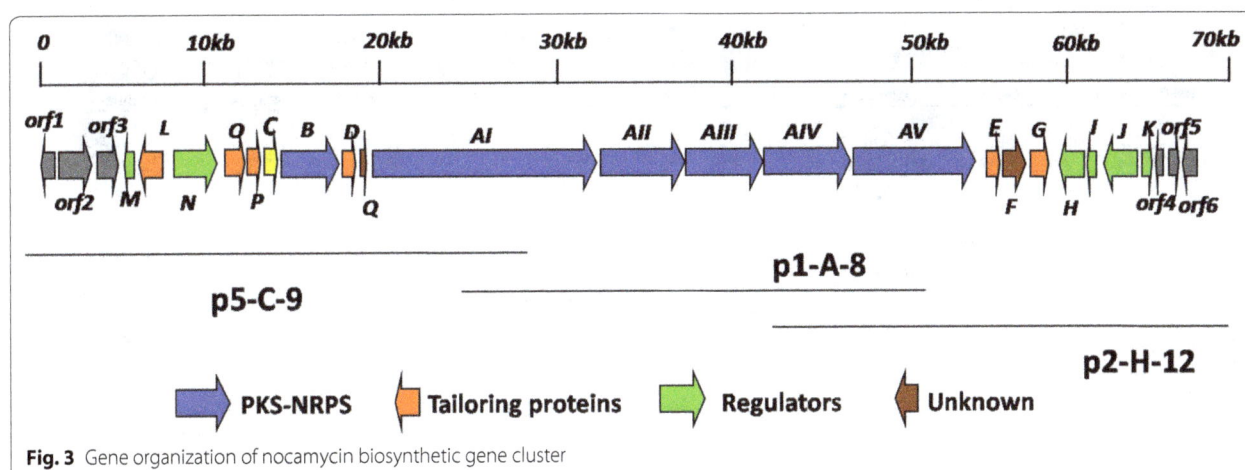

Fig. 3 Gene organization of nocamycin biosynthetic gene cluster

Table 3 Deduced functions of genes in the nocamycin biosynthetic gene cluster

Gene	Length (amino acids)	Closest similar protein accession number, origin, identity/similarity (%)	Deduced function
Orf1	306	KOX27604, *Saccharothrix* sp. NRRL B-16348	Short-chain dehydrogenase
Ofr2	664	KOX27605.1, *Saccharothrix* sp. NRRL B-16348, 81/86%	Helicase
Orf3	455	WP_053719771, *Saccharothrix* sp. NRRL B-16348	Hypothetical protein
NcmM	225	SCD95979, *Streptomyces* sp. PalvLS-984, 68/79%	Transcriptional regulator
NcmL	511	EJI98707, *Rhodococcus* sp. JVH1, 50/60%	Monooxygenase
NcmN	911	AFI57028 (QmnRg4), *Amycolatopsis orientalis*, 43/55%	LuxR family regulator
NcmO	412	AFI57027 (QmnO), *Amycolatopsis orientalis*, 48/63%	Cytochrome P450 oxidase
NcmP	287	ACN29714 (NokK), *Nonomuraea longicatena*, 44/55%	Carboxylate O-methyltransferase
NcmC	272	ADY38535 (TrdC), *Streptomyces* sp. SCSIO1666, 45/58%	Dieckmann cyclase
NcmB	1119	ADY38536 (TrdD), *Streptomyces* sp. SCSIO1666, 56/66%	Non-ribosomal peptide synthetase
NcmD	271	AII10529, *Rhodococcus opacus*, 45/60%	Short-chain dehydrogenase
NcmQ	120	EJY55702, *Alicyclobacillus hesperidum* URH17-3-68, 41/55%	Glyoxalase/bleomycin resistance protein
NcmAI	4915	CBA11584 (SlgA1), *Streptomyces lydicus*, 54/64%	Type I polyketide synthase
NcmAII	1786	EHY88978, *Saccharomonospora azurea* NA-128, 54/65%	Type I polyketide synthase
NcmAIII	1554	CCF23202, *Streptomyces hygroscopicus*, 60/69%	Type I polyketide synthase
NcmAIV	1786	CCF23202.1, *Streptomyces hygroscopicus*, 56/66%	Type I polyketide synthase
NcmAV	2679	AEP40935.1, *Nocardiopsis* sp. FU40, 50/60%	Type I polyketide synthase
NcmE	273	ADC79643 (TamE), *Streptomyces* sp. 307-9, 60/76%	Glycoside hydrolase
NcmF	487	ADY38538 (TrdF), *Streptomyces* sp. SCSIO1666, 50/61%	Prenyltransferase
NcmG	397	ADZ45320(Mur7), *Streptomyces* sp. NRRL 30471, 51/64%	Cytochrome P450 oxidase
NcmH	543	CAH10178 (ChaT1), *Streptomyces chartreusis*, 42/60%	Multiple drug transporter
NcmI	197	KKZ83567, Rhizobium phaseoli Ch24-10, 41/61%	PadR family transcriptional regulator
NcmJ	713	WP_037345636, *Scisionella* sp. SE31, 69/79%	AAA family ATPase
NcmK	213	ADY38543(TrdK) *Streptomyces* sp. SCSIO1666, 49/64%	TetR family transcriptional regulator
Orf4	110	GAT66653, Planomonospora sphaerica, 43/54%	Ohr subfamily peroxiredoxin
Orf5	232	GAT10151, *Mycobacterium novocastrense*, 64/73%	Ubiquinone biosynthesis methyltransferase UbiE
Orf6	322	KDO05396, Amycolatopsis mediterranei, 68/77%	(2Fe–2S) ferredoxin

involved in tirandamycin biosynthetic pathway [14]. TrdC and its analogues SlgC, KirHI have been determined as Dieckmann cyclases, and they catalyze the formation of tetramic acid or pyridone moiety [21]. Bioinformatics analyses revealed that NcmC also possesses the characteristic catalytic traid Cys-Asp-His (Additional

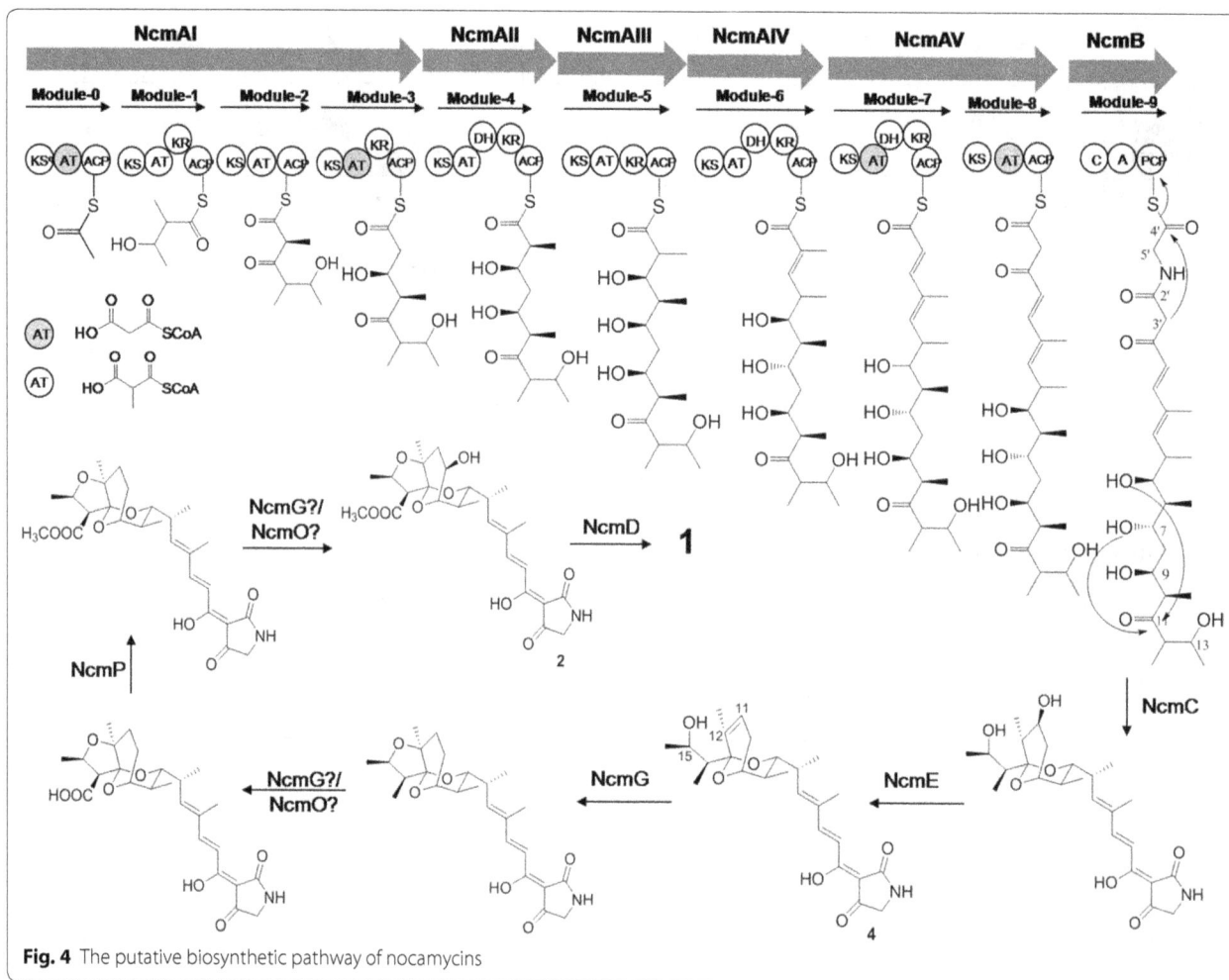

Fig. 4 The putative biosynthetic pathway of nocamycins

file 1: Figure S4). Thus, in nocamycin biosynthesis pathway, NcmC is proposed to be responsible for the PK-NRP chain releasing and catalyze the formation of tetramic acid moiety.

Genes involved in post-tailoring steps

After linear chain released from PKS-NRPS and formation of tetramic acid moiety, several post tailoring processes including oxolane ring system, C-10 hydroxyl/ketone group, C-14 methoxycarbonyl group are required to synthesis nocamycin I. Within the identified gene cluster, there are six genes encoding two cytochrome P450 monooxygenase (*ncmO* and *ncmG*), one monooxygenase (*ncmL*), one carboxylate O-methyltransferase (*ncmP*), one short chain dehydrogenase (*ncmD*) and one glycoside hydrolase (*ncmE*) are likely to be involved in these steps.

The glycoside hydrolase NcmE shows identity to TrdE (60% identity/76% similarity) involved in tirandamycin biosynthesis [14]. In tirandamycin pathway, TrdE functions as a dehydratase and it is responsible for the

formation of C11–C12 double bond [17]. Thus, we propose that NcmE is a dehydratase and it catalyzes the formation of C11–C12 double bond.

Both cytochrome P450 monooxygenases NcmG and NcmO possess the highly conserved heme-binding domain (GXXXCXG), K-helix (EEXLL) and oxygen binding region (Additional file 1: Figure S5) [32, 33]. NcmG shows similarity to Mur7 (51% identity/64% similarity) involved in muraymycin biosynthesis [34]. NcmO shows similarity to QnmO (48% identity/63% similarity) involved in quartromicin biosynthesis [35]. Since at least three oxidative tailoring steps, including the formation of tetrahydrofuran fused in bicyclic ketal structure, C-10 hydroxyl and C-14 carboxyl group are required, NcmG or NcmO is proposed to be bifunctional. Sequence alignments analysis revealed that NcmO and NcmG are distinct from the cytochrome P450 oxidases TrdI/TamI, SlgO1 and SlgO2 involved in tirandamycin or streptolydigin biosynthetic pathway [10, 15] (Additional file 1: Figure S6). The reason for this phenomenon may attribute

Table 4 ^{1}H and ^{13}CNMR spectroscopic data for nocamycin III (4) and nocamycin IV (5)

Position	4[a]		5[b]	
	δ_C type	δ_H mult. (J in Hz)	δ_C type	δ_H mult. (J in Hz)
1	175.5		Not observed	
2	116.6	7.17, d (15.7)	Not observed	7.33, d (14.2)
3	150.3	7.60, d (15.7)	Not observed	7.56, d (15.6)
4	135.3		136.5	
5	144.5	6.05, d (10.2)	144.0	6.04, d (9.4)
6	34.6	2.89, m	35.7	2.97, m
7	78.2	3.66, dd (11.2, 2.0)	79.3	3.75, dd (11.3, 1.6)
8	35.2	2.0, m	36.7	1.95, m
9	70.7	4.0, t (6.3)	71.7	4.04, t (6.2)
10	23.9	1.98, m; 2.4, m	24.4	2.16, m; 2.43, m
11	125.9	5.84, d (3.5)	125.8	6.23, d (4.2)
12	130.8		135.7	
13	101.3		100.8	
14	43.7	1.94, m	45.2	1.89, dd (14.1, 7.0)
15	68.9	4.29, dq (8.5, 6.3)	69.0	4.36, dq (12.5, 6.3)
16	20.7	1.23, d (6.3)	20.6	1.22, d (6.3)
17	12.4	1.92, s	12.6	1.94, s
18	17.0	1.07, d (7.0)	17.7	1.09, d (6.8)
19	13.2	0.71, d (6.9)	13.4	0.79, d (6.9)
20	18.0	1.62, s	62.2	3.96, d (13.1); 4.07, d (14.9)
21	11.8	0.79, d (7.0)	11.2	0.85, d (6.9)
1′				
2′	176.7		Not observed	
3′	Not observed		Not observed	
4′	192.8		Not observed	
5′	51.7	3.84, s	52.0	3.79, s

[a] Measured in CDCl$_3$

[b] Measured in MeOD

to the different oxidative modification in bicyclic ketal structure of nocamycin, streptolydigin and tirandamycin.

Within nocamycin gene cluster, only *ncmP* encodes for a SAM-dependent carboxylate *O*-methyltransferase and it shows identity to NokK (44% identity/55% similarity) and NivG (43% identity/53% similarity). Both NokK and NivG are proposed to catalyze methyl esterification of the carboxylate group in biosynthesis of K-252a and nivetetracyclates, respectively [36, 37]. Hence, it should be evident that NcmP serves as the best candidate responsible for methyl esterification during nocamycin biosynthetic pathway.

NcmL shows similarities to monooxygenase from a series of actinomycetes. BLAST analysis revealed that NcmL displays FAD-binding domain (pfam01494). Unlike bicovalent flavinylation protein TrdL/TamL involved in tirandamycin biosynthetic pathway, NcmL has no conserved His and Cys dual active site residues that distributed in 8α-histidyl and 6-S-cysteinyl FAD linked monooxygenase family (Additional file 1: Figure S7) [15, 16]. To investigate the function of NcmL in nocamycin biosynthesis, the gene *ncmL* was inactivated (Additional file 1: Figure S2). The fermentation broth of *ΔncmL* mutant strain was analyzed by HPLC (Fig. 2III). The results revealed that the titer of nocamycin I and nocamycin II in *ΔncmL* deletion strain is identical to that in wild type, indicating NcmL is not involved in nocamycin biosynthesis.

The putative product of *ncmD* shows identity to a series of short chain dehydrogenase (SDR) family oxidoreductase originated from various bacteria. NcmD shares the Rossmann fold NAD-binding motif and characteristic NAD-binding and catalytic sequence patterns [38]. NcmD shows closet similarity to BatM (40% identity/56% similarity) which was proposed to catalyze the conversion from hydroxyl to ketone in C-17 position during kalimantacin/batumin-related polyketide antibiotic biosynthesis [39]. Thus, NcmD is proposed to be the candidate to convert hydroxy to ketone in C-10 position.

Genes involved in regulation, resistance and unknown functions

Five genes related to regulation and resistance are easily discerned from the nocamycin biosynthetic gene cluster. *NcmN* encodes for a LuxR family regulator and it shows similarity to a series of regulators from different actinomycetes, including QmnRg4 (43% identity/55% similarity) from *Amycolatopsis orientalis* involved in quartromicin biosynthesis and TamH (39% identity/52% similarity) involved in tirandamycin biosynthesis [14, 35]. The characteristic C-terminal helix-turn-helix (HTH) DNA binding domain signature and a N-terminal ATP-binding domain represented by discernible Walker A (GxxGxGK) and Walker B (R/K-X(7-8)-H(4)-D) motifs present in all members of this family of regulatory proteins are found in NcmN [40]. NcmJ is similar to AAA family ATPase from different actinomycetes. AAA family ATPases are present in all kingdoms and they are often involved in DNA replication, repair, recombination and transcription [41]. NcmJ contains the Walker A and Walker B motifs, which is the hallmark of ATP-binding domain in these proteins [41]. NcmI encodes a

PadR family transcriptional regulator and it shows similarities to several PadR-like proteins of unknown function from different actinomycetes. PadR-like proteins is a quite recently identified family of regulatory proteins, named after the phenolic acid decarboxylation repressor of *Bacillus subtilis* [42, 43]. The hallmark of this family transcriptional regulator is a highly conserved N-terminal winged helix-turn-helix (HTH) domain with about 80–90 residues [44, 45], which is also found in NcmI. NcmK encodes for a TetR family transcriptional regulator and it shows identity to TrdK (49% identity/64% similarity) involved in tirandamycin biosynthesis [14]. The characteristic N-terminal helix-turn-helix (HTH) DNA binding domain signature (pfam00440) presented in all members of this family of regulatory proteins has been found in NcmK.

NcmH, a major facilitator superfamily (MFS) transporter, shows identity to ChaT1 (42% identity/60% similarity) from *Streptomyces chartreusis* involved in antitumor agent chartreuse in biosynthesis pathway, is a candidate protein for resistance [46]. NcmQ is similar to the proteins belong to glyoxalase/bleomycin resistance protein/dioxygenase superfamily. The exact role of NcmQ in nocamycin biosynthesis is unclear and we assume that NcmQ is likely involved in resistance.

The deduced product of *ncmF* shows similarity to a series of prenyltransferase, including TrdF (50% identity/61% similarity) involved in tirandamycin biosynthesis and SlgF (51% identity/60% similarity) involved in streptolydigin biosynthesis, respectively [10, 14]. Previous studies on TrdF and SlgF demonstrated that both the proteins show no relationship with tirandamycin or streptolydigin biosynthesis [10, 14]. Thus, we hypothesize that NcmF maybe not involved in nocamycin biosynthesis.

Inactivation of *ncmG* and isolation the new derivatives from the mutant strain

Cytochrome P450 oxidases are often play important roles in post-tailoring steps during antibiotic biosynthesis. Generally, oxygenation modification is a vital approach to improve bioactivity of parent molecule. To obtain more nocamycin derivatives, we inactivated *ncmG* by λ-RED/ET technology and generated Δ*ncmG* mutant strain *S. syringae* MoS1003 (Additional file 1: Figure S3). HPLC analysis revealed that *S. syringae* MoS1003 abolished nocamycin I and nocamycin II production completely and two new peaks with similar characteristic UV absorption to these of nocamycin I and nocamycin II are detected (Fig. 2II). Then, A 15-L scale fermentation of Δ*ncmG* mutant strain led the purification of nocamycin III and nocamycin IV. The structures of nocamycin III and nocamycin IV were determined by multiple

spectroscopy data analyses. Both nocamycin III and IV are new nocamycin derivatives. Compared to nocamyin I and II, nocamycin III and IV show less oxidative modification, lacking of tetrahydrofuran ring, C-10 and C-21 modification.

The molecular formula of nocamycin III (**4**) is $C_{25}H_{35}NO_6$ ($m/z = 445.25$), which was determined by HR–ESI–MS ([M − H]$^-$ $m/z = 444.2422$, [M + H]$^+$ $m/z = 446.2548$, [M + Na]$^+$ $m/z = 468.2354$) (Additional file 1: Figure S8). Comparisons of the ^{1}H and ^{13}C NMR spectroscopic data of nocamycin III to those of nocamycin I (Bu-2313B) suggested that they share a similar structure. Complete spectral data including COSY, HSQC, and HMBC spectra were also acquired (Additional file 1: Figures S10–S16), thereby allowing full assignments of the ^{1}H and ^{13}C signals (Table 4). Comparisons of the ^{1}H and ^{13}C NMR data for nocamycin I and nocamycin III revealed that the tetrahydrofuran ring is not closed and a $\Delta^{11,12}$ double bond is apparent in nocamycin III. HMBC correlations from H-20 to C-11, C-12, and C-13, and the COSY correlations of H-10/H-11 further substantiated these assignments. Additionally, H-15 was shifted to δ_H 4.29 due to the ring opening, relative to the same position of the cyclic form. A keto group in nocamycin I was replaced by a methylene group (δ_H, 1.98, 2.4; δ_C 23.9) at C-10 in **4**, which was confirmed by the HMBC correlations from H-8, H-9, and H-11 to C-10 and from the COSY correlations of H-9/H-10α, and of H-10β/H-11. Another obvious difference observed from the ^{1}H and ^{13}C spectroscopic data was the absence of a –COOCH$_3$ in **4** compared to that of nocamycin I. In turn, a methyl group (δ_H, 0.79, δ_C 11.8) was found to be attached at C-14. Cross peak of H-14/H-21 in the COSY spectrum and the HMBC correlations from H-21 to C-13, C-14 and C-15 further confirmed this assignment. Inspection of other NMR data for nocamycin III revealed other structural elements are identical to those of nocamycin I. Consequently, the structure of nocamycin III was elucidated as shown in Fig. 5.

Nocamycin IV (**5**) was isolated as a yellowish amorphous solid. Its molecular formula was determined as $C_{25}H_{35}NO_7$ ($m/z = 461.24$) by HR–ESI–MS ([M − H]$^-$ $m/z = 460.2355$, [M + H]$^+$ $m/z = 462.2503$, [M + Na]$^+$ $m/z = 484.2303$) (Additional file 1: Figure S9), 16 mass units greater than that of nocamycin III, indicating one more oxygen atom than that of nocamycin III. Complete spectral data including COSY, HSQC, and HMBC spectra were also acquired (Additional file 1: Figures S17–S21), thereby allowing full assignments of the ^{1}H and ^{13}C signals (Table 4). It shared a similar structure to that of nocamycin III, except that a methyl signal at δ_H 1.62 was disappeared and an oxygen-bearing methylene signal at δ_H 3.96 and 4.07 occurred. Key HMBC correlations from

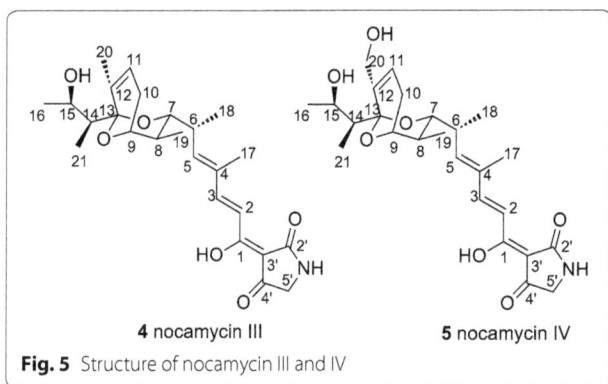

Fig. 5 Structure of nocamycin III and IV

H-20 to C-12 and from H-11 to C-12 further confirmed the location of the –CH$_2$OH group at C-12. Thus, the structure of nocamycin IV was elucidated as 20-hydroxy-nocamycin III (Fig. 5).

Discussion

In this study, the gene cluster responsible for nocamycin biosynthesis identified from S. syringae consists of 21 ORFs: 12 coding for structural proteins, seven involved in regulator and resistance and two with unknown function. Like the reported biosynthetic gene clusters of tirandamycin and streptolydigin, a hybrid PKS-NRPS mechanism is employed to assemble the chain PK-NPR backbone by co-linearity rule [10, 13, 14]. The core structure of nocamycin is bicyclic ketal unit and tetramic acid moiety. To date, tetramic acid structure has been identified in numerous natural products and four phylogenetically different family enzymes have been characterized to catalyze the tetramic acid formation through Dieckmann cyclisation reaction [21, 47–50]. In previous report, TrdC and its homologous protein SlgL have been characterized as Dieckmann cyclases to catalyze the formation of tetramic acid moiety in tirandamycin and streptolydigin biosynthetic pathway, respectively [21]. Thus, it is plausible to assume that NcmC, the homologous protein to TrdC, is employed to generate tetramic acid moiety through Dieckmann cyclisation during nocamycins biosynthetic pathway [21].

Formation of bicyclic ketal ring represents the most intriguing issue of nocamycin family natural products, which is not fully understood. In our previous study, an abnormal DH at module 3 in tirandamycin PKS was proposed to be involved in spiroketal structure formation [14]. Comparing with tirandamycin and streptolydigin gene clusters, it is important to notice that all the three gene clusters possess a similar unexpected DH domain with conserved active motif in the corresponding PKS. This abnormal DH domain at module 4 are likely not to catalyze the dehydration reaction to afford C-10 and C-11 double bond because the C-11 hydroxy group is

absolutely required for the C-13 spiroketal group formation and no nocamycin derivatives possess C-10 and C-11 double bond have been identified. Recently, linear 7,13,9,13-diseco-tirandamycin derivative tirandamycin K, a shunt pathway product in tirandamycin pathway, was isolated from Streptomyces sp. 307-9 and its P450 monooxygenase disruption mutant strain [51]. C-9 hydroxyl in tirandamycin K clearly indicates that DH3-catalyzed dehydration can be avoided, and it also provides evidence to support the mechanism that DH3 is involved in bicyclic ketal formation [51]. Due to the high similarity in polyketide structure and domains organization of PKS between tirandamycin, nocamycin and streptolydigin gene clusters, the abnormal DH catalytic mechanisms are likely to be common spiroketalization mechanisms in these three pathways.

Based on bioinformatics and genetic engineering data, post tailoring steps of nocamycin can be predicted as follows (Fig. 4). Firstly, the earliest intermediate released from the PCP protein possesses a hydroxyl group in C-11 position, NcmE catalyze the dehydration process to afford nocamycin III. Next, nocamycin III undergoes several oxidative and one methyl esterification steps to produce nocamycin I. At last, NcmD catalyzes the dehydrogenation process to afford nocamycin II. Comparisons of gene clusters of tirandamycin and nocamycin revealed an interesting phenomenon that the post-tailoring enzymes involved in modification of similar structure are varied. In tirandamycin biosynthetic pathway, a FAD-dependent dehydrogenase TrdL/TamL is responsible for the conversion from C-10 hydroxyl to C-10 ketone [15, 16]. In our initial hypothesis, a TrdL/TamL homologous protein is predicted to be responsible for the same process, however, no TrdL/TamL homologous protein has been observed within the gene cluster. Although NcmL shows FAD-binding domain, it lacks the conserved bicovalent FAD linked active sites to that in TamL/TrdL [15, 16]. Meanwhile, the gene inactivation experiments revealed that NcmL shows no relationship to nocamycin biosynthesis, and this result also indicates that diversified modification mechanism occurred in this class of natural products. Overview the gene cluster, the short-chain dehydrogenase NcmD is the best candidate to catalyze the last C-10 dehydrogenation step in nocamycin biosynthetic pathway. The complex oxidative modifications including formation of fused oxolane ring system in bicyclic ketal moiety and the conversion from methyl group to carboxyl are intriguing issues, and the two cytochrome P450 oxidase NcmG and NcmO are expected to be involved in these steps. Two new derivatives nocamycin III and nocamycin IV lacking of closed tetrahydrofuran ring from ΔncmG mutant strain indicates NcmG's involvement in the formation of the fused oxolane ring system. In terms of oxolane ring system formation, four different biosynthetic routes have been envisioned [52–55]. The mechanism of tetrahydrofuran ring in nocamycin is proposed to be similar to that in nonactin biosynthesis pathway [52]. NcmG is likely to catalyze conjugate addition of C-15 hydroxyl groups to the adjacent C-11 and C-12 alkenyl moiety to form oxolane ring (Fig. 4). We notice that the C-20 hydroxyl in nocamycin IV is similar to C-18 hydroxyl in tirandamycin B. In tirandamycin biosynthetic pathway, a multifunctional cytochrome P450 TamI has been verified to be responsible for the formation of C-18 hydroxyl group [15]. However, C-20 hydroxylation modification is not required in nocamycin biosynthetic pathway (Fig. 4). Thus, we hypothesize that nocamycin IV is probably a shunt product in nocamycins biosynthetic pathway and an oxidase located elsewhere of the genome can catalyze the hydroxylation process. Considerations of several oxidative modifications are required to afford nocamycin II, one of NcmG and NcmO is potentially responsible for more than one oxidative tailoring steps. Elucidation of the exact roles of NcmG and NcmO and the timing of modification in nocamycin biosynthesis is our ongoing project.

Up to now, the biosynthetic gene clusters responsible for streptolydigin, tirandamycin and nocamycin biosynthesis have been identified. Comparisons of the three gene clusters will help us deeply understand the biosynthetic mechanisms of this small class of natural products. The genetic insights and elucidations of enzyme function will facilitate us to rationally generate new derivatives with improved pharmacological property by manipulating biosynthetic pathway.

Conclusion

The nocamycin I and II, bearing a tricyclic ketal moiety, belong to acyl tetramic acid natural products and they display broad antimicrobial activity. In this report, we identify nocamycins biosynthetic gene cluster from rare actinomycete *Saccharothrix syringae*, which provides us the genetic insights into nocamycins biosynthesis and enzyme candidates for several intriguing biochemical transformations. Inactivation of cytochrome P450 monoxygenase NcmG led to isolation of two novel nocamycin derivatives from the mutant strain. Based on gene cluster data and new derivatives isolated from gene inactivation mutant strain, a putative biosynthetic pathway of nocamycin is proposed. These findings provide insights into further investigation of nocamycin biosynthetic mechanism, and also set the stage to rationally engineer new nocamycin derivatives via manipulating biosynthetic pathway.

Additional file

Additional file 1: Table S1. Conserved motifs from PKS modules in nocamycin gene cluster. **Figure S1.** Inactivation of *ncmB* by gene disruption. **Figure S2.** Inactivation of *NcmL* by gene disruption. **Figure S3.** Inactivation of *NcmG* by gene disruption. **Figure S4.** Multiple sequence alignment of NcmC and its homologous protein sequence. **Figure S5.** Multiple sequence alignments of the cytochrome P450 domains of NcmG and NcmO. **Figure S6.** Unrooted phylogenetical tree of NcmO and NcmG with TrdI, SlgO1, SlgO2 and other cytochrome P450s. **Figure S7.** Multiple sequence alignments of NcmL with confirmed proteins that contain biocovently linked FAD cofactor. **Figure S8.** HR–ESI–MS of nocamycin III (**4**). **Figure S9.** HR–ESI–MS of nocamycin IV (**5**). **Figure S10.** 1H NMR (700 MHz) spectrum of compound 4 in CDCl3. **Figure S11.** 13C NMR (176 MHz) spectrum of compound **4** in CDCl3. **Figure S12.** DEPT 135 spectrum of compound **4** in $CDCl_3$. **Figure S13.** 1H–1H COSY spectrum of compound **4** in $CDCl_3$. **Figure S14.** HSQC spectrum of compound 4 in CDCl3. **Figure S15.** HMBC spectrum of compound 4 in CDCl3. **Figure S16.** NOESY spectrum of compound **4** in CDCl3. **Figure S17.** 1H NMR (500 MHz) spectrum of compound **5** in MeOD. **Figure S18.** 13C NMR (125 MHz) spectrum of compound **5** in MeOD. **Figure S19.** 1H–1H COSY spectrum of compound **5** in MeOD. **Figure S20.** HSQC spectrum of compound 5 in MeOD. **Figure S21.** HMBC spectrum of compound 5 in MeOD.

Abbreviations

ORF: open reading frame; NRPS: nonribosomal peptide synthetases; PKS: polyketide synthases; A domain: adenylation domain; PCP: peptidyl carrier protein; C domain: condensation domain; KS: ketosyntheatase; AT: acyltransferase; ACP: acyl carrier protein; DH: dehydratase; KR: ketoreductase; Apr^R: apramycin resistant phenotype; Kan^S: kanamycin sensitive phenotype; HPLC: high-performance liquid chromatography; HTH: helix-turn-helix; M: module; MS: mass spectrometry; NMR: nuclear magnetic resonance; TFA: trifluoroacetic acid; Amp: ampicillin; Apr: apramycin; Kan: kanamycin; Cml: chloroamphenicol; TMP: trimethoprim.

Authors' contributions

XM and QW conceived and designed the project. XM, CS, CG, YZ and JJ conducted and analyzed the data, XM, JJ and QW wrote the paper. All authors read and approved the final manuscript.

Author details

[1] Shandong Key Laboratory of Applied Mycology, School of Life Sciences, Qingdao Agricultural University, Qingdao 266109, China. [2] CAS Key Laboratory of Tropical Marine Bio-resources and Ecology, Guangdong Key Laboratory of Marine Materia Medica, RNAM Center for Marine Microbiology, South China Sea Institute of Oceanology, Chinese Academy of Sciences, 164 West Xingang Rd., Guangzhou 510301, China.

Acknowledgements

We are so grateful to Agricultural Research Service Culture Collection (NRRL) for providing S. syringae NRRL B-16468. Research in our laboratory was supported by Grants from the National Natural Science Foundation of China (Nos. 31300063, 31300005), Natural Science Foundation of Shandong Province (NO. ZR2013CL020), Research Foundation for Advanced Talents of Qingdao Agricultural University (No. 631301) and Open Research Foundation of CAS Key Laboratory of Tropical Marine Bio-resources and Ecology (No. LMB141007).

Competing interests

The authors declare that they have no competing interests.

References

1. Gauze G, Sveshnikova M, Ukholina R, Komarova G, Bazhanov V. Formation of a new antibiotic, nocamycin, by a culture of *Nocardiopsis syringae* sp. nov. Antibiotiki. 1977;22:483–6.
2. Brazhnikova M, Konstantinova N, Potapova N, Tolstykh I. Physicochmemical characteristics of the new antineoplastic antibiotic, nocamycin. Antibiotiki. 1977;22:486–9.
3. Horváth G, Brazhnikova MG, Konstantinova NV, Tolstykh IV, Potapova NP. The structure of nocamycin, a new antitumor antibiotic. J Antibiot (Tokyo). 1979;32:555–8.
4. Tsunakawa M, Toda S, Okita T, Hanada M, Nakagawa S, Tsukiura H, Naito T, Kawaguchi H. Bu-2313, a new antibiotic complex active against anaerobes II. Structure determination of Bu-2313 A and B. J Antibiot. 1980;33:166–72.
5. Tsukiura H, Tomita K, Hanada M, Kobaru S, Tsunakawa M, Fujisawa M, Fujisawa K, Kawaguchi H. Bu-2313, a new antibiotic complex active against anaerobes I. production, isolation and properties of Bu-2313 A and B. J Antibiot. 1980;33(2):157–64.
6. Bansal M, Dhawan V, Thadepalli H. In vitro activity of Bu-2313B against anaerobic bacteria. Chemotharapy. 1982;28:200–3.
7. Toda S, Nakagawa S, Naito T. Bu-2313, a new antibiotic complex active against anaerobes III. Semi-synthesis of Bu-2313 A and B. J Antibiot. 1980;33:166–71.
8. Tuske S, Sarafianos SG, Wang X, Hudson B, Sineva E, Mukhopadhyay J, Birktoft JJ, Leroy O, Ismail S, Clark AD Jr, Dharia C, Napoli A, Laptenko O, Lee J, Borukhov S, Ebright RH, Arnold E. Inhibition of bacterial RNA polymerase by streptolydigin: stabilization of a straight-bridge-helix active-center conformation. Cell. 2005;122:541–52.
9. Temiakov D, Zenkin N, Vassylyeva MN, Perederina A, Tahirov TH, Kashkina E, Savkina M, Zorov S, Nikiforov V, Igarashi N, Matsugaki N, Wakatsuki S, Severinov K, Vassylyev DG. Structural basis of transcription inhibition by antibiotic streptolydigin. Mol Cell. 2005;19:655–66.
10. Olano C, Gómez C, Pérez M, Palomino M, Pineda-Lucena A, Carbajo RJ, Braña AF, Méndez C, Salas JA. Deciphering biosynthesis of the RNA polymerase inhibitor streptolydigin and generation of glycosylated derivatives. Chem Biol. 2009;16:1031–44.
11. Carlson JC, Li S, Burr DA, Sherman DH. Isolation and characterization of tirandamycins from a marine-derived *Streptomyces* sp. J Nat Prod. 2009;72:2076–9.
12. Horna DH, Gómez C, Olano C, Palomino-Schätzlein M, Pineda-Lucena A, Carbajo RJ, Braña AF, Méndez C, Salas JA. Biosynthesis of the RNA polymerase inhibitor streptolydigin in *Streptomyces lydicus*: tailoring modification of 3-methyl-aspartate. J Bacteriol. 2011;193:2647–51.
13. Carlson JC, Fortman JL, Anzai Y, Li S, Burr DA, Sherman DH. Identification of the tirandamycin biosynthetic gene cluster from *Streptomyces* sp. 307-9. ChemBioChem. 2010;11:564–72.
14. Mo X, Wang Z, Wang B, Ma J, Huang H, Tian X, Zhang S, Zhang C, Ju J. Cloning and characterization of the biosynthetic gene cluster of the bacterial RNA polymerase inhibitor tirandamycin from marine-derived *Streptomyces* sp. SCSIO1666. Biochem Biophys Res Commun. 2011;406:341–7.
15. Carlson JC, Li S, Gunatilleke SS, Anzai Y, Burr DA, Podust LM, Sherman DH. Tirandamycin biosynthesis is mediated by co-dependent oxidative enzymes. Nat Chem. 2011;3:628–33.
16. Mo X, Huang H, Ma J, Wang Z, Wang B, Zhang S, Zhang C, Ju J. Characterization of TrdL as a 10-hydroxy dehydrogenase and generation of new analogues from a tirandamycin biosynthetic pathway. Org Lett. 2011;13:2212–5.
17. Mo X, Ma J, Huang H, Wang B, Song Y, Zhang S, Zhang C, Ju J. Δ(11,12) double bond formation in tirandamycin biosynthesis is atypically catalyzed by TrdE, a glycoside hydrolase family enzyme. J Am Chem Soc. 2002;134:2844–7.
18. Gómez C, Horna DH, Olano C, Palomino-Schätzlein M, Pineda-Lucena A, Carbajo RJ, Braña AF, Méndez C, Salas JA. Amino acid precursor supply in the biosynthesis of the RNA polymerase inhibitor streptolydigin by *Streptomyces lydicus*. J Bacteriol. 2011;193:4214–23.
19. Gómez C, Olano C, Méndez C, Salas JA. Three pathway-specific regulators control streptolydigin biosynthesis in *Streptomyces lydicus*. Microbiology. 2012;158:2504–14.
20. Gómez C, Horna DH, Olano C, Méndez C, Salas JA. Participation of putative glycoside hydrolases SlgC1 and SlgC2 in the biosynthesis of streptolydigin in *Streptomyces lydicus*. Microb Biotechnol. 2012;5:663–7.
21. Gui C, Li Q, Mo X, Qin X, Ma J, Ju J. Discovery of a new family of Dieckmann cyclases essential to tetramic acid and pyridone-based natural products biosynthesis. Org Lett. 2015;17:628–31.
22. Macneil DJ, Gewain KM, Ruby CL, Dezeny G, Gibbons PH, Macneil T. Analysis of *Streptomyces-avermitilis* genes required for avermectin biosynthesis utilizing a novel integration vector. Gene. 1992;111:61–8.
23. Datsenko KA, Wanner BL. One-step inactivation of chromosomal genes in *Escherichia coli* K-12 using PCR products. Proc Natl Acad Sci USA. 2000;97:6640–5.
24. Gust B, Chandra G, Jakimowicz D, Yuqing T, Bruton CJ, Chater KF. Lambda red-mediated genetic manipulation of antibiotic-producing streptomyces. Adv Appl Microbiol. 2004;54:107–28.
25. Paget MSB, Chamberlin L, Atrih A, Foster SJ, Buttner MJ. Evidence that the extracytoplasmic function sigma factor sigma(e) is required for normal cell wall structure in *Streptomyces coelicolor* A3(2). J Bacteriol. 1999;181:204–11.
26. Kieser T, Bibb MJ, Buttner MJ, Chater KF, Hopwood DA. Practical streptomyces genetics. Norwich: John Innes foundation; 2000.
27. Weber T, Blin K, Duddela S, Krug D, Kim HU, Bruccoleri R, Lee SY, Fischbach MA, Müller R, Wohlleben W, Breitling R, Takano E, Medema MH. antiSMASH 3.0—a comprehensive resource for the genome mining of biosynthetic gene clusters. Nucleic Acids Res. 2015;43:W237–43.
28. Caffrey P. Conserved amino acid residues correlating with ketoreductase stereospecificity in modular polyketide synthases. ChemBioChem. 2003;4:654–7.
29. Haydock SF, Aparicio JF, Molnar I, Schwecke T, Khaw LE, Konig A, Marsden AFA, Galloway IS, Staunton J, Leadlay PF. Divergent sequence motifs correlated with the substrate-specificity of (methyl)malonyl-coa-acyl carrier protein transacylase domains in modular polyketide synthases. FEBS Lett. 1995;374(2):246–8.
30. Reeves CD, Murli S, Ashley GW, Piagentini M, Hutchinson CR, McDaniel R. Alteration of the substrate specificity of a modular polyketide synthase acyltransferase domain through site-specific mutations. Biochemistry. 2001;40:15464–70.
31. Bevitt DJ, Cortes J, Haydock SF, Leadlay PF. 6-Deoxyerythronolide-B synthase from *Saccharopolyspora erythraea*: cloning of the structural gene, sequence-analysis and inferred domain-structure of the multifunctional enzyme. Eur J Biochem. 1992;204:39–49.
32. Nagano S, Cupp-Vickery JR, Poulos TL. Crystal structures of the ferrous dioxygen complex of wild-type cytochrome P450eryF and its mutants, A245S and A245T: investigation of the proton transfer system in P450eryF. J Biol Chem. 2005;280:22102–7.
33. Parajuli N, Basnet DB, Chan Lee H, Sohng JK, Liou K. Genome analyses of *Streptomyces peucetius* ATCC 27952 for the identification and comparison of cytochrome P450 complement with other Streptomyces. Arch Biochem Biophys. 2004;425:233–41.
34. Tang GL, Cheng YQ, Shen B. Leinamycin biosynthesis revealing unprecedented architectural complexity for a hybrid polyketide synthase and nonribosomal peptide synthetase. Chem Biol. 2004;11:33–45.
35. He HY, Pan HX, Wu LF, Zhang BB, Chai HB, Liu W, Tang GL. Quartromicin biosynthesis: two alternative polyketide chains produced by one polyketide synthase assembly line. Chem Biol. 2012;19:1313–23.
36. Chiu HT, Chen YL, Chen CY, Jin C, Lee MN, Lin YC. Molecular cloning, sequence analysis and functional characterization of the gene cluster for biosynthesis of K-252a and its analogs. Mol BioSyst. 2009;5:1180–91.
37. Chen C, Liu X, Abdel-Mageed WM, Guo H, Hou W, Jaspars M, Li L, Xie F, Ren B, Wang Q, Dai H, Song F, Zhang L. Nivetetracyclates A and B: novel compounds isolated from *Streptomyces niveus*. Org Lett. 2013;15:5762–5.
38. Kavanagh KL, Jörnvall H, Persson B, Oppermann U. Medium- and short-chain dehydrogenase/reductase gene and protein families: the SDR superfamily: functional and structural diversity within a family of metabolic and regulatory enzymes. Cell Mol Life Sci. 2008;65:3895–906.
39. Mattheus W, Masschelein J, Gao LJ, Herdewijn P, Landuyt B, Volckaert G, Lavigne R. The kalimantacin/batumin biosynthesis operon encodes a self-resistance isoform of the FabI bacterial target. Chem Biol. 2010;17:1067–71.
40. Walker JE, Saraste M, Runswick MJ, Gay NJ. Distantlyrelated sequences in the alpha- and beta-subunits of ATP synthase, myosin, kinases and other ATP-requiring enzymes and a common nucleotide binding fold. EMBO J. 1982;1:945–51.

41. Iyer LM, Leipe DD, Koonin EV, Aravind L. Evolutionary history and higher order classification of AAA + ATPases. J Struct Biol. 2004;146:11–31.

42. Barthelmebs L, Lecomte B, Diviès C, Cavin JF. Inducible metabolism of phenolic acids in *Pediococcus pentosaceus* is encoded by an autoregulated operon which involves a new class of negative transcriptional regulator. J Bacteriol. 2000;182:6724–31.

43. Gury J, Barthelmebs L, Tran NP, Diviès C, Cavin JF. Cloning, deletion, and characterization of PadR, the transcriptional repressor of the phenolic acid decarboxylase-encoding padA gene of *Lactobacillus plantarum*. Appl Environ Microbiol. 2004;70:2146–53.

44. De Silva RS, Kovacikova G, Lin W, Taylor RK, Skorupski K, Kull FJ. Crystal structure of the virulence gene activator AphA from *Vibrio cholerae* reveals it is a novel member of the winged helix transcription factor superfamily. J Biol Chem. 2005;280:13779–83.

45. Madoori PK, Agustiandari H, Driessen AJ, Thunnissen AM. Structure of the transcriptional regulator LmrR and its mechanism of multidrug recognition. EMBO J. 2009;28:156–66.

46. Xu Z, Jakobi K, Welzel K, Hertweck C. Biosynthesis of the antitumor agent chartreusin involves the oxidative rearrangement of an anthracyclic polyketide. Chem Biol. 2005;12:579–88.

47. Blodgett JA, Oh DC, Cao S, Currie CR, Kolter R, Clardy J. Common biosynthetic origins for polycyclic tetramate macrolactams from phylogenetically diverse bacteria. Proc Natl Acad Sci USA. 2010;107:11692–7.

48. Wu Q, Wu Z, Qu X, Liu W. Insights into pyrroindomycin biosynthesis reveal a uniform paradigm for tetramate/tetronate formation. J Am Chem Soc. 2012;134:17342–5.

49. Sims JW, Schmidt EW. Thioesterase-like role for fungal PKS-NRPS hybrid reductive domains. J Am Chem Soc. 2008;130:11149–55.

50. Liu X, Walsh CT. Cyclopiazonic acid biosynthesis in *Aspergillus* sp.: characterization of a reductase-like R* domain in cyclopiazonate synthetase that forms and releases cyclo-acetoacetyl-ʟ-tryptophan. Biochemistry. 2009;48:8746–57.

51. Zhang X, Zhong L, Du L, Chlipala GE, Lopez PC, Zhang W, Sherman DH, Li S. Identification of an unexpected shunt pathway product provides new insightsinto tirandamycin biosynthesis. Tetrahedron Lett. 2016;57:5919–23.

52. Woo AJ, Strohl WR, Priestley ND. Nonactin biosynthesis: the product of nonS catalyzes the formation of the furan ring of nonactic acid. Antimicrob Agents Chemother. 1999;43:1662–8.

53. Demydchuk Y, Sun Y, Hong H, Staunton J, Spencer JB, Leadlay PF. Analysis of the tetronomycin gene cluster: insights into the biosynthesis of a polyether tetronate antibiotic. ChemBioChem. 2008;9:1136–45.

54. Bode HB, Zeeck A. Biosynthesis of kendomycin: origin of the oxygen atoms and further investigations. J Chem Soc Perkin Trans. 2000;1:2665–70.

55. Richter ME, Traitcheva N, Knüpfer U, Hertweck C. Sequential asymmetric polyketide heterocyclization catalyzed by a single cytochrome P450 monooxygenase (AurH). Angew Chem Int Ed. 2008;47:8872–5.

5

Engineering redox homeostasis to develop efficient alcohol-producing microbial cell factories

Chunhua Zhao[1,2], Qiuwei Zhao[1], Yin Li[1] and Yanping Zhang[1]*

Abstract

The biosynthetic pathways of most alcohols are linked to intracellular redox homeostasis, which is crucial for life. This crucial balance is primarily controlled by the generation of reducing equivalents, as well as the (reduction)-oxidation metabolic cycle and the thiol redox homeostasis system. As a main oxidation pathway of reducing equivalents, the biosynthesis of most alcohols includes redox reactions, which are dependent on cofactors such as NADH or NADPH. Thus, when engineering alcohol-producing strains, the availability of cofactors and redox homeostasis must be considered. In this review, recent advances on the engineering of cellular redox homeostasis systems to accelerate alcohol biosynthesis are summarized. Recent approaches include improving cofactor availability, manipulating the affinity of redox enzymes to specific cofactors, as well as globally controlling redox reactions, indicating the power of these approaches, and opening a path towards improving the production of a number of different industrially-relevant alcohols in the near future.

Keywords: Redox homeostasis, Metabolic engineering, Alcohol, Cofactor engineering, Glutathione, Reducing equivalent

Background

Due to the increasing concerns surrounding limited fossil resources and environmental problems, there has been much interest in the microbial production of chemicals and fuels from renewable resources. Alcohols such as ethanol, 1,3-propanediol, butanol, isobutanol, 2,3-butanediol and 1,4-butanediol, can be used as important platform chemicals or biofuels [1]. Since they are bulk products, the demand for most of these compounds is highly cost-sensitive. To meet this challenge, the microbial cell factories for producing alcohols must be engineered to increase the titer, yield and productivity of target products as much as possible.

Since wild-type microorganisms do not allow the production of industrially relevant alcohols with high enough efficiency, many efforts have been undertaken to improve their production by systems metabolic engineering [2]. To develop microbial strains that maximize the titer, yield and productivity of the target products, intracellular metabolic fluxes must be optimized using various molecular and high-throughput techniques, including, but not limited to: selecting the best biosynthesis genes [3], overexpressing rate-limiting enzymes, fine-tuning the expression of pathway enzymes [4], reinforcing the direct biosynthesis route [5–7], deleting or down-regulating competing pathways [8, 9], as well as deactivating degradation and utilization pathways or removing feedback regulation [10].

Most of the recent successful systems metabolic engineering examples of the development of alcohol-producing microorganisms focused on directly engineering enzymes of the metabolic pathways in question. However, in addition to the activity of enzymes involved in the pathway itself, the metabolic flux also depends on the concentrations of precursors and cofactors in the cells [11]. Since most alcohol production pathways comprise redox reactions, their production efficiency depends

*Correspondence: zhangyp@im.ac.cn
[1] CAS Key Laboratory of Microbial Physiological and Metabolic Engineering, Institute of Microbiology, Chinese Academy of Sciences, No. 1 West Beichen Road Chaoyang District, Beijing 100101, China
Full list of author information is available at the end of the article

on the availability of cofactors. The cofactors in question are usually some type of reducing equivalents, such as NADH and NADPH, which usually act as carriers of electrons generated from substrate oxidation. Under aerobic conditions, the electrons provided by NAD(P)H are commonly ultimately accepted by O_2 [12], whereby NAD(P)H is converted to its oxidized form. Therefore, since alcohol production is generally performed under anaerobic conditions, the strains maintain their cellular redox balance mainly through the reactions of central metabolism, which are significantly different from aerobic microbial metabolism. By decreasing the amounts of acid-forming enzymes and/or enhancing the butanol synthetic pathway genes expression in the non-sporulating, non-solventogenic *Clostridium acetobutylicum* strain M5, Sillers et al. [13] demonstrated the rigidity of intracellular electron balance. Thus, in order to sustain growth and metabolism, the metabolic network must be tweaked to maintain the redox balance in the cells [14].

Currently, the primary feedstocks used in the biological production of alcohols are sugarcane, sugar beet, maize (corn) and sorghum, due to their low price and wide availability in the market [15]. These feedstocks mainly provide fermentable sugars, which are easily metabolized by the production strains, generating NADH, NADPH, ferredoxin and other reducing equivalents that are needed in the alcohol biosynthetic pathways. However, due to the unfavorable stoichiometry of available electrons from a substrate such as glucose [16], the maximum theoretical yields for alcohols are mostly lower than 0.5 g/g, with the exception of ethanol, at 0.51 g/g [17]. Furthermore, in addition to alcohol synthesis, there are many other pathways that are competing for reducing equivalents, especially in anaerobes, such as hydrogen production [14]. Actually, due to the imbalances between the generation of reducing equivalents from substrates and their oxidation by redox enzymes in the alcohol biosynthesis pathways, the carbon metabolic flux of substrates is generally distributed unfavorably between alcohol biosynthesis and other competing pathways [18–20]. This leads to a much lower yield of the target alcohol from sugars in the actual production process.

Therefore, to improve alcohol production, and especially the yield that can be achieved from cheap substrates, cellular redox homeostasis must be manipulated to avoid a possible limitation of reducing equivalents. In this article, we review recent advances in accelerating the production of alcohols by engineering microbial redox homeostasis, including providing sufficient amounts of needed cofactors, improving the affinity of key enzymes to the available reducing equivalents, manipulating the intracellular electron transport chain, and other approaches for engineering the cellular redox balance.

Improving the availability of required cofactors to enhance cofactor-dependent alcohol production

Targeted regulation of enzymes or genes involved in the target pathway is often the first step in metabolic engineering of microbes for alcohol production. However, once the enzyme levels are no longer limiting, cofactor availability can become the main bottleneck for cofactor-dependent redox reactions [21]. Nicotinamide adenine dinucleotide (NAD) functions as a cofactor in over 300 oxidation–reduction reactions and regulates various enzymes and genetic processes [21]. The NADH/NAD$^+$ cofactor pair also plays a major role in microbial catabolism [22]. Due to their role as co-substrates, the concentration of cofactors, together with other substrates, determines the rate of enzymatic reactions and therefore the flux of the corresponding pathway. Many strategies have thus been developed to improve the availability of cofactors such as NADH and NADPH, and successfully applied to enhance the microbial production of various alcohols.

Fine-tuning of genes expression in alcohol biosynthetic pathway to enhance NAD(P)H competitiveness

Usually, there would be more than one enzyme involved in the alcohol synthetic pathway. Thus a proper proportion of these enzymes especially the NAD(P)H-dependent one is of crucial role. Fine-tuning of gene expression through manipulation of mRNA stability [23], modulation of the ribosome binding site (RBS) [24], codon optimization [25] and other approaches [26, 27] can be benefit for the redox balance in alcohol-producing cells.

Fine-tuning of *GRE3* which is strictly NADPH-dependent expression could be more useful to reduce xylitol formation and increase ethanol production from xylose in *Saccharomyces cerevisiae* [28, 29]. Meanwhile, fine-tuned overexpression of xylulokinase in *S. cerevisiae* could lead to improved fermentation of xylose to ethanol [29] and fine-tuning of NADH oxidase could decrease byproduct accumulation in *S. cerevisiae* [30]. Sun et al. engineered a 1,2,4-butanetriol-producing *Escherichia coli* and fine-tuned the expression of *yjhG* and *mdlC*. The relative strain BW-026 increased 1,2,4-butanetriol titer by 71.4% [4]. Recently, Ohtake et al. [31] engineered a high titer butanol-producing *E. coli* strain by fine-tuning of *adhE2* which is NADH-dependent. The authors believed a CoA imbalance problem was solved improving the butanol production. On the other hand, the redox balance was also further achieved as *adhE2* is responsible for two steps consuming NADH in butanol synthetic pathway.

Blocking of competing NADH-withdrawing pathways to redirect metabolic flux towards the target alcohols

In many microorganisms, and most production strains, glycolysis is the key upstream pathway in the fermentation process from sugars to alcohols, with pyruvate as the node linking different directions of carbon flow. Concomitantly with the generation of pyruvate, a net two NADH molecules are generated from one glucose molecule [32]. To return this reduced cofactor to its oxidized state, oxidative phosphorylation or anaerobic fermentation is implemented to generate ATP or reduced byproducts, respectively [18]. In *E. coli*, lactate, ethanol, succinate, amino acids, and some other chemicals can be derived from pyruvate [or phosphoenolpyruvate (PEP)], consuming NADH under anaerobic conditions [33]. Hence, a direct approach to provide more NADH for alcohol formation is to block the pathways competing for it.

Lactate can be directly generated from pyruvate and NADH with no additional intermediate reactions, thus making it a very competitive byproduct that needs to be removed. Berríos-Rivera et al. [19] showed that an *ldh*− genotype increased the synthesis of 1,2-propanediol (1,2-PDO) in *E. coli*, which was considered an NADH-limited system. This work manipulated the NADH/NAD+ pool by eliminating the competing lactate pathway, which provided a more reducing environment for alcohol production [19]. Likewise, Zhang et al. inactivated the *aldA* gene encoding ALDH, an enzyme that competes with 1,3-propanediol (1,3-PDO) oxidoreductase for NADH in *Klebsiella pneumoniae*, to produce higher amounts of 1,3-PDO. By this manipulation, the product titer was increased by 33% compared with the control strain, and the yield of 1,3-PDO from glycerol was increased from 0.355 to 0.699 mol/mol, reaching an astonishing 97.1% of the maximal theoretical yield [34]. Similar effects were found in the engineered butanol-producing strains. By deleting the main competing NADH-withdrawing pathway genes in *E. coli*, including *adhE* for ethanol, *ldhA* for lactate, and *frdBC* for succinate, butanol production was significantly improved, leading to a doubling of the titer. After additionally blocking other byproduct pathways, the final butanol titer of the resulting strain increased by 133% [20, 35, 36].

An approach guided by in silico metabolic engineering of *E. coli* for direct production of 1,4-butanediol (1,4-BDO) also led to a strategy of eliminating pathways which compete for reducing power [37, 38]. Similarly, Fu et al. pointed out that although the deletion of *ldh* did not increase the metabolic flux towards the 2,3-butanediol (2,3-BDO) pathway, it increased the NADH/NAD+ ratio for further conversion of acetoin to 2,3-BDO, underscoring that NADH availability was the key factor for 2,3-BDO production [39].

Increasing the total level of NAD to accelerate alcohols production

The total level of NAD (NAD+ and NADH) is strictly controlled in microorganisms through specific regulatory mechanisms [40]. A de novo pathway and a pyridine nucleotide salvage pathway was found in *E. coli* to maintain its total intracellular NADH/NAD+ pool [40]. Berríos-Rivera et al. found that the nicotinic acid phosphoribosyltransferase, encoded by the *pncB* gene, can catalyze the formation of a precursor of NAD. Consequently, they overexpressed the *pncB* gene from *Salmonella typhimurium* to increase the total level of NAD. Anaerobic tube experiments showed that the strains overexpressing *pncB* had higher biomass and increased ethanol/acetate ratios [40]. Jawed et al. [41] also performed this *pncB*-overexpressing method in a *Klebsiella* HQ-3 strain and observed increased production and yield of H2. Along with H2, 2,3-BDO and ethanol titers were improved as well due to the increased availability of NADH [41]. Another study showed enhancement of succinate production by expressing nicotinic acid phosphoribosyltransferase gene *pncB* [42]. Although it is not alcohol related, succinate is a reducing chemical which makes it a valuable reference.

Regeneration of NAD(P)H to increase the availability of its reduced form to accelerate alcohol production

In addition to the total NAD(P) pool, the ratio of the reduced to the oxidized form will determine the reaction activity. Reduced cofactors (NADH, NADPH, reduced ferredoxin) are needed to provide electrons for the reduction of precursors to alcohols [43]. Therefore, efficient regeneration of NAD(P)H is crucial for optimal production of alcohols, especially in anaerobic fermentations.

Several enzymatic methods have been developed for the regeneration of NADH [44]. By overexpressing the NAD+-dependent formate dehydrogenase (FDH) from *Candida boidinii* in *E. coli*, the maximum yield of NADH was doubled from 2 to 4 mol NADH/mol glucose consumed [21]. Compared with the control strain, the ethanol to acetate (Et/Ac) ratio of the engineered strain containing heterologous FDH increased dramatically, by nearly 30-fold. What makes it even more interesting is the observation that the increased availability of NADH induced the production of ethanol even in the presence of O2, and the amount of ethanol was dependent on the amount of added formate [21]. This approach was also demonstrated to be effective for improving the Et/Ac ratio in minimal medium [22]. Similarly, the *fdh* gene was introduced into *Klebsiella oxytoca*. Interestingly, in said case both the oxidative and the reductive metabolism of glycerol was enhanced [45]. Results indicated that the engineered strain OF-1 produced more 1,3-propanediol, ethanol, and lactate than the control strain, as a result of increased NADH availability. The molar yield of 1,3-PDO was 17.3% higher than that of the control strain [45]. Using the same formate/formate dehydrogenase NADH regeneration system, the target pathways of (2S,3S)-2,3-butanediol [46] and butanol [47, 48] were effectively coupled to the NADH driving force, respectively, and the product titers were also improved significantly.

In addition to fine-tuning *fdh1* expression levels, it was demonstrated that the intracellular redox state could be modulated by anaerobically activating the pyruvate dehydrogenase (PDH) complex. The engineered strain showed the highest reported butanol productivity from glucose in *E. coli* (0.26 g/L/h) [35]. It indicated a new approach to improve the availability of NADH.

In spite of NADH, there are strategies reported on NADPH regeneration for alcohols or reduced chemicals production. Verho et al. expressed a discovered *GDP1* gene coding an $NADP^+$-dependent D-glyceraldehyde-3-phosphate dehydrogenase for ethanol fermentation in *S. cerevisiae* [49]. The *GDP1*-overexpressed strain produced ethanol with a higher rate and yield than the control strain. Combining with the deletion of *ZWF1* (coding glucose-6-phosphate dehydrogenase for NADPH and CO_2 generation) for redox balance, the resulting strain produced 11% more ethanol and 69% less xylitol which is the main byproduct in xylose fermentation [49]. Furthermore, glucose dehydrogenases from different microorganisms were also used for NADPH regeneration [50, 51]. Eguchi et al. used a glucose dehydrogenase cloned from *Gluconobacter scleroides* for recycling of cofactor NADPH in vitro [50], while Xu et al. cloned a glucose dehydrogenase gene *gdh* from *Bacillus megaterium* to regenerate NADPH in vitro and in vivo [51]. A recent study also reported an approach to enhancing NADPH supply by overexpressing glucose-6-phosphate dehydrogenase [52]. These examples demonstrated the possibility of engineering the regeneration of NADPH for efficient alcohol production.

In addition to the purely bio-catalytic regeneration of NADH and NADPH, electricity-driven NAD(P)H regeneration and direct electron transfer are rapidly being developed and have been applied experimentally for CO_2 fixation in the recent 5 years [53–56]. These studies focused on the delivery of electrons from electrodes to the cells to supply reducing power, which in turn can be used for alcohol production [57]. CO_2 is an oxidizing compound which requires large amounts of energy and reducing power to be fixed into organics. In nature, cyanobacteria and higher plants use NADPH to fix CO_2 in the Calvin cycle [58, 59]. Li et al. [53] designed an integrated electro-microbial process to convert CO_2 into formate, which was further turned into NADH by formate dehydrogenase. The generated NADH was used for isobutanol synthesis in *Ralstonia eutropha*. About 846 mg/L isobutanol was produced, indicating the tantalizing possibility of microbial electrosynthesis of alcohols. Torella

et al. [55] reported a hybrid microbial water-splitting catalyst system which was similar to natural photosynthesis. In this system, water was electrolyzed by electricity for the supply of reduced cofactors (NADPH) with the help of hydrogenases, and CO_2 was fixed through the Calvin cycle in an engineered *R. eutropha* strain using the obtained NADPH. Using this system, 216 mg/L isopropanol was synthesized with high selectivity [55].

In the above content, we listed some approaches of improving the availability of needed cofactors for alcohol production and described each approach respectively. However, these approaches are not always separately employed in metabolic engineering for alcohol production. Blocking of competing NADH-withdrawing pathways was usually accompanied by introduction of NADH regeneration systems [35]. Analogously, fine-tuning of gene expression may connect with introduction of NADH regeneration systems in alcohols synthetic pathway [31, 48]. Additionally, the strategy of increasing the total level of NAD can conceivably be combined with the introduction of an NADH regeneration system to exert an even stronger effect [60]. Therefore, in systems metabolic engineering of alcohol production, different kinds of cofactor engineering approaches could be considered and combined.

Manipulating the affinity of key redox enzymes for NADH or NADPH to improve alcohol production

In cells, various redox enzymes prefer different reducing equivalents. NAD(H) and its phosphate form NAD(P)H play major roles in metabolic processes of all living beings [21]. In microorganisms, over 400 redox enzymes have a high affinity to NAD(H) and another 400 ones have a high affinity to NADP(H), they are dependent on NAD(H) and NADP(H), respectively [38, 61]. In addition, some redox enzymes are dependent on ferredoxin, the flavin nucleotides flavin-adenine dinucleotide (FAD) and flavin mononucleotide (FMN), heme, pyrroquinoline quinone (PQQ) or other cofactors [38, 62]. As shown in Fig. 1, NADH and NADPH can be generated from different pathways in microbes. In any case, the electron balance must be satisfied and thus reduced electron carriers, like NADH and NADPH, must be re-oxidized, mostly via the reduction of substrates to alcohols, or the formation of H_2 and/or other reductive metabolites [43]. Commonly, electrons are transferred between the reduced and oxidized forms of the cofactor, the corresponding redox enzyme and the reactants, forming a redox cycle. However, it is also possible that some of the proteins mediate the exchange of electrons between NADH, NADPH, ferredoxin and other reducing equivalents. Sometimes, the types of reducing equivalents generated from the available substrates are not fit for the redox enzymes that re-oxidize

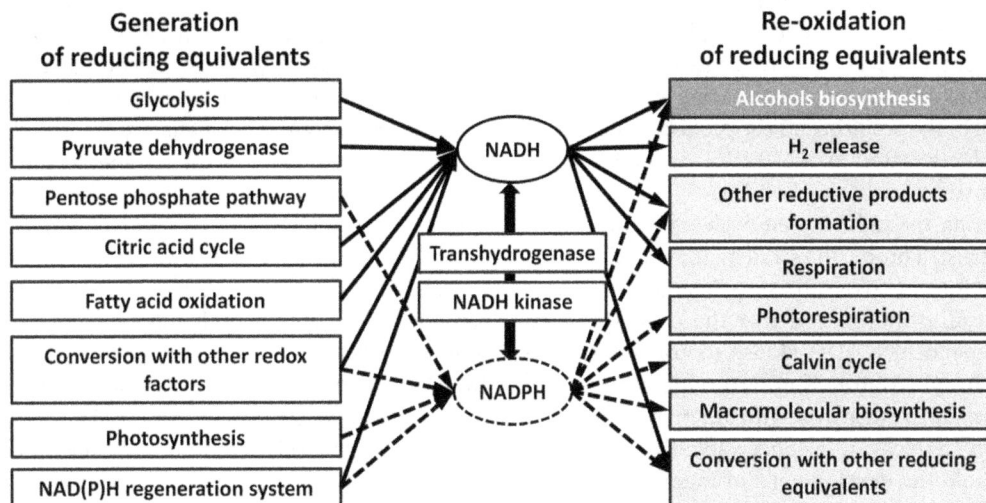

Fig. 1 Common NAD(P)H-dependent metabolic pathways in microbes. *Dashed arrow line*: NADPH; *solid arrow line*: NADH

the necessary cofactors [11, 63]. Thus, to meet the redox requirements for alcohol biosynthesis, it is necessary to construct novel redox cycles and therefore to achieve new redox homeostasis. Recently, many attempts have been made to change the affinity of key redox enzymes for different types of reducing equivalents, or to interconvert the reducing equivalents between different types.

Switching the redox enzymes' affinity from one type of reducing equivalent to another to efficiently couple alcohol production to cellular redox homeostasis

As described above, NAD(H) is the most abundant reducing equivalent in most bacteria and yeasts. Consequently, many efforts have been made to change the preferential affinity of redox enzymes from NADPH to NADH. For example, using xylose as a feedstock to produce ethanol in *S. cerevisiae* has attracted much attention, and it was found that the ethanol yield was far below the theoretical maximum because of imbalanced coenzyme utilization [63]. An NADPH-preferring xylose reductase (XR) and a strictly NAD^+-dependent xylitol dehydrogenase (XDH) caused the cofactor imbalance, leading to a low yield. Consequently, researchers employed structure-guided site-directed mutagenesis to change the coenzyme preference of *Candida tenuis* XR from NADPH in the wild-type enzyme to NADH [63, 64]. The strain harboring the resulting XR double mutant showed a 42% enhanced ethanol yield (0.34 g/g) compared to the reference strain harboring wild-type XR, in anaerobic bioconversions of xylose [63]. Likewise, the NADH preference of *Pichia stipitis* XR could also be altered by site-directed mutagenesis [65]. An engineered XR with the point mutation K270R was combined with

the capability of xylose utilization, and the resulting *S. cerevisiae* gave an ethanol yield of 0.39 g/g and a titer of 25.3 g/L, which was 18 and 51% higher than the reference strain, respectively [65].

Generally, NADH is the preferred electron donor for redox enzymes in most organisms, but some are capable of efficiently generating NADPH. An example of this are photosynthetic cyanobacteria [66]. For these microbes, the use of NADPH-dependent enzymes can be beneficial for alcohol production. Lan and Liao introduced the butanol pathway into *Synechococcus elongatus* PCC 7942 by exchanging the NAD^+-dependent enzymes with $NADP^+$-dependent ones, enabling them to consume the NADPH generated through photosynthesis [67]. By coupling the pathway with an ATP-driven step, the cyanobacterial strain successfully produced 29.9 mg/L butanol, increasing butanol production by fourfold [67]. Interestingly, by introducing an additional NADPH-consuming isopropanol synthetic pathway, the photosynthesis of *Synechocystis* sp. PCC 6803 was improved by about 50%, due to the immediate re-oxidation of NADPH that was generated from the photoreaction. At the same time, 226.9 mg/L of isopropanol was produced by this engineered strain [68]. Considering their ethanol-producing potential [57], cyanobacteria may well become the most cost-effective alcohol producing microbial cell factory in the future [69].

Dai et al. [70] introduced a single secondary alcohol dehydrogenase into *C. acetobutylicum* to consume NADPH for isopropanol production which switches ABE fermentation to a higher level IBE fermentation. The clostridial butanol synthesis pathway utilizes both NADH and reduced ferredoxin as sources of reducing power [71].

In order to couple the NADH driving force to the butanol pathway, a *trans*-enoyl-CoA reductase (Ter) was chosen to replace the butyryl-CoA dehydrogenase complex (Bcd-EtfAB), and thus to balance the reducing power in the form of NADH. The resulting strain produced 1.8 g/L of butanol in 24 h compared to only 0.1 g/L generated by an equivalent construct harboring Bcd-EtfAB [48].

Interconverting the reducing equivalents between different types is also a promising strategy to meet the redox requirements for the biosynthesis of target products. Panagiotou et al. demonstrated that the overexpression of an ATP-dependent NADH kinase to convert NADH into NADPH had a positive effect on growth efficiency in *Aspergillus nidulans*. Since aspergilli are major players in industrial biotechnology, it is conceivable that this strategy could enable the development of many new strains capable of generating the important reducing power in the form of NADPH, which is crucial for efficient production of metabolites and enzymes in large-scale fermenters [72]. In some cases, NADPH is needed directly for the production of target chemicals. For this purpose, researchers have genetically engineered an *E. coli* strain to increase the availability of NADPH by replacing the native NAD$^+$-dependent glyceraldehyde-3-phosphate dehydrogenase (GAPDH) with an NADP$^+$-dependent GAPDH from *C. acetobutylicum*. This resulted in the generation of 2 mol of NADPH, instead of NADH, per mole of glucose consumed [11].

Taking an approach that is different from engineering the affinity for natural cofactors, recently, Zhao et al. created artificial redox systems which depend on nicotinamide flucytosine dinucleotide and showed excellent activity with the NAD-dependent malic enzyme [73]. This opens a new avenue for engineering bioorthogonal redox systems for a wide variety of applications in systems and synthetic biology, which could also be implemented in alcohol production [38].

Engineering of key enzymes to improve their affinity for NAD(P)H and decrease the redox requirements for alcohol production

During the production of alcohols, some cofactor-dependent key enzymes are often rate-limiting, which is obviously unfavorable. Ingram et al. found more effective enzymes utilizing NADH in ethanol production. Alcohol dehydrogenase II and pyruvate decarboxylase from *Zymomonas mobilis* were expressed at high levels in *E. coli*, resulting in increased cell growth and the production of ethanol as the principal fermentation product from glucose [18].

In addition to substituting intrinsic enzymes with more efficient ones, direct engineering of target enzymes to improve their affinity for specific cofactors is also a practical way to increase the product titer of various

alcohols. Directed evolution which is a method for protein engineering and protein evolution mimicking natural selection has often been performed to engineer the characteristics of target enzymes [74]. Bastian et al. engineered an NADH-dependent IlvC by directed evolution, and coupled it with an engineered *Lactococcus lactis* AdhA in the isobutanol pathway. The K_m value of this IlvC variant for NADH was dramatically decreased from 1080 to 30 µM, which was even lower than the K_m of its native substrate NADPH which is 40 µM. At the same time, the engineered AdhA also showed increased affinity for NADH, with a change in K_m value from 11.7 to 1.7 mM. Strains carrying the two engineered enzymes improved the yield to practically 100% of the theoretical limit under anaerobic conditions using glucose as feedstock [75].

Structure-based rational design is also an important approach used to engineer enzymes. Meng et al. recently engineered the affinity of a D-lactate dehydrogenase for NADH and NADPH. Based on computational design and analysis, the wild-type NADH-dependent D-lactate dehydrogenase from *Lactobacillus delbrueckii* was rationally mutated to increase its affinity for both NADPH and NADH. The mutant enzyme was able to super-efficiently utilize both NADPH and NADH as cofactors [76]. This study is not directly related to alcohol production, yet it may provide useful reference points.

Engineering the cellular redox environment at a global level to benefit alcohol production

As described above, fermentations for alcohol production are mostly performed under anaerobic conditions. In the presence of sufficient O_2, most industrial organisms use active respiration to re-oxidize NADH and O_2 is usually used as the final electron acceptor. Furthermore, under some conditions, O_2 can lead to the production of free radicals from the electron transport chain, which can cause severe stress to microorganisms [77]. This in turn can indirectly hinder alcohol production. Some strategies have been reported to improve alcohol production by blocking O_2-mediated NADH oxidation and strengthening the redox balance [78, 79].

Manipulating respiratory levels to redirect the electron transport chain towards aerobic formation of alcohols

Under aerobic conditions, cells produce large amounts of ATP through respiration, and grow rapidly, but alcohol production is inhibited due to a lack of NADH. Zhu et al. reported a smart strategy to limit respiratory levels, allowing the formation of reduced chemicals such as ethanol even under fully aerobic conditions. By knocking out the *ubiCA* genes, which encode two critical enzymes for ubiquinone synthesis and therefore respiration in *E. coli*, and by supplementing external coenzyme Q1, the respiratory level was manipulated so that up to 80% of the carbon atoms from glycerol were converted into ethanol [78]. It thus demonstrated that NADH (redox) partitioning between energy generation in the electron transport chain (respiration) and the use of NADH for reduction of metabolic intermediates could be precisely controlled.

In addition to genetically manipulating respiratory levels to redirect the electron transport chain, electron carriers based on artificial chemicals have also been used to direct electron flow. Stoichiometric network analysis revealed that NAD(P)H that was lost from the fermentation in the form of H_2 limited the yield of butanol, and led to the accumulation of acetone. By using methyl viologen as an electron carrier to divert the electron flow away from H_2 production, the NAD(P)H supply was reinforced, which increased butanol yields by 37.8%, along with strongly diminished acetone production [79].

Introduction of glutathione to enhance the thiol redox balance and accelerate alcohol biosynthesis

In addition to its direct participation in NAD(P) H-dependent reactions, these cofactors also play a prominent role in the physiological functions linked to microbial growth and metabolism. As the storage molecules of reducing power, NADH and NADPH provide most of the electrons that reverse O_2-dependent thiol oxidation, constituting the thiol redox system, together with the glutathione (GSH, L-γ-glutamyl-L-cysteinylglycine) and thioredoxin pathways [12], which control intracellular redox homeostasis. Correspondingly, the microbial thiol redox system, including GSH, is presumed to affect the NADH and NADPH availability and therefore control the flux of NAD(P)H-dependent pathways.

GSH is the most abundant non-protein thiol, and is widely distributed in living organisms [80]. It plays important roles in many physiological and metabolic processes, including thiol redox homeostasis, protein stabilization, antioxidation, stresses tolerance and provision of electrons to reductive enzymes via NADPH [81–83]. The biosynthesis of GSH involves two consecutive enzymatic reactions, catalyzed either by the two separate enzymes γ-glutamylcysteine synthetase (γ-GCS, encoded by *gshA*) and GSH synthetase (GS, encoded by *gshB*), or by a bifunctional γ-glutamate-cysteine ligase/GSH synthetase (GshF). By over-expressing the *gshAB* genes from *E. coli*, GSH biosynthetic capability was introduced into *C. acetobutylicum* DSM 1731, and the resulting strain produced 14.8 g/L butanol, which was 37% higher than its wild-type parent. The engineered strain also exhibited improved tolerance to aeration and butanol [84]. This strategy was also applied in the butanol-producing

strain *C. acetobutylicum* ATCC 824. By expressing the *gshAB* genes from *E. coli* in the *adc* locus, butanol production in the engineered strain 824*adc::gsh* was increased by 59%, reaching 8.3 g/L [85].

Engineering the redox-sensitive transcription factor Rex to control NADH/NAD+ homeostasis in order to manipulate alcohol biosynthesis

Anaerobic microbes, such as *C. acetobutylicum*, have evolved a number of strategies to cope with the oxidative stress from reactive oxygen species and molecular O_2. In addition to the protection provided by GSH, it was found that reducing equivalents directly participate in the defense against oxidative stress in *Clostridium* by reducing O_2 and oxygen free radicals, which favorably shifts the cellular redox balance [14, 86]. Interestingly, the redox-sensing transcriptional repressor Rex has recently been found to play a role in the solventogenic shift of *C. acetobutylicum* [87]. Rex is composed of two domains, an N-terminal winged-helix DNA-binding domain and a C-terminal Rossmann-like domain involved in NADH binding and subunit dimerization. The DNA-binding activity of Rex protein is modulated by the ratio of NADH to NAD+ [88, 89]. By systematically investigating the Rex regulons in 11 diverse clostridial species, Zhang et al. [14] suggested that Rex plays an important role in maintaining NADH/NAD+ homeostasis. This indicates a possible method to improve NADH-dependent alcohol production in clostridia.

Conclusions and perspectives

The main recent advances on engineering redox homeostasis to accelerate alcohol biosynthesis, from the viewpoints of cofactors availability, enzyme affinity to cofactors and global redox regulation, have been summarized in this article. A number of approaches, as reviewed here, demonstrate the power of redox homeostasis to improve alcohol production. The strategy of improving the availability of the required cofactors can increase both the titers and yields of the desired alcohols to different extents. Although the productivity data are usually not indicated, an increased titer mostly also indicate increased productivity [34, 35, 46]. Manipulating the affinity of key redox enzymes for NADH or NADPH is an effective strategy to meet the specific cofactor requirements for alcohol biosynthesis and yield improving [63, 65, 75]. Globally engineered cellular redox state benefited the microbes' tolerance to serious stresses, and therefore indirectly benefited the production of alcohols [78, 84, 87]. By employing these approaches, the alcohol production improvements were truly profound in certain cases, and are reflected by the final titers, yields and productivities (Table 1).

Table 1 Strategies for engineering redox homeostasis and its effects on alcohols production

Strategy	Specific approach	Target product	Main effects	Yield	Productivity	Ref.
			Titer			
Improving the availability of cofactors						
Fine-tuning of NAD(P)H-dependent gene	Fine-tuning of *sjhG* and *mdlC*	1,2,4-Butanetriol	Increased by 71.4%	NR	NR	[4]
	Fine-tuning of *adhE2*	Butanol	Increased from 15 to 18.3 g/L	NR	NR	[31]
Blocking NADH-competing pathways	Knock out *ldh*	1,2-Propanediol	Increased from 1.08 to 1.30 g/L anaerobically, from 1.10 to 1.40 g/L microaerobically	Increased by 43% anaerobically, by 67% microaerobically	NR	[19]
	Knock out *aldA*	1,3-Propanediol	Increased from 698.6 to 927.6 mM	NR	Increased by 33%	[34]
	Knock out *adh*, *ldh* and *frd*	Butanol	Increased from 141 to 274 mg/L	NR	NR	[20]
	Knock out *mdh*	1,4-Butanediol	Increased from ~3 to ~8 mM	NR	NR	[37]
Increasing total NAD level	Overexpress *pncB*	Ethanol	Increased from 11.50 to 28.58 mM	NR	NR	[40]
Introducing NAD(P)H regeneration systems	Overexpress *fdh1*	Ethanol	Increased from ~15 to ~175 mM	NR	NR	[21]
	Overexpress *fdh1*	Ethanol	Increased from 52.20 to 117.77 mM	Increased from 0.72 to 1.33 mol/mol	NR	[22]
	Overexpress *fdh*	1,3-Propanediol	NR	Increased by 17.3%	NR	[45]
	Overexpress *fdh*	2,3-Butanediol	Increased from 16.1 to 17.8 g/L	Increased from 82.5 to 91.8%	Increased by 33%	[46]
	Activate pyruvate dehydrogenase, fine-tune express *fdh1*	Butanol	Increased from 5.02 to 6.8 g/L	NR	Increased by 136%	[35]
Manipulating affinity of redox enzymes for NAD(P)H						
	Overexpress *GDP1*	Ethanol	Increased from 90 to 100 mM	Increased from 18 to 41%	NR	[49]
	Electrically regenerate NADH	Isobutanol	Produced 846 mg/L	NR	NR	[53]
	Electrically regenerate NADPH	Isopropanol	Produced 216 mg/L	NR	NR	[55]
Switching the affinity from one type to another	Mutate XR (NADPH to NADH)	Ethanol	NR	Increased from 0.24 to 0.34 g/g	NR	[63]
	Mutate XR (NADPH to NADH)	Ethanol	Increased from 16.7 to 25.3 g/L	Increased from 0.33 to 0.38 g/g	NR	[65]
	Introduce NADPH-preferring enzymes in *Synechococcus*	Butanol	Increased from 6.4 to 29.9 mg/L	NR	NR	[67]
	Replace *bcd-etfAB* with *ter*	Butanol	Increased from 0.1 to 1.8 g/L	NR	NR	[48]
Improving affinity for NAD(P)H	Introduce alcohol dehydrogenase I and pyruvate decarboxylase genes from *Z. mobilis*	Ethanol	Increased from 18 to 750 mM	NR	NR	[18]
	Increase affinities of IlvC and AdhA for NADH	Isobutanol	Increased from 1 to 13.4 g/L	Increased from 53 to 100% of the theoretical yield	Increased by 38–88%	[75]
Globally engineering cellular redox balance						
Manipulating respiratory levels	Knock out *ubiCA* and supply coenzyme Q1	Ethanol	NR	Increased from 0.48 to 0.80 mol/mol aerobically	NR	[78]
Introducing glutathione	Overexpress *gshAB*	Butanol	Increased from 10.8 to 14.8 g/L	NR	NR	[84]
Engineering redox-sensitive transcription factor Rex	Inactivate *rex*	Ethanol, Butanol	Increased from ~20 to ~120 mM and increased from 60 to 120 mM, respectively	NR	NR	[87]

NR not reported

Redox homeostasis engineering may play an important role in developing alcohol-producing microbial cell factories, yet it is not omnipotent. Firstly, it is hard to quantify the exact impact of cofactor manipulation on reducing equivalents as some unknown formats of reducing equivalents exist not only NAD(P)H, $FADH_2$, etc. [90, 91]. Consequently, some strategies could be useless or bring burden to the cells, and sometimes may even be harmful to the cell hosts [92]. Secondly, the cellular redox state is dynamically changed and cannot be monitored in real time, which makes it difficult to completely understand the whole process of alcohol production. Thirdly, there are other redox relevant enzymes except for alcohol synthetic pathway enzymes. These enzymes may have physiological function shifting the cell to another metabolic pattern after the above approaches were adopted [93].

Although rapidly advancing, the tools and methods of systems metabolic engineering still await more exciting developments for controlling the metabolic fluxes and energy/redox requirements in the context of maximizing product titer, yield and productivity. Since traditional cofactor engineering might not be sufficient to meet the demand for higher titer, yield and productivity of target products, future work will have to use systems and synthetic biology approaches in order to further understand the redox systems of typical industrially relevant bacteria. In addition, the product yield is always limited by the provided substrate (including co-substrate) due to the stoichiometry of available electrons from a substrate [16]. Engineering of redox homeostasis made it possible to close to the maximal theoretical yield, but it was hardly to obtain a yield beyond the limits from substrate. Reports on other target chemicals have also provided certain reference points for future engineering of redox homeostasis. Feedstocks which are more reduced than glucose may be suitable for the production of alcohols such as glycerol [5] and sorbitol [22], but also fatty acids [17]. Additionally, extracellular redox potential (ORP) was validated as an effective parameter that controls the anaerobic microbial production of 1,3-propanediol [94] and butanol [95]. In the future, improving the metabolic flux towards target products by controlling extracellular ORP could be employed in some reactions which are difficult to conduct, especially ones that need very low redox potentials [96, 97].

Abbreviations
NADH: reduced nicotinamide adenine dinucleotide; NAD^+: oxidized nicotinamide adenine dinucleotide; NADPH: reduced nicotinamide adenine dinucleotide phosphate; $NADP^+$: oxidized nicotinamide adenine dinucleotide phosphate; RBS: ribosome binding site; PEP: phosphoenolpyruvate; 1,2-PDO: 1,2-propanediol; 1,3-PDO: 1,3-propanediol; 1,4-BDO: 1,4-butanediol; 2,3-BDO: 2,3-butanediol; FDH: formate dehydrogenase; Et/Ac: ethanol to acetate; PDH: pyruvate dehydrogenase; FAD: flavin adenine dinucleotide; FMN: flavin

mononucleotide; PQQ: pyrroquinoline quinone; XR: xylose reductase; XDH: xylitol dehydrogenase; GAPDH: glyceraldehyde-3-phosphate dehydrogenase; GSH: glutathione; γ-GCS: γ-glutamylcysteine synthetase; GS: GSH synthetase; ORP: redox potential.

Authors' contributions
YZ developed the concept and edited the paper. CZ, QZ and YL wrote the paper. All authors read and approved the final manuscript.

Author details
[1] CAS Key Laboratory of Microbial Physiological and Metabolic Engineering, Institute of Microbiology, Chinese Academy of Sciences, No. 1 West Beichen Road Chaoyang District, Beijing 100101, China. [2] University of Chinese Academy of Sciences, Beijing 100049, China.

Acknowledgements
Not applicable.

Competing interests
The authors declare that they have no competing interests.

Funding
This work was supported by the Key Research Program of the Chinese Academy of Sciences (ZDRW-ZS-2016-3), and the National Natural Science Foundation of China (31470231). Yanping Zhang is supported by the Youth Innovation Promotion Association CAS (No. 2014076).

References
1. Dai ZJ, Dong HJ, Zhang YP, Li Y. Elucidating the contributions of multiple aldehyde/alcohol dehydrogenases to butanol and ethanol production in *Clostridium acetobutylicum*. Sci Rep. 2016; 6.
2. Jang YS, Lee J, Malaviya A, Seung DY, Cho JH, Lee SY. Butanol production from renewable biomass: rediscovery of metabolic pathways and metabolic engineering. Biotechnol J. 2012;7(2):186–98.
3. Mann MS, Lutke-Eversloh T. Thiolase engineering for enhanced butanol production in *Clostridium acetobutylicum*. Biotechnol Bioeng. 2013;110(3):887–97.
4. Sun L, Yang F, Sun HB, Zhu TC, Li XH, Li Y, Xu ZH, Zhang YP. Synthetic pathway optimization for improved 1,2,4-butanetriol production. J Ind Microbiol Biotechnol. 2016;43(1):67–78.
5. Ammar EM, Wang ZQ, Yang ST. Metabolic engineering of *Propionibacterium freudenreichii* for n-propanol production. Appl Microbiol Biotechnol. 2013;97(10):4677–90.
6. Jang YS, Lee JY, Lee J, Park JH, Im JA, Eom MH, Lee J, Lee SH, Song H, Cho JH, Seung DY, Lee SY. Enhanced butanol production obtained by reinforcing the direct butanol-forming route in *Clostridium acetobutylicum*. mBio. 2012;3(5):e00314–12.
7. Runguphan W, Keasling JD. Metabolic engineering of *Saccharomyces cerevisiae* for production of fatty acid-derived biofuels and chemicals. Metab Eng. 2014;21:103–13.
8. Tummala SB, Welker NE, Papoutsakis ET. Design of antisense RNA constructs for downregulation of the acetone formation pathway of *Clostridium acetobutylicum*. J Bacteriol. 2003;185(9):1923–34.
9. Jiang Y, Xu CM, Dong F, Yang YL, Jiang WH, Yang S. Disruption of the acetoacetate decarboxylase gene in solvent-producing *Clostridium acetobutylicum* increases the butanol ratio. Metab Eng. 2009;11(4):284–91.
10. Cho C, Choi SY, Luo ZW, Lee SY. Recent advances in microbial production of fuels and chemicals using tools and strategies of systems metabolic engineering. Biotechnol Adv. 2015;33(7):1455–66.

11. Martinez I, Zhu JF, Lin H, Bennett GN, San KY. Replacing *Escherichia coli* NAD-dependent glyceraldehyde 3-phosphate dehydrogenase (GAPDH) with a NADP-dependent enzyme from *Clostridium acetobutylicum* facilitates NADPH dependent pathways. Metab Eng. 2008;10(6):352–9.

12. Toledano MB, Kumar C, Le Moan N, Spector D, Tacnet F. The system biology of thiol redox system in *Escherichia coli* and yeast: differential functions in oxidative stress, iron metabolism and DNA synthesis. FEBS Lett. 2007;581(19):3598–607.

13. Sillers R, Chow A, Tracy B, Papoutsakis ET. Metabolic engineering of the non-sporulating, non-solventogenic *Clostridium acetobutylicum* strain M5 to produce butanol without acetone demonstrate the robustness of the acid-formation pathways and the importance of the electron balance. Metab Eng. 2008;10(6):321–32.

14. Zhang L, Nie X, Ravcheev DA, Rodionov DA, Sheng J, Gu Y, Yang S, Jiang W, Yang C. Redox-responsive repressor Rex modulates alcohol production and oxidative stress tolerance in *Clostridium acetobutylicum*. J Bacteriol. 2014;196(22):3949–63.

15. Gerbens-Leenes P, Hoekstra A. The water footprint of sweeteners and bioethanol from sugar cane, sugar beet and maize. Value of Water Research Report Series No. 38, UNESCO-IHE and University of Twente, Delft and Enschede, The Netherlands. 2009.

16. Erickson LE, Selga SE, Viesturs UE. Application of mass and energy balance regularities to product formation. Biotechnol Bioeng. 1978;20(10):1623–38.

17. Dellomonaco C, Rivera C, Campbell P, Gonzalez R. Engineered respiro-fermentative metabolism for the production of biofuels and biochemicals from fatty acid-rich feedstocks. Appl Environ Microbiol. 2010;76(15):5067–78.

18. Ingram LO, Conway T, Clark DP, Sewell GW, Preston JF. Genetic engineering of ethanol production in *Escherichia coli*. Appl Environ Microbiol. 1987;53(10):2420–5.

19. Berríos-Rivera SJ, San KY, Bennett GN. The effect of carbon sources and lactate dehydrogenase deletion on 1,2-propanediol production in *Escherichia coli*. J Ind Microbiol Biotechnol. 2003;30(1):34–40.

20. Atsumi S, Cann AF, Connor MR, Shen CR, Smith KM, Brynildsen MP, Chou KJY, Hanai T, Liao JC. Metabolic engineering of *Escherichia coli* for 1-butanol production. Metab Eng. 2008;10(6):305–11.

21. Berríos-Rivera SJ, Bennett GN, San KY. Metabolic engineering of *Escherichia coli*: Increase of NADH availability by overexpressing an NAD(+)-dependent formate dehydrogenase. Metab Eng. 2002;4(3):217–29.

22. Berríos-Rivera SJ, Sanchez AM, Bennett GN, San KY. Effect of different levels of NADH availability on metabolite distribution in *Escherichia coli* fermentation in minimal and complex media. Appl Microbiol Biotechnol. 2004;65(4):426–32.

23. Carrier TA, Keasling JD. Library of synthetic 5′ secondary structures to manipulate mRNA stability in *Escherichia coli*. Biotechnol Prog. 1999;15(1):58–64.

24. Park YS, Seo SW, Hwang S, Chu HS, Ahn J-H, Kim T-W, Kim D-M, Jung GY. Design of 3′-untranslated region variants for tunable expression in *Escherichia coli*. Biochem Biophys Res Commun. 2007;356(1):136–41.

25. Jana S, Deb J. Strategies for efficient production of heterologous proteins in *Escherichia coli*. Appl Microbiol Biotechnol. 2005;67(3):289–98.

26. Jarboe LR, Zhang X, Wang X, Moore JC, Shanmugam K, Ingram LO. Metabolic engineering for production of biorenewable fuels and chemicals: contributions of synthetic biology. Biomed Res Int. 2010;2010:761042.

27. Nevoigt E, Kohnke J, Fischer CR, Alper H, Stahl U, Stephanopoulos G. Engineering of promoter replacement cassettes for fine-tuning of gene expression in *Saccharomyces cerevisiae*. Appl Environ Microbiol. 2006;72(8):5266–73.

28. Kuhn A, van Zyl C, van Tonder A, Prior BA. Purification and partial characterization of an aldo-keto reductase from *Saccharomyces cerevisiae*. Appl Environ Microbiol. 1995;61(4):1580–5.

29. Matsushika A, Inoue H, Kodaki T, Sawayama S. Ethanol production from xylose in engineered *Saccharomyces cerevisiae* strains: current state and perspectives. Appl Microbiol Biotechnol. 2009;84(1):37–53.

30. Hou J, Suo F, Wang C, Li X, Shen Y, Bao X. Fine-tuning of NADH oxidase decreases byproduct accumulation in respiration deficient xylose metabolic *Saccharomyces cerevisiae*. BMC Biotechnol. 2014;14:13.

31. Ohtake T, Pontrelli S, Laviña WA, Liao JC, Putri SP, Fukusaki E. Metabolomics-driven approach to solving a CoA imbalance for improved 1-butanol production in *Escherichia coli*. Metab Eng. 2017;41:135–43.

32. Katz J, Wood HG. The use of glucose-C^{14} for the evaluation of the pathways of glucose metabolism. J Biol Chem. 1960;235(8):2165–77.

33. Ingram LO, Aldrich HC, Borges ACC, Causey TB, Martinez A, Morales F, Saleh A, Underwood SA, Yomano LP, York SW, Zaldivar J, Zhou SD. Enteric bacterial catalysts for fuel ethanol production. Biotechnol Prog. 1999;15(5):855–66.

34. Zhang YP, Li Y, Du CY, Liu M, Cao ZA. Inactivation of aldehyde dehydrogenase: a key factor for engineering 1,3-propanediol production by *Klebsiella pneumoniae*. Metab Eng. 2006;8(6):578–86.

35. Lim JH, Seo SW, Kim SY, Jung GY. Model-driven rebalancing of the intracellular redox state for optimization of a heterologous *n*-butanol pathway in *Escherichia coli*. Metab Eng. 2013;20:49–55.

36. Saini M, Chen MH, Chiang C-J, Chao Y-P. Potential production platform of *n*-butanol in *Escherichia coli*. Metab Eng. 2015;27:76–82.

37. Yim H, Haselbeck R, Niu W, Pujol-Baxley C, Burgard A, Boldt J, Khandurina J, Trawick JD, Osterhout RE, Stephen R, Estadilla J, Teisan S, Schreyer HB, Andrae S, Yang TH, Lee SY, Burk MJ, Van Dien S. Metabolic engineering of *Escherichia coli* for direct production of 1,4-butanediol. Nat Chem Biol. 2011;7(7):445–52.

38. Chen XL, Li SB, Liu LM. Engineering redox balance through cofactor systems. Trends Biotechnol. 2014;32(6):337–43.

39. Fu J, Huo GX, Feng LL, Mao YF, Wang ZW, Ma HW, Chen T, Zhao XM. Metabolic engineering of *Bacillus subtilis* for chiral pure meso-2,3-butanediol production. Biotechnol Biofuels. 2016;9(1):1.

40. Berríos-Rivera SJ, San KY, Bennett GN. The effect of NAPRTase overexpression on the total levels of NAD, the NADH/NAD$^+$ ratio, and the distribution of metabolites in *Escherichia coli*. Metab Eng. 2002;4(3):238–47.

41. Jawed M, Pi J, Xu L, Zhang H, Hakeem A, Yan Y. Enhanced H_2 production and redirected metabolic flux via overexpression of *fhlA* and *pncB* in *Klebsiella* HQ-3 strain. Appl Biochem Biotechnol. 2016;178(6):1113–28.

42. Ma J, Gou D, Liang L, Liu R, Chen X, Zhang C, Zhang J, Chen K, Jiang M. Enhancement of succinate production by metabolically engineered *Escherichia coli* with co-expression of nicotinic acid phosphoribosyltransferase and pyruvate carboxylase. Appl Microbiol Biotechnol. 2013;97(15):6739–47.

43. Wang YP, San KY, Bennett GN. Cofactor engineering for advancing chemical biotechnology. Curr Opin Biotech. 2013;24(6):994–9.

44. Wong CH, Whitesides GM. Enzyme-catalyzed organic synthesis: regeneration of deuterated nicotinamide cofactors for use in large-scale enzymatic synthesis of deuterated substances. J Am Chem Soc. 1983;105(15):5012–4.

45. Zhang YP, Huang ZH, Du CY, Li Y, Cao ZA. Introduction of an NADH regeneration system into *Klebsiella oxytoca* leads to an enhanced oxidative and reductive metabolism of glycerol. Metab Eng. 2009;11(2):101–6.

46. Wang Y, Li LX, Ma CQ, Gao C, Tao F, Xu P. Engineering of cofactor regeneration enhances (2S,3S)-2,3-butanediol production from diacetyl. Sci Rep. 2013;3:2643.

47. Wang SH, Zhang YP, Dong HJ, Mao SM, Zhu Y, Wang RJ, Luan GD, Li Y. Formic acid triggers the "acid crash" of acetone-butanol-ethanol fermentation by *Clostridium acetobutylicum*. Appl Environ Microbiol. 2011;77(5):1674–80.

48. Shen CR, Lan EI, Dekishima Y, Baez A, Cho KM, Liao JC. Driving forces enable high-titer anaerobic 1-butanol synthesis in *Escherichia coli*. Appl Environ Microbiol. 2011;77(9):2905–15.

49. Verho R, Londesborough J, Penttilä M, Richard P. Engineering redox cofactor regeneration for improved pentose fermentation in *Saccharomyces cerevisiae*. Appl Environ Microbiol. 2003;69(10):5892–7.

50. Eguchi T, Kuge Y, Inoue K, Yoshikawa N, Mochida K, Uwajima T. NADPH regeneration by glucose dehydrogenase from *Gluconobacter scleroides* for l-leucovorin synthesis. Biosci Biotechnol Biochem. 1992;56(5):701–3.

51. Xu Z, Jing K, Liu Y, Cen P. High-level expression of recombinant glucose dehydrogenase and its application in NADPH regeneration. J Ind Microbiol Biotechnol. 2007;34(1):83–90.

52. Xue J, Balamurugan S, Li D-W, Liu Y-H, Zeng H, Wang L, Yang W-D, Liu J-S, Li H-Y. Glucose-6-phosphate dehydrogenase as a target for highly efficient fatty acid biosynthesis in microalgae by enhancing NADPH supply. Metab Eng. 2017;41:212–21.

53. Li H, Opgenorth PH, Wernick DG, Rogers S, Wu T-Y, Higashide W, Malati P, Huo Y-X, Cho KM, Liao JC. Integrated electromicrobial conversion of CO_2 to higher alcohols. Science. 2012;335(6076):1596.

54. Jourdin L, Grieger T, Monetti J, Flexer V, Freguia S, Lu Y, Chen J, Romano M, Wallace GG, Keller J. High acetic acid production rate obtained by microbial electrosynthesis from carbon dioxide. Environ Sci Technol. 2015;49(22):13566–74.

55. Torella JP, Gagliardi CJ, Chen JS, Bediako DK, Colon B, Way JC, Silver PA, Nocera DG. Efficient solar-to-fuels production from a hybrid microbial-water-splitting catalyst system. Proc Natl Acad Sci USA. 2015;112(8):2337–42.

56. Jourdin L, Freguia S, Flexer V, Keller J. Bringing high-rate, CO₂-based microbial electrosynthesis closer to practical implementation through improved electrode design and operating conditions. Environ Sci Technol. 2016;50(4):1982–9.

57. Schievano A, Sciarria TP, Vanbroekhoven K, De Wever H, Puig S, Andersen SJ, Rabaey K, Pant D. Electro-fermentation—merging electrochemistry with fermentation in industrial applications. Trends Biotechnol. 2016;34(11):866–78.

58. Gong FY, Cai Z, Li Y. Synthetic biology for CO₂ fixation. Sci China Life Sci. 2016;59(11):1106–14.

59. Raines CA. The Calvin cycle revisited. Photosynth Res. 2003;75(1):1–10.

60. San KY, Bennett GN, Berrios-Rivera SJ, Vadali RV, Yang Y-T, Horton E, Rudolph FB, Sariyar B, Blackwood K. Metabolic engineering through cofactor manipulation and its effects on metabolic flux redistribution in Escherichia coli. Metab Eng. 2002;4(2):182–92.

61. Foster JW, Park Y, Penfound T, Fenger T, Spector M. Regulation of NAD metabolism in Salmonella typhimurium: molecular sequence analysis of the bifunctional nadR regulator and the nadA-pnuC operon. J Bacteriol. 1990;172(8):4187–96.

62. Duine JA. Cofactor diversity in biological oxidations: implications and applications. Chem Rec. 2001;1(1):74–83.

63. Petschacher B, Nidetzky B. Altering the coenzyme preference of xylose reductase to favor utilization of NADH enhances ethanol yield from xylose in a metabolically engineered strain of Saccharomyces cerevisiae. Microb Cell Fact. 2008;7(1):9.

64. Petschacher B, Leitgeb S, Kavanagh KL, Wilson DK, Nidetzky B. The coenzyme specificity of Candida tenuis xylose reductase (AKR2B5) explored by site-directed mutagenesis and X-ray crystallography. Biochem J. 2005;385:75–83.

65. Bengtsson O, Hahn-Hägerdal B, Gorwa-Grauslund MF. Xylose reductase from Pichia stipitis with altered coenzyme preference improves ethanolic xylose fermentation by recombinant Saccharomyces cerevisiae. Biotechnol Biofuels. 2009;2(1):1.

66. Choi YJ, Lee J, Jang YS, Lee SY. Metabolic engineering of microorganisms for the production of higher alcohols. mBio. 2014;5(5):e01524-14.

67. Lan EI, Liao JC. ATP drives direct photosynthetic production of 1-butanol in cyanobacteria. Proc Natl Acad Sci USA. 2012;109(16):6018–23.

68. Zhou J, Zhang FL, Meng HK, Zhang YP, Li Y. Introducing extra NADPH consumption ability significantly increases the photosynthetic efficiency and biomass production of cyanobacteria. Metab Eng. 2016;38:217–27.

69. Zhou J, Zhang HF, Zhang YP, Li Y, Ma YH. Designing and creating a modularized synthetic pathway in cyanobacterium Synechocystis enables production of acetone from carbon dioxide. Metab Eng. 2012;14(4):394–400.

70. Dai ZJ, Dong HJ, Zhu Y, Zhang YP, Li Y, Ma YH. Introducing a single secondary alcohol dehydrogenase into butanol-tolerant Clostridium acetobutylicum Rh8 switches ABE fermentation to high level IBE fermentation. Biotechnol Biofuels. 2012;5(1):44.

71. Li F, Hinderberger J, Seedorf H, Zhang J, Buckel W, Thauer RK. Coupled ferredoxin and crotonyl coenzyme a (CoA) reduction with NADH catalyzed by the butyryl-CoA dehydrogenase/Etf complex from Clostridium kluyveri. J Bacteriol. 2008;190(3):843–50.

72. Panagiotou G, Grotkjær T, Hofmann G, Bapat PM, Olsson L. Overexpression of a novel endogenous NADH kinase in Aspergillus nidulans enhances growth. Metab Eng. 2009;11(1):31–9.

73. Ji DB, Wang L, Hou SH, Liu WJ, Wang JX, Wang Q, Zhao ZK. Creation of bioorthogonal redox systems depending on nicotinamide flucytosine dinucleotide. J Am Chem Soc. 2011;133(51):20857–62.

74. Arnold FH. Directed evolution: creating biocatalysts for the future. Chem Eng Sci. 1996;51(23):5091–102.

75. Bastian S, Liu X, Meyerowitz JT, Snow CD, Chen MM, Arnold FH. Engineered ketol-acid reductoisomerase and alcohol dehydrogenase enable anaerobic 2-methylpropan-1-ol production at theoretical yield in Escherichia coli. Metab Eng. 2011;13(3):345–52.

76. Meng HK, Liu P, Sun HB, Cai Z, Zhou J, Lin JP, Li Y. Engineering a D-lactate dehydrogenase that can super-efficiently utilize NADPH and NADH as cofactors. Sci Rep. 2016;6:24887.

77. Knappe J, Sawers G. A radical-chemical route to acetyl-CoA: the anaerobically induced pyruvate formate-lyase system of Escherichia coli. FEMS Microbiol Rev. 1990;6(4):383–98.

78. Zhu JF, Sanchez A, Bennett GN, San KY. Manipulating respiratory levels in Escherichia coli for aerobic formation of reduced chemical products. Metab Eng. 2011;13(6):704–12.

79. Liu D, Chen Y, Li A, Ding FY, Zhou T, He Y, Li BB, Niu HQ, Lin XQ, Xie JJ, Chen XC, Wu JL, Ying HJ. Enhanced butanol production by modulation of electron flow in Clostridium acetobutylicum B3 immobilized by surface adsorption. Bioresour Technol. 2013;129:321–8.

80. Meister A, Anderson ME. Glutathione. Annu Rev Biochem. 1983;52:711–60.

81. Flohé L. The glutathione peroxidase reaction: molecular basis of the antioxidant function of selenium in mammals. Curr Top Cell Regul. 1985;27:473–8.

82. Penninckx M. A short review on the role of glutathione in the response of yeasts to nutritional, environmental, and oxidative stresses. Enzyme Microb Technol. 2000;26(9–10):737–42.

83. Li Y, Wei GY, Chen J. Glutathione: a review on biotechnological production. Appl Microbiol Biotechnol. 2004;66(3):233–42.

84. Zhu LJ, Dong HJ, Zhang YP, Li Y. Engineering the robustness of Clostridium acetobutylicum by introducing glutathione biosynthetic capability. Metab Eng. 2011;13(4):426–34.

85. Hou XH, Peng WF, Xiong L, Huang C, Chen XF, Chen XD, Zhang WG. Engineering Clostridium acetobutylicum for alcohol production. J Biotechnol. 2013;166(1):25–33.

86. Kawasaki S, Sakai Y, Takahashi T, Suzuki I, Niimura Y. O₂ and reactive oxygen species detoxification complex, composed of O₂-responsive NADH: rubredoxin oxidoreductase-flavoprotein A2-desulfoferrodoxin operon enzymes, rubperoxin, and rubredoxin, in Clostridium acetobutylicum. Appl Environ Microbiol. 2009;75(4):1021–9.

87. Wietzke M, Bahl H. The redox-sensing protein Rex, a transcriptional regulator of solventogenesis in Clostridium acetobutylicum. Appl Microbiol Biotechnol. 2012;96(3):749–61.

88. Brekasis D, Paget MSB. A novel sensor of NADH/NAD(+) redox poise in Streptomyces coelicolor A3(2). EMBO J. 2003;22(18):4856–65.

89. Wang E, Bauer MC, Rogstam A, Linse S, Logan DT, von Wachenfeldt C. Structure and functional properties of the Bacillus subtilis transcriptional repressor Rex. Mol Microbiol. 2008;69(2):466–78.

90. Barnes HJ, Arlotto MP, Waterman MR. Expression and enzymatic activity of recombinant cytochrome P450 17 alpha-hydroxylase in Escherichia coli. Proc Natl Acad Sci USA. 1991;88(13):5597–601.

91. Wargacki AJ, Leonard E, Win MN, Regitsky DD, Santos CNS, Kim PB, Cooper SR, Raisner RM, Herman A, Sivitz AB. An engineered microbial platform for direct biofuel production from brown macroalgae. Science. 2012;335(6066):308–13.

92. Heux S, Cachon R, Dequin S. Cofactor engineering in Saccharomyces cerevisiae: expression of a H₂O-forming NADH oxidase and impact on redox metabolism. Metab Eng. 2006;8(4):303–14.

93. De Felipe FL, Kleerebezem M, de Vos WM, Hugenholtz J. Cofactor engineering: a novel approach to metabolic engineering in Lactococcus lactis by controlled expression of NADH oxidase. J Bacteriol. 1998;180(15):3804–8.

94. Du CY, Yan H, Zhang YP, Li Y, Cao ZA. Use of oxidoreduction potential as an indicator to regulate 1,3-propanediol fermentation by Klebsiella pneumoniae. Appl Microbiol Biotechnol. 2006;69(5):554–63.

95. Wang SH, Zhu Y, Zhang YP, Li Y. Controlling the oxidoreduction potential of the culture of Clostridium acetobutylicum leads to an earlier initiation of solventogenesis, thus increasing solvent productivity. Appl Microbiol Biotechnol. 2012;93(3):1021–30.

96. Ragsdale SW. Pyruvate ferredoxin oxidoreductase and its radical intermediate. Chem Rev. 2003;103(6):2333–46.

97. Li B, Elliott SJ. The Catalytic Bias of 2-Oxoacid: ferredoxin oxidoreductase in CO₂: evolution and reduction through a ferredoxin-mediated electrocatalytic assay. Electrochim Acta. 2016;199:349–56.

6

Type IIs restriction based combinatory modulation technique for metabolic pathway optimization

Lijun Ye[1,2], Ping He[1,2,3], Qingyan Li[1,2], Xueli Zhang[1,2*] and Changhao Bi[1,2*]

Abstract

Background: One of the most important research subjects of metabolic engineering is pursuing a balanced metabolic pathway, which is the basis of an efficient cell factory. In this work, we dedicated to develop a simple and efficient technique to modulate expression of multiple genes simultaneously, and select for the optimal regulation pattern.

Results: A Type IIs restriction based combinatory modulation (TRCM) technique was designed and established in the research. With this technique, a plasmid library containing variably regulated *mvaE*, *mvaS*, *mvaK₁*, *mvaD* and *mvaK₂* of the mevalonate (MVA) pathway were obtained and transformed into *E. coli* DXS37-IDI46 to obtain a β-carotene producer library. The ratio of successfully assembled plasmids was determined to be 35%, which was increased to 100% when color based pre-screening was applied. Representative strains were sequenced to contain diverse RBSs as designed to regulate expression of MVA pathway genes. A relatively balanced MVA pathway was achieved in *E. coli* cell factory to increase the β-carotene yield by two fold. Furthermore, the approximate regulation pattern of this optimal MVA pathway was illustrated.

Conclusions: A TRCM technique for metabolic pathway optimization was designed and established in this research, which can be applied to various applications in terms of metabolic pathway regulation and optimization.

Keywords: Metabolic pathway optimization, Type IIs restriction, β-carotene, MVA pathway, Terpene

Background

As the development of Synthetic Biology and Metabolic Engineering, various microbial cell factories have been developed for producing value-added chemical compounds. However, engineering of cell metabolism often disturbs the metabolic network,triggers energetic and objective inefficiency inside the cell, and impedes cell metabolism [1]. Hence, one of the most important research subjects of metabolic engineering is pursuing the balanced metabolic network and pathways. Techniques have been developed to analyze metabolic pathways, including genome-scale models and C^{13}-metabolic flux analysis. And there have been strategies developed to relieve the metabolic burden, such as enhancing respiration, co-utilizing nutrient resources, decoupling cell growth with production phases, and dynamic regulatory systems [1]. As for a specific metabolic pathway, gene expression level is the key effector of the pathway efficiency [2]. Lower expression of genes decreases metabolic pathway flux, while overexpressed genes may over-consume building blocks and cause cells metabolic burden [3]. Furthermore, imbalanced pathway may cause accumulation of pathway intermediates, some of which may even be toxic and jeopardize cell growth [4].

Due to the complexity of metabolic network in organisms and difficulty to precisely control expression of certain gene, it is almost impossible to rationally design and construct an optimized metabolic pathway. In most metabolic engineering projects, one common way was to

*Correspondence: bi_ch@tib.cas.cn; zhang_xl@tib.cas.cn
[1] Tianjin Institute of Industrial Biotechnology, Chinese Academy of Sciences, Tianjin 300308, People's Republic of China
Full list of author information is available at the end of the article

modulate gene expression one by one [5]. With this strategy, the possibility to achieve an optimized regulation pattern is very low. A better solution was to analyze all possible expression levels of pathway genes in a combinatorial fashion. With this strategy, Yin et al. constructed a plasmid library containing the possible combinations of gene regulation patterns [6]. A similar strategy was employed by Xu et al. to optimize fatty acid pathway. Plasmids of various copy numbers were used to carry expression genes for the first round of optimization, which was followed by fine tuning expression with four RBS elements [7]. However, in their work, regulatory parts were limited and the plasmids were exhaustively constructed one by one, which limited the experiment outcome and made the process laborious. The same group also established a BioBrick based method with specially designed restriction adapters. Genes with regulatory elements could be iteratively integrated into the ePathBrick vectors to create a diversified expression library [8]. Similarly, Zelcbuch et al. created a plasmid library construction method to "span high-dimensional expression space" [9]. In their methods, the libraries were constructed with multiple rounds of plasmid construction, which made the practice very time consuming. In the work of Lee et al., a combinatorial library was established by Gibson assembly based method in one reaction [10, 11]. However, only five regulatory parts were employed, which decreased the diversity of the combinatorial library. Based on the extensive researches and great progress achieved by fellow researchers in this subject, we were able to develop a convenient method for constructing complex combinatorial expression library, which was aimed for optimizing a metabolic pathway with maximal outcome and minimal lab hours.

β-carotene, one kind of isoprenoids, is one of the strongest antioxidant in nature [12], and has tremendous potential in healthcare and pharmaceutical industries [13, 14]. Isoprenoids are all derived from two five-carbon building blocks, isopentenyl diphosphate (IPP) and dimethylallyl diphosphate (DMAPP), which are synthesized either by the mevalonate (MVA) pathway in eukaryotes, archaea, and some bacteria or 2-C-methyl-d-erythritol-4-phosphate (MEP) pathway in other prokaryotes and plastids in plants (Fig. 1) [13–15]. In MVA pathway, two acetyl-CoA are condensed into one atetoacetyl-CoA, which is then reduced into 3-hydroxy-3-methyl-glutaryl-coenzyme A (HMG-CoA). The CoA group is released from HMG-CoA to form MVA, which is phosphorylated into mevalonate-5-phosphate, and then mevalonate-5-diphosphate. This compound is coverted into IPP, the common precursor of isoprenoids (Fig. 1). The heterologous MVA pathways have been introduced into E. coli to improve precursor supply of IPP and

DMAPP [16–20]. Isoprenoid production was improved by employment of the bottom portion of MVA pathway derived from Streptococcus pneumoniae and supplementation of MVA in culture [18, 20]. Lycopene production of E. coli with only native MEP pathway was increased twofold with introduction of the whole MVA pathway from Streptomyces sp. CL190 [19]. However, overexpression of mevalonate pathway genes was reported to inhibit cell growth. Pitera et al. found that accumulation of MVA pathway intermediate HMG-CoA inhibited cell growth, which was caused by overexpression of atoB, mvaS and hmg1 [4]. Mevalonate kinase (MK), encoded by erg12, was identified as another rate-limiting enzyme when MVA pathway was adopted for amorphadiene production in E. coli [21]. Thus, the MVA pathway has to be expressed in an optimized and balanced status to benefit isoprenoid cell factories, otherwise an unbalanced MVA pathway would impede the growth and production. In this work, an MVA pathway optimized by combinatorial expression library technique was to be introduced into MEP pathway dependent E. coli cell factory for improving β-carotene production (Fig. 2).

Methods
Strains, medium and growth conditions
Strains used in this study are listed in Table 1. During strain construction, cultures were grown aerobically at 30 or 37 °C in Luria broth (per liter: 10 g Difco tryptone, 5 g Difco yeast extract and 5 g NaCl) and fermentation medium (per liter: 10 g Difco yeast extract, 15 g glycerol, 10.5 g $K_2HPO_4 \cdot 3H_2O$, 6 g $(NH_4)_2HPO_4$, 5 g $MgSO_4 \cdot 7H_2O$, 1.84 g citric acid monohydrate and 10 mL microelements solution; Microelements solution per liter: 10 g $FeSO_4 \cdot 7H_2O$, 5.25 g $ZnSO_4 \cdot 7H_2O$, 3.0 g $CuSO_4 \cdot 5H_2O$, 0.5 g $MnSO_4 \cdot 4H_2O$, 0.23 g $Na_2B_4O_7 \cdot 10H_2O$, 2.0 g $CaCl_2$ and 0.1 g $(NH_4)_6Mo_7O_{24}$). For β-carotene production, single colonies were picked from LB plate and inoculated into 15×100 mm tubes containing 4 mL LB with or without 34 mg/L chloramphenicol, and grown at 30 °C and 250 rpm overnight. Seed culture was subsequently inoculated into 100 ml flask containing 10 mL fermentation medium at an initial OD_{600} of 0.05, with or without 34 mg/L chloramphenicol, and grown at 30 °C and 250 rpm. After growth for 24 h, cells were collected for measurement of β-carotene production.

Genes, vector and primers
MVA pathway genes mvaE, mvaS, mvaK₁, mvaD, and mvaK₂ were amplified from genomic DNA of Enterococcus faecalis CGMCC No.1.2135 using primer set Ga2-R1-EfmvaE-F/Ga2-R1-EfmvaE-R, Ga3-R1-EfmvaS-F/Ga3-R1-EfmvaS-R, and from genomic DNA of Streptococcus pneumoniae CGMCC No.1.8722 using primer

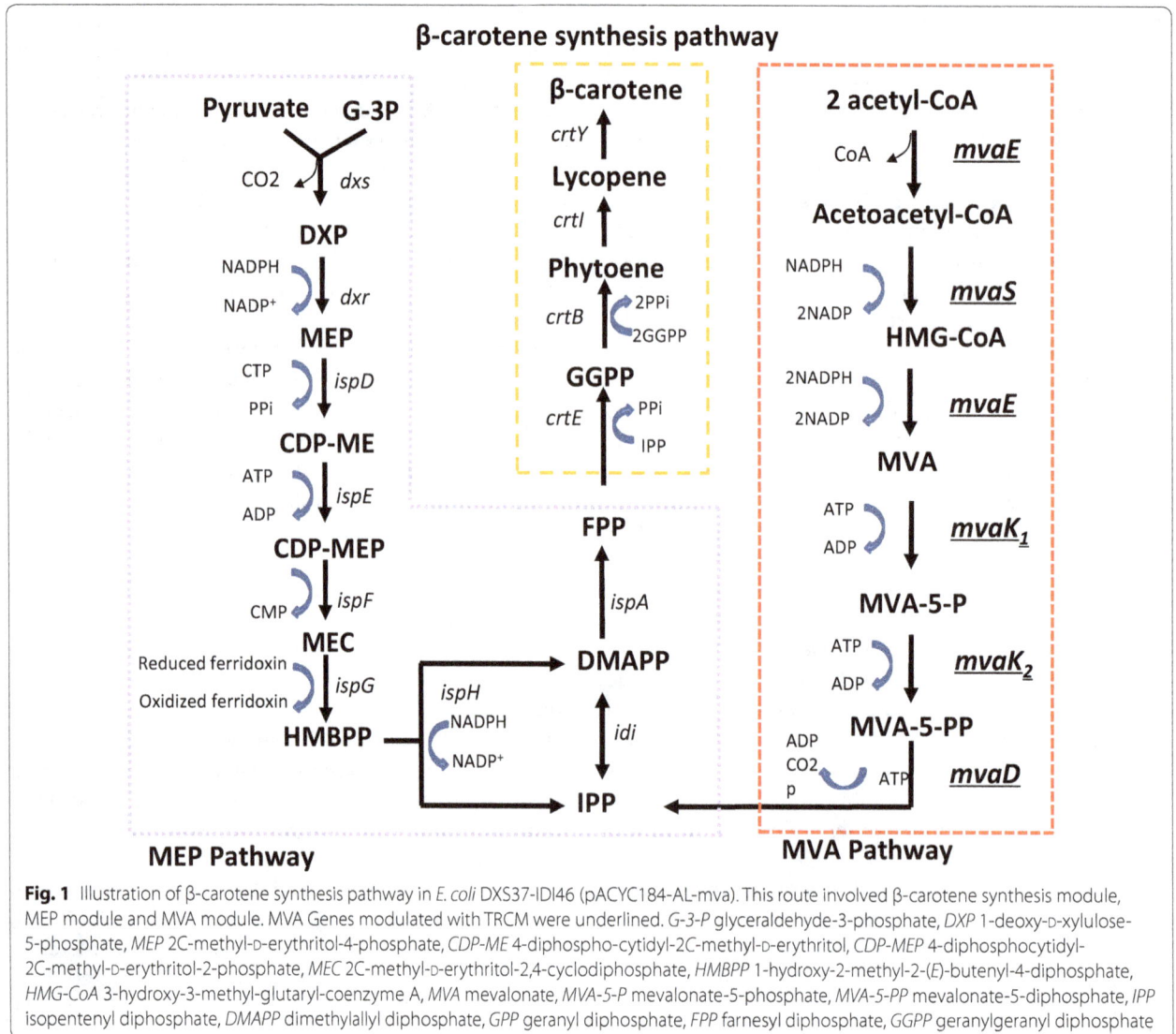

Fig. 1 Illustration of β-carotene synthesis pathway in *E. coli* DXS37-IDI46 (pACYC184-AL-mva). This route involved β-carotene synthesis module, MEP module and MVA module. MVA Genes modulated with TRCM were underlined. *G-3-P* glyceraldehyde-3-phosphate, *DXP* 1-deoxy-D-xylulose-5-phosphate, *MEP* 2C-methyl-D-erythritol-4-phosphate, *CDP-ME* 4-diphospho-cytidyl-2C-methyl-D-erythritol, *CDP-MEP* 4-diphosphocytidyl-2C-methyl-D-erythritol-2-phosphate, *MEC* 2C-methyl-D-erythritol-2,4-cyclodiphosphate, *HMBPP* 1-hydroxy-2-methyl-2-(*E*)-butenyl-4-diphosphate, *HMG-CoA* 3-hydroxy-3-methyl-glutaryl-coenzyme A, *MVA* mevalonate, *MVA-5-P* mevalonate-5-phosphate, *MVA-5-PP* mevalonate-5-diphosphate, *IPP* isopentenyl diphosphate, *DMAPP* dimethylallyl diphosphate, *GPP* geranyl diphosphate, *FPP* farnesyl diphosphate, *GGPP* geranylgeranyl diphosphate

set Ga46-R1-SpmvaK1-F/Ga46-R1-SpmvaK1-R, Ga7-R1-SpmvaD-F/Ga7-R1-SpmvaD-R, and Ga8-R1-SpmvaK2-F/Ga8-R1-SpmvaK2-R respectively (Additional file 1: Table S1). The DNA fragments used for assembly were gel purified and designated as Ga2-mvaE, Ga3-mvaS, Ga46-mvaK1, Ga7-mvaD and Ga8-mvaK2 (Fig. 1b). Vector Fragment Ga91-184A was amplified from pACYC184-PgadA-RFP, and subjected to *Dpn*I digestion (10 U, 16 h, 37 °C) and gel purification. PCR was performed with PrimeSTAR® HS DNA Polymerase (Takara) with primers purchased from GENEWIZ (Suzhou, China). All assembly primers were designed with optimized linkers for Type IIs restriction enzyme based assembly, in which forward primers for amplification of genes were embedded with an RBS library at 5′ ends. Primers used in this study are summarized in Additional file 1: Table S1.

Construction of *mva* operon variants using TRCM

DNA fragments were assembled by Golden Gate DNA assembly method [22, 23]. 100 nanogram vector fragment Ga91-184A and equimolar amount of PCR amplified genes Ga2-mvaE, Ga3-mvaS, Ga46-mvaK1, Ga7-mvaD and Ga8-mvaK2 were mixed in 20 μL Golden Gate reaction solution with 1 μL *Bsa*I-HF, 1 μL T4 ligase (New England Biolabs, Ipswich, MA) and 1× T4 ligase buffer. The reaction was carried out in a thermocycler using the following program: 37 °C for 5 min, 37 °C for 5 min (step 2), 16 °C for 10 min (step 3), step 2 and 3 for 20 cycles, 16 °C for 20 min, 37 °C for 30 min, 75 °C for 6 min, and 4 °C for hold. After PCR, 0.5 μL plasmid-safe nuclease (Epicenter), and 1 μL of 25 mM ATP was added to the reaction, which was incubated at 37 °C for 15 min. 1.5 μL of the resultant reaction solution were transformed into 80 μL competent DXS37-IDI46 cells to obtain the library [24].

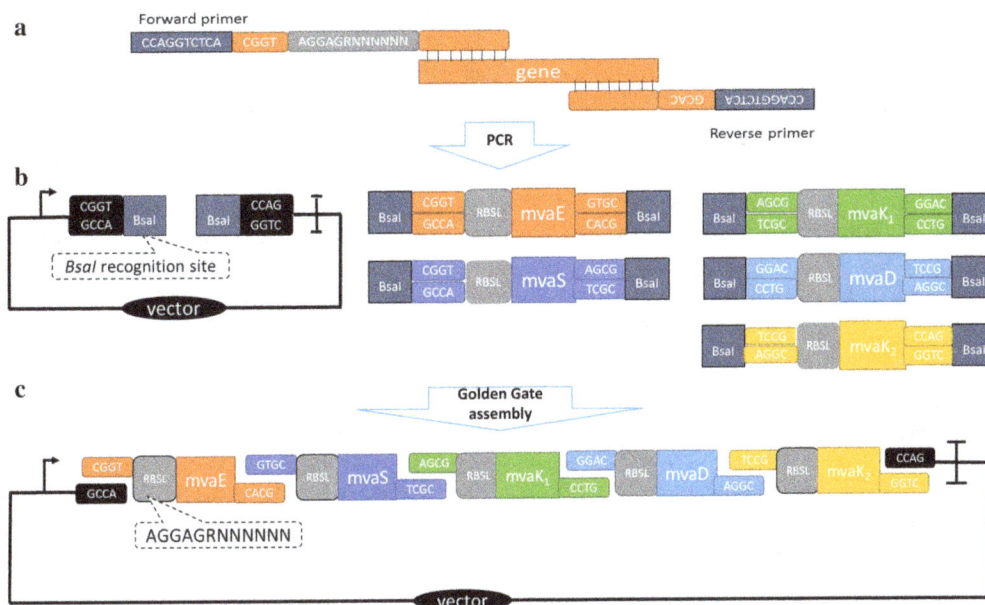

Fig. 2 Modulation and optimization of MVA pathway with TRCM technique. **a** Primer design for obtaining assembly parts containing degenerate RBS sequences. Primers for amplification of MVA genes were embedded with *Bsal* recognition site, optimized four bp linkers, and random RBS sequence AGGAGRNNNNNN. **b** TRCM DNA parts for assembly. One ready-made vector part was used to assemble with five library parts. **c** Combinatory expression library of MVA pathway. The plasmid library contained all five genes regulated with degenerate RBSs in various combinations

Table 1 Strains and plasmids in this study

Strains/plasmids	Relevant characteristics	Source/notes
Strains		
E. coli DH5α	F⁻*endA1thi-1 recA1 relA1 gyrA96deoRΦ80dlacΔ(lacZ) M15 Δ (lacZYA-argF) U169hsdR17 (r$_K^-$, m$_K^+$) λ⁻supE44 phoA*	Invitrogen
CGMCC 1.2135	*Enterococcus faecalis* wild-type	CGMCC
CGMCC 1.8722	*Streptococcus Pneumoniae* wild-type	CGMCC
Dxs37-Idi46	ATCC 8739, *ldhA*::M1-12::*crtEXYIB*::*ldhA*, M1-37::*dxs*, M1-46::*idi*	Laboratory stock
Plasmids		
pACYC184-PgadA-RFP	*E. coli* expression vector derived from pACYC184, promoter of *gadA*, RFP, cat	Laboratory stock
pACYC184-AL-mva	plasmid library of combinatorically regulated MVA pathway, derived from pACYC184-PgadA-RFP	This study

In order to determine whether the MVA pathway genes were successfully incorporated into vector backbone, recombinant clones were subjected to colony PCR analysis. PCR primers were designed to amplify the region from *mvaE* (the first gene on plasmid) to *mvaK₂* (the last gene on plasmid) by primers fE-JF and pK2-JR (Additional file 1: Table S1), which had a product size of 3.8 Kbp. A master mix with 1× Es Taq MasterMix (CWBio, Peking, China) and 0.4 μM forward and reverse primer (Additional file 1: Table S1) was prepared, and 20 μL master mix was dispensed into each PCR tubes. Colonies were directly transferred from LB agar plates into the PCR tubes with sterile toothpicks. The PCR cycling was started with an initial denaturation temperature at 94 °C for 10 min, followed by 30 cycles (94 °C, 30 s; 61 °C, 30 s; and 72 °C, 2 min) and one fill-up cycle (72 °C, 2 min). The PCR products were analyzed on agarose–TAE gels.

Measurement of β-carotene titer

Production of β-carotene was quantified by measuring absorption of acetone-extracted β-carotene at 453 nm as previously reported [24]. A standard curve was obtained by measuring OD_{453} of β-carotene standard samples (Cat. No. C4582, Sigma, USA) with varied concentrations using a Shimadzu UV-2550 spectrophotometer (Shimadzu, Kyoto, Japan). The results represented the mean ± standard deviation (S.D.) of three independent experiments. Dry cell weight (DCW) was calculated based on optical density at 600 nm (1 OD_{600} = 323 mg DCW).

Calculation of MVA pathway genes RBS strength of strains from TRCM libraries

RBS sequences of *mvaE*, *mvaS*, *mvaK₁*, *mvaD*, and *mvaK₂* in representative strains were obtained by PCR and DNA sequencing. Their theoretical RBS strength characterized by the value of translation initiation rate was calculated with the RBS Calculator [25, 26]. The RBS sequence diversity of the combinatory library was analyzed with the Weblogo software [27].

Results and discussions
Design of a Type IIs restriction based combinatory modulation technique (TRCM) for metabolic pathway optimization

With the purpose of developing a simple technique to modulate and optimize expression of multiple genes simultaneously, we designed a Type IIs restriction based combinatory modulation technique (TRCM) for metabolic pathway optimization as illustrated in Fig. 2. Variably regulated genes were obtained by PCR amplification with extended primers, in which degenerate RBS nucleotides were embedded at the 5′ ends. Specifically designed linkers for Type IIs restriction enzymes were also imbedded in the primers to ensure the assembly pattern and efficiency.

Type IIs restriction based Golden Gate [23] was employed as DNA assembly method in this work, which has several advantages. First, there is no PCR reaction involved in the assembling process, which reduces the possibility of mutation compared with other PCR based assembly methods such as Gibson and CPEC [10]; second, the ligase facilitated irreversible ligation greatly improves assembly efficiency compared with homologous arm based method [23]. With this method, gene parts of a pathway were assembled with a vector part to form an expression plasmid. Since each gene part was constructed to carry a collection of regulatory parts, a combinatory plasmid library with variably regulated pathway genes was created, which was subsequently transformed into dedicated host to be screened and selected for strains carrying optimized pathways.

This technique was designed with the modularized strategy to be as simple as possible. The vector part was ready-made for all reactions, providing a stable plasmid backbone. By incorporation of fixed linkers and regulatory elements in primers for amplification of genes, the only variable parts of this method were the actual PCR primer sequences of pathway genes (Fig. 2a).

Development and application of TRCM for MVA pathway optimization

Our lab has constructed a few E. coli β-carotene producers, such as DXS37-IDI46 and CAR001, by modulating several key genes of the MEP pathway module, the pentose phosphate pathway (PPP) module, the ATP module and the tricarboxylic acid cycle (TCA) module [24]. In this work, a heterologous MVA pathway optimized with TRCM technique was introduced into DXS37-IDI46 for further improving its β-carotene production (Fig. 2).

Since MVA pathway upstream genes $mvaS$, $mvaE$ from Enterococcus faecalis and downstream genes $mvaK_1$, $mvaK_2$, $mvaD$ from Streptococcus pneumoniae were reported to be successfully expressed in E. coli, they were selected to be used in this research [17]. As designed with the TRCM technique, primers for PCR amplification of vector and MVA genes were embedded with BsaI recognition sites GGTCTC and specific four bp linkers, in order to be assembled in sequence. These linkers were rationally designed and experimentally tested to enable efficient assembly of DNA parts regardless of condition change (Fig. 2a).

To create a library of differently regulated genes, the RBS sequences of each gene were degenerated. For this purpose, forward primers of MVA genes Ga2-R1-EfmvaE-F, Ga3-R1-EfmvaS-F, Ga46-R1-SpmvaK1-F, Ga7-R1-SpmvaD-F and Ga8-R1-SpmvaK2-F were embedded with the random RBS sequence AGGAGRNNNNNN behind the 4 bp linkers. The starting code ATG of each gene was located behind the six Ns, which was also the starting point of actual PCR primers (Fig. 1a).

Gene parts, which carried front and back linkers for assembly in sequence, were obtained with PCR amplification. In Golden Gate assembly reaction, PCR amplified $mvaS$, $mvaE$, $mvaK_1$, $mvaK_2$ and $mvaD$ parts were mixed with the ready-made vector part Ga91-184A (Fig. 1b). After reaction, a plasmid library was created, which had differentially regulated MVA genes in various combinations. Theoretically, all patterns of differently expressed MVA pathway could be obtained in such a library (Fig. 1c). With this simple method, we achieved the goal of spanning high-dimensional expression space [9].

β-carotene production was improved with TRCM optimized MVA pathway

To select for the optimally expressed MVA pathways, Golden Gate reaction solution containing plasmid library was transformed into the β-carotene producer strain DXS37-IDI46 (Table 1). After electroporation, transformed cells were plated on solid LB with 34 mg/L chloramphenicol to select for plasmid bearing transformants. Orange colored colonies with various intensity appeared on the plates as expected after incubation overnight (Fig. 3a), which indicated their different capacity of β-carotene production. Colonies were randomly picked and analyzed by colony PCR to determine the ratio of successful assembly, for which PCR primers were designed to amplify the region from $mvaE$ to $mvaK_2$ with primers fE-JF and pK2-JR (Additional file 1: Table S1), with an expected product size of 3.8 Kbp. As illustrated in Fig. 3a of TAE gel electrophoresis, 22 of 63 screened colonies had PCR bands of the correct size, which indicated a 35% successful assembly ratio of the synthetic operon. To

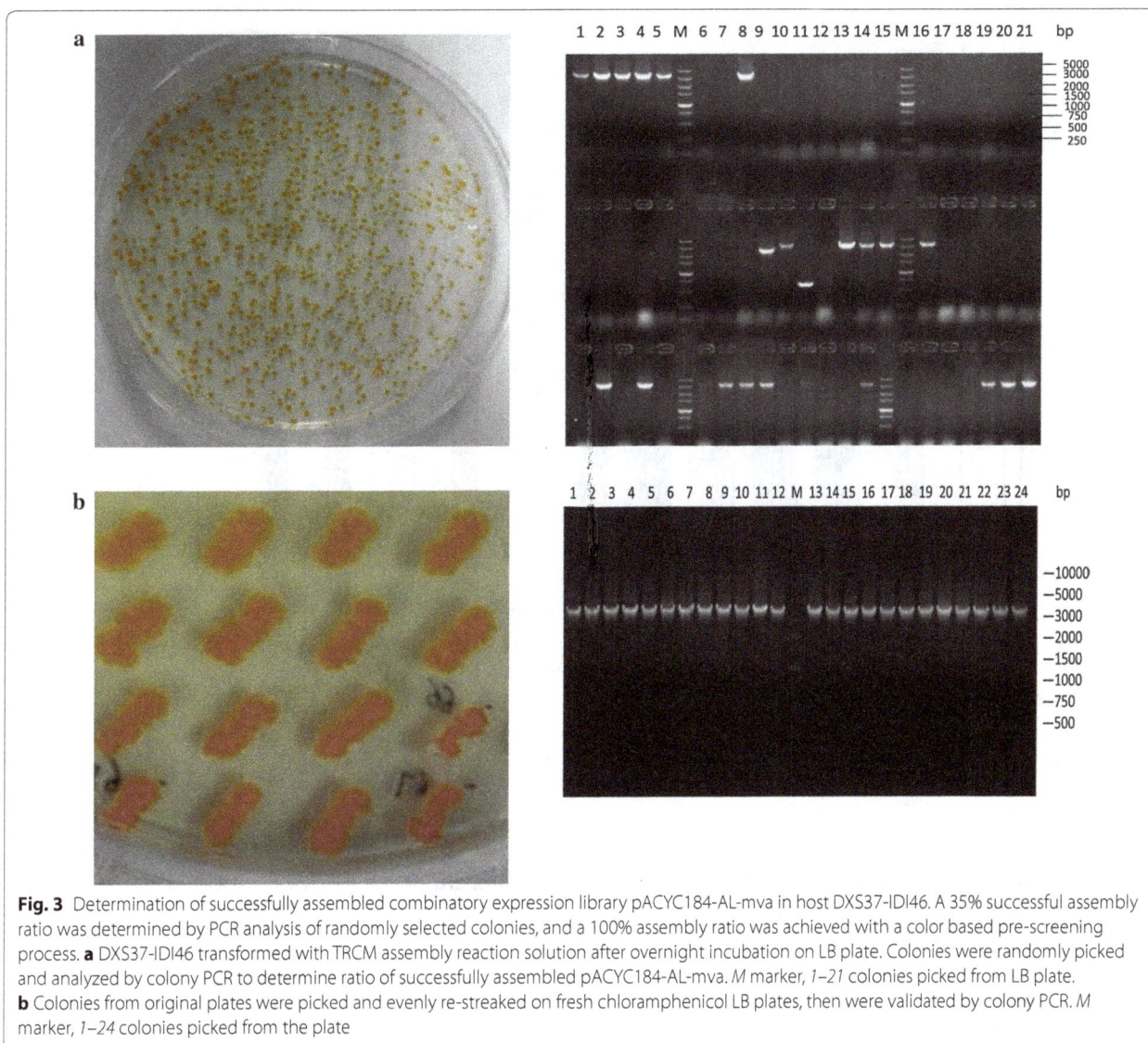

Fig. 3 Determination of successfully assembled combinatory expression library pACYC184-AL-mva in host DXS37-IDI46. A 35% successful assembly ratio was determined by PCR analysis of randomly selected colonies, and a 100% assembly ratio was achieved with a color based pre-screening process. **a** DXS37-IDI46 transformed with TRCM assembly reaction solution after overnight incubation on LB plate. Colonies were randomly picked and analyzed by colony PCR to determine ratio of successfully assembled pACYC184-AL-mva. *M* marker, *1–21* colonies picked from LB plate. **b** Colonies from original plates were picked and evenly re-streaked on fresh chloramphenicol LB plates, then were validated by colony PCR. *M* marker, *1–24* colonies picked from the plate

determine this ratio of strains with increased β-carotene production, transformants were picked and evenly re-streaked on fresh chloramphenicol LB plates, which was to improve the reliability of color based β-carotene production screening (Fig. 3b). Colonies with deeper orange color were selected for PCR analysis. The gel electrophoresis (Fig. 3b) indicated a positive ratio of 100%, since all 24 strains gave the correct PCR products. The results indicated a decent assembly ratio of 35% was achieved by TRCM technique with modulation of five genes simultaneously, and a higher successful ratio could be achieved with a simple color based pre-screening process.

To select strains with significantly improved β-carotene producing capacity, colonies with deeper color were cultured aerobically, and the β-carotene titer was measured. Ten representative strains ALV104, ALV131, ALV100, ALV108, ALV20, ALV63, ALV25, ALV133, ALV23, and

ALV145 were determined to have improved β-carotene production to various extent in comparison with DXS37-IDI46, as illustrated in Fig. 4. Strain ALV145 had the highest yield of 11.17 ± 0.82 mg/g, which was a 96.0% increase compared with the parent strain DXS37-IDI46. The significant improvement indicated that an efficient cell factory with optimized metabolic pathway could be obtained by the simple TRCM technique.

A combinatorial expression library with five genes regulated by diverse RBSs was obtained with TRCM technique

To analyze the degenerated RBS sequences regulating MVA gene, PCR was used to amplify regions containing the RBS sequences of MVA operon genes in representative strains. The PCR products were sequenced subsequently to obtain the RBS sequence information, which

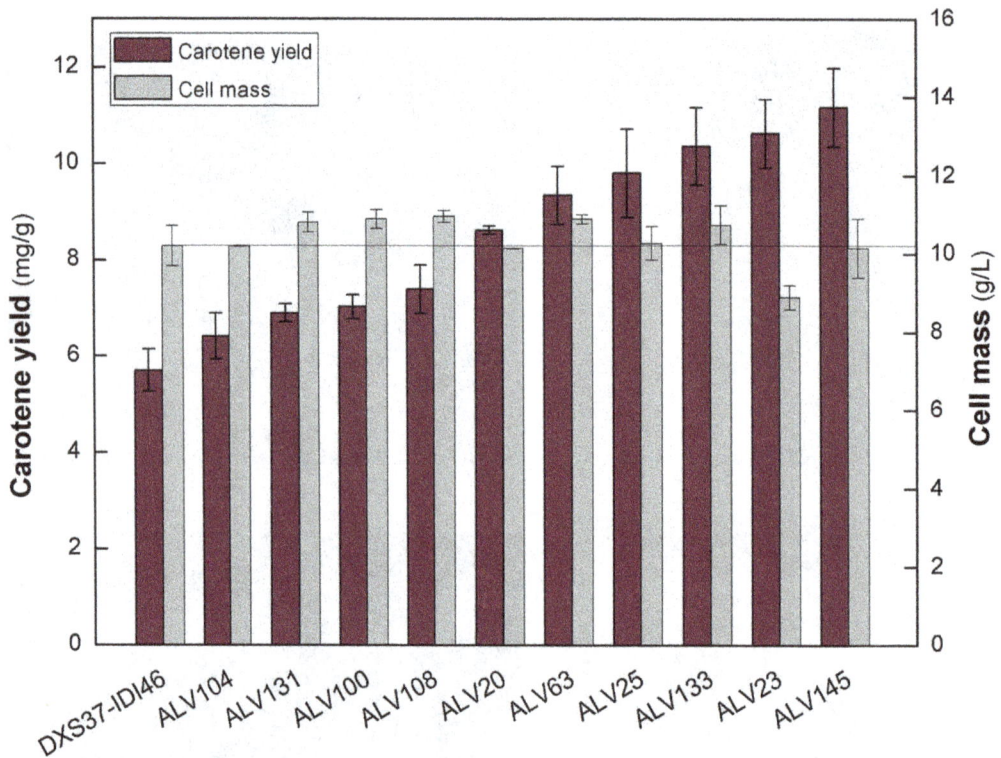

Fig. 4 β-carotene yield and cell mass of representative strains from the combinatory expression library DXS37-IDI46 (pACYC184-AL-mva). Strains include ALV104, ALV131, ALV100, ALV108, ALV20, ALV63, ALV25, ALV133, ALV23, ALV145 and parent strain DXS37-IDI46. Three repeats were performed for each strain, and the *error bars* represented standard deviation

was summarized in Table 2. As designed, a combinatorial expression library with the five MVA genes regulated by diverse RBSs was obtained. For each strain, RBSs of the five genes were all different; and for each gene, all ten RBSs were not same either. Diversity of the degenerate RBS sequence RNNNNNN from combinatory library was further analyzed with the Weblogo software [27]. A good but not great diversity was obtained for each of the five MVA genes, probably due to the low sample number, which was only ten for one gene. However, when all fifty RBS sequences were subjected to calculation, a logo with very high diversity was achieved (Additional file 2: Figure S1). In addition, among the fifty sequenced RBSs, ratio of the highest calculated RBS strength to the lowest was 183,455/92 [25, 26], which indicated an expression dynamic range of about 2000-fold. The results proved that TRCM technique process did not create significant bias, and were able to generate quite diverse combinatory expression library.

An approximate expression pattern of optimal MVA pathway was illustrated

To illustrate a general and approximate expression pattern of an efficient MVA pathway in *E. coli* cell factories,

RBS strength of the MVA genes was analyzed with the RBS Calculator [25, 26]. Calculated RBS strength was represented by the translation initiation rate as listed in Table 2. The calculated RBS strength was by no means a very accurate measurement of the expression status of *mvaE*, *mvaS*, *mvaK₁*, *mvaD* and *mvaK₂*, however, could give a good estimation of a general trend of the optimized expression status of MVA pathway. For better illustration, the highest RBS strength for each gene was defined as 1, and a relative RBS strength of the ten strains for this gene was calculated accordingly (Fig. 5). The ten strains are lined up along the X axis according to their β-carotene yield. MVA gene expression pattern of ALV145 strain with the highest yield was obvious, that all genes were regulated to a medium level. In contrast, some of the inefficient strains had one or more MVA genes fell to very low expression level, or one or more genes reached the highest level. It was reported that some MVA pathway intermediates were toxic, for example HMG-CoA, accumulation of which affected cell growth and pathway efficiency [4]. Besides, Mevalonate kinase (MK), encoded by *mvaK₁*, was identified as a rate-limiting enzyme [21]. Thus in an optimized MVA pathway, *mvaE* should be coordinately expressed with *mvaS* to avoid HMG-CoA

Table 2 MVA gene RBS sequences with their calculated strength of representative strains from the combinatory expression library DXS37-IDI46 (pACYC184-AL-mva)

Strains	The genes RBS of MVA pathway										Relative yield (%)[b]
	mvaE		mvaS		mvaK1		mvaK2		mvaD		
	Sequence	Calculated strength[a]	Sequence	Calculated strength	Sequence	calculated strength	Sequence	Calculated strength	Sequence	Calculated strength	
ALV104	AGGAGGCTAGAA	17988	AGGAGAGTTTTA	16327	AGGAGAGTCTTT	12463	AGGAGGGGGGTGC	30731	AGGAGATTCCTG	92	112.5
ALV131	AGGAGGCTTTGG	8370	AGGAGGGTGGGGG	162464	AGGAGGTATTTC	183455	AGGAGGTTGCGT	90504	AGGAGAAATGGGC	612	120.9
ALV100	AGGAGGGTTTGA	9158	AGGAGGGTGGGG	72267	AGGAGATTGTGC	7513	AGGAGGGTGTCC	29379	AGGAGAAATGGGG	447	123.2
ALV108	AGGAGGTTGGTG	8110	AGGAGAAATTGG	2064	AGGAGGGGGGGGT	35173	AGGAGGGGGGGTT	15646	AGGAGAGTTTCG	259	129.6
ALV20	AGGAGGGGGCTA	15575	AGGAGACTTTGT	7615	AGGAGGGCGCCT	6080	AGGAGGTTTGCT	103587	AGGAGGTATGCT	3238	151.2
ALV63	AGGAGGAGGTGT	33777	AGGAGGCATGGG	44050	AGGAGAAACGGC	30183	AGGAGGGGGTCTA	14690	AGGAGGTTGCAT	5813	163.9
ALV25	AGGAGACACAAT	10817	AGGAGGGCGGTT	15646	AGGAGGGCTTTT	25668	AGGAGGTGC	1258	AGGAGGCTTTCGT	559	171.9
ALV133	AGGAGAGTTTAC	25607	AGGAGGACGAGG	17119	AGGAGGGCTCCT	22126	AGGAGGGCGGGC	12493	AGGAGATCGGCC	92	181.8
ALV23	AGGAGGGAGTAA	50645	AGGAGGTGGGTG	148479	AGGAGAGGGTCC	8332	AGGAGGGATGCG	11418	AGGAGGTGCCAT	1870	186.3
ALV145	AGGAGGGATTGT	33777	AGGAGGAGCTCT	57705	AGGAGGTGGTGT	160285	AGGAGGGGGGTGC	30731	AGGAGGTCTTGC	1258	196.0

a Calculated Strength was represented by the translation initiation rate calculated by RBS Library Calculator [9, 17]

b Relative yield compared with their parent strain DXS37-IDI46

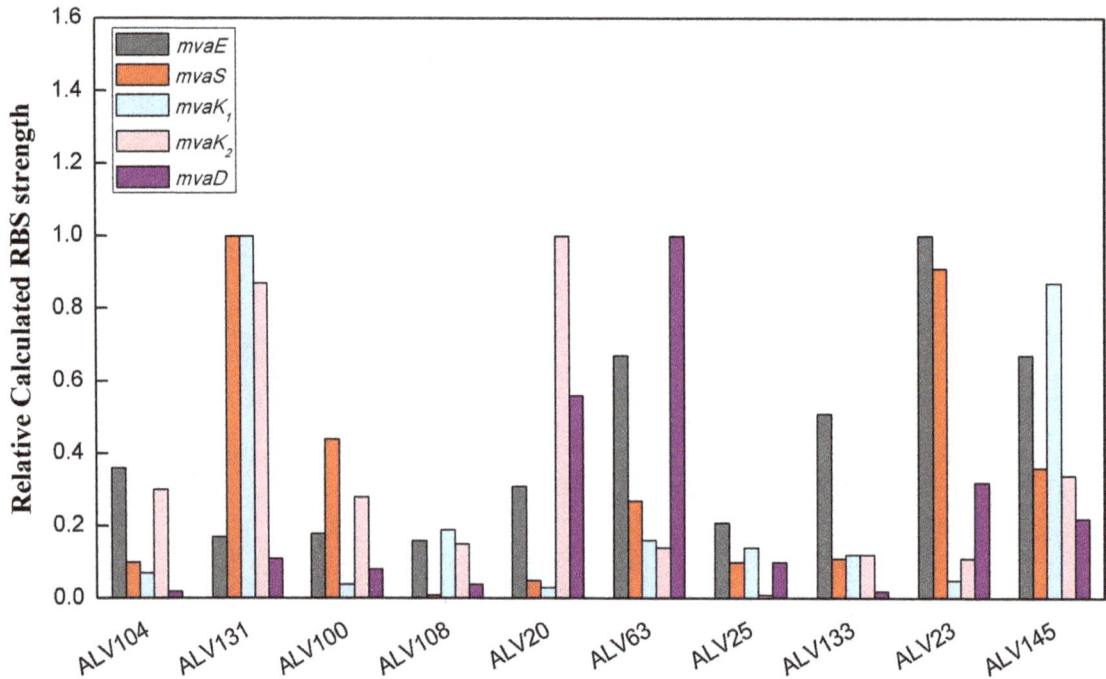

Fig. 5 The relative calculated RBS strength of MVA genes in representative strains from library DXS37-IDI46 (pACYC184-AL-mva). The highest RBS strength for each gene was defined as 1, and a relative RBS strength of the ten strains for this gene was calculated accordingly. The ten strains are lined up along the *X* axis according to their β-carotene yield from low to high

accumulation, and a higher expression of $mvaK_1$ is desired. It was found by the expression analysis, except ALV131, most of the inefficient strains did not follow these two rules.

Analysis of the representative strains indicated that an efficient MVA pathway contained genes expressed at a medium level, among which *mvaE* coordinately expresses with *mvaS* to avoid HMG-CoA accumulation, and a higher expression level of $mvaK_1$ is beneficial.

Conclusion

A TRCM technique was designed and established in the research, which could be easily applied to various applications in terms of metabolic pathway regulation and optimization. An optimized MVA pathway was constructed with TRCM to increase β-carotene yield of *E. coli* cell factory by twofold, and the optimal regulation pattern of MVA pathway was analyzed and illustrated.

Abbreviations
LB: lysogeny broth; RBS: ribosome-binding site; DCW: dry cell weight; ATP: adenosine triphosphate.

Authors' contributions
YL and HP planned and performed experiments, analyzed and interpreted the data. LQ, BC and ZX supervised the study, designed experiments and interpreted the results. YL wrote the manuscript. All authors read and approved the final manuscript.

Author details
[1] Tianjin Institute of Industrial Biotechnology, Chinese Academy of Sciences, Tianjin 300308, People's Republic of China. [2] Key Laboratory of Systems Microbial Biotechnology, Chinese Academy of Sciences, Tianjin 300308, People's Republic of China. [3] School of Pharmacy, East China University of Science and Technology, Shanghai 200237, People's Republic of China.

Acknowledgements
We would like to thank the two great reviewers this manuscript is lucky to have, who took a lot of time and made tremendous efforts to give very precise criticism and very constructive suggestions to help make this manuscript better.

Competing interests
The authors declare that they have no competing interests.

Funding
This research was supported by grants from National High Technology Research and Development Program of China (2015AA020202), Tianjin Key Technology RD program of Tianjin Municipal Science and Technology Commission (Y5M2121111), National Natural Science Foundation of China (31522002), Natural Science Foundation of Tianjin (15JCYBJC49400), and Chinese Academy of Sciences (NN-CAS) Research Fund (NNCAS-2015-2).

References

1. Wu G, Yan Q, Jones JA, Tang YJ, Fong SS, Koffas MA. Metabolic burden: cornerstones in synthetic biology and metabolic engineering applications. Trends Biotechnol. 2016;34:652–64.
2. Scott M, Gunderson CW, Mateescu EM, Zhang Z, Hwa T. Interdependence of cell growth and gene expression: origins and consequences. Science. 2010;330:1099–102.
3. Glick BR. Metabolic load and heterologous gene expression. Biotechnol Adv. 1995;13:247–61.
4. Pitera DJ, Paddon CJ, Newman JD, Keasling JD. Balancing a heterologous mevalonate pathway for improved isoprenoid production in *Escherichia coli*. Metab Eng. 2007;9:193–207.
5. Lu J, Tang J, Liu Y, Zhu X, Zhang T, Zhang X. Combinatorial modulation of galP and glk gene expression for improved alternative glucose utilization. Appl Microbiol Biotechnol. 2011;93:2455–62.
6. Yin L, Zhao J, Chen C, Hu X, Wang X. Enhancing the carbon flux and NADPH supply to increase L-isoleucine production in *Corynebacterium glutamicum*. Biotechnol Bioprocess Eng. 2014;19:132–42.
7. Xu P, Gu Q, Wang W, Wong L, Bower AG, Collins CH, Koffas MA. Modular optimization of multi-gene pathways for fatty acids production in *E. coli*. Nat Commun. 2013;4:1409.
8. Xu P, Vansiri A, Bhan N, Koffas MA. ePathBrick: a synthetic biology platform for engineering metabolic pathways in *E. coli*. ACS Synth Biol. 2012;1:256–66.
9. Zelcbuch L, Antonovsky N, Bar-Even A, Levin-Karp A, Barenholz U, Dayagi M, Liebermeister W, Flamholz A, Noor E, Amram S, et al. Spanning high-dimensional expression space using ribosome-binding site combinatorics. Nucleic Acids Res. 2013;41:e98.
10. Gibson DG, Young L, Chuang R-Y, Venter JC, Hutchison CA, Smith HO. Enzymatic assembly of DNA molecules up to several hundred kilobases. Nat Methods. 2009;6:343–5.
11. Lee ME, Aswani A, Han AS, Tomlin CJ, Dueber JE. Expression-level optimization of a multi-enzyme pathway in the absence of a high-throughput assay. Nucleic Acids Res. 2013;41:10668–78.
12. Ajikumar PK, Tyo K, Carlsen S, Mucha O, Phon TH, Stephanopoulos G. Terpenoids: opportunities for biosynthesis of natural product drugs using engineered microorganisms. Mol Pharm. 2008;5:167–90.
13. Das A, Yoon SH, Lee SH, Kim JY, Oh DK, Kim SW. An update on microbial carotenoid production: application of recent metabolic engineering tools. Appl Microbiol Biotechnol. 2007;77:505–12.
14. Lee PC, Schmidt-Dannert C. Metabolic engineering towards biotechnological production of carotenoids in microorganisms. Appl Microbiol Biotechnol. 2002;60:1–11.
15. Yadav VG, De Mey M, Giaw Lim C, Kumaran Ajikumar P, Stephanopoulos G. The future of metabolic engineering and synthetic biology: towards a systematic practice. Metab Eng. 2012;14:233–41.
16. Martin VJ, Pitera DJ, Withers ST, Newman JD, Keasling JD. Engineering a mevalonate pathway in *Escherichia coli* for production of terpenoids. Nat Biotechnol. 2003;21:796–802.
17. Yoon SH, Lee SH, Das A, Ryu HK, Jang HJ, Kim JY, Oh DK, Keasling JD, Kim SW. Combinatorial expression of bacterial whole mevalonate pathway for the production of beta-carotene in *E. coli*. Biotechnol. 2009;140:218–26.
18. Yoon SH, Lee YM, Kim JE, Lee SH, Lee JH, Kim JY, Jung KH, Shin YC, Keasling JD, Kim SW. Enhanced lycopene production in *Escherichia coli* engineered to synthesize isopentenyl diphosphate and dimethylallyl diphosphate from mevalonate. Biotechnol Bioeng. 2006;94:1025–32.
19. Vadali RV, Fu Y, Bennett GN, San KY. Enhanced lycopene productivity by manipulation of carbon flow to isopentenyl diphosphate in *Escherichia coli*. Biotechnol Prog. 2005;21:1558–61.
20. Yoon S-H, Park H-M, Kim JE, Lee SH, Choi MS, Kim JY, Oh DK, Keasling JD, Kim SW. Increased β-carotene production in recombinant *Escherichia coli* harboring an engineered isoprenoid precursor pathway with mevalonate addition. Biotechnol Prog. 2007;23:599–605.
21. Anthony JR, Anthony LC, Nowroozi F, Kwon G, Newman JD, Keasling JD. Optimization of the mevalonate-based isoprenoid biosynthetic pathway in *Escherichia coli* for production of the anti-malarial drug precursor amorpha-4,11-diene. Metab Eng. 2009;11:13.
22. Liang J, Chao R, Abil Z, Bao Z, Zhao H. FairyTALE: a high-throughput TAL effector synthesis platform. ACS Synth Biol. 2014;3:67–73.
23. Engler C, Marillonnet S. Generation of families of construct variants using golden gate shuffling. Methods Mol Biol. 2011;729:167–81.
24. Zhao J, Li Q, Sun T, Zhu X, Xu H, Tang J, Zhang X, Ma Y. Engineering central metabolic modules of *Escherichia coli* for improving beta-carotene production. Metab Eng. 2013;17:42–50.
25. Salis HM, Mirsky EA, Voigt CA. Automated design of synthetic ribosome binding sites to control protein expression. Nat Biotechnol. 2009;27:946–50.
26. Espah Borujeni A, Channarasappa AS, Salis HM. Translation rate is controlled by coupled trade-offs between site accessibility, selective RNA unfolding and sliding at upstream standby sites. Nucleic Acids Res. 2014;42:2646–59.
27. Crooks GE, Hon G, Chandonia JM, Brenner SE. WebLogo: a sequence logo generator. Genome Res. 2004;14:1188–90.

7

Scaling up and scaling down the production of galactaric acid from pectin using *Trichoderma reesei*

Toni Paasikallio, Anne Huuskonen and Marilyn G. Wiebe[*] [iD]

Abstract

Background: Bioconversion of D-galacturonic acid to galactaric (mucic) acid has previously been carried out in small scale (50–1000 mL) cultures, which produce tens of grams of galactaric acid. To obtain larger amounts of biologically produced galactaric acid, the process needed to be scaled up using a readily available technical substrate. Food grade pectin was selected as a readily available source of D-galacturonic acid for conversion to galactaric acid.

Results: We demonstrated that the process using *Trichoderma reesei* QM6a Δ*gar1 udh* can be scaled up from 1 L to 10 and 250 L, replacing pure D-galacturonic acid with commercially available pectin. *T. reesei* produced 18 g L^{-1} galactaric acid from food-grade pectin (yield 1.00 g [g D-galacturonate consumed]$^{-1}$) when grown at 1 L scale, 21 g L^{-1} galactaric acid (yield 1.11 g [g D-galacturonate consumed]$^{-1}$) when grown at 10 L scale and 14 g L^{-1} galactaric acid (yield 0.77 g [g D-galacturonate consumed]$^{-1}$) when grown at 250 L scale. Initial production rates were similar to those observed in 500 mL cultures with pure D-galacturonate as substrate. Approximately 2.8 kg galactaric acid was precipitated from the 250 L culture, representing a recovery of 77% of the galactaric acid in the supernatant. In addition to scaling up, we also demonstrated that the process could be scaled down to 4 mL for screening of production strains in 24-well plate format. Production of galactaric acid from pectin was assessed for three strains expressing uronate dehydrogenase under alternative promoters and up to 11 g L^{-1} galactaric acid were produced in the batch process.

Conclusions: The process of producing galactaric acid by bioconversion with *T. reesei* was demonstrated to be equally efficient using pectin as it was with D-galacturonic acid. The 24-well plate batch process will be useful screening new constructs, but cannot replace process optimisation in bioreactors. Scaling up to 250 L demonstrated good reproducibility with the smaller scale but there was a loss in yield at 250 L which indicated that total biomass extraction and more efficient DSP would both be needed for a large scale process.

Keywords: Galactaric acid, Mucic acid, D-Galacturonic acid, Pectin, *Trichoderma reesei*, Scale-up, Scale-down

Background

There is considerable interest in replacing chemicals derived from petroleum with bio-derived chemicals, i.e. chemicals obtained by bioconversion and/or chemical conversion from renewable biological resources, primarily plants. Both galactaric and glucaric acids have been identified as substrates for chemical conversion to adipic acid, furandicarboxylic acid (which is being developed as a substitute for terephthalic acid) and other platform chemicals, including anhydrides, diesters and diallyls which can be used in the synthesis of higher value products [1–3]. Both compounds can be prepared by nitric acid oxidation of the corresponding monosaccharide (D-glucose or D-galactose) [4]. Although biotechnological conversions may be preferable to nitric acid oxidation, biotechnological conversion of D-glucose to glucaric acid continues to be challenging [5]. However, galactaric acid has been produced from D-galacturonic acid using

*Correspondence: Marilyn.Wiebe@vtt.fi
VTT Technical Research Centre of Finland Ltd., P.O. Box 1000, 02044 Espoo, Finland

genetically modified *Escherichia coli* and *Trichoderma reesei* at concentrations of 10 [6] to 20 [7] g L^{-1}, which are high enough concentrations for precipitation from the culture broth, making biotechnologically derived galactaric acid available for assessment in further chemical conversions [6]. The genetically modified *E. coli* or *T. reesei* convert D-galacturonic acid to galactaric acid by the action of a uronate dehydrogenase (UDH), which is expressed in a strain in which normal D-galacturonic acid metabolism has been disrupted [6, 8]. In the case of *T. reesei*, deletion of the gene encoding NADPH-dependent D-galacturonate reductase (*gar1*), the first step in the fungal pathway for D-galacturonic acid metabolism, is sufficient to disrupt the pathway and expression of the *A. tumefaciens udh* gene results in a strain which produces galactaric acid [8].

Although chemically produced galactaric acid is commercially available, biotechnologically produced galactaric acid will contain different impurities and the effect of these on specific chemical reactions needs to be assessed to provide feedback on the purity requirements and consequent implications for downstream processing (DSP) in the biotechnological process. For this reason it may be necessary to produce substantial quantities of galactaric acid using biotechnology, even though the process is not ready for commercialisation.

The substrate for biotechnological production of galactaric acid is D-galacturonic acid, which can be obtained by hydrolysis of pectin. Zhang et al. [6] demonstrated that galactaric acid could be produced from enzymatically hydrolysed sugar beet pulp, but noted that the yield of galactaric acid on sugar beet pulp was relatively low (0.14 g galactaric acid per g sugar beet pulp), reflecting the high concentration of other compounds in the pulp, particularly D-glucose and L-arabinose. D-Galacturonic acid is currently available only at high prices (more than €3000 per kg) and is thus not suitable for producing kilogram amounts of galactaric acid for chemical testing. However, food grade pectins are commercially available at prices in the range of €10–100 per kg, making them a reasonable source for preparation of large amounts of galactaric acid. It should be noted that food grade pectin may be diluted with additives such as sucrose or glucose. Having received a request to provide a 2 kg sample of biotechnologically produced galactaric acid for use in chemical reactions, we decided to use pectin as the source of D-galacturonic acid in scaling up the process of galactaric acid production with *T. reesei Δgar1 udh*. An industrial process would be expected to use sugar beet pulp, citrus waste and other pectin-rich waste streams or crude extractions of pectin from these sources, which would not compete with the food use of current pectin production, but these were not readily available at the scale needed. *T. reesei Δgar1 udh* was used in scaling up the process since it has already been demonstrated to produce up to 20 g L^{-1} galactaric acid from D-galacturonic acid [7]. Since *T. reesei* does not hydrolyse pectin [9] and to facilitate sterilisation of viscous pectin solutions, enzymatic pre-hydrolysis of the pectin was necessary. An alternative approach would be to develop a consolidated process using a production strain such as *Aspergillus niger*, which produces native pectinases. However, *A. niger* can metabolise galactaric acid [8] and a strain in which this pathway has been disrupted has only recently become available [10]. A production process with *A. niger* has not yet been developed.

In addition to scaling up the process of producing galactaric acid to provide galactaric acid for chemical reactions, we were also interested in providing a scaled down process, which would enable the screening of new strains [11]. Galactaric acid production by *T. reesei* has previously been demonstrated in flask cultures [8], but at much lower concentrations than can be obtained in bioreactors [7]. Running and Bansal [12] demonstrated that 24-well plates can have as good or better oxygen transfer as shaken flasks, depending on the shaking regime applied, and 24-well plates are increasingly being used for the cultivation of filamentous fungi [13–15]. It is therefore useful to assess whether galactaric acid production by *T. reesei* could be scaled down for production in 24-well plates using pectin as substrate.

In this paper we describe the production of galactaric acid by *T. reesei* VTT D-161646 from enzyme hydrolysed pectin at 1, 10 and 250 L scales. We also demonstrate that the process can be scaled down to 4 mL for strain screening, for example in considering the effectiveness of alternative promoters for expression of the uronate dehydrogenase gene, as shown here.

Results

Scaling down production of galactaric acid in 24-well plates

Three transformants of *T. reesei* were generated in which the uronate dehydrogenase (*udh*) gene was expressed under different promoters and these were cultivated in 24-well plates to assess the suitability of the 24-well plate format for galactaric acid production. All transformants which contained the *udh* gene produced galactaric acid in the 24-well plates. *T. reesei* VTT D-161646 produced 8.8 g L^{-1} galactaric acid from 10.7 g D-galacturonic acid in 6 days and 10.5 g L^{-1} from 20 g L^{-1} pectin (Sigma), hydrolysed to give 11.6 g L^{-1} D-galacturonic acid, demonstrating that galactaric acid could be produced in the 24-well plates (Fig. 1; Table 1). No galactaric acid was produced by M122, the control strain lacking *udh*, nor in wells which had not been

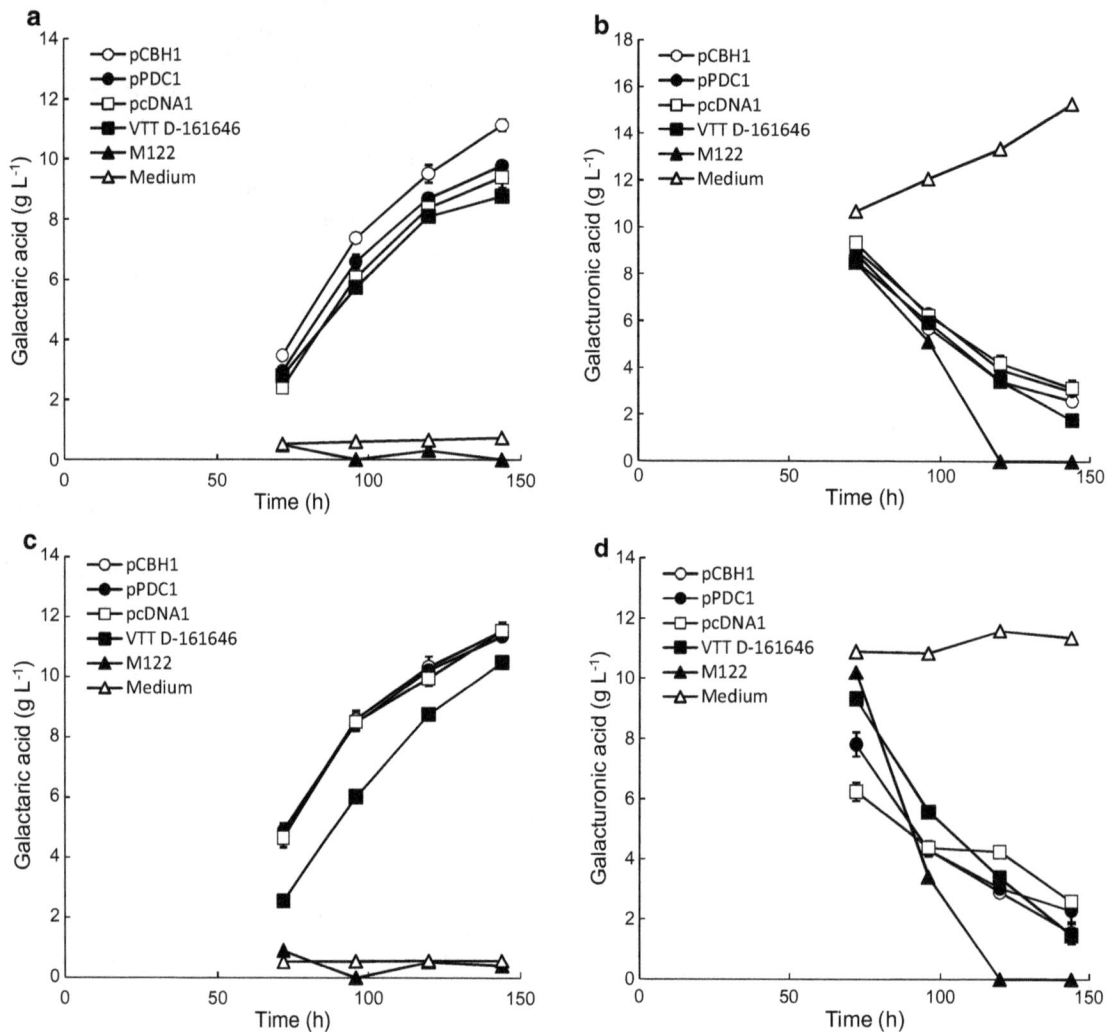

Fig. 1 Galactaric acid production and D-galacturonic acid consumption in 24-well plates. *T. reesei* Δ*gar1 udh* strains, in which the *udh* gene was under control of the *CBH1*, *PDC1*, *cDNA1* or *GPDA* (VTT D-161646) promoters, and M122, which contains *gar1* and does not contain *udh*, were grown in 4 mL medium containing pure D-galacturonate (**a**, **b**) or hydrolysed pectin (**c**, **d**) as substrate in 24-well plates. All values have been adjusted for evaporation, except those for the medium in wells with pure D-galacturonic acid. Values for p*CBH1*, p*PDC1* and p*cDNA1* represent mean ± sem for n = 3. Where not seen, *error bars* were smaller than the *symbols*

inoculated (Fig. 1). The increase in concentration of D-galacturonic acid (Fig. 1b) and lactose in the uninoculated wells provided evidence of the extent of evaporation in the wells and this data was used to calculate the actual concentrations of galactaric acid produced in other wells. Hydrolysis of the pectin was not complete within 72 h, so changes in the concentrations of D-galacturonic acid and lactose in the well with medium containing D-galacturonate were used to adjust for evaporation in the pectin containing wells.

Each of the new transformants produced similar amounts of galactaric acid to VTT D-161646 and to each other (Fig. 1; Table 1). The strain expressing *udh*

under control of the CBH1 (cellobiohydrolase I) promoter produced significantly more (p < 0.05) galactaric acid (11.2 ± 0.2 g L^{-1}) from D-galacturonic acid than the other three strains, but there were no significant differences (p > 0.05) in the amounts produced from Sigma pectin (Fig. 1). The yield of galactaric acid on D-galacturonic acid was between 0.87 and 1.00 g g^{-1} in the 24-well plates when D-galacturonic acid was used as the substrate, and between 0.76 and 0.90 g g^{-1} when pectin was used as substrate (Table 1). Production rates between 72 and 120 h were 0.11–0.13 g L^{-1} h^{-1} with either pure D-galacturonic acid or with pectin as the source of D-galacturonic acid (Table 1).

Table 1 Galactaric acid production by *T. reesei* expressing *udh* under promoters p*CBH1*, p*cDNA1*, p*PDC1* or p*GPDA*

Strain	pCBH1	pcDNA1	pPDC1	pGPDA[a]
24-well plates—D-galacturonate				
Galactaric acid (g L^{-1})	11.1 ± 0.2	9.4 ± 0.4	9.8 ± 0.1	8.8
Galactaric acid production rate (72–120 h) (g l^{-1} h^{-1})	0.13 ± 0.01	0.12 ± 0.00	0.12 ± 0.00	0.11
Yield galactaric/D-galacturonic (g g^{-1})	0.95 ± 0.02	0.87 ± 0.02	0.89 ± 0.00	0.89
24-well plates—pectin				
Galactaric acid (g L^{-1})	11.6 ± 0.2	11.5 ± 0.3	11.4 ± 0.1	10.5
Galactaric acid production rate (72–120 h) (g L^{-1} h^{-1})	0.12 ± 0.04	0.11 ± 0.00	0.11 ± 0.00	0.13
Yield galactaric/D-galacturonic (g g^{-1})	0.84 ± 0.03	0.90 ± 0.03	0.89 ± 0.03	0.76
0.5 L bioreactors—D-galacturonate				
Galactaric acid (g L^{-1})	15.7	13.6	16.0	19.7
Galactaric acid production rate (0–72 h) (g L^{-1} h^{-1})	0.22	0.21	0.09	0.19
Yield galactaric/D-galacturonic (g g^{-1})	0.90	0.82	1.11	0.85
Maximum biomass (g L^{-1})	9.5 ± 0.2	9.0 ± 0.3	6.5 ± 0.5	7.9 ± 0.3

Galactaric acid production (titre, rate of production and yield on D-galacturonic acid) in 24-well plates and in 0.5 L bioreactors using either D-galacturonic acid or pectin (Sigma) as substrate, with lactose as co-substrate. Maximum biomass in bioreactors was observed at 50 (pCBH1, pcDNA1, pGPDA) or 72 (pPDC1) h after which time biomass decreases (see also Fig. 3). Values are mean ± sem for three replicates

[a] Strain VTT D-161646

In lactose-D-galacturonic acid fed-batch cultures (0.5 L, using the optimised process described by [7]) transformants expressing *udh* under the *gpdA* (glyceraldehyde-3-phosphate dehydrogenase, strain VTT D-161646), *CBH1*, *cDNA1* (unidentified hypothetical protein), and *PDC1* (pyruvate decarboxylase) promoters produced 20, 16, 14 and 16 g L^{-1} galactaric acid, respectively (Table 1). Yields were 0.85, 0.90, 0.82 and 1.11 g galactaric [g D-galacturonate consumed]$^{-1}$, respectively. However, the strain producing galactaric acid under the *pdc1* promoter grew slower than the other strains, producing less biomass, and the initial production rate (between 0 and 72 h) of D-galactaric acid (0.09 g L^{-1} h^{-1}) was slower than that of the other three strains (0.19–0.20 g L^{-1} h^{-1}, Table 1).

Production of galactaric acid from pectin at 1, 10 and 250 L scale

Strain VTT D-161646 was used in the scaling up experiments since production conditions have been developed specifically for this strain [7] and there was no improvement in galactaric acid production with the new strains under these conditions. In order to scale up production of galactaric acid, D-galacturonic acid was replaced with pectin, using either Sigma or Meridianstar pectin (Table 2). A process scheme for the production of galactaric acid from pectin is shown in Fig. 2. The 10 L culture was carried out using Sigma pectin while waiting for the Meridianstar pectin to arrive. Because of the high D-glucose content in the Meridianstar pectin it was necessary to adjust the process at 1 L scale to assess the extent to which the high D-glucose content would repress

Table 2 Carbohydrate composition (% dry matter) of pectin

Pectin	Sigma P1935	Meridianstar rapid set
D-Galacturonate	65.6	43.0
D-Glucose	1.7	35.5
L-Arabinose	1.2	1.5
D-Galactose + D-xylose	11	4

Pectin was hydrolysed by addition of 0.5–1.0 mL L^{-1} Pectinex Ultra with 0.1 mL L^{-1} Pectinex Smash and incubation at 40 °C. D-Galactose and D-xylose were not separated on the HPLC column used

galactaric acid production. Therefore both 1 and 250 L cultures were carried out using Meridianstar pectin.

When Sigma pectin replaced D-galacturonic acid in the *T. reesei* galactaric acid production process, 21 g L^{-1} galactaric acid was produced with a yield of 1.11 g galactaric acid [g D-galacturonic acid]$^{-1}$, assuming that the pectin had been fully hydrolysed. The yield of galactaric acid on pectin was 0.73 g g^{-1}. The production rate during the first 140 h (0.14 g L^{-1} h^{-1}) was comparable to that observed previously with pure D-galacturonic acid during the same time interval (0.15 g L^{-1} h^{-1}; Fig. 3) [7] and the process generally showed good reproducibility with the 500 mL scale production from D-galacturonic acid. No accumulation of D-galacturonic acid or other carbohydrates (lactose, glucose, galactose or arabinose) was observed during the feeding phase. Biomass production on pectin with lactose was a bit higher than on D-galacturonic acid with lactose, even though the lactose concentration had been reduced to take into account the

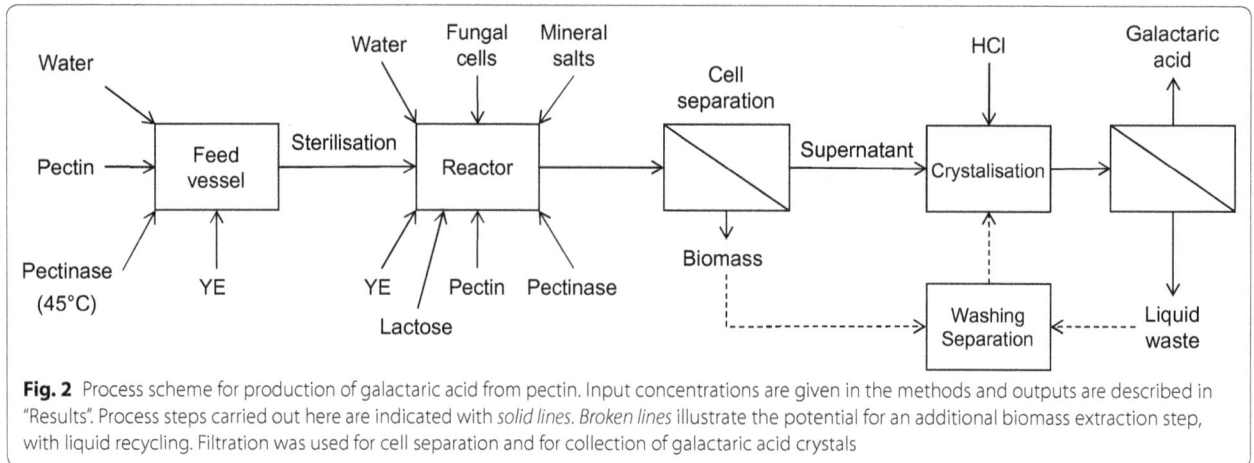

Fig. 2 Process scheme for production of galactaric acid from pectin. Input concentrations are given in the methods and outputs are described in "Results". Process steps carried out here are indicated with *solid lines*. *Broken lines* illustrate the potential for an additional biomass extraction step, with liquid recycling. Filtration was used for cell separation and for collection of galactaric acid crystals

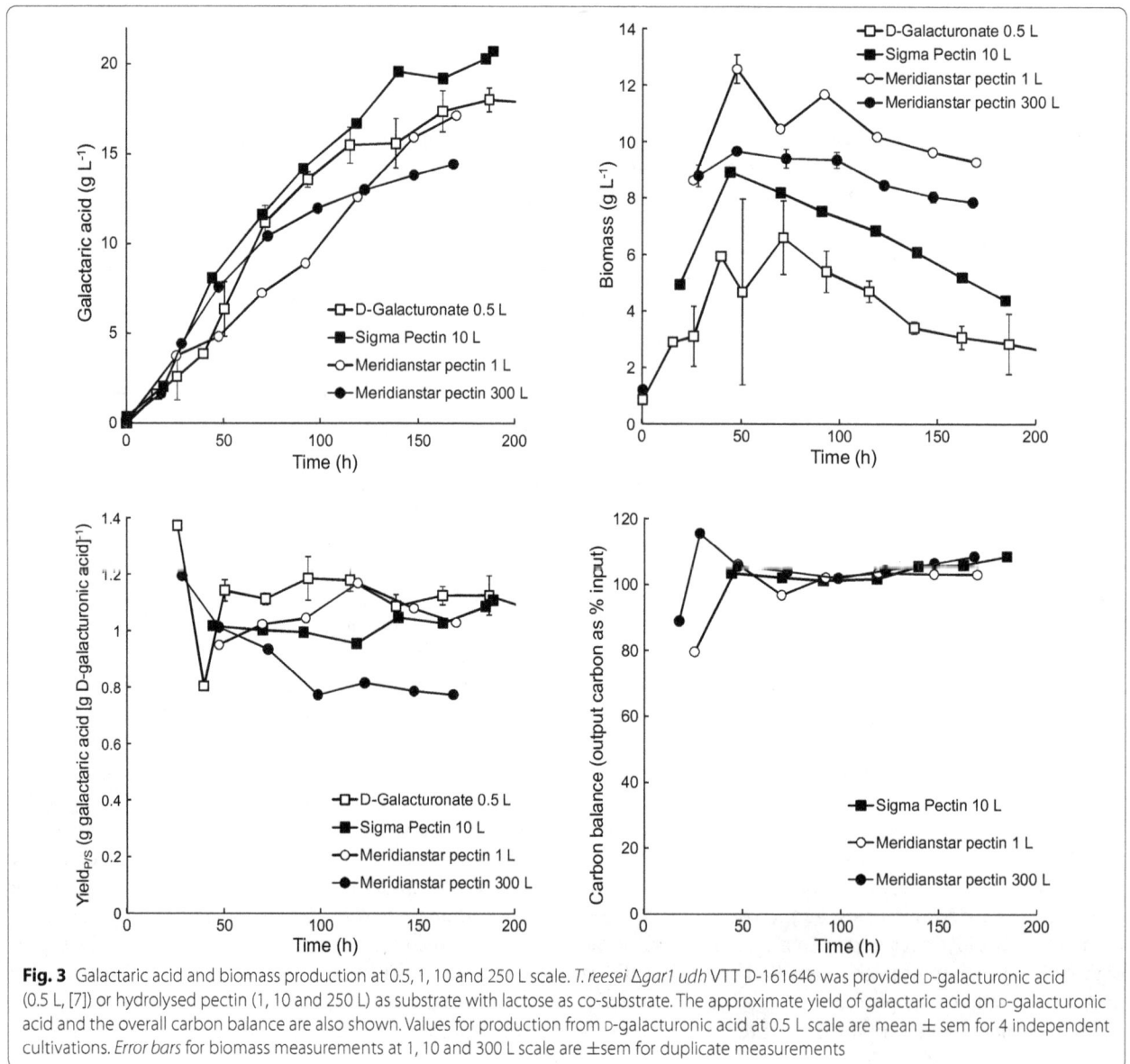

Fig. 3 Galactaric acid and biomass production at 0.5, 1, 10 and 250 L scale. *T. reesei Δgar1 udh* VTT D-161646 was provided D-galacturonic acid (0.5 L, [7]) or hydrolysed pectin (1, 10 and 250 L) as substrate with lactose as co-substrate. The approximate yield of galactaric acid on D-galacturonic acid and the overall carbon balance are also shown. Values for production from D-galacturonic acid at 0.5 L scale are mean ± sem for 4 independent cultivations. *Error bars* for biomass measurements at 1, 10 and 300 L scale are ±sem for duplicate measurements

carbohydrates introduced with the pectin (Fig. 3). D-Glucose remained limiting throughout the feeding phase.

The culture broth was collected at the end of the cultivation and the biomass separated by filtration, generating 8.8 L clear liquid and 282 g wet biomass. Precipitating and collecting the crystals of galactaric acid resulted in 154 g galactaric acid, representing 85% of the galactaric acid produced (Table 3).

Using Meridianstar Rapid Set pectin, 19 g L^{-1} galactaric acid was produced with a yield of 1.00 g galactaric acid [g D-galacturonic acid]$^{-1}$, assuming complete hydrolysis of the pectin (Fig. 3). The yield of galactaric acid on Meridianstar pectin was 0.41 g g^{-1}. Feeding was not started until all D-glucose and lactose had been consumed, but this resulted in D-galacturonic acid limitation during the batch phase and reduced the initial galactaric acid production rate. The initial production rate (0–70 h) was only 0.10 g L^{-1} h^{-1}, but during the feeding phase (90–150 h) it increased to 0.13 g L^{-1} h^{-1}. Although there was no accumulation of D-glucose during the feeding phase, there was initial accumulation of D-galacturonic acid (up to 4.3 g L^{-1}) in the culture supernatant and the feed rate was reduced to allow more complete conversion to D-galactaric acid. More biomass was produced when *T. reesei* VTT D-161646 was grown on Meridianstar pectin than on Sigma pectin and the biomass concentration was maintained during the feeding phase (Fig. 3). Approximately 89% of the galactaric acid was recovered by precipitation in acid (Table 3).

Based on the results at the 1 L scale, *T. reesei* VTT D-161646 was grown in a 300 L bioreactor and 256 L of 14 g L^{-1} galactaric acid was produced. The initial (0–73 h) production rate of 0.15 g L^{-1} h^{-1} was only sustained for 72 h, after which time both the rate of galactaric acid production and the yield of galactaric acid on D-galacturonic acid decreased. The final yield was 0.77 g galactaric acid [g D-galacturonic acid]$^{-1}$ or 0.32 g [g pectin]$^{-1}$ (Fig. 3). Crystals were observed among the mycelia under the microscope and were also present in the foam which had been produced in the bioreactor. After precipitation, collection and drying, 2.81 kg galactaric acid was obtained, or 73% of the amount present in the supernatant (Table 3). Biomass was not extracted.

Discussion

The process of producing galactaric acid using *T. reesei* was scalable both up and down, with little loss in productivity, even with the substitution of pectin for D-galacturonic acid. Pectin was found to be a suitable substrate for galactaric acid production, although for *T. reesei* this required the addition of commercial pectinases. Addition of pectinases also facilitated sterilisation of the pectin feed, so even though *T. reesei* could be engineered to express a full range of pectinases to reduce the cost of enzyme addition, a fully consolidated process may not be desirable. The only other adjustment required for the use of pectin was a reduction in the amount of added co-substrate, since co-substrates were provided with the pectin, particularly in

Table 3 Process parameters for galactaric acid production at 0.5, 1, 10 and 250 L

	D-Galacturonic acid[a]	Meridianstar pectin	Sigma pectin	Meridianstar pectin
Reactor volume (L)	0.5	1	10	250
Process time (h)	242	215	185	168
Inputs				
Lactose (g L^{-1})	28	10	14	9
Galacturonic acid (g L^{-1})	21			
Pectin (g L^{-1})		48.4	29.2	45.9
Pectinase(s) (mL L^{-1})		1	1	2
Fungal cells (g L^{-1})	1	1	1	1
Air (vvm)	1.4	0.5	0.5	0.5
Output				
Galactaric acid (g L^{-1})	21 ± 2	18	21	15
Yield (g [g galacturonate]$^{-1}$)	1.0 ± 0.2	1.0	1.1	0.8
Yield (g [g pectin]$^{-1}$)		0.41	0.73	0.32
Biomass (g L^{-1})[b]	4 ± 1	9 ± 0.1	4 ± 0.1	8 ± 0.2
Product recovery (kg)		0.014	0.15	2.8
Product recovery (%)	68 ± 7	89	85	73

[a] Data from [7]. Values are mean ± sem for four cultivations

[b] Biomass measurements are mean ± sem for 2–3 replicates, except for 0.5 L cultivations (mean ± sem for 4 cultivations), at end of cultivation

the case of the food-grade pectin, which contained a high concentration of D-glucose. Earlier results had indicated that D-glucose-limited fed-batch culture was less effective than lactose-limited fed-batch culture for galactaric acid production with *T. reesei* VTT D-161646 [7], but this did not appear to be the case when D-glucose was supplied with hydrolysed pectin in the feed.

Pectin was also used in the 24-well plates, with no apparent problems with mixing. Evaporation could be a problem at this scale, however, as seen in the increasing concentration of substrates in uninoculated wells, and needed to be taken into account. Production from hydrolysed pectin was comparable to that observed when pure D-galacturonic acid was provided. Production of galactaric acid was much higher in these 4 mL batch cultures than has previously been reported for flask cultures with filamentous fungi (around 4 g L^{-1}; [8, 10]). The medium in the 24-well plates contained spent grain extract, which was expected to induce the CBH1 promoter [16], but not necessarily other promoters. Spent grain extract also provides organic nitrogen, resulting in a richer medium, which may contribute to the good conversion of D-galacturonic acid into galactaric acid. Kuivanen et al. [10] recently found that providing yeast extract and peptone to a *A. niger* galactaric acid producing strain, resulted in improved galactaric acid production, although the amounts were still low.

Production of galactaric acid by three new strains was compared in the 24-well plates, but production of all strains was found to be similar. This suggested that the amount of uronate dehydrogenase produced under the control of each of the promoters tested (p*CBH1*, pc*DNA1*, p*PDC1* and the original p*GPDA*) was adequate to catalyse the conversion of D-galacturonic acid to galactaric acid at the rate at which it was taken up under these conditions. The effectiveness of these strains in producing galactaric acid was confirmed in 0.5 L bioreactors, however, the bioreactor cultures also revealed a slow growth (and production) phenotype for the strain expressing *udh* under the *PDC1* promoter, which was not observed in the microtiter plates. The *PDC* promoter is expected to be particularly well induced when D-glucose is the main carbon source, and conversion of D-galacturonic to galactaric acid may be better with D-glucose as co-substrate [17], rather than lactose, as used here. The provision of better aeration and substrate-limited feeding in the bioreactors may also contribute to the phenotype.

The yield of galactaric acid on D-galacturonate was close to the theoretical yield of 1.08 g g^{-1} in the 1 and 10 L cultures, but in the 250 L culture the yield decreased to 0.8 g galactaric acid [g D-galacturonate consumed]$^{-1}$ after 72 h. The carbon balance (Fig. 3) indicated that all of the input carbon was accounted for in the output.

This suggests that galactaric acid crystals may have been included with inadequately washed biomass in the dry weight determination, or that some D-galacturonate contributed to CO_2 production through an unknown pathway. Up to around 7% galactaric acid was expected to be removed with the biomass, but is expected to be washed off during rinsing for the dry weight determination. However, when harvesting the 250 L culture we observed a decrease in galactaric acid concentration during the filtration process, which indicated that galactaric acid crystals were trapped as more biomass accumulated in the filter. The concentration of galactaric acid was determined after filtration. Large scale extraction of the biomass was not possible at this time, but it is clear that this would be needed to obtain the maximum galactaric acid output from a scaled up process.

In spite of the low yield at the 250 L scale, around 3.8 kg galactaric acid was produced and 2.8 kg was recovered by acid precipitation. Poor mixing and inadequate cooling in the vessels used for the crystallisation step will have contributed to the recovery of only 73% of the galactaric acid. This was adequate for supplying material for further chemical conversions, but highlights the need for further DSP development.

Conclusions

The process of producing galactaric acid by bioconversion with *T. reesei* was demonstrated to be equally efficient using hydrolysed pectin as D-galacturonic acid as the substrate, and D-glucose present with the pectin did not inhibit the process. Scaling up to 250 L demonstrated good reproducibility with the smaller scale but there was a loss in yield at 250 L which indicated that both total biomass extraction and more efficient DSP would be needed for a large scale process. It would also be necessary to obtain higher production titres and production rates before a biotechnological process for producing galactaric acid would be considered for commercial production. However, the low solubility of galactaric acid [7] made it feasible to produce kilogram amounts of galactaric acid with the currently available strains so that assessment on purity requirements and further chemical transformations can proceed in parallel with further strain engineering. The 24-well plate batch process will be useful in screening new constructs, but cannot replace process optimisation in bioreactors.

Methods
Strains

Trichoderma reesei QM6a (VTT D-071262T, ATCC13631), M122 (RutC-30 *mus53-*), M1123 (M122 *pyr4-*) and VTT D-161646 were obtained from VTT's strain collections. Spore suspensions were prepared by cultivating the fungus on potato-dextrose agar (BD, Sparks, Maryland, USA) for 5–7 days, after which the spores were harvested, suspended in a buffer containing 0.8% NaCl, 0.025% Tween-20 and 20% glycerol, filtered through cotton and stored at −80 °C. *Saccharomyces cerevisiae* strain H3488 (FY834, obtained from Jay C. Dunlap) was used to clone expression construct plasmids. *Escherichia coli* TOP10 (Invitrogen) was used for propagation of the plasmids.

Expression vectors

Expression plasmids contained the 5′ and 3′ flanking regions of *T. reesei* D-galacturonic acid reductase (*gar1*, [18]), *T. reesei* promoter (*cbh1*, *cDNA1* or *pdc1*), *Agrobacterium tumefasciens* UDH gene, *Aspergillus trpC* terminator, *T. reesei pyr4* selection marker with a loop-out fragment for marker removal, and the pRS426 plasmid as vector backbone.

Plasmid pRS426 (ATCC77107), used in cloning expression construct plasmids in yeast, was obtained from Jay C. Dunlap. For cloning, pRS426 was digested with restriction enzymes *Eco*RI and *Xho*I. A plasmid containing *A. tumefasciens* D-galacturonic acid dehydrogenase (*udh* gene, GenBank no. BK006462.1) codon optimised for *Aspergillus niger*, [8] was digested with *Nco*I and *Sac*II to release the *udh* gene from the vector backbone. The 5′ and 3′ flank (1572 and 1500 bp) fragments of *gar1*, the three promoters (*cbh1*, *cDNA1* and *pdc1*) and 285 bp of the *gar1* 3′ flank as a direct repeat for looping out the selection marker were obtained by PCR using strain QM6a as template. The *Aspergillus trpC* terminator (773 bp) was obtained by PCR from a plasmid containing the element. PCR amplification was performed with a KAPA HiFi HotStart ReadyMix PCR

kit (KAPABiosystems) using the primers listed in Table 4. The selection marker *pyr4* was obtained from an existing plasmid with *Not*I digestion. All PCRs and digestions were separated with agarose gel electrophoresis and correct fragments were extracted with a gel extraction kit (Qiagen).

The appropriate purified DNA fragments were transformed to *S. cerevisiae* FY834 using the method described in Gietz and Woods [19]. The plasmids obtained through homologous recombination reactions were isolated from *S. cerevisiae*, amplified in *E. coli* and checked by restriction enzyme digestions and sequencing.

Strain generation

Expression plasmids (10 µg) were digested with *Mss*I, resulting in approximately 6 µg of released expression cassette from each. To generate 3′ single stranded overhangs and thus to improve transformation efficiency, digestion mixtures were further treated with T7 exonuclease [20] and used without further purification.

Strain M1123 was transformed with each of the three expression constructs and positive transformants were selected on minimal medium as described previously [21, 22]. Transformants were sub-cultured onto agar-solidified minimal medium containing 1 mL L^{-1} Triton X-100. The isolates were screened by PCR to verify correct integration at the *gar1* locus and a few correctly integrated transformants were purified as single spore isolates. Deletion of *gar1* was verified by PCR.

Media

The low phosphate medium described by [7] was used for 24-well plate, flask and bioreactor cultivations, with D-galacturonic acid, pectin and lactose provided as

Table 4 Primers used to generate fragments for cloning by PCR amplification

Product	Primer	Sequence	Product size (bp)
gar1 5′flank (tre22004)	MU01_gar1_5f_for	GGTAACGCCAGGGTTTTCCCAGTCACGACGGTTTAAACTTATATCCACCGTGTCCCAG	1610
	MU02_gar1_5f_rev	GACGCAGTTGTTTGAGCAAC	
trpC terminator (Aspergillus)	MU03_trpCt_for	GTGGATAACCCCATCTTCAAGCAGTCCTGAGATCCACTTAACGTTACTGAAATCA	803
	MU04_trpCt_rev	GAGTGGAGATGTGGAGTGGG	
gar1 3′direct repeat (tre22004)	MU05_gar1_3dr_for	TGTGTAAGCGCCCACTCCACATCTCCACTCGGCGCGCCTTGCATTGGTCAGAGCGGTA	361
	MU06_gar1_3dr_rev	CGAGAGCAGAGCAGCAGTAGTCGATGCTAGGCGGCCGCCCCGACTTGGAGAAGCTCGTC	
gar1 3′flank (tre22004)	MU07_gar1_3f_for	CCAGCTGCGATTGATGTGTATCTTTGCATGGCGGCCGCTTGCATTGGTCAGAGCGGTA	1576
	MU08_gar1_3f_rev	AGCGGATAACAATTTCACACAGGAAACAGCGTTTAAACAAGCAGTGGATGACTTGCTG	
cbh1 promoter (tre123989)	MU09_cbh1p_for	AGGTTAGTAGGTTGCTCAAACAACTGCGTCGGCCGGCCTGTGGCAACAAGAGGCCAGA	1646
	MU10_cbh1p_rev	GGCAGCGCCGGTGACGAGCAGGCGCTTCATGATGCGCAGTCCGCGGTTGA	
cDNA1 promoter (tre123515)	MU11_cDNA1p_for	AGGTTAGTAGGTTGCTCAAACAACTGCGTCGGCCGGCCGAATTCGGTCTGAAGGACGT	1231
	MU12_cDNA1p_rev	GGCAGCGCCGGTGACGAGCAGGCGCTTCATGTTGAGAGAAGTTGTTGGATTGA	
pdc1 promoter (tre121534)	MU13_pdc1p_for	AGGTTAGTAGGTTGCTCAAACAACTGCGTCGGCCGGCCAAAGGAGGGAGCATTCTTCG	1370
	MU14_pdc1p_rev	GGCAGCGCCGGTGACGAGCAGGCGCTTCATGATTGTGCTGTAGCTGCGCT	

The gene identifiers (tre-numbers) refer to Joint Genome Institute *T. reesei* assembly release version 2.0

carbon sources, as indicated. Medium for 24-well plate cultivations contained 15 g L^{-1} lactose with 10 g L^{-1} D-galacturonate or 10 g L^{-1} lactose with 20 g L^{-1} pectin (Sigma P1935, Table 2). Immediately before inoculation, 1 mL L^{-1} Pectinex Ultra SP-L (Novozymes) was added to pectin-containing medium. All 24-well plate medium also contained 100 mM PIPPS and 1 g L^{-1} spent grain extract, with the pH adjusted to 5.5.

Pre-cultures for bioreactors were grown in low phosphate medium with 20 g L^{-1} lactose and 1 g L^{-1} yeast extract.

Media for bioreactors contained 15 g L^{-1} lactose and 7.5 g L^{-1} Sigma pectin or 9 g L^{-1} lactose and 11 g L^{-1} Meridianstar Rapid set pectin (Table 2) in the batch phase, plus 3 g L^{-1} yeast extract. Filter sterilised Pectinex Ultra SP-L (1 mL L^{-1}) was added to the reactors after sterilisation at 1 and 10 L scale, but unsterilized enzyme was added before sterilisation for the 250 L pilot. Feed for the fed-batch cultivations contained 1 g L^{-1} yeast extract and either 50 g L^{-1} Sigma pectin with 6 g L^{-1} lactose or 78 g L^{-1} Meridianstar pectin. Pectin in the feed was hydrolysed by the addition of 1.0 mL L^{-1} Pectinex Ultra SP-L with 0.1 mL L^{-1} Pectinex Smash (Novozymes) and incubation at ~45 °C, with agitation, prior to sterilisation. Hydrolysis was allowed to proceed until the viscosity had been reduced to allow adequate sterilisation. The D-galacturonic acid concentration was assessed by HPLC and if hydrolysis was not yet complete, additional sterile Pectinex Ultra SP-L was added after sterilisation to ensure the availability of D-galacturonic acid for the process.

Culture conditions

Twenty-four-well plates (Whatman 10 mL round bottom Uniplate) contained 4 mL medium per well, with either 10 g L^{-1} D-galacturonate or 20 g L^{-1} pectin (Sigma), which was enzymatically hydrolysed simultaneously with the cultivation, as the substrate and lactose as the carbon source for growth. Wells were inoculated with approximately 10^5 spores in 10 µL suspension (~2.5 × 10^4 spores mL^{-1} final concentration). One well for each medium was not inoculated and used as a control. Three wells were inoculated with each new transformant (expressing *udh* under the *CBH1*, *PDC1* and *cDNA1* promoters) and one well each with M122 (negative control) and VTT D-161646. The plates were covered with adhesive, breathable rayon fibre film for culture plates (VWR) and incubated at 28 °C, 800 rpm (Infors HT Microtron, Switzerland), 85% relative humidity for up to 7 days. Samples (100 µL) were taken daily after 72 h using wide-mouth 200 µL pipette tips and either diluted and analysed immediately or stored at −20 °C. Sample size was minimised

for the 24-well plates to ensure adequate culture volume for continued incubation.

Pre-cultures for bioreactor inocula were started in Erlenmeyer flasks (500–2000 mL containing 20% volume medium) inoculated with ~1 × 10^5 spores mL^{-1} (final concentration) to provide a 10% v/v inoculum for 1, 10 or 32 L reactors. The 32 L reactor (New Brunswick 32 L BioFlo 510, Eppendorf) was used to provide inoculum for the 300 L reactor. Flasks were incubated at 30 °C with 200 rpm agitation for approximately 45 h before being transferred directly to bioreactors to provide an initial biomass of about 1 g L^{-1}.

Bioreactor cultures were grown in Multifors (Infors HT, Switzerland, max. working volume 0.5 L), Biostat Q (Sartorius AG, Germany, 1 L max. working volume), Biostat C (Sartorius AG, Germany, 10 L max. working volume), BioFlo 510 (Eppendorf) or New Brunswick Scientific IF400 (Eppendorf, 350 L max. working volume) bioreactors. Cultures were maintained at 35 °C, with 500–900 rpm (0.5 L), 350–500 rpm (1 L), 400 rpm (10 L), 200 rpm (32 L) or 125–200 rpm (350 L reactor) agitation and approximately 0.5 volume gas (volume culture)$^{-1}$ min^{-1} (vvm). Culture pH was kept constant at pH 4.0 by the addition of sterile 2–4 M NaOH or 1 M H$_3$PO$_4$. Gas concentration (CO$_2$, O$_2$, N$_2$ and Ar) was analysed continuously using a Prima Pro Process mass spectrometer (Thermo Scientific, UK) calibrated with 3% CO$_2$ in Ar, 5% CO$_2$ with 0.99% Ar and 15% O$_2$ in N$_2$, 20% O$_2$ plus 20% Ar in N$_2$, and 0.04% ethanol in N$_2$. Fed-batch cultures were provided feed at a constant rate after at least 5 g L^{-1} biomass had been produced. The start of feeding for the 1 L culture was delayed to allow complete utilisation of the lactose, but this appeared to reduce productivity and the feed for the 250 L culture (using the same substrate) was started with about 2 g L^{-1} lactose still present in the medium, as was the 10 L culture.

Chemical analyses

The concentrations of D-glucose, lactose, D-galactose, D-galacturonic acid, and galactaric acid were determined by HPLC using a Fast Acid Analysis Column (100 × 7.8 mm, BioRad Laboratories, Hercules, CA) linked to an Aminex HPX-87H organic acid analysis column (300 × 7.8 mm, BioRad Laboratories) with 5 mM H$_2$SO$_4$ as eluent and a flow rate of 0.3 or 0.5 mL min^{-1}. The column was maintained at 55 °C. Peaks were detected using a Waters 410 differential refractometer and a Waters 2487 dual wavelength UV (210 nm) detector.

Samples were diluted with eluent to give expected concentrations of galactaric acid between 1 and 3 g L^{-1} and heated at 100 °C for 1 h to solubilise crystals of galactaric acid prior to HPLC analysis. Samples from 24-well plates were diluted without separation of biomass and supernatant, heated at 100 °C for 0.5–1 h, then centrifuged at room temperature to remove biomass.

Downstream processing

Culture broth was collected from the 1, 10 and 250 L cultures and the mycelium separated from the supernatant by filtration. At the 1 and 10 L scale, the mycelium was filtered through thick cleaning cloth under vacuum. The 250 L culture was warmed to 45 °C before biomass was separated from the liquid using Seitz K300 depth filter sheets (Pall Corporation) in a Seitz 40–30A4 filter press to obtain approximately 250 L filtrate and 20 kg wet biomass. The supernatant was transferred during filtration to 2 large vessels (165 and 85 L) which were cooled via cold jackets operated with water circulated from a chiller to give temperatures between 7 and 12 °C. One and 10 L volumes were stored at 4 °C. HCl was added to all culture supernatants to reduce the pH to between 1.5 and 2.2. Supernatants with precipitating galactaric acid crystals were mixed periodically by shaking (1 and 2 × 5 L) or stirring with a paddle (165 and 85 L) until it was convenient to collect the crystals of galactaric acid.

Galactaric acid crystals were collected by filtration through Whatman GF/A filter paper. For large volumes, the crystals were first allowed to settle to the bottom of the container and the clear liquid decanted to a clean vessel. The decanted liquid was expected to contain some galactaric acid and was returned to the cold to allow further crystal formation, precipitation and collection. The process was repeated twice for the 1 L culture supernatant and four times for the 10 and 250 L volumes.

Biomass determination

Mycelia were collected by filtration through Whatman GF/B filters under vacuum and washed twice with an equal or greater volume of distilled H_2O. Mycelia were dried to a constant weight at 105 °C.

Authors' contributions

AH carried out all work related to scaling down the process (transformations and 24-well plate cultivations). TP carried out the scaling up cultivations. MW designed the study, supervised the scaling-up and carried out HPLC analysis and DSP. MW and AH drafted the manuscript. All authors read and approved the final manuscript.

Acknowledgements

We thank Merja Aarnio for technical assistance.

Competing interests

The authors declare that they have no competing interests.

Funding

This research was supported by the VTT BioEconomy Programme.

References

1. Boussie TR, Dias EL, Fresco ZM, Murphy VJ, Shoemaker J, Archer R, et al. Adipic acid composition. 2011; Patent WO2011/109051 A1.
2. Thomas D, Asikainen M, Harlin A. Method for producing muconic acids and furans from aldaric acids. 2015; Patent WO2015189481 A1.
3. Lavilla C, Alla A, Martínez de Ilarduya A, Benito E, García-Martín MG, Galbis JA, Muñoz-Guerra S. Carbohydrate-based polyesters made from bicyclic acetalized galactaric acid. Biomacromolecules. 2011;12:2642–52.
4. Kiely DE, Kirk RHS. Method of oxidation using nitric acid. 2010; Patent US7692041 B2.
5. Liu Y, Gong X, Wang C, Du G, Chen J, Kang Z. Production of glucaric acid from myo-inositol in engineered *Pichia pastoris*. Enzym Microb Technol. 2016;91:8–16.
6. Zhang H, Li X, Su X, Ang EL, Zhang Y, Zhao H. Production of adipic acid from sugar beet residue by combined biological and chemical catalysis. ChemCatChem. 2016;8:1500–6.
7. Barth D, Wiebe MG. Enhancing fungal production of galactaric acid. Appl Microbiol Biotechnol. 2017;101:4033–40.
8. Mojzita D, Wiebe M, Hilditch S, Boer H, Penttilä M, Richard P. Metabolic engineering of fungal strains for conversion of D-galacturonate to meso-galactarate. Appl Environ Microbiol. 2010;76:169–75.
9. Wiebe MG, Mojzita D, Hilditch S, Ruohonen L, Penttilä M. Bioconversion of D-galacturonate to keto-deoxy-L-galactonate (3-deoxy-L-threo-hex-2-ulosonate) using filamentous fungi. BMC Biotechnol. 2010;10:63.
10. Kuivanen J, Wang Y-MJ, Richard P. Engineering *Aspergillus niger* for galactaric acid production: elimination of galactaric acid catabolism by using RNA sequencing and CRISPR/Cas9. Microb Cell Fact. 2016;15:210.
11. Formenti LR, Nørregaard A, Bolic A, Hernandez DQ, Hagemann T, Heins AL, Larsson H, Mears L, Mauricio-Iglesias M, Krühne U, Gernaey KV. Challenges in industrial fermentation technology research. Biotechnol J. 2014;9:727–38.
12. Running JA, Bansal K. Oxygen transfer rates in shaken culture vessels from Fernbach flasks to microtiter plates. Biotechnol Bioeng. 2016;113:1729–35.
13. Linde T, Hansen NB, Lübeck M, Lübeck PS. Fermentation in 24-well plates is an efficient screening platform for filamentous fungi. Lett Appl Microbiol. 2014;59:224–30.
14. Dana CM, Dotson-Fagerstrom A, Roche CM, Kal SM, Chokhawala HA, Blanch HW, Clark DS. The importance of pyroglutamate in cellulase Cel7A. Biotechnol Bioeng. 2014;111:842–7.
15. Landowski CP, Mustalahti E, Wahl R, Croute L, Sivasiddarthan D, Westerholm-Parvinen A, Sommer B, Ostermeier C, Helk B, Saarinen J, Saloheimo M. Enabling low cost biopharmaceuticals: high level interferon alpha-2b production in *Trichoderma reesei*. Microb Cell Fact. 2016;15:104.
16. Nguyen EV, Imanishi SY, Haapaniemi P, Yadav A, Saloheimo M, Corthals GL, Pakula TM. Quantitative site-specific phosphoproteomics of *Trichoderma reesei* signaling pathways upon induction of hydrolytic enzyme production. J Proteome Res. 2016;15:457–67.
17. Li J, Wang J, Wang S, Xing M, Yu S, Liu G. Achieving efficient protein expression in *Trichoderma reesei* by using strong constitutive promoters. Microb Cell Fact. 2012;11:84.
18. Kuorelahti S, Kalkkinen N, Penttilä M, Londesborough J, Richard P. Identification in the mold *Hypocrea jecorina* of the first fungal D-galacturonic acid reductase. Biochemistry. 2005;44:11234–40.
19. Gietz RD, Woods RA. Transformation of yeast by lithium acetate/single-stranded carrier DNA/polyethylene glycol method. Guid Yeast Genet Mol Cell Biol Pt B. 2002;350:87–96.
20. Korppoo A. Development of the CRISPR/Cas9 method for use in *T. reesei*. University of Helsinki; 2017. https://helda.helsinki.fi/bitstream/handle/10138/175220/developm.pdf?sequence=1. Accessed 19 Mar 2017.
21. Gruber F, Visser J, Kubicek CP, Degraaff LH. The development of a heterologous transformation system for the cellulolytic fungus *Trichoderma reesei* based on a *pyrg*-negative mutant strain. Curr Genet. 1990;18:71–6.
22. Penttilä M, Nevalainen H, Rättö M, Salminen E, Knowles J. A versatile transformation system for the cellulolytic filamentous fungus *Trichoderma reesei*. Gene. 1987;61:155–64.

8

Cell foundry with high product specificity and catalytic activity for 21-deoxycortisol biotransformation

Shuting Xiong[1,2†], Ying Wang[1,2†], Mingdong Yao[1,2], Hong Liu[1,2], Xiao Zhou[1,2], Wenhai Xiao[1,2*] and Yingjin Yuan[1,2]

Abstract

Background: 21-deoxycortisol (21-DF) is the key intermediate to manufacture pharmaceutical glucocorticoids. Recently, a Japan patent has realized 21-DF production via biotransformation of 17-hydroxyprogesterone (17-OHP) by purified steroid 11β-hydroxylase CYP11B1. Due to the less costs on enzyme isolation, purification and stabilization as well as cofactors supply, whole-cell should be preferentially employed as the biocatalyst over purified enzymes. No reports as so far have demonstrated a whole-cell system to produce 21-DF. Therefore, this study aimed to establish a whole-cell biocatalyst to achieve 21-DF transformation with high catalytic activity and product specificity.

Results: In this study, *Escherichia coli* MG1655(DE3), which exhibited the highest substrate transportation rate among other tested chassises, was employed as the host cell to construct our biocatalyst by co-expressing heterologous CYP11B1 together with bovine adrenodoxin and adrenodoxin reductase. Through screening CYP11B1s (with mutagenesis at N-terminus) from nine sources, *Homo sapiens* CYP11B1 mutant (G25R/G46R/L52 M) achieved the highest 21-DF transformation rate at 10.6 mg/L/h. Furthermore, an optimal substrate concentration of 2.4 g/L and a corresponding transformation rate of 16.2 mg/L/h were obtained by screening substrate concentrations. To be noted, based on structural analysis of the enzyme-substrate complex, two types of site-directed mutations were designed to adjust the relative position between the catalytic active site heme and the substrate. Accordingly, 1.96-fold enhancement on 21-DF transformation rate (to 47.9 mg/L/h) and 2.78-fold improvement on product/by-product ratio (from 0.36 to 1.36) were achieved by the combined mutagenesis of F381A/L382S/I488L. Eventually, after 38-h biotransformation in shake-flask, the production of 21-DF reached to 1.42 g/L with a yield of 52.7%, which is the highest 21-DF production as known.

Conclusions: Heterologous CYP11B1 was manipulated to construct *E. coli* biocatalyst converting 17-OHP to 21-DF. Through the strategies in terms of (1) screening enzymes (with N-terminal mutagenesis) sources, (2) optimizing substrate concentration, and most importantly (3) rational design novel mutants aided by structural analysis, the 21-DF transformation rate was stepwise improved by 19.5-fold along with 4.67-fold increase on the product/byproduct ratio. Eventually, the highest 21-DF reported production was achieved in shake-flask after 38-h biotransformation. This study highlighted above described methods to obtain a high efficient and specific biocatalyst for the desired biotransformation.

Keywords: Synthetic biology, Whole-cell biocatalysis, Product specificity, Catalytic activity, 21-Deoxycortisol, CYP11B1

*Correspondence: wenhai.xiao@tju.edu.cn
†Shuting Xiong and Ying Wang contributed equally to this work
[1] Key Laboratory of Systems Bioengineering (Ministry of Education), Tianjin University, No. 92, Weijin Road, Nankai District, Tianjin 300072, People's Republic of China
Full list of author information is available at the end of the article

Background

Manufacturing steroid compounds with therapeutic usage and commercial value is a successful example to apply biocatalysts in large-scale industrial processes [1–4]. The high regio- and stereo-selectivity make bioconversion more advantageous than chemical synthesis of steroid molecules with complex structures [5, 6]. Considering the costs involved in enzyme isolation, purification and stabilization as well as cofactors [such as NAD(P)H] supply, whole-cell as biocatalysts is better than purified enzyme to some extent [5, 6]. As one of steroids, 21-deoxycortisol (21-DF) is the key intermediate for pharmaceutical glucocorticoids with anti-inflammatory effect. For instance, it can be used as the substrate to chemosynthesize cortisol via a matured and environmental friendly process [7]. However, the accessibility of 21-DF in bulks remains a great challenge due to catalytic selectivity for 11β-hydroxylation, which has restricted the large-scale application of 21-DF in associated steroid industry. In 2006, a Japan patent [8] realized 21-DF transformation only through one step catalysis by purified adrenodoxin (Adx), adrenodoxin reductase (AdR) and CYP11B1. However, no reports as known has demonstrated a whole-cell system for 21-DF transformation. Therefore, this study aims to establish a whole-cell biocatalyst of CYP11B1 to achieve the desired conversion (Fig. 1) with high catalytic activity and product specificity.

Cytochrome P450s are highly involved in biocatalysts [9–16], catalyzing a broad range of reactions including steroid hydroxylation [1, 17–20]. For manipulation of heterologous P450 in an appropriate host cell, many approaches have been successfully applied to improve the catalytic activity and soluble expression level of the target protein [9, 21–25]. And some P450s exhibiting relaxed substrate specificities [26–29], probably leading to a concerted oxidation reaction without structural selectivity [30]. Therefore, poor product specificity

brought by these P450s became a main obstacle [31, 32]. In order to alleviate by-product accumulation, site-direct mutagenesis within P450s was applied to alter product profiles, in which those successful mutations were obtained in irrational [33–35], semi-rational [27, 36–38] or even rational design [39–41]. Compared with other approaches, rational design not only avoids laborious and time-consuming operation, but also directly builds up the correlation between mutant structures and catalytic mechanism. Majority of the hot spot positions for site-directed mutagenesis were within or surround the active site cavity [39–41], which aimed to make relatively minor alternations on substrate-recognition sites and substrate-access channels to strengthen the targeted substrate binding and reposition the compound of interest [37, 42]. In order to increase the proportion of perillyl alcohol, Seifert et al. [43] once designed mutations to reshape the substrate binding cavity of CYP102A1 by selectively exposing the C7 of (4R)-limonene towards the activated heme oxygen. For our targeted protein CYP11B1, Schiffer et al. [44] once reported that CYP11B1-dependent biocatalyst transformed spironolactone into three isomers with hydroxyl group at C11, C18 and C19, respectively, which might due to the relative position between the potential hydroxylation sites and the catalytic active site heme. Thus, in order to concentrate our target product 21-DF, it will be promising solution to regulate the relative position between the substrate and the catalytic active site heme through the mutagenesis especially at essential residues associated to heme.

In this study, a whole-cell biocatalyst of CYP11B1 was established to transform 17-OHP into 21-DF. Through screening five host candidates with unique characteristics to fit steroid conversion, E. coli MG1655(DE3), which exhibited the highest substrate transportation rate, was employed as the host cell to construct our biocatalyst. The key enzyme CYP11B1 has already been successfully

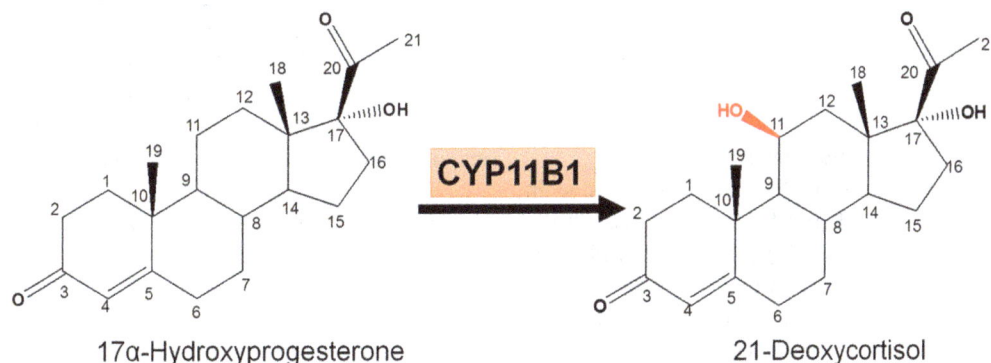

Fig. 1 The paradigm of 21-deoxycortisol (21-DF) biosynthetic pathway from 17α-hydroxyprogesterone (17-OHP). CYP11B1, which is responsible for 11β-hydroxylation, was highlighted in *red*

expressed in both eukaryotes (such as *Saccharomyces cerevisiae* [45] and *Schizosaccharomyces pombe* [46]) and prokaryotes (such as *E. coli* [11]) to achieve the conversion of 11-deoxycortisol to cortisol, those approaches to improve the expression level and catalytic activity of CYP11B1 would provide valuable information to assist our work. Notably, in order to improve the catalytic activity as well as product specificity of our biocatalyst, a series of novel mutations were designed based on structural analysis of the enzyme-substrate complex to regulate the relative position between the substrate 17-OHP and the catalytic active site heme in CYP11B1. 19.5-fold increase on 21-DF transformation rate along with 4.67-fold enhancement on product/by-product ratio were totally achieved via screening CYP11B1s from diverse sources, N-terminal mutagenesis, optimizing substrate concentration and rational design site-directed mutations in the activity pocket of CYP11B1. Eventually, the production of 21-DF reached to 1.42 g/L with a yield of 52.7%, which is the highest 21-DF production as known. This study not only provides promising methods to optimize heterologous P450s by rational design, but also sets a good example of manipulation the key enzyme to systematically improve the efficiency and catalytic selectivity of the whole-cell biocatalyst to accomplish the desired reaction.

Methods

Strains and cultivation

Escherichia coli DH5α was used to construct and maintain plasmids, while *E. coli* C43(DE3) [47] and MG1655(DE3) [48] were attempted to be employed for steroids conversion. All the *E. coli* strains were cultured in Luria–Bertani (LB) medium at 37 °C. When needed, 100 μg/mL ampicillin and 34 μg/mL chloramphenicol were supplemented into the medium. In the meanwhile, three model organisms, i.e. *S. cerevisiae* CEN.PK2, *Yarrowia lipolytica* ATCC201249 [49] and *Mycobacterium smegmatis* mc^{2}155 [50] were also attempted to convert steroids. *S. cerevisiae* and *Y. lipolytica* strains were grown in YPD medium [51] at 30 °C, while *M. smegmatis* strain was cultured in Middlebrook 7H9 broth (Difco) [52] at 37 °C.

Plasmids construction and mutagenesis

Primers and plasmids used in this study were listed in Additional file 1: Tables S1 and S2, respectively. All the heterologous genes including *Cyp11B1*, *AdR*, and truncated *Adx* (Adx_{4-108}) were codon optimized (Additional file 1: Table S3) and synthesized by GENEWIZ (Suzhou, China). These genes were delivered as pUC57-simple serious plasmids (Additional file 1: Table S2). All the Adx-CPY11B1-AdR expression plasmids were constructed

based on vector pET-21a-YX, which was derived from vector pET-21a (Novagen, Germany) by Cao et al. [53] through adding a SpeI site between the BamHI and BglI site of the initial plasmid. Therefore, genes *Cyp11B1*, Adx_{4-108} and *AdR* could be assembled into pET-21a-YX via BioBrick™ strategy [54]. To be noted, as shown in Additional file 1: Figure S1, genes Adx_{4-108}, *AdR* and *Cyp11B1* were recovered from pUC57-simple series plasmids by NdeI/SpeI digestion and inserted into the same sites of pET21a-YX, obtaining pXST-01 (pET21a-Adx_{4-108}), pXST-02 (pET21a-*AdR*) and pXST-03–11 (pET21a-*Cyp11B1*), respectively. Then gene *Cyp11B1* was cut from the corresponding pET21a-*Cyp11B1* plasmids by XbaI/BamHI and inserted into the BamHI/SpeI sites of pXST-01, generating plasmids pXST-21–29 (pET21a-Adx_{4-108}-*Cyp11B1*). After that, gene *AdR* was cut from plasmid pXST-02 by XbaI/BamHI and inserted into the BamHI/SpeI sites of plasmids pET21a-Adx_{4-108}-*Cyp11B1* to construct the final Adx_{4-108}-CPY11B1-AdR co-expression plasmids (pXST-39–47, Table 1) for steroids conversion.

A modified overlap extension PCR (OE-PCR) was employed for mutagenesis in CYP11B1. As illustrated in Additional file 1: Figure S2, mutated residue(s) were introduced into the tails of matched primers OE-1 and OE-2, while NdeI and SpeI restriction sties were introduced into 5′ end of primers OE-F and OE-R, respectively. The products of the first round of PCR with primer pairs OE-F/OE-1 and OE-2/OE-R separately were used as the templates of the second round of PCR with primer pairs OE-F/OE-R. Then the final amplification product was digested by NdeI/SpeI for constructing the final Adx-CPY11B1-AdR co-expression plasmid via the method described above.

Substrate feeding assay

The substrate-feeding assay was carried out in 250 mL flasks with 50 mL appropriate medium according to organism specie. The substrate 17-OHP was supplemented into the culture at the very beginning of the cultivation from a stock solution in 50% (m/v) (2-Hydroxypropyl)-γ-cyclodextrin. The concentration of 17-OHP was 36 mg/mL in the stock solution and diluted to 720 μg/mL in final reaction solution. During this experiment, samples were taken every 8 h.

Whole-cell biocatalysis in shake-flask

In this study, whole-cell biocatalysis was separated into two periods as protein expression and steroid conversion, respectively. All the experiments involving in these two processes were conducted by *E. coli* MG1655(DE3). This strain was kindly provided by Professor Kristala L. J. Prather from Massachusetts Institute of Technology

Table 1 Strains used in this study

Strain	Description	Source
E. coli C43(DE3)	F^- *ompT gal hsdSB (rB- mB-) dcm lon λ*	[47]
E. coli MG1655(DE3)	MG1655 *ΔendA ΔrecA* (DE3), transformation strain	[48]
S. cerevisiae	MATα, *his3Δ1 leu2Δ0 lys2Δ0 ura3Δ0*	This lab
Y. lipolytica	MATA, *ura3-302, leu2-270, lys8-11, pex17-ha*	[49]
M. smegmatis mc^{2}155	*ept-1 mc26* mutant efficient for electroporation	[50]
Ec02040101	*E. coli* MG1655(DE3) with plasmids pETXST39 (pET21a-*Adx$_{4-108}$-Cyp11B1*_Hs-*AdR*) and pGro7	This study
Ec02040102	*E. coli* MG1655(DE3) with plasmids pETXST40 (pET21a-*Adx$_{4-108}$-Cyp11B1*_Bt-*AdR*) and pGro7	This study
Ec02040103	*E. coli* MG1655(DE3) with plasmids pETXST41 (pET21a-*Adx$_{4-108}$-Cyp11B1*_Rn-*AdR*) and pGro7	This study
Ec02040104	*E. coli* MG1655(DE3) with plasmids pETXST42 (pET21a-*Adx$_{4-108}$-Cyp11B1*_Oa-*AdR*) and pGro7	This study
Ec02040105	*E. coli* MG1655(DE3) with plasmids pETXST43 (pET21a-*Adx$_{4-108}$-Cyp11B1*_Sh-*AdR*) and pGro7	This study
Ec02040106	*E. coli* MG1655(DE3) with plasmids pETXST44 (pET21a-*Adx$_{4-108}$-Cyp11B1*_Pa-*AdR*) and pGro7	This study
Ec02040107	*E. coli* MG1655(DE3) with plasmids pETXST45 (pET21a-*Adx$_{4-108}$-Cyp11B1*_Cco-*AdR*) and pGro7	This study
Ec02040108	*E. coli* MG1655(DE3) with plasmids pETXST46 (pET21a-*Adx$_{4-108}$-Cyp11B1*_Mo-*AdR*) and pGro7	This study
Ec02040109	*E. coli* MG1655(DE3) with plasmids pETXST47 (pET21a-*Adx$_{4-108}$-Cyp11B1*_Sa-*AdR*) and pGro7	This study
Ec02040110	*E. coli* MG1655(DE3) with plasmids pETXST48 (pET21a-*Adx$_{4-108}$-Cyp11B1*_Hs$_{G25R/G46R/L52M}$-*AdR*) and pGro7	This study
Ec02040111	*E. coli* MG1655(DE3) with plasmids pETXST49 (pET21a-*Adx$_{4-108}$-Cyp11B1*_Bt$_{G25R}$-*AdR*) and pGro7	This study
Ec02040112	*E. coli* MG1655(DE3) with plasmids pETXST50 (pET21a-*Adx$_{4-108}$-Cyp11B1*_Rn$_{G25R/G46R}$-*AdR*) and pGro7	This study
Ec02040113	*E. coli* MG1655(DE3) with plasmids pETXST51 (pET21a-*Adx$_{4-108}$-Cyp11B1*_Oa$_{G25R/G46R/V52M}$-*AdR*) and pGro7	This study
Ec02040114	*E. coli* MG1655(DE3) with plasmids pETXST52 (pET21a-*Adx$_{4-108}$-Cyp11B1*_Sh$_{T26R/S55R}$-*AdR*) and pGro7	This study
Ec02040115	*E. coli* MG1655(DE3) with plasmids pETXST53 (pET21a-*Adx$_{4-108}$-Cyp11B1*_Pa$_{G25R/G46R/A52M}$-*AdR*) and pGro7	This study
Ec02040116	*E. coli* MG1655(DE3) with plasmids pETXST54 (pET21a-*Adx$_{4-108}$-Cyp11B1*_Cco$_{Y20R/W38R/N44M}$-*AdR*) and pGro7	This study
Ec02040117	*E. coli* MG1655(DE3) with plasmids pETXST55 (pET21a-*Adx$_{4-108}$-Cyp11B1*_Mo$_{A30R/S50R/D56M}$-*AdR*) and pGro7	This study
Ec02040118	*E. coli* MG1655(DE3) with plasmids pETXST56 (pET21a-*Adx$_{4-108}$-Cyp11B1*_Sa$_{A30R/V41R/L47M}$-*AdR*) and pGro7	This study
Ec02040119	*E. coli* MG1655(DE3) with plasmids pETXST57 (pET21a*Adx$_{4-108}$-Cyp11B1*_Hs$_{G25R/G46R/L52M/R384A}$-*AdR*) and pGro7	This study
Ec02040120	*E. coli* MG1655(DE3) with plasmids pETXST58 (pET21a- *Adx$_{4-108}$-Cyp11B1*_Hs $_{G25R/G46R/L52M/R110A}$-*AdR*) and pGro7	This study
Ec02040122	*E. coli* MG1655(DE3) with plasmids pETXST59 (pET21a-*Adx$_{4-108}$-Cyp11B1*_Hs $_{G25R/G46R/L52M/F381A/L382S}$-*AdR*) and pGro7	This study
Ec02040123	*E. coli* MG1655(DE3) with plasmids pETXST60 (pET21a-*Adx$_{4-108}$-Cyp11B1*_Hs $_{G25R/G46R/L52M/F381A/L382T}$-*AdR*) and pGro7	This study
Ec02040124	*E. coli* MG1655(DE3) with plasmids pETXST61 (pET21a-*Adx$_{4-108}$-Cyp11B1*_Hs $_{G25R/G46R/L52M/I488L}$-*AdR*) and pGro7	This study
Ec02040125	*E. coli* MG1655(DE3) with plasmids pETXST62 (pET21a-*Adx$_{4-108}$-Cyp11B1*_Hs $_{G25R/G46R/L52M/F381A/L382S/I488L}$-*AdR*) and pGro7	This study

(MIT), USA. Before the whole-cell biocatalysis, the particular Adx-CPY11B1-AdR co-expression plasmid together with chaperone vector pGro7 (Takara, Japan) were transformed into *E. coli* MG1655(DE3) by electroporation. Plasmid pGro7 contains two chaperone genes *groEL* and *groES* to assist protein folding [55]. Then the recombinant strains were selected from LB agar plates supplemented with 100 μg/mL ampicillin and 34 μg/mL chloramphenicol.

The procedures for protein expression and steroids conversion were modified according to Schiffer et al. [11]. For protein expression, the recombinant strains for whole-cell biocatalysis were grown in 4 mL TB medium (12 g/L peptone, 24 g/L yeast extract, 0.4% (v/v) glycerol, 4.62 g/L KH_2PO_4, 25 g/L K_2HPO_4) with 100 μg/mL ampicillin and 34 μg/mL chloramphenicol, shaking at 37 °C, 250 rpm overnight. After that, the preculture was transferred into 50 mL fresh TB medium with the same antibiotics and grown at 37 °C, 250 rpm until the OD_{600} reached 0.5–0.7. Then the temperature and shaking speed were reduced to 25 °C and 200 rpm, respectively. After that 1 mM IPTG, 1 mM the heme precursor δ-aminolevulinic acid, 4 mg/mL L-arabinose and 50 μg/mL ampicillin were added into the medium to initialize protein expression. The protein expression period would last for 21 h.

For steroids conversion with resting cells, the cultures from protein expression process were harvested by centrifugation (8000 rpm, at 4 °C for 5 min). Then the cells were washed twice by 50 mM potassium phosphate buffer (pH 7.4) and resuspended by the steroids conversion solution containing 50 mM potassium phosphate buffer (pH 7.4), 2% (v/v) glycerol, 1 mM IPTG, 1 mM δ-aminolevulinic acid, 4 mg/mL L-arabinose, 100 μg/mL ampicillin and 34 μg/mL chloramphenicol. The cell density (OD_{600}) in this solution maintained in 75–85. The substrate 17-OHP was supplemented into the solution with appropriate concentration. Biotransformation was performed at 27.5 °C. Samples were taken at the defined time points. According to the data in Additional file 1: Figures S4, S5a and S6, the data at 12 and 18 h were all in the linear range of all the conversion. Therefore, the maximal conversion rate (mg/L/h) was determined as $(P_{18}-P_{12})/(t_{18}-t_{12})$, where P is for product (21-DF) concentration (mg/L) and t is for time (h).

Extraction and analysis of steroids

The standards of 17-OHP and 21-DF were purchased from Sigma (Sigma-Aldrich, USA). The procedures for extracting and analyzing the related steroids were modified according to the methods for 11-deoxycortisol and cortisol [11]. To be specific, 10 mL samples were extracted by 10 mL dichloromethane twice. After centrifugation at 8000 rpm for 5 min, the organic solvent containing steroids on the bottom was collected and then evaporated by nitrogen blow. Then products were dissolved in methanol and analyzed by high-performance liquid chromatography system (HPLC, Waterse2695, Waters Corp., USA) equipped with a BDS HYPERSIL C18 column (150 mm × 4.6 mm, 5 μm, Thermo Scientific) and a Photodiode array detector (Waters 2996). The signals were detected at 240 nm. 70% (v/v) methanol–water (containing 0.1% formic acid) was chosen as the mobile phase with a flow rate at 1 mL/min and the column temperature was set at 40 °C. Semi-quantitative analysis of the unidentified by-product was applied based on the standard curve of 21-DF.

Bioinformatics and structural analysis of CYP11B1

The amino acid sequences of CYP11B1 from diversity species were aligned by ClustalW with default settings [56]. In the meanwhile, in order to analyze and design the complex structure of CYP11B1_Hs mutant1 (G25R/G46R/L52M) with the substrate, three-dimensional structure models were developed by Swiss-Model (http://swissmodel.expasy.org/) with the high sequence homology (>90% identify) and high-resolution crystal structure of CYP11B2 (the aldosterone synthase from *Homo sapiens* with ligand, PDB accession 4dvq-F) as the template. The structure models were subjected to energy minimization by Swiss-Pdb Viewer (http://spdbv.vital-it.ch/). Afterwards enzyme-ligand docking was performed by AutoDockVina program [57]. And the docking studies were run with 17-OHP as the ligand for above-mentioned structure model. The 17-OHP structure files (ligand) were retrieved from ZINC site [58]. Docking cluster analysis was performed in the AutoDockVina program environment, and clusters were characterized by binding energy (in kilocalories per mole). Establishment of docking models was also followed by energy minimization. The built complex structural analysis was done with Pymol software [59]. The mutation at specific amino acid site was also introduced by this software, which allowed exploration of the spatial and molecular interactions among amino acids.

Results and discussion

Selected the host cell with the strongest capability of substrate transportation

Engineered yeast and *E. coli* strains have been reported to realize the conversion of 11-deoxycortisol to cortisol by CYP11B1 [60]. Meanwhile, *Y. lipolytica* strains are good at absorbing hydrophobic materials as well as accumulating lipid bodies that can store less polar metabolites and avoid cell burden [61]. *Mycobacterium* strains were widely applied to convert steroid compounds in pharmaceutical industry. In sum, the strains belonging to these species process distinguishing characteristics that might suit our desired biocatalyst for 21-DF production. In this study, the model organism from these species, i.e. *E. coli* MG1655(DE3), *E. coli* C43(DE3), *S. cerevisiae* CEN.PK2, *Y. lipolytica* ATCC201349 and *M. smegmatis* mc²155, were selected and their transportion capabilities of the substrate 17-OHP across the biocatalyst membrane were initially characterized. Through measuring the extracellular 17-OHP concentration at each time point, it was demonstrated that 17-OHP concentration was decreased significantly only in the media culturing *Y. lipolytica* ATCC201249 or *E. coli* MG1655(DE3) (Fig. 2a). And the most dramatic decline in extracellular 17-OHP concentration was found in the medium with *E. coli* MG1655(DE3) (Fig. 2b), indicating this strain exhibited the strongest ability to transport 17-OHP. As previous study demonstrated that endogenous protein GCY1 could convert 17-OHP into 17α,20α-dihydroxypregn-4-ene-3-one in *S. cerevisiae* [45], and *E. coli* protein Akr is the isoenzyme of GCY1 (http://www.kegg.jp). Therefore, whether the substrate 17-OHP was metabolized in *E. coli* MG1655(DE3) should be investigated. As shown in Fig. 2b, it was observed that the sums of extracellular and intracelluar 17-OHP amount were almost equal at each time point during the cultivation of *E. coli* MG1655(DE3), suggesting 17-OHP is stable during this period and strain MG1655(DE3) could not metabolize 17-OHP. Thus, *E. coli* MG1655(DE3) was employed as the host cell to construct the whole-cell biocatalysis in this study.

Construct of a *E. coli* biocatalyst to convert 17-OHP into 21-DF

In order to realize the reaction from 17-OHP to 21-DF by biocatalyst, heterologous CYP11B1 as well as its redox patterns (Adx and AdR) from *Bos taurus* were all codon

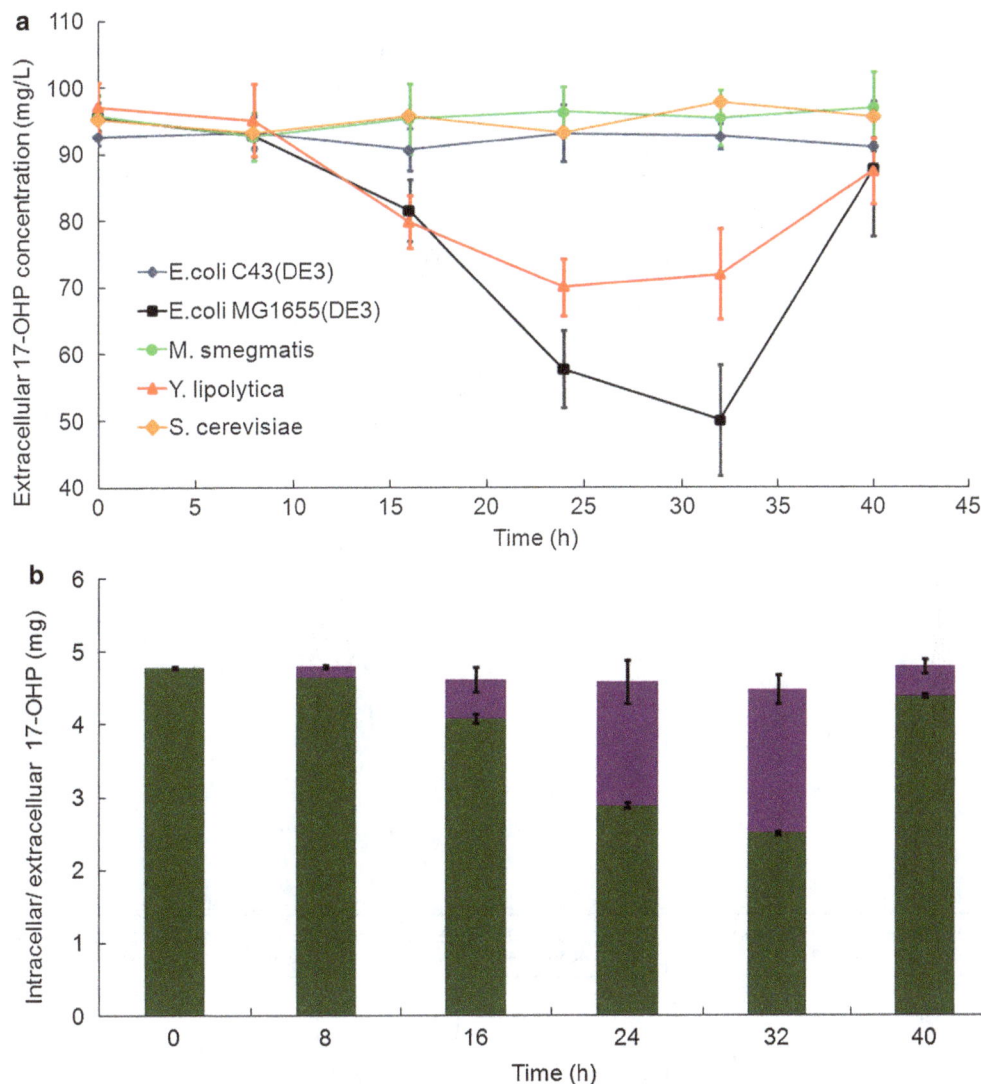

Fig. 2 Selecting the host cell processing high substrate transportation ability. **a** Extracellular substrate (17-OHP) concentration for *E. coli* C43(DE3) (*dark blue*), *E. coli* MG1655(DE3) (*black*), *S. cerevisiae* CEN.PK2 (*orange*), *Y. lipolytica* ATCC201249 (*red*) and *M. smegmatis* mc^{2}155 (*green*) over the time course. **b** The intracelluar (*purple*) and extracelluar (*green*) substrate amounts for *E. coli* MG1655(DE3) over the time course. The substrate 17-OHP was supplemented into the culture at the very beginning of the cultivation. Samples were taken every 8 h

optimized and introduced into *E. coli* MG1655(DE3). Adx was truncated from R1 to T3 (Adx4–108) to improve the efficiency of electron transfer [62]. As illustrated in Fig. 3a, all these three encoding genes were assembled into one cluster and carried by a T7 RNA polymerase-based vector. According to Nishihara et al. [63], bacterial chaperonin GroEL and its cofactor GroES, were also introduced into this system and co-expressed with the heterologous proteins to ensure proper folding. CYP11B1 from *Coprinopsis cinerea okayama* (CYP11B1_Cco), was initially chosen to establish the whole-cell system, generating strain Ec02040107. According to Schiffer et al. [11],

the whole-cell biocatalysis procedure was separated into two periods as protein expression and steroid conversion, respectively. Transformation of 17-OHP was conducted with resting cells to avoid indole accumulation under glucose depletion [64], since indole might inhibit the activity of P450 [65]. The extract from the steroids conversion solution with strain Ec02040107 was analyzed by HPLC. As shown in Fig. 3b, 45.3 mg/L 21-DF (at 3.5 min, the same retention time as 21-DF standard) was successfully detected after 24-h transformation, indicating conversion of 17-OHP to 21-DF was succeeded here. However, there was plenty of the substrate (17-OHP, at 7.0 min)

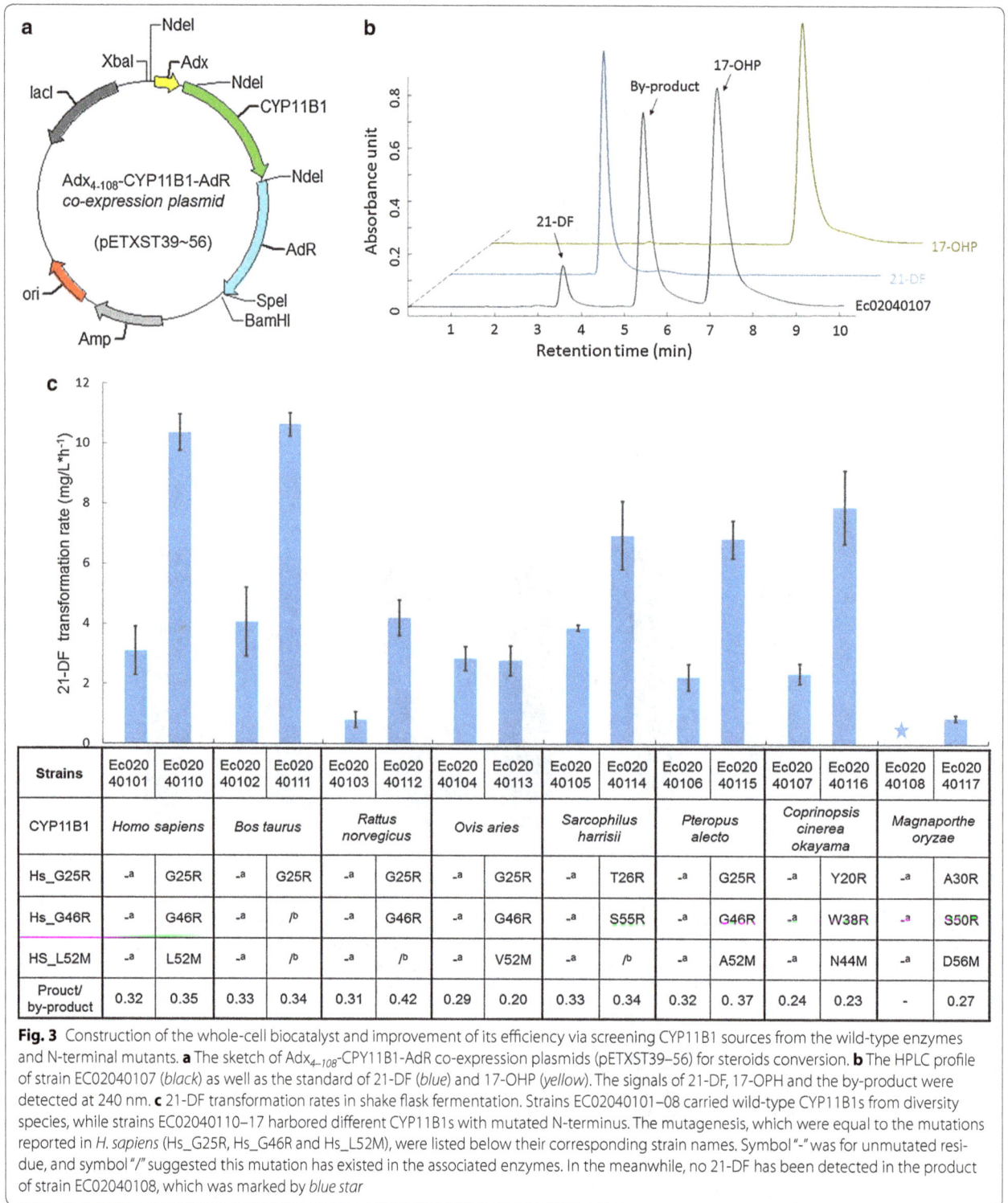

Fig. 3 Construction of the whole-cell biocatalyst and improvement of its efficiency via screening CYP11B1 sources from the wild-type enzymes and N-terminal mutants. **a** The sketch of Adx_{4-108}-CPY11B1-AdR co-expression plasmids (pETXST39–56) for steroids conversion. **b** The HPLC profile of strain EC02040107 (*black*) as well as the standard of 21-DF (*blue*) and 17-OHP (*yellow*). The signals of 21-DF, 17-OPH and the by-product were detected at 240 nm. **c** 21-DF transformation rates in shake flask fermentation. Strains EC02040101–08 carried wild-type CYP11B1s from diversity species, while strains EC02040110–17 harbored different CYP11B1s with mutated N-terminus. The mutagenesis, which were equal to the mutations reported in *H. sapiens* (Hs_G25R, Hs_G46R and Hs_L52M), were listed below their corresponding strain names. Symbol "-" was for unmutated residue, and symbol "/" suggested this mutation has existed in the associated enzymes. In the meanwhile, no 21-DF has been detected in the product of strain EC02040108, which was marked by *blue star*

remaining here, and one kind of unidentified by-product (5.5 min) accumulated dramatically in the extracts (Fig. 3b).

Improve biocatalyst efficiency via screening CYP11B1 sources, N-terminal mutagenesis and optimizing substrate concentration

Screening enzymes from diverse sources has been wildly applied and proved to be a promising strategy to overproduce desired compounds from single carbon sources [66–70]. Here, this method was also used to improve the conversion rate of our biocatalyst. Besides CYP11B1_Cco, eight other wild-type CYP11B1s from *H. sapiens* (CYP11B1_Hs), *Bos taurus* (CYP11B1_Bt), *Rattus norvegicus* (CYP11B1-_Rn), *Ovis aries* (CYP11B1_Oa), *Sarcophilus harrisii* (CYP11B1_Sh), *Pteropus alecto* (CYP11B1_Pa), *Magnaporthe oryzae* (CYP11B1_Mo) and *Sorex araneus* (CYP11B1_Sa) were selected to construct the whole-cell system accordingly, obtaining strains Ec02040101–09. And N-terminus modification (drastic truncation or site-directed mutagenesis) in P450s can impact their catalytic activity as well as expression level [71]. Schiffer et al. [11] once reported that mutagenesis at G23R of CYP11B1_Hs increased its expression level as well as maintained its catalytic activity for the reaction from 11-deoxycortisol to cortisol in *E. coli*. It was also reported mutations G46R [11, 72] and L52M [46] could enhance the activity of CYP11B1_Hs. Since mutations effected in one enzyme can be transposed to its homologous proteins [40], therefore these mutations were also introduced into all the CYP11B1s selected in this study. By aligning the protein sequences of these enzymes, the amino acid residues equal to G23, G46 and L52 within CYP11B1_Hs were identified in all the selected CYP11B1 (Additional file 1: Figure S3a) and then were mutated accordingly, obtaining strains Ec02040110–18. Strains Ec02040101–09 containing the wild-type CYP11B1s and strains Ec02040110–18 harboring the mutated CYP11B1s from diverse species were cultured in shake flask to determine the corresponding transformation rate for 21-DF. As a result, the introduced mutations significantly increased the transformation rates except ones in CYP11B1_Oa (Fig. 3c). And stains Ec02040110 and Ec02040111 processing mutated CYP11B1_Hs and CYP11B1_Bt achieved the highest transformation rate at 10.36 and 10.63 mg/L/h, respectively. Because CYP11B1_Hs mutant G25R/G46R/L52M (CYP11B1_Hs mutant1) achieved a little higher product/by-product ratio (0.35) than mutated CYP11B1_Bt (0.34), CYP11B1_Hs mutant1 was employed for further study.

For the reaction from 11-deoxycortisol to cortisol, Schiffer et al. [11] once reported that increasing substrate concentration led to enhancement on CYP11B1 activity.

Here, the effect of substrate concentration on our biotransformation (from 17-OHP to 21-DF) was also characterized under different 17-OHP concentrations from 0.72 to 7.2 g/L (Additional file 1: Figure S5a). As shown in Additional file 1: Figure S5b, higher 17-OHP concentrations resulted in firstly increases and then decreases on the maximal conversion rate. And the highest transformation rate of 16.2 mg/L/h was achieved under the 17-OHP concentration at 2.4 g/L (Additional file 1: Figure S5b). This concentration was then applied in the following optimization process.

Improve the product specificity via regulating the relative position between the catalytic active site heme and the substrate

In this study, regardless of CYP11B1 species, there was always an unidentified by-product accumulated that was much more than our desired product 21-DF in the fermentation (Additional file 1: Figure S4). For instance, the product/by-product ratio of the control strain Ec02040110 with CYP11B1_Hs mutant1 was 0.35 (Fig. 3c). In order to improve the product specificity of CYP11B1, the structural model of CYP11B1_Hs mutant1 with 17-OHP was generated and analyzed. As shown in Fig. 4a, the Fe ion of heme in the catalytic active site of CYP11B1 adjoined C11 site of the substrate with a long distance of 4.13 Å, which implied the weak catalytic activity for hydroxylation on C11 site of 17-OHP. That would result in poor product conversion or selectivity of 11-hydroxylation, which was supported by the low product/by-products ratio achieved in our study (Fig. 3c). In order to improve the catalytic activity and selectivity of 11-hydroxylation, the relative position between the catalytic active site heme and the substrate was directionally regulated via designing and introducing two kinds of site-directed mutations (Type I and II) in the activity pocket of CYP11B1, according to the structures of enzyme-substrate complex (Fig. 4).

Type I mutagenesis was proposed to move the Fe ion of heme close to C11 site in the direction as the orange arrow indicated in Fig. 4a. The structural analysis revealed that R110 as well as R384 were essential for structural stability and catalysis function of heme (Fig. 4b). These two residues were also highly conserved among the nine selected CYP11B1s (Additional file 1: Figure S3b). When R110 and R384 were separately mutated into alanine within CYP11B1_Hs mutant1 in the control strain Ec02040110, the corresponding transformation rate for 21-DF decreased by 100 and 90.1%, respectively (Fig. 4d), indicating that weakening the interaction with the carboxyl of heme would decay the catalytic activity of CYP11B1 for generating the targeted product 21-DF. By contrast, if L382 was substituted by the residue with

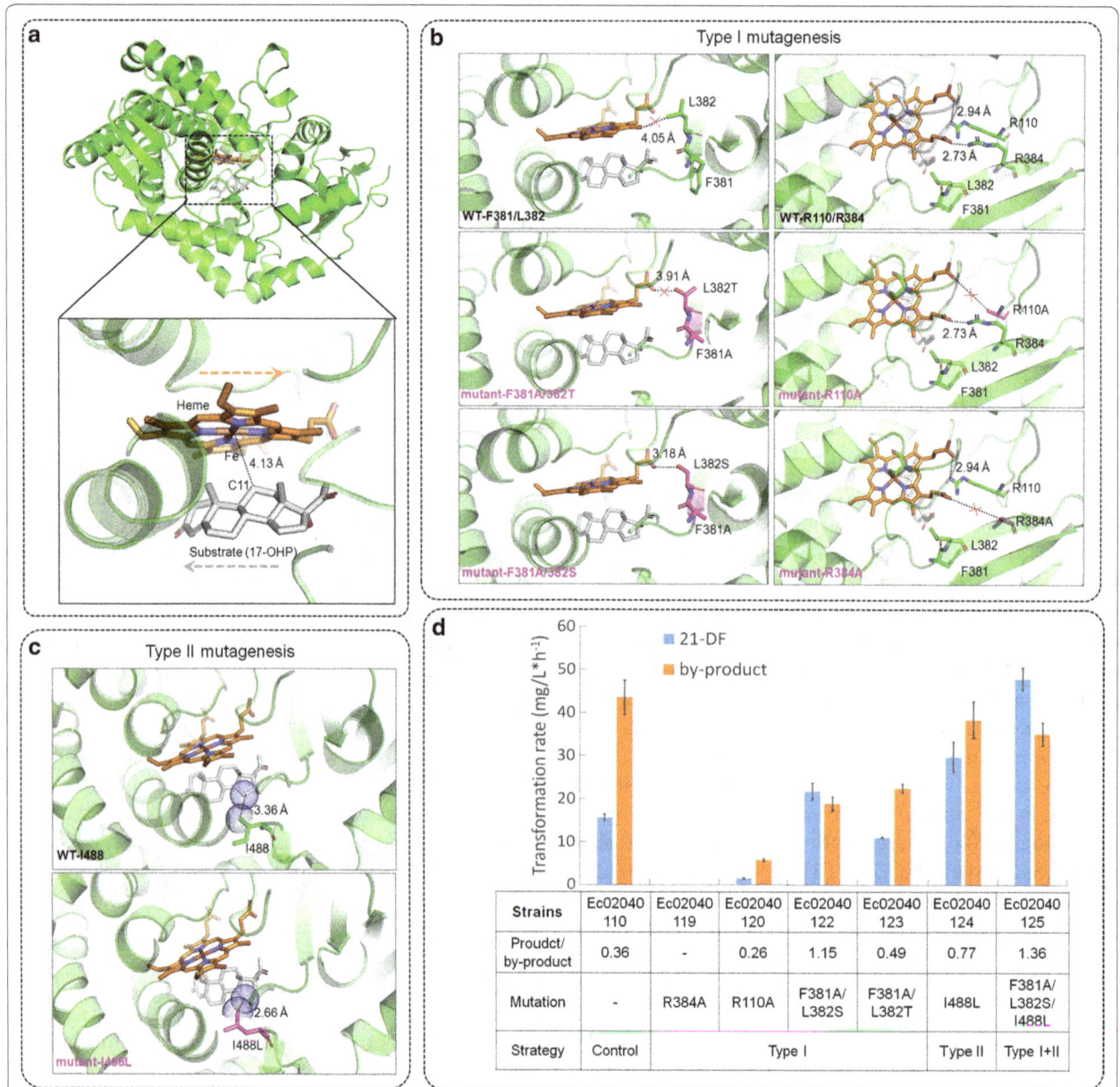

Fig. 4 Improve the targeted product specificity by mutagenesis to adjust the relative position between the catalytic active site heme and the substrate. Heme in the catalytic active site of CYP11B1 was represented by *orange*, while the substrate 17-OHP was represented by *grey*. All the mutated residues were highlighted in *purple*. **a** The structural model of CYP11B1_Hs mutant1/17-OHP complex. The *orange* and *grey arrows* indicated the moving direction of the heme and the substrate, respectively. **b** The strategy for type I mutagenesis including F381A/L382S, F381A/L382T, R110A and R384A. **c** The strategy for type II mutagenesis as I488L. **d** The transformation rate of 21-DF and product/by-product ratio achieved by the designed mutagenesis

hydroxyl group in the side chain (i.e. serine or threonine), it would provide additional hydrogen bond interaction with the carboxyl of heme to draw the heme close to the C11 side of the substrate. Besides this mutagenesis site L382, mutation F381A was also introduced to enhance the effect of mutation L382 by the flexibility of alanine. As a result, compared with the data from the control strain, mutant F381A/L382S improved the product/by-product ratio and the transformation rate by 219 and by 39.1%, respectively (Fig. 4d). This outcome confirmed our design on type I mutagenesis which mainly improve the specificity of CYP11B1 to our targeted product. However, mutant F381A/L382T enhanced the product/by-product ratio only by 36.1%, and decreased the transformation rate for 21-DF by 28.8% (Fig. 4d). This unexpected result might due to that the methyl in the side chain of residue T382 forced the orientation of its adjacent hydroxyl group to be opposite to our desired direction. Under the circumstances, no additional interaction with heme was formed and this mutation probably damaged the functional structure of CYP11B1 (Fig. 4b). Thus, only mutant F381A/L382S could be applied latter.

Type II mutagenesis aimed to strengthen the substrate binding of CYP11B1 as well as to move the C11 site of substrate toward the Fe ion of heme in the direction as the grey arrow indicated in Fig. 4a. The complex structural model demonstrated that the side chain in residue I488 formed a hydrophobic force with the substrate (Fig. 4c). If I488 was substituted by a residue with larger hydrophobic group in the side chain (such as leucine), the distance between this residue and the substrate was shortened (e.g. from 3.36 to 2.66 Å for I488L), leading to an enhanced hydrophobic force which might improve substrate binding and slightly shift C11 site of 17-OHP towards the Fe ion of heme (Fig. 4c). As illustrated in Fig. 4d, compared with control strain Ec02040110, mutant I488L improved the transformation rate and the product/by-product ratio for 21-DF by 91.0 and by 114%, respectively. These data corroborated type II mutagenesis that improved the target product specificity as well as the catalytic activity of CYP11B1. Moreover, substrate binding to a P450 usually results in the binding spectra occurring from a shift from low-spin FeIII to high-spin FeIII upon binding of the substrate [73, 74]. Therefore, in order to confirm that mutant I488L improved the substrate binding of CYP11B1 besides the bioinformatical data, it is necessary to determine the binding spectra of CYP11B1 to 17-OHP as well as to compare the KS value of the mutant to the wild-type enzyme in future study.

Eventually, the mutations exhibiting positive effect (i.e. F381A/L382S from type I mutagenesis and I488L from type II mutagenesis) were all incorporated into the CYP11B1_Hs mutant1, obtaining CYP11B1_Hs mutant2. As a result, 1.96-fold enhancement on 21-DF transformation rate (to 47.9 mg/L/h) and 2.78-fold improvement on product/by-product ratio (to 1.36) were achieved in strain Ec02040125 harboring CYP11B1_Hs mutant2 (Fig. 4d). After 38-h biotransformation in shake-flask, the production of 21-DF reached to 1.42 g/L with a yield of 52.7%, which was the highest 21-DF production as known.

Conclusions

In our study, *E. coli* MG1655(DE3) exhibited the highest up-taking rate of 17-OHP than other tested host cells. Then the biotransformation of 17-OHP to 21-DF was successfully achieved by establishing a whole-cell catalyst via incorporating heterologous CYP11B1 along with bovine Adx and AdR in *E. coli* MG1655(DE3). The effects of CYP11B1 with N-terminal mutagenesis from diverse sources and substrate concentration were investigated to stepwise increase 21-DF transformation rate by 343 and 56.4%, respectively. To be noted, in order to improve the catalytic activity as well as the product specificity by adjusting the relative position between the substrate 17-OHP and the catalytic active site heme in CYP11B1, two types of novel mutations were designed based on structural analysis of the enzyme-substrate complex. 1.96-fold enhancement on 21-DF transformation rate (to 47.9 mg/L/h) and 2.78-fold improvement on product/by-product ratio (to 1.36) were further achieved by the combination of two types of mutations. Eventually, the highest reported 21-DF production (1.42 g/L) along with a yield of 52.7% was obtained in strain Ec02040125 harboring CYP11B1_Hs mutant2 (G25R/G46R/L52M/F381A/L382S/I488L). This study offers a good example of whole cell biocatalyst system for efficient steroids biotransformation. It also provides important insights that can guide the audiences to regulate catalytic acidity as well as product specificity of other P450s.

Additional file

Additional file 1: Table S1. Oligonucleotides used in this study. **Table S2.** Plasmids used in this study. **Table S3.** The Codon-optimized sequences of *CY11B1, Adx* and *AdR* involved in this study. **Figure S1.** Schematic representation of the constructing strategies for Adx$_{4-108}$-CYP11B1-AdR co-expression plasmid (pETXST39–56). **Figure S2.** Schematic representation of the mutagenesis strategy applied in this study. **Figure S3.** Alignment of CYP11B1 s from nine species screened in this study. **Figure S4.** Conversion of 17-OHP to 21-DF by biocatalysts harboring different wild-type and mutated CYP11B1s from diversity species. **Figure S5.** Optimizing substrate concentration for higher biocatalyst efficiency. **Figure S6.** Biotransformation of 17-OHP to 21-DF by strains Ec02040126 and Ec02040110 during the time course.

Abbreviations
21-DF: 21-deoxycortisol; 17-OHP: 17α-hydroxyprdgesterone; CYP11B1: 11β-hydroxylase; AdR: adrenodoxin reductase; Adx: adrenodoxin; Ho: *Homo sapiens*; Bt: *Bos taurus*; Oa: *Ovis aries*; Sh: *Sarcophilus harrisii*; Pa: *Pteropus Alecto*; Cco: *Coprinopsis cinerea okayama*; Mo: *Magnaporthe oryzae*, Sa: *Sorex araneus*.

Authors' contributions
SX, YW, WX and YY conceived of the study. YW and WX participated in design and coordination of the study. SX carried out the strain construction. SX and WX carried out the whole-cell biocatalysis experiments. MY carried out the protein analysis. HL and XZ participated in HPLC analysis. YW and MY helped to draft the manuscript. WX supervised the whole research and revised the manuscript. All authors read and approved the final manuscript.

Author details
¹ Key Laboratory of Systems Bioengineering (Ministry of Education), Tianjin University, No. 92, Weijin Road, Nankai District, Tianjin 300072, People's Republic of China. ² SynBio Research Platform, Collaborative Innovation Center of Chemical Science and Engineering (Tianjin), School of Chemical Engineering and Technology, Tianjin University, Tianjin 300072, People's Republic of China.

Acknowledgements

The authors are grateful for Prof. Kristala L. Jones Prather from MIT, USA to kindly provided *E. coli* strain MG1655(DE3). This study was supported by the financial support from the International S&T Cooperation Program of China (2015DFA00960), the National Natural Science Foundation of China (21621004, 21390203 and 31570088), the Ministry of Science and Technology of China ("973" Program: 2014CB745100) and Innovative Talents and Platform Program of Tianjin (16PTSYJC00050).

Competing interests

The authors declare that they have no competing interests.

Funding

The International S&T Cooperation Program of China (2015DFA00960), the National Natural Science Foundation of China (21621004, 21390203 and 31570088), the Ministry of Science and Technology of China ("973" Program: 2014CB745100) and Innovative Talents and Platform Program of Tianjin (16PTSYJC00050).

References

1. Garcia JL, Uhia I, Galan B. Catabolism and biotechnological applications of cholesterol degrading bacteria. Microb Biotechnol. 2012;5:679–99.
2. Alessi P, Cortesi A, Kikic I, et al. Particle production of steroid drugs using supercritical fluid processing. Ind Eng Chem Res. 1996;35:4718–26.
3. Flickinger MC. Encyclopedia of industrial biotechnology: bioprocess, bioseparation, and cell technology. New York: Wiley; 2010.
4. Robinson T, Singh D, Nigam P. Solid-state fermentation: a promising microbial technology for secondary metabolite production. Appl Microbiol Biotechnol. 2001;55:284–9.
5. Fernandes P, Cruz A, Angelova B, et al. Microbial conversion of steroid compounds: recent developments. Enzyme Microb Tech. 2003;32:688–705.
6. Brixius-Anderko S, Schiffer L, Hannemann F, et al. A CYP21A2 based whole-cell system in *Escherichia coli* for the biotechnological production of premedrol. Microb Cell Fact. 2015;14:135.
7. Fang WM, Tang SH. The method to prepare hydrocortisone. CN, CN201110201729.1; 2012.
8. Ogawa J, Takatori k. Production method of 11-beta hydroxysteroid via a novel steroid 11-beta hydroxylase. JP, JP/2006/124179; 2006.
9. Ringle M, Khatri Y, Zapp J, et al. Application of a new versatile electron transfer system for cytochrome P450-based *Escherichia coli* whole-cell bioconversions. Appl Microbiol Biotechnol. 2013;97:7741–54.
10. Cook DJ, Finnigan JD, Cook K, et al. Chapter five—cytochromes p450: history, classes, catalytic mechanism, and industrial application. Adv Protein Chem Struct. 2016;105:105–26.
11. Schiffer L, Anderko S, Hobler A, et al. A recombinant CYP11B1 dependent *Escherichia coli* biocatalyst for selective cortisol production and optimization towards a preparative scale. Microb Cell Fact. 2015;14:25.
12. Li A, Ilie A, Sun Z, et al. Whole-cell-catalyzed multiple regio- and stereoselective functionalizations in cascade reactions enabled by directed evolution. Angew Chem Int Ed Engl. 2016;55:12026–9.
13. Hu S, Huang J, Mei L, et al. Altering the regioselectivity of cytochrome P450 BM-3 by saturation mutagenesis for the biosynthesis of indirubin. J Mol Catal B-Enzym. 2010;67:29–35.
14. Bodkin MJ. Why don't we see a greater uptake of computational chemistry approaches by the medicinal chemistry community? Future Med Chem. 2012;4:1889–91.
15. Urlacher VB, Girhard M. Cytochrome P450 monooxygenases: an update on perspectives for synthetic application. Trends Biotechnol. 2006;30:26.
16. Sakaki T. Practical application of cytochrome P450. Biol Pharm Bull. 2012;35:844–9.
17. Zehentgruber D, Dragan CA, Bureik M, et al. Challenges of steroid biotransformation with human cytochrome P450 monooxygenase CYP21 using resting cells of recombinant *Schizosaccharomyces* pombe. J Biotechnol. 2010;146:179–85.
18. Kille S, Zilly FE, Acevedo JP, et al. Regio- and stereoselectivity of P450-catalysed hydroxylation of steroids controlled by laboratory evolution. Nat Chem. 2011;3:738–43.
19. Makino T, Katsuyama Y, Otomatsu T, et al. Regio- and stereospecific hydroxylation of various steroids at the 16α position of the D ring by the *Streptomyces griseus* cytochrome P450 CYP154C3. Appl Environ Microb. 2014;80:1371–9.
20. Bracco P, Janssen DB, Schallmey A. Selective steroid oxyfunctionalisation by CYP154C5, a bacterial cytochrome P450. Microb Cell Fact. 2013;12:95.
21. Ichinose H, Wariishi H. High-level heterologous expression of fungal cytochrome P450s in *Escherichia coli*. Biochem Biophys Res Commun. 2013;438:289–94.
22. Ichinose H, Hatakeyama M, Yamauchi Y. Sequence modifications and heterologous expression of eukaryotic cytochromes P450 in *Escherichia coli*. J Biosci Bioeng. 2015;120:268–74.
23. Carmichael AB, Wong LL. Protein engineering of *Bacillus megaterium* CYP102. The oxidation of polycyclic aromatic hydrocarbons. Eur J Biochem. 2001;268:3117–25.
24. Ichinose H, Wariishi H. Heterologous expression and mechanistic investigation of a fungal cytochrome P450 (CYP5150A2): involvement of alternative redox partners. Arch Biochem Biophys. 2012;518:8–15.
25. Pan Y, Abd-Rashid BA, Ismail Z, et al. Heterologous expression of human cytochromes P450 2D6 and CYP3A4 in *Escherichia coli* and their functional characterization. Protein J. 2011;30:581–91.
26. Jung ST, Lauchli R, Arnold FH. Cytochrome P450: taming a wild type enzyme. Curr Opin Biotechnol. 2011;22:809–17.
27. Bommarius AS, Blum JK, Abrahamson MJ. Status of protein engineering for biocatalysts: how to design an industrially useful biocatalyst. Curr Opin Chem Biol. 2011;15:194–200.
28. O'Reilly E, Koehler V, Flitsch SL, et al. Cytochromes P450 as useful biocatalysts: addressing the limitations. Chem Commun (Camb). 2011;47:2490–501.
29. Roiban GD, Reetz MT. Expanding the toolbox of organic chemists: directed evolution of P450 monooxygenases as catalysts in regio- and stereoselective oxidative hydroxylation. Chem Commun (Camb). 2014;46:2208–24.
30. Meunier B, de Visser SP, Shaik S. Mechanism of oxidation reactions catalyzed by cytochrome p450 enzymes. Chem Rev. 2004;104:3947–80.
31. Rea V, Kolkman AJ, Vottero E, et al. Active site substitution A82W improves the regioselectivity of steroid hydroxylation by cytochrome P450 BM3 mutants as rationalized by spin relaxation nuclear magnetic resonance studies. Biochemistry. 2012;51:750–60.
32. Cha GS, Ryu SH, Ahn T, et al. Regioselective hydroxylation of 17beta-estradiol by mutants of CYP102A1 from *Bacillus megaterium*. Biotechnol Lett. 2014;36:2501–6.
33. Gumulya Y, Sanchis J, Reetz MT. Many pathways in laboratory evolution can lead to improved enzymes: how to escape from local minima. ChemBioChem. 2012;13:1060.
34. Agudo R, Roiban GD, Reetz MT. Achieving regio- and enantioselectivity of P450-catalyzed oxidative CH activation of small functionalized molecules by structure-guided directed evolution. ChemBioChem. 2012;13:1465–73.
35. Hunter DJ, Behrendorff JB, Johnston WA, et al. Facile production of minor metabolites for drug development using a CYP3A shuffled library. Metab Eng. 2011;13:682–93.
36. Edgar S, Li FS, Qiao K, et al. Engineering of taxadiene synthase for improved selectivity and yield of a key taxol biosynthetic intermediate. ACS Synth Biol. 2017;6:201–5.

37. Nguyen KT, Virus C, Gunnewich N, et al. Changing the regioselectivity of a P450 from C15 to C11 hydroxylation of progesterone. ChemBioChem. 2012;13:1161–6.

38. Zhang K, Shafer BM, Demars MD, et al. Controlled oxidation of remote sp(3) C–H bonds in artemisinin via P450 catalysts with fine-tuned regio- and stereoselectivity. J Am Chem Soc. 2012;134:18695–704.

39. Roiban GD, Reetz MT. Expanding the toolbox of organic chemists: directed evolution of P450 monooxygenases as catalysts in regio- and stereoselective oxidative hydroxylation. Chem Commun (Camb). 2015;51:2208–24.

40. Behrendorff JB, Huang W, Gillam EM. Directed evolution of cytochrome P450 enzymes for biocatalysis: exploiting the catalytic versatility of enzymes with relaxed substrate specificity. Biochem J. 2015;467:1–15.

41. Zorn K, Oroz-Guinea I, Brundiek H, et al. Engineering and application of enzymes for lipid modification, an update. Prog Lipid Res. 2016;63:153–64.

42. Dietrich M, Do TA, Schmid RD, et al. Altering the regioselectivity of the subterminal fatty acid hydroxylase P450 BM-3 towards gamma- and delta-positions. J Biotechnol. 2009;139:115–7.

43. Seifert A, Antonovici M, Hauer B, et al. An efficient route to selective bio-oxidation catalysts: an iterative approach comprising modeling, diversification, and screening, based on CYP102A1. ChemBioChem. 2011;12:1346–51.

44. Schiffer L, Muller AR, Hobler A, et al. Biotransformation of the mineralocorticoid receptor antagonists spironolactone and canrenone by human CYP11B1 and CYP11B2: characterization of the products and their influence on mineralocorticoid receptor transactivation. J Steroid Biochem Mol Biol. 2016;163:68–76.

45. Szczebara FM, Chandelier C, Villeret C, et al. Total biosynthesis of hydrocortisone from a simple carbon source in yeast. Nat Biotechnol. 2003;21:143–9.

46. Hakki T, Zearo S, Dragan CA, et al. Coexpression of redox partners increases the hydrocortisone (cortisol) production efficiency in CYP11B1 expressing fission yeast Schizosaccharomyces pombe. J Biotechnol. 2008;133:351–9.

47. Miroux B, Walker JE. Over-production of proteins in Escherichia coli: mutant hosts that allow synthesis of some membrane proteins and globular proteins at high levels. J Mol Biol. 1996;260:289–98.

48. Tseng HC, Harwell CL, Martin CH, et al. Biosynthesis of chiral 3-hydroxyvalerate from single propionate-unrelated carbon sources in metabolically engineered E. coli. Microb Cell Fact. 2010;9:96.

49. Papanikolaou S, Muniglia L, Chevalot I, et al. Yarrowia lipolytica as a potential producer of citric acid from raw glycerol. J Appl Microbiol. 2002;92:737–44.

50. Snapper SB, Melton RE, Mustafa S, et al. Isolation and characterization of efficient plasmid transformation mutants of Mycobacterium smegmatis. Mol Microbiol. 1990;4:1911–9.

51. Su W, Xiao WH, Wang Y, et al. Alleviating Redox imbalance enhances 7-dehydrocholesterol production in engineered Saccharomyces cerevisiae. PLoS ONE. 2015;10:e0130840.

52. Kendall SL, Burgess P, Balhana R, et al. Cholesterol utilization in mycobacteria is controlled by two TetR-type transcriptional regulators: KstR and KstR2. Microbiology. 2010;156:1362–71.

53. Cao YX, Xiao WH, Liu D, et al. Biosynthesis of odd-chain fatty alcohols in Escherichia coli. Metab Eng. 2015;29:113–23.

54. Shetty RP, Endy D, Knight TF Jr. Engineering BioBrick vectors from BioBrick parts. J Biol Eng. 2008;2:5.

55. Hayer-Hartl M, Bracher A, Hartl FU. The GroEL-GroES chaperonin machine: a nano-cage for protein folding. Trends Biochem Sci. 2016;41:62–76.

56. Thompson JD, Higgins DG, Gibson TJ. CLUSTAL W: improving the sensitivity of progressive multiple sequence alignment through sequence weighting, position-specific gap penalties and weight matrix choice. Nucleic Acids Res. 1994;22:4673–80.

57. Trott O, Olson AJ. AutoDock Vina: improving the speed and accuracy of docking with a new scoring function, efficient optimization, and multithreading. J Comput Chem. 2010;31:455–61.

58. Irwin JJ, Sterling T, Mysinger MM, et al. ZINC: a free tool to discover chemistry for biology. J Chem Inf Model. 2012;52:1757–68.

59. Delano WL. The PyMOL molecular graphics system. San Carlos: DeLano Scientific; 2002.

60. Bureik M, Lisurek M, Bernhardt R. The human steroid hydroxylases CYP1B1 and CYP11B2. Biol Chem. 2002;383:1537–51.

61. Du HX, Xiao WH, Wang Y, et al. Engineering Yarrowia lipolytica for campesterol overproduction. PLoS ONE. 2016;11:e0146773.

62. Ewen KM, Ringle M, Bernhardt R. Adrenodoxin—a versatile ferredoxin. IUBMB Life. 2012;64:506–12.

63. Nishihara K, Kanemori M, Kitagawa M, et al. Chaperone coexpression plasmids: differential and synergistic roles of DnaK-DnaJ-GrpE and GroEL-GroES in assisting folding of an allergen of Japanese cedar pollen, Cryj2, in Escherichia coli. Appl Environ Microbiol. 1998;64:1694–9.

64. Yanofsky C, Horn V, Gollnick P. Physiological studies of tryptophan transport and tryptophanase operon induction in Escherichia coli. J Bacteriol. 1991;173:6009–17.

65. Brixius-Anderko S, Hannemann F, Ringle M, et al. An indole-deficient Escherichia coli strain improves screening of cytochromes P450 for biotechnological applications. Biotechnol Appl Biochem. 2016. doi: 10.1002/bab.1488. [Epub ahead of print].

66. Wang R, Gu X, Yao M, et al. Engineering of β-carotene hydroxylase and ketolase for astaxanthin overproduction in Saccharomyces cerevisiae. Front Chem Sci Eng. 2017;11:89–99.

67. Chen Y, Xiao W, Wang Y, et al. Lycopene overproduction in Saccharomyces cerevisiae through combining pathway engineering with host engineering. Microb Cell Fact. 2016;15:113.

68. Chai F, Wang Y, Mei X, et al. Heterologous biosynthesis and manipulation of crocetin in Saccharomyces cerevisiae. Microb Cell Fact. 2017;16:54.

69. Jiang GZ, Yao MD, Wang Y, et al. Manipulation of GES and ERG20 for geraniol overproduction in Saccharomyces cerevisiae. Metab Eng. 2017;41:57–66.

70. Zhang Y, Wang Y, Yao M, et al. Improved campesterol production in engineered Yarrowia lipolytica strains. Biotechnol Lett. 2017. doi: 10.1007/s10529-017-2331-4. [Epub ahead of print].

71. Zelasko S, Palaria A, Das A. Optimizations to achieve high-level expression of cytochrome P450 proteins using Escherichia coli expression systems. Protein Expr Purif. 2013;92:77–87.

72. Kagawa N, Cao Q, Kusano K. Expression of human aromatase (CYP19) in Escherichia coli by N-terminal replacement and induction of cold stress response. Steroids. 2003;68:205–9.

73. Isin EM, Guengerich FP. Substrate binding to cytochromes P450. Anal Bioanal Chem. 2008;392:1019–30.

74. Schenkman JB, Sligar SG, Cinti DL. Substrate interaction with cytochrome P-450. Pharmacol Ther. 1981;12:43–71.

9

Reversible bacterial immobilization based on the salt-dependent adhesion of the bacterionanofiber protein AtaA

Shogo Yoshimoto, Yuki Ohara, Hajime Nakatani and Katsutoshi Hori* ⓘ

Abstract

Background: Immobilization of microbial cells is an important strategy for the efficient use of whole-cell catalysts because it simplifies product separation, enables the cell concentration to be increased, stabilizes enzymatic activity, and permits repeated or continuous biocatalyst use. However, conventional immobilization methods have practical limitations, such as limited mass transfer in the inner part of a gel, gel fragility, cell leakage from the support matrix, and adverse effects on cell viability and catalytic activity. We previously showed a new method for bacterial cell immobilization using AtaA, a member of the trimeric autotransporter adhesin family found in *Acinetobacter* sp. Tol 5. This approach is expected to solve the drawbacks of conventional immobilization methods. However, similar to all other immobilization methods, the use of support materials increases the cost of bioprocesses and subsequent waste materials.

Results: We found that the stickiness of the AtaA molecule isolated from Tol 5 cells is drastically diminished at ionic strengths lower than 10 mM and that it cannot adhere in deionized water, which also inhibits cell adhesion mediated by AtaA. Cells immobilized on well plates and polyurethane foam in a salt solution were detached in deionized water by rinsing and shaking, respectively. The detached cells regained their adhesiveness in a salt solution and could rapidly be re-immobilized. The cells expressing the *ataA* gene maintained their adhesiveness throughout four repeated immobilization and detachment cycles and could be repeatedly immobilized to polyurethane foam by a 10-min shake in a flask. We also demonstrated that both bacterial cells and a support used in a reaction could be reused for a different type of reaction after detachment of the initially immobilized cells from the support and a subsequent immobilization step.

Conclusions: We invented a unique reversible immobilization method based on the salt-dependent adhesion of the AtaA molecule that allows us to reuse bacterial cells and supports by a simple manipulation involving a deionized water wash. This mitigates problems caused by the use of support materials and greatly helps to enhance the efficiency and productivity of microbial production processes.

Keywords: Immobilization, Whole-cell catalyst, Trimeric autotransporter adhesin, *Acinetobacter*

Background

Although microbial cells are expected to provide environmentally friendly production processes as whole-cell biocatalysts with highly effective and selective reactivity under ordinary temperatures and pressures [1–5], their

disadvantages, such as complicated handling due to their fragility, a requirement for costly product separation processes, and easy inactivation at unsuitable temperatures, pH, and substrate and product concentrations, raise production costs and hinder their commercial use in chemical production processes. To expand the use of microbial production processes in the industrial sector, it is important to improve the efficiency of the entire process, from an upstream process including strain development to a

*Correspondence: khori@chembio.nagoya-u.ac.jp
Department of Biomolecular Engineering, Graduate School of Engineering, Nagoya University, Furo-cho, Chikusa-ku, Nagoya 464-8603, Japan

downstream process including product separation [6]. Recently, a systems metabolic engineering approach targeting an upstream process received considerable attention by aiming to develop a novel biosynthetic pathway producing high-value products and/or improve their productivity in microbial cells [6–11]. As for the downstream process, cell immobilization is important because it simplifies product separation, enables the cell concentration to be increased, stabilizes the enzymatic activity, and permits repetitive or continuous use of precious and expensive biocatalysts [12–15]. Conventional methods for cell immobilization are gel entrapment, covalent bonding to solid surfaces, cross-linkage, and physical adsorption [16, 17]. These methods, however, have practical limitations, such as limited mass transfer in the inner part of a gel [18, 19], gel fragility, cell leakage from the support matrix, and adverse effects on cell viability and catalytic activity [12].

We previously invented a method for bacterial cell immobilization using the adhesive protein AtaA found in *Acinetobacter* sp. Tol 5 [20–22], which belongs to the trimeric autotransporter adhesin (TAA) family [23]. Although AtaA shares a fibrous architecture consisting of an N-terminus—passenger domain (PSD) containing head and stalk domains—transmembrane anchor (TM)—C-terminus with TAA family members [24], which usually bind to target biotic surfaces, AtaA uniquely confers nonspecific high adhesiveness to both abiotic and biotic surfaces on bacterial cells transformed with its gene. Large amounts of growing, resting, even lyophilized transformant cells can be quickly and firmly immobilized onto any material surfaces selected according to the application [25]. Cells immobilized directly on surfaces through AtaA are not embedded in extracellular polymeric substances with mass transfer limitations, show enhanced tolerance [22], increase chemical reaction rates, and can be repeatedly used in reactions without inactivation [25]. However, similar to all other immobilization methods, the use of support materials increases the cost of bioprocesses and subsequent waste materials. These might be inevitable problems as long as support materials are used in the immobilization process. A way to minimize these drawbacks should be developed so as to, for example, reduce the amount of support materials, use inexpensive materials or waste materials, and reuse support materials.

AtaA is a homotrimer of polypeptides comprising 3630 amino acids. In a previous study, we developed a method to isolate its PSD, which is secreted to the bacterial cell surface through the TM and is responsible for biological functions, by genetically introducing a recognition site for human rhinovirus 3C (HRV 3C) protease [26]. Specific cleavage by the protease reaps AtaA PSD

nanofibers 225 nm in length from the cell surface. This enables biochemical and biophysical analyses of the purified huge AtaA PSD in the native molecular state. Here, we demonstrate a new phenomenon: AtaA PSD cannot adhere to surfaces in deionized water (dH$_2$O). Based on this molecular property of AtaA, we developed a unique method for the reversible immobilization of bacterial cells, which can solve the problems caused by the use of support materials.

Results

Effect of ionic strength on the adhesive property of the AtaA molecule

To investigate the adhesive property of the AtaA molecule, AtaA PSD was isolated by the enzymatic reaping method from a Tol 5 derivative strain, 4140, transformed with p3CAtaA [26, 27]. KCl solutions (50 µL) of various concentrations containing 5 µg/mL of the purified AtaA PSD were incubated in 96-well polystyrene (PS) and glass plates at 28 °C for 2 h. The protein solution was removed from each well using a micropipette and the well was rinsed three times with 200 µL of phosphate-buffered saline containing 0.05% Tween-20 (PBS-T). The AtaA PSD adsorbed to the well plates was assessed by ELISA. Interestingly, we found that the amounts of AtaA PSD molecules adsorbed onto surfaces of hydrophobic PS and hydrophilic glass dropped sharply at ionic strengths lower than 10 mM, with the molecules hardly adhering in dH$_2$O, despite their high adhesiveness at higher ionic strengths (Fig. 1a). AtaA PSD cannot be considered to be denatured in dH$_2$O because AtaA PSD has high structural stability [26, 28]. Indeed, AtaA PSD molecules isolated from cells by the enzymatic reaping method mentioned above were dissolved in dH$_2$O, and subsequently KCl solution was added to the adherence assay to attain the final ionic strengths intended. Furthermore, by using a quartz crystal microbalance (QCM), which enables the quantification of molecules adhered to its quartz crystal sensor chip as a frequency shift, we confirmed that the adhesiveness of AtaA PSD can be recovered in a salt solution by adding KCl salt to fresh water. AtaA PSD molecules did not adhere to a gold-coated sensor chip of QCM in dH$_2$O, but started adhering to the chip immediately after KCl solution was added (Fig. 1b). Evidently, AtaA PSD is not denatured in dH$_2$O and recovers its adhesiveness in a salt solution.

Effect of the ionic strength on bacterial cell adhesion mediated by AtaA

The identification of the ionic strength-dependent stickiness of the AtaA PSD prompted us to examine whether or not bacterial cell adhesion mediated by AtaA also depends on ionic strength. Tol 5 cells were grown,

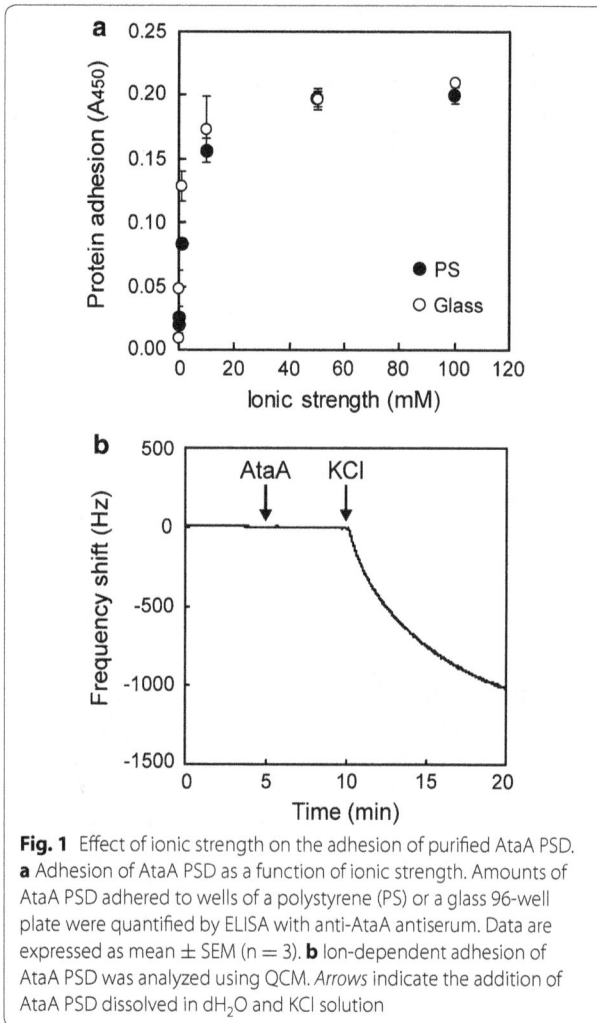

Fig. 1 Effect of ionic strength on the adhesion of purified AtaA PSD. **a** Adhesion of AtaA PSD as a function of ionic strength. Amounts of AtaA PSD adhered to wells of a polystyrene (PS) or a glass 96-well plate were quantified by ELISA with anti-AtaA antiserum. Data are expressed as mean ± SEM (n = 3). **b** Ion-dependent adhesion of AtaA PSD was analyzed using QCM. *Arrows* indicate the addition of AtaA PSD dissolved in dH$_2$O and KCl solution

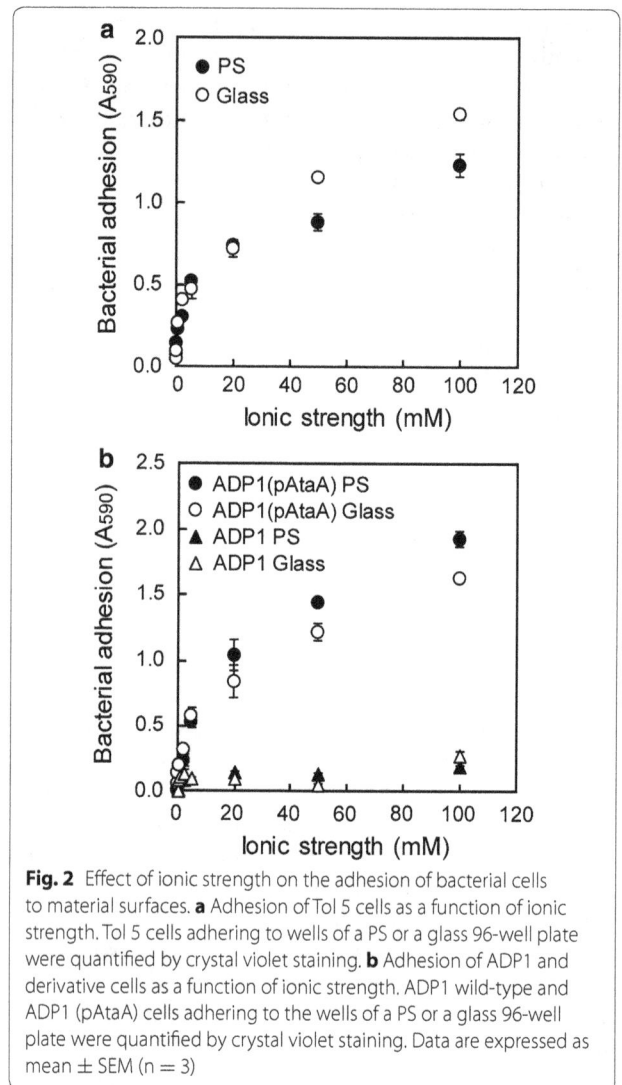

Fig. 2 Effect of ionic strength on the adhesion of bacterial cells to material surfaces. **a** Adhesion of Tol 5 cells as a function of ionic strength. Tol 5 cells adhering to wells of a PS or a glass 96-well plate were quantified by crystal violet staining. **b** Adhesion of ADP1 and derivative cells as a function of ionic strength. ADP1 wild-type and ADP1 (pAtaA) cells adhering to the wells of a PS or a glass 96-well plate were quantified by crystal violet staining. Data are expressed as mean ± SEM (n = 3)

harvested, washed with dH$_2$O, suspended in dH$_2$O and KCl solutions of various concentrations at an OD$_{660}$ of 0.5, and placed into 96-well plates. After a 2-h incubation at 28 °C without shaking, the cell suspension was removed from each well by a micropipette and the well was rinsed three times with 200 μL of dH$_2$O or KCl solution of each concentration using a micropipette. The cells immobilized onto the well surfaces were quantified by crystal violet staining. At ionic strengths higher than 20 mM, a large amount of Tol 5 cells was immobilized onto PS and glass surfaces, although the amount gradually increased as the ionic strength increased (Fig. 2a). However, Tol 5 cell adhesion dropped at ionic strengths lower than 5 mM for both PS and glass surfaces, and Tol 5 cells were unable to adhere to either surface in dH$_2$O. To confirm that such ionic strength-dependent adhesion of bacterial cells can be decisively attributed to the adhesive properties of the AtaA molecule, the cell adhesion

of *Acinetobacter baylyi* ADP1 and its transformant with *ataA*, ADP1 (pAtaA), to PS and glass surfaces was examined at various ionic strengths by the same procedure used for Tol 5 cells. ADP1 cells expressing AtaA showed high adhesiveness to both PS and glass surfaces at ionic strengths higher than 20 mM with a gradual increase in adhesion with ionic strength, whereas wild-type ADP1 was hardly immobilized at any ionic strength (Fig. 2b). However, at an ionic strength lower than 5 mM, even ADP1 cells expressing AtaA showed the same diminished adhesion as Tol 5 cells, adhering to neither PS surface nor glass surface in dH$_2$O. Therefore, the adhesion profiles of Tol 5 cells and ADP1 (pAtaA) cells at various ionic strengths directly reflect the properties of the AtaA molecule.

Cell detachment and reversible immobilization using AtaA

Our finding that the cell adhesion mediated by AtaA is inhibited by dH_2O prompted us to examine the ability of a simple dH_2O wash to detach bacterial cells already immobilized on material surfaces. After immobilization of ADP1 (pAtaA) cells onto the well surfaces in 100 mM KCl solution as described above, the wells were rinsed with 200 µL of dH_2O or 100 mM KCl solution using a micropipette. This washing step was repeated three times. Thereafter, the cells still immobilized on the well surfaces were quantified by crystal violet staining. Most of the cells washed with dH_2O were detached from both PS and glass surfaces, whereas the cells washed with 100 mM KCl solution were retained on the surfaces (Fig. 3).

We also confirmed the ability of dH_2O to detach ADP1 (pAtaA) cells previously immobilized on a polyurethane foam support, which is often used in bioprocesses as a support. The cells were immobilized onto 1-cm^3 pieces of polyurethane foam support. A piece of the support with the immobilized cells was transferred into fresh 100 mM KCl solution, gently rinsed, picked up with tweezers, and shaken in dH_2O or 100 mM KCl solution for video recording. When shaken in dH_2O, the immobilized cells immediately began to detach and an increase in the turbidity of the surrounding H_2O solution was observed. An additional movie file shows this in more detail (see Additional file 1). In contrast, the immobilized cells were not detached at all by being shaken in 100 mM KCl solution (see Additional file 2).

Next, we tried to repeat the immobilization and detachment of bacterial cells expressing AtaA. ADP1 (pAtaA) cells were repeatedly subjected to the immobilization/detachment process; cells suspended in 100 mM KCl solution were immobilized onto the PS well surface by a 2-h incubation (immobilization process) and subsequently detached by the dH_2O wash using the same procedure described above (detachment process). The detached cells were collected by centrifugation, resuspended in 100 mM KCl solution at an OD_{660} of 0.5, and placed into the new well for the next immobilization cycle. This immobilization/detachment process was repeated four times. As shown in Fig. 4a, the detached cells showed the same adhesion ability as the fresh cells used for the first adherence assay, and the cell adhesiveness did not decrease throughout four immobilization/detachment cycles. The cells finally detached were subjected to flow cytometry to quantify the amount of AtaA displayed on the ADP1 (pAtaA) cell surface. This revealed that the amount of AtaA molecules on the cell surface did not decrease even after the fourth detachment compared with that on fresh cells before the first immobilization (Fig. 4b), suggesting that AtaA was not impaired throughout the repeated detachment manipulations by washing with dH_2O. The ADP1 (pAtaA) cells immobilized in each immobilization/detachment cycle were subjected to an esterase activity assay involving the addition of a reaction buffer containing a substrate directly to the well. The cell-bound esterase activity of the immobilized cells on the PS surface was maintained at the same level throughout the four cycles (Fig. 4c), suggesting that the repeated immobilization/detachment cycle also did not deteriorate the integrity of the surface of ADP1 (pAtaA) cells.

We also examined whether or not ADP1 (pAtaA) cells can be reversibly immobilized onto polyurethane foam support. Six pieces of the polyurethane foam (a 1 cm cube) were placed into a 30-mL cell suspension of ADP1 (pAtaA) at an OD_{660} of 2.0 in a 100-mL Erlenmeyer flask and shaken at 115 rpm for immobilization of the bacterial cells. The percentage of immobilized cells over time is shown in Fig. 4d ("1st"). More than 90% of the cells were immobilized within 10 min (immobilization process). Subsequently, the supernatant was discarded by decantation and 30 mL of dH_2O was poured into the flask, which was then shaken at 115 rpm for 5 min. This washing step was repeated three times (detachment process). The detached cells from each washing step were collected by centrifugation and resuspended in 100 mM KCl solution at an OD_{660} of 2.0. Six fresh pieces of the polyurethane foam were placed into the cell suspension for the next immobilization cycle. This immobilization/detachment cycle was also repeated four times. The time profiles of the immobilization were similar throughout the four

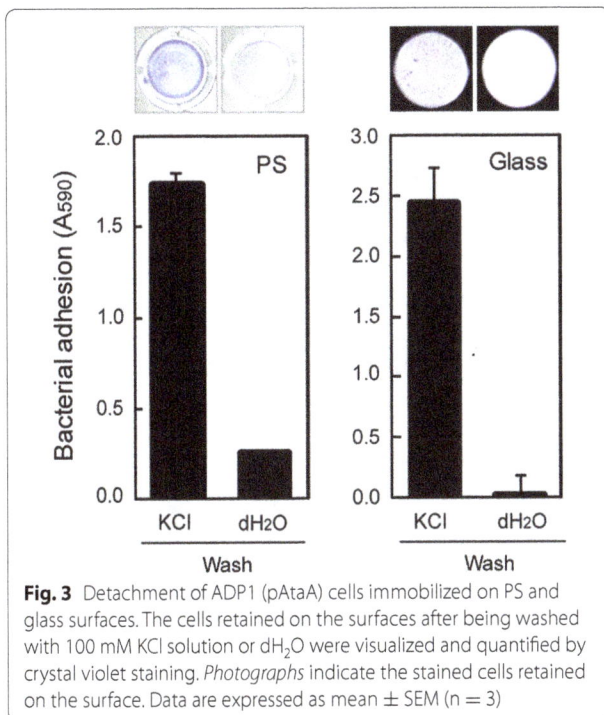

Fig. 3 Detachment of ADP1 (pAtaA) cells immobilized on PS and glass surfaces. The cells retained on the surfaces after being washed with 100 mM KCl solution or dH_2O were visualized and quantified by crystal violet staining. *Photographs* indicate the stained cells retained on the surface. Data are expressed as mean ± SEM (n = 3)

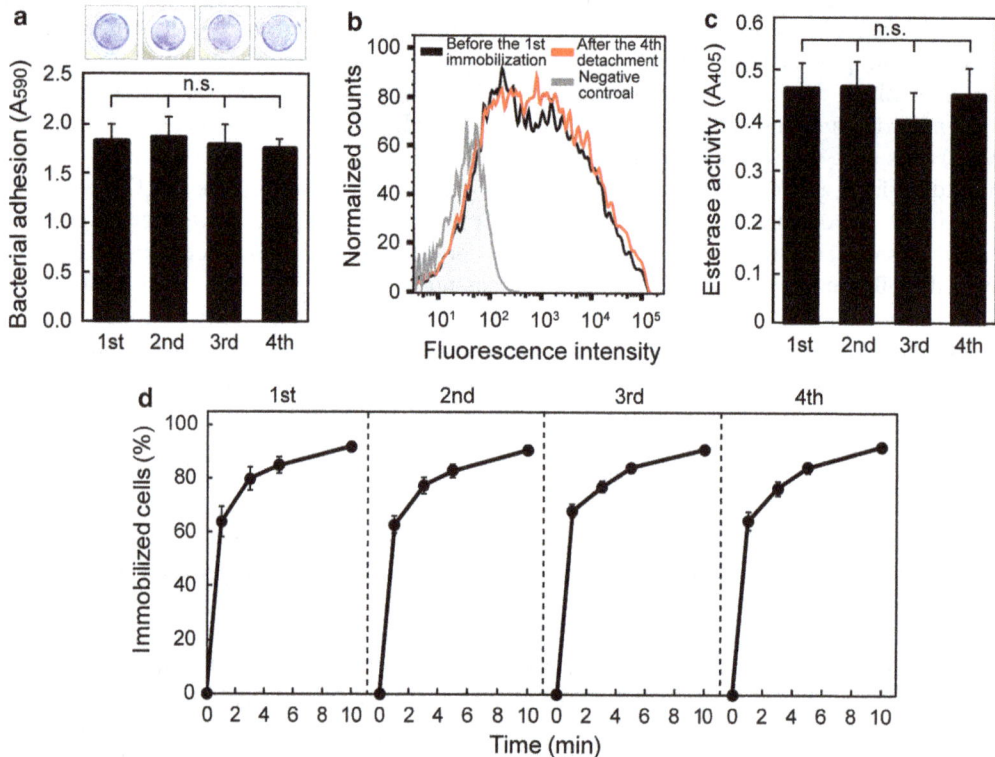

Fig. 4 Re-immobilization of ADP1 (pAtaA) cells. **a** Repetition of the immobilization/detachment cycle of ADP1 (pAtaA) cells. The cells were immobilized onto well surfaces of a PS plate in 100 mM KCl solution for 2 h, detached by washing with dH$_2$O from the PS plate, collected by centrifugation, resuspended in the KCl solution, added to fresh wells, and incubated for re-immobilization for 2 h. Immobilized cells were visualized and quantified by crystal violet staining. This process was repeated four times. *Photographs* indicate the stained cells immobilized on the surface at each cycle. **b** AtaA molecules on the ADP1 (pAtaA) cell surface before and after the repetition of the immobilization/detachment cycle were analyzed by flow cytometry. The *black* and *red lines* show surface-displayed AtaA before the first immobilization and after the fourth cycle of the repetition, respectively. The wild type of ADP1 without AtaA molecules was subjected to the same analysis as a negative control. **c** Esterase activities of ADP1 (pAtaA) cells immobilized on the plate wells in **a**. **d** Time course of the immobilization ratio of ADP1 (pAtaA) cells onto the polyurethane support in each immobilization/detachment cycle. In the respective cycles, the cells were immobilized onto polyurethane foam in 100 mM KCl solution in a flask with shaking and the immobilization ratio was calculated from a decrease in the OD$_{660}$ of the cell suspension. Data are expressed as mean ± SEM (n = 3). *n.s.* not significant (ANOVA, $P > 0.1$)

cycles; more than 90% of ADP1 (pAtaA) cells detached from the polyurethane were re-immobilized onto the fresh polyurethane foam within 10 min (Fig. 4d).

Thus, we have succeeded in developing a novel method for the reversible immobilization of bacterial cells, which enables the reuse of cells without impairment, by means of AtaA expression and a simple manipulation involving a dH$_2$O wash. Other conventional immobilization methods are unsuitable for the development of a reversible process without support destruction or cell inactivation.

Reuse of bacterial cells in a different type of reaction

To show the merit of our reversible immobilization method, we attempted to demonstrate the reusability of bacterial cells for different types of reactions, an ester hydrolysis and toluene degradation, using the schema shown in Fig. 5a. At first, Tol 5 cells were immobilized onto three pieces of the polyurethane foam support in a 100-mL Erlenmeyer flask containing the cells suspended in 30 mL of 100 mM KCl solution (OD$_{660}$ = 1.0) by shaking at 115 rpm at 28 °C for 30 min. The cells loosely attaching to the support were removed by dipping them into 100 mM KCl solution and gentle squeezing. A piece of the support with the immobilized Tol 5 cells was placed into esterase reaction buffer in a test tube. After a 10-min incubation of the test tube at 28 °C, the reaction buffer turned from colorless to yellow due to the 4-nitrophenol produced by esterase on the immobilized Tol 5 cells (Fig. 5b). Next, three pieces of the support used in the esterase reaction were collected and washed in 100 mL of dH$_2$O in a 500-mL Erlenmeyer flask by shaking at 115 rpm for 5 min. This washing step was repeated three times. The detached cells from each washing step were collected by centrifugation, resuspended in 30 mL

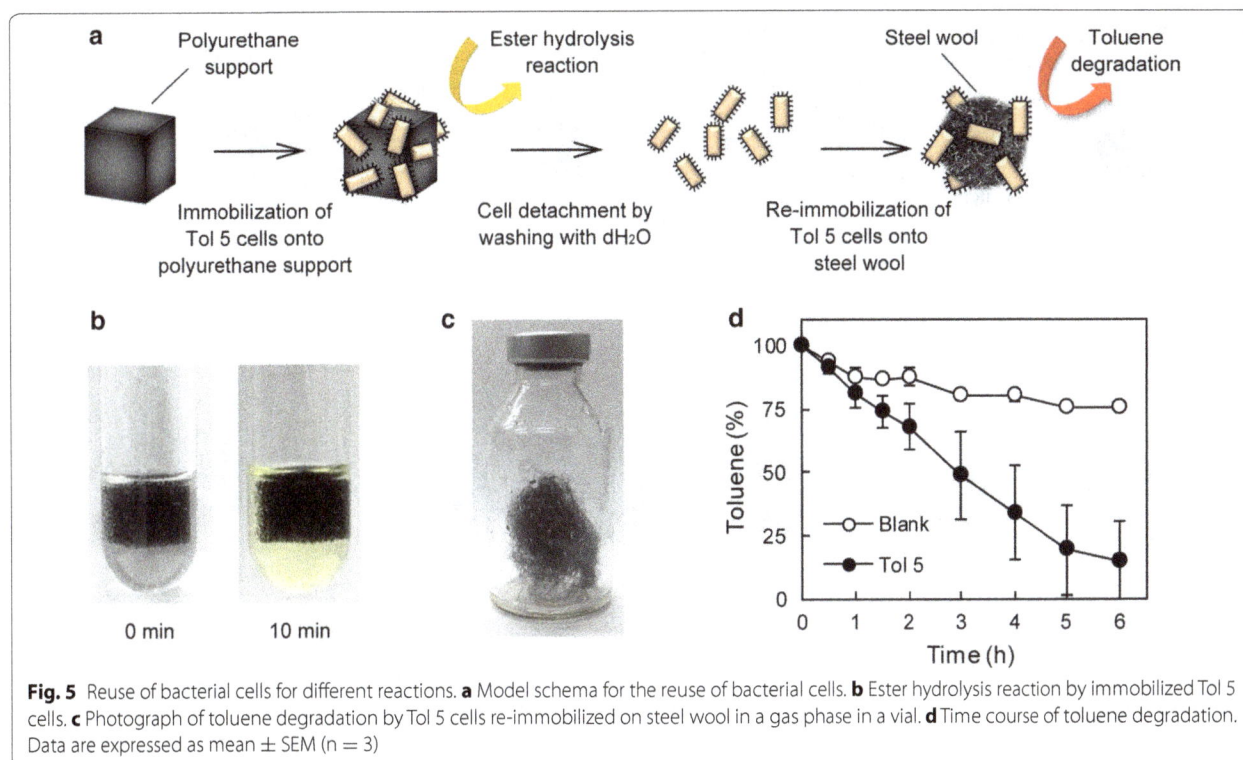

Fig. 5 Reuse of bacterial cells for different reactions. **a** Model schema for the reuse of bacterial cells. **b** Ester hydrolysis reaction by immobilized Tol 5 cells. **c** Photograph of toluene degradation by Tol 5 cells re-immobilized on steel wool in a gas phase in a vial. **d** Time course of toluene degradation. Data are expressed as mean ± SEM (n = 3)

of basal salt (BS) medium in a 100-mL Erlenmeyer flask, and re-immobilized onto 300 mg of steel wool support, which is not eroded by organic solvents, by shaking at 115 rpm for 1 h.

For the induction of toluene-degrading gene expression, the steel wool support with the re-immobilized Tol 5 cells was picked up, touched with paper towel to remove extra water, transferred into a 25-mL vial, and incubated at 28 °C for 1 day under a toluene atmosphere. After this induction step, the immobilized cells on the steel wool support were subjected to a toluene-degradation reaction in a gas phase (Fig. 5c). The reaction was started by injecting 1 μL of toluene into the vial, and thereafter the toluene concertation of the gas in the vial was quantified by gas chromatography–mass spectrometry (GC/MS) and its time-dependent decrease was plotted (Fig. 5d). The cells immobilized on the steel wool support linearly degraded toluene for 5 h and thereafter the degradation rate lowered following the first-order reaction kinetics that depends on the toluene concentration. The slight decrease in the toluene concentration in the control vial without the bacterial cells (blank) suggests adsorption of toluene onto a butyl rubber septum on the cap or solubilization of toluene in a small amount of water from the wetting support. Thus, our reversible immobilization method uniquely allows us to reuse bacterial cells for different types of chemical reactions after detachment from

a support, re-immobilization, and an appropriate induction or reactivation for second reaction.

Reuse of supports for a different type of reaction

To show the further merit of our reversible immobilization method, we attempted to demonstrate the reusability of the support used for the cell immobilization using the schema shown in Fig. 6a. Three pieces of the polyurethane foam support were placed into 30 mL of the suspension of resting ADP1 (pAtaA) cells in 100 mM KCl solution (OD_{660} = 1.0) and shaken in a 100-mL Erlenmeyer flask at 115 rpm at 30 °C for 30 min. The cells loosely attaching to the support were removed by dipping them into 100 mM KCl solution and gentle squeezing. A piece of the support with the immobilized ADP1 (pAtaA) cells was placed into an esterase reaction buffer in a test tube. After a 10-min incubation of the test tube at 28 °C, the reaction buffer turned from colorless to yellow due to the 4-nitrophenol produced by esterase on the immobilized ADP1 (pAtaA) cells, as with the Tol 5 cells (Fig. 6b).

Three pieces of the support used in the reaction were collected and washed in 100 mL of dH_2O in a 500-mL Erlenmeyer flask by shaking at 115 rpm for 10 min. This washing step was repeated three times to thoroughly remove ADP1 (pAtaA) cells. Subsequently, three pieces of the used support were transferred into 30 mL of a

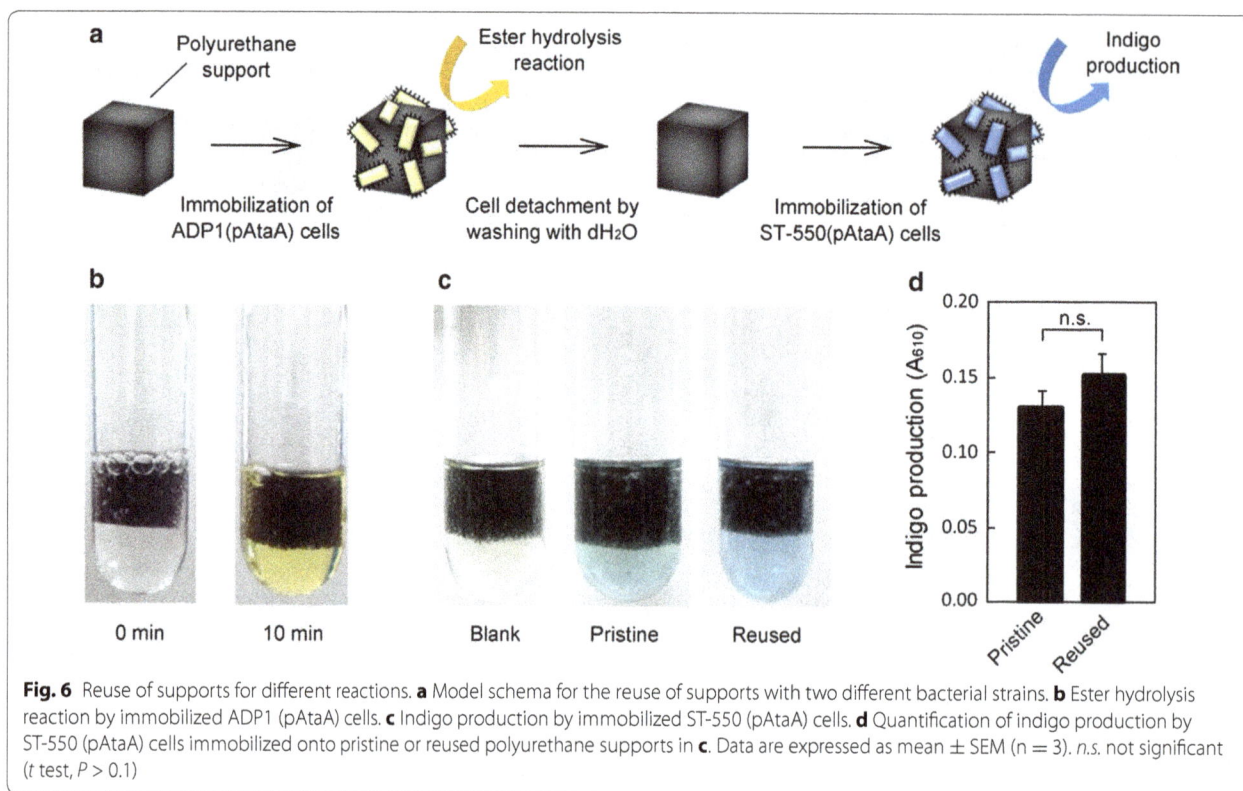

Fig. 6 Reuse of supports for different reactions. **a** Model schema for the reuse of supports with two different bacterial strains. **b** Ester hydrolysis reaction by immobilized ADP1 (pAtaA) cells. **c** Indigo production by immobilized ST-550 (pAtaA) cells. **d** Quantification of indigo production by ST-550 (pAtaA) cells immobilized onto pristine or reused polyurethane supports in **c**. Data are expressed as mean $\pm$ SEM (n = 3). *n.s.* not significant (*t* test, $P > 0.1$)

suspension of resting cells of *Acinetobacter* sp. ST-550 transformed with *ataA*, ST-550 (pAtaA) [22], which has the ability to produce indigo from indole using its phenol hydroxylase [29, 30], in 100 mM KCl solution ($OD_{660} = 1.0$) and shaken in a 100-mL Erlenmeyer flask at 115 rpm at 30 °C for 30 min for cell immobilization. Pristine pieces of the polyurethane foam support were also subjected to the cell immobilization for a control experiment. The cells loosely attaching to the support were removed by dipping them into 100 mM KCl solution and gentle squeezing. Each piece of the support with the immobilized ST-550 (pAtaA) cells was placed into an indigo reaction solution and incubated for 12 h. Indigo produced by the immobilized ST-550 (pAtaA) was extracted with *N*,*N*-dimethylformamide (DMF). Figure 6c shows the solution extracted from each support. The quantity of indigo produced from each support is shown in Fig. 6d. The productivity with the reused support was similar to that with the pristine support, implying that the reused support retained its capability for bacterial immobilization after the immobilization/detachment process. Thus, our reversible immobilization method uniquely allows us to reuse supports, even for different chemical reactions, via the immobilization of different bacterial cells.

Discussion

Immobilization of biocatalysts simplifies product separation, stabilizes biocatalysts, and enables the repeated or continuous use of biocatalysts, which are typically expensive to produce [12, 13]. However, they are usually discarded, together with the supports, after a reaction. In protein immobilization, many techniques for reversible immobilization of enzymes (e.g. lipase, amyloglucosidase, glucoamylase, and aminoacylase) have been studied to enable the regeneration and reuse of support materials [31–35]. Additionally, with regard to bacterial cell immobilization, reversibility should be beneficial. However, the reuse of gel supports is impossible after their use in entrapment immobilization, which is most frequently employed for bacterial cells. Biofilm reactors are also used in the production of valuable compounds, such as alcohols and organic acids, not just in wastewater treatment [15, 36, 37]. These bioreactors use biofilms formed on support materials as immobilized microbial cells [15, 38]. Once biofilms are formed, it is difficult to completely detach them from supports by simple treatments. Therefore, when catalytic activity decreases or a chemical reaction has to be switched for another one, the support with biofilms would be discarded and a new biofilm would be reconstructed on the fresh support. However, a long startup time is required to rebuild an active biofilm.

In this study, we found that the stickiness of the AtaA molecule is drastically diminished at a lower ionic strength and is completely lost in dH$_2$O (Fig. 1). Cell adhesion mediated by AtaA also depends on ionic strength in the same manner as the AtaA molecule, and even bacterial cells previously adhered to supports through AtaA can be detached in dH$_2$O (Figs. 2, 3). Based on this adhesion property, we have established a reversible immobilization method for microbial cells (Fig. 4) and demonstrated the reuse of both cells and supports by means of this reversible immobilization (Figs. 5, 6).

Cells immobilized with AtaA can be detached from supports by a simple manipulation involving a dH$_2$O wash and active cells can be quickly immobilized onto the same previously used support. Because this method is not based on the characteristics of support materials but the unique adhesion property of AtaA, various materials can be employed as reusable supports. For example, supports that have a structure with pores, fibers, or slits for a large surface area and are formed in a combined unit or integrated into a reactor vessel might be used.

In our new reversible immobilization method, both cells and supports can be reused for different types of chemical reactions. Three patterns can be considered about reused processes; (1) used cells are re-immobilized onto a new support, (2) fresh cells are immobilized onto a used support, and (3) used cells are re-immobilized onto a used support. In other words, one of or both of cells and a support are reused. It is expensive to grow bacterial cells on a medium containing many kinds of chemicals, such as nutrients, inducers, and antibiotics, using energy for sterilization, agitation, aeration, and temperature control. In this study, we demonstrated that bacterial cells can be reused for a different type of chemical reaction after a simple induction or reactivation step. We can choose a different support material that is suitable for the subsequent reaction. For example, we used polyurethane foam for the first reaction of ester hydrolysis in a buffer and steel wool for the second reaction of toluene degradation in a gas phase. In addition, the polyurethane foam can also be reused for a different reaction, such as indigo production, after immobilization of a different bacterial strain. The reuse of support material mitigates problems caused by the use of support materials, such as the cost and waste of support materials. Reversible microbial cell immobilization would make bioreactors and bioprocesses simpler, more efficient, more cost-effective, and more convenient.

Conclusions

In summary, we found that the stickiness of isolated AtaA PSD and cell adhesion mediated by AtaA are drastically diminished in deionized water and that deionized water even detaches bacterial cells previously adhered to

support in a salt solution. Using this phenomenon, we invented a unique reversible immobilization method that allows us to reuse bacterial cells and supports for different chemical reactions by a simple manipulation involving a dH$_2$O wash. This method for the immobilization of bacterial cells using AtaA would make bioprocesses more cost-effective and enhance their commercial use for environmentally friendly chemical productions.

Methods
Bacterial strains and culture conditions
The bacterial strains used in this study are detailed in Table 1. These bacterial strains were grown as described previously [20].

Enzyme-linked immunosorbent assay (ELISA)
AtaA PSD was isolated and purified from the Tol 5 derivative strain 4140 [27] harboring p3CAtaA plasmid (Tol 5 4140 (p3CAtaA)), as described previously [26]. Twenty-five microliters of the purified AtaA PSD (10 µg/mL) and an equal volume of KCl solution were added to each well of 96-well PS plates (353072; Becton, Dickinson and Company, NJ) or 96-well glass plates (FB-96; Nippon Sheet Glass Company, Ltd., Tokyo, Japan) and incubated at 28 °C for adsorption. After a 2-h incubation, the solution was removed by a micropipette and each well was gently rinsed three times by removing and replacing 200 µL of PBS-T. Then, 200 µL of PBS-T containing 2% skim milk was added to each well and incubated for 30 min for blocking. Thereafter, the protein adsorbed to the well surface was treated with anti-AtaA$_{699-1014}$ antiserum [20] at a 1:10,000 dilution in PBS-T for 30 min and subsequently with a HRP-conjugated anti-rabbit antibody at

Table 1 Bacterial strains and plasmids used in this study

Strains or plasmids	Description	Reference
Acinetobacter		
sp. Tol 5	Wild type strain	[21]
sp. Tol 5 4140	Unmarked Δ*ataA* mutant of Tol 5	[27]
sp. Tol 5 4140 (p3CAtaA)	Tol 5 4140 harboring p3CAtaA plasmid	[26]
baylyi ADP1	Wild type strain: ATCC 33305	[40]
baylyi ADP1 (pAtaA)	ADP1 harboring pAtaA plasmid	[20]
sp. ST-550	Indigo productive strain	[30]
sp. ST-550 (pAtaA)	ST-550 harboring pAtaA plasmid	[22]
Plasmids		
pARP3	*E. coli-Acinetobacter* shuttle expression vector, *araC*-P$_{BAD}$, Gm^r, Amp^r	[20]
pAtaA	*ataA*-expression vector, pARP3::*ataA*	[20]
p3CAtaA	*3CataA*-expression vector, pARP3::*3CataA*	[26]

a 1:10,000 dilution in PBS-T for 30 min. Finally, the wells were rinsed five times with PBS-T, and 100 μL of substrate solution [ELISA POD Substrate TMB Solution (Popular); Nacalai Tesque, Kyoto, Japan] was added, followed by a 15-min incubation at room temperature. The reaction was stopped by the addition of 100 μL of 1.0 M H_2SO_4 and the absorbance at 450 nm (A_{450}) was measured using a microplate reader (ARVO X3; PerkinElmer, Inc., MA).

Quartz crystal microbalance

The adhesiveness of AtaA PSD was measured using a QCM system (AFFINIX Q8; ULVAC, Kanagawa, Japan) as described previously [26] with a slight modification. A gold-coated electrode (QCM01S-01; ULVAC) was cleaned with piranha solution (H_2SO_4:30% H_2O_2 = 3:1) and equilibrated with 98 μL of dH_2O at 25 °C. After the equilibration, 1 μL of AtaA PSD solution (0.1 mg/mL) was added and the frequency change was measured for 5 min. Subsequently, 1 μL of 100 mM KCl solution (final concentration = 1 mM) was added and the measurement was continued.

Cell attachment and detachment assays using well plates

For a cell attachment assay, the harvested cells were washed three times with dH_2O and suspended in KCl solutions of various concentrations or dH_2O at an OD_{660} of 0.5. The cell suspensions (200 μL each) were placed into wells of 96-well PS or glass plates. After a 2-h incubation at 28 °C without shaking, the cell suspensions were removed by a micropipette and each well was rinsed three times with 200 μL of KCl solution of each concentration or dH_2O. The immobilized cells were stained with 200 μL of 0.1% crystal violet solution for 15 min. After three rinses with 200 μL of KCl solution of each concentration or dH_2O, the dye was eluted from the cells with 200 μL of 70% ethanol, and the absorbance at 590 nm (A_{590}) of the elution was measured by a microplate reader.

For a cell detachment assay, each well was rinsed three times with 200 μL of 100 mM KCl solution or dH_2O using a micropipette. The remaining cells were quantified by crystal violet staining as described above. For re-attachment, the cells detached by rinsing each well with dH_2O were collected by centrifugation, resuspended in 100 mM KCl solution at an OD_{660} of 0.5, and added to a new well for the next immobilization cycle.

Immobilization of bacterial cells onto support materials

Polyurethane foam with a specific surface area of 37.5 cm^2/cm^3 (CFH-30; Inoac Corporation, Nagoya, Japan) in the shape of a cube (1 cm^3) was used as a sponge support. The steel wool used in this study was the same as that previously used [25], which was purchased from Handy Crown (Tokyo, Japan).

To immobilize bacterial cells onto the polyurethane foam support, cells were suspended in 100 mM KCl solution in a 100-mL Erlenmeyer flask. The value of the OD_{660} was adjusted to 2.0 for the visualization of the cell detachment from the support and for the analysis of the time profile of cell immobilization or to 1.0 for use in chemical reactions. Pieces of the support were placed into the cell suspension and shaken at 115 rpm at 28 or 30 °C for 10–30 min. For analysis of the time profile of cell immobilization, the OD_{660} of the cell suspension was measured periodically. The immobilization ratio of the cells was calculated from the following equation:

Immobilization ratio (%)

$$= \left(OD_{660\ initial} - OD_{660\ after\ shaking}\right) / OD_{660\ initial} \times 100.$$

To immobilize bacterial cells onto the steel wool support, the cells detached from the three pieces of the polyurethane foam support used for the esterase reaction were resuspended in 30 mL of BS medium [39] in a 100-mL Erlenmeyer flask. Into this cell suspension, 300 mg of the steel wool support was placed and shaken at 115 rpm at 28 °C for 1 h.

Flow cytometry

Bacterial cells before and after the cell immobilization/detachment test were resuspended in PBS containing 4% paraformaldehyde and incubated at room temperature for 15 min. The samples were washed with PBS and treated with anti-$AtaA_{699-1014}$ antiserum diluted 1:10,000 in PBS. After a 1-h incubation at room temperature, the samples were washed twice with NET buffer (150 mM NaCl, 5 mM EDTA, 50 mM Tris–HCl, 0.05% Triton X-100, pH 7.6) and treated with Alexa Fluor 488-conjugated anti-rabbit antibody (Cell Signaling Technology, MA) diluted 1:500 in NET buffer for 30 min. Finally, the samples were resuspended in dH_2O, and the fluorescence was measured by FACS Canto II (Becton, Dickinson and Company, NJ).

Chemical reactions by immobilized bacteria

For the measurement of cell-bound esterase activity, cells immobilized on plate wells were reacted with 1.9 mM 4-nitrophenyl butyrate (4-NPB) in 200 μL reaction buffer (1.1% Triton X-100, 50 mM 3,3-dimethylglutaric acid, 50 mM Tris, 50 mM 2-amino-2-methyl-1,3-propanediol) at 28 °C for 30 min. Triton X-100 was eliminated from the reaction buffer when the esterase activity of Tol 5 was measured. The A_{405} of 4-nitrophenol produced by the reaction was measured by a microplate reader. To measure the esterase activity of cells immobilized on the polyurethane support, a piece of the support with the immobilized cells was placed into 3 mL of the reaction buffer in a test tube and incubated at 28 °C for 10 min.

For indigo production, a piece of polyurethane foam support with immobilized cells was transferred into a 3-mL reaction solution (1 mM indole, 1% DMF, 40 mM potassium phosphate buffer, pH 7.0) in a test tube and incubated at 30 °C for 12 h. Indigo produced from indole was extracted with DMF, and its concentration was determined using A_{610} measurement.

For toluene degradation, bacterial cells immobilized on steel wool were placed in a 25-mL vial capped by a butyl rubber septum. The reaction was started by injecting 1 μL of toluene into the vial using a gastight syringe (MS-GFN100; ITO Corporation, Fuji, Japan). The toluene concentration was measured using GC/MS, which comprised a GC system (GC7820A; Agilent Technologies, Santa Clara, CA) coupled to a MS detector (MSD 5975; Agilent Technologies), equipped with a Rtx-200 capillary column (30 m × 0.32 mm × 0.5 μm; RESTEC, Bellefonte, PA). Gas sample (50 μL) was taken from a vial and injected into the GC/MS system using the gastight syringe. The split ratio and the flow rate of helium were set at 10:1 and 2 mL/min, respectively. The operating program started with an isocratic step at 90 °C for 1 min, followed by temperature ramping of 25 °C/min to the final temperature at 120 °C, then ion fragments of m/z = 65 and m/z = 91 were monitored on the selected ion monitoring mode. The peak, which showed toluene, was detected at a retention time of 1.4 min. The degradation ratio of toluene was calculated from the ratio of the peak area at each time to that at 0 min.

Additional files

Additional file 1. Detachment of adhered ADP1 (pAtaA) cells from a polyurethane support by washing in dH_2O. The polyurethane supports were put into the ADP1 (pAtaA) cell suspension in 100 mM KCl solution and were incubated at 28 °C with shaking at 115 rpm. After incubation for 10 min, the supports were retrieved and lightly washed with 100 mM KCl solution. Then, the supports were picked up with tweezers and shaken in 100 mL of dH_2O for 1 min. Note that the white grains in the dH_2O are detached cells.

Additional file 2. Detachment of adhered ADP1 (pAtaA) cells from a polyurethane support by washing in 100 mM KCl solution. The polyurethane supports were put into the ADP1 (pAtaA) cell suspension in 100 mM KCl solution and were incubated at 28 °C with shaking at 115 rpm. After incubation for 10 min, the supports were retrieved and lightly washed with 100 mM KCl solution. Then, the supports were picked up with tweezers and shaken in 100 mL of 100 mM KCl solution for 1 min.

Abbreviations
TAA: trimeric autotransporter adhesion; PSD: passenger domain; TM: transmembrane anchor; HRV 3C: human rhinovirus 3C; dH_2O: deionized water; ELISA: enzyme-linked immunosorbent assay; HRP: horseradish peroxidase; POD: peroxidase; QCM: quartz crystal microbalance; 4-NPB: 4-nitrophenyl butyrate; BS: basal salt; GC/MS: gas chromatography–mass spectrometry; DMF: N,N-dimethylformamide; PS: polystyrene.

Authors' contributions
The manuscript was written through contributions of SY, YO, HN and KH. All authors read and approved the final manuscript.

Acknowledgements
Not applicable.

Competing interests
A patent application has been filed in relation to this work (JP2014/056966).

Funding
This work was supported by the Advanced Low Carbon Technology Research and Development Program (ALCA) of the Japan Science and Technology Agency (JST) and by the Program for Leading Graduate Schools "Integrative Graduate Education and Research in Green Natural Sciences", MEXT, Japan.

References
1. Taher E, Chandran K. High-rate, high-yield production of methanol by ammonia-oxidizing bacteria. Environ Sci Technol. 2013;47:3167–73.
2. Fukuda H, Hama S, Tamalampudi S, Noda H. Whole-cell biocatalysts for biodiesel fuel production. Trends Biotechnol. 2008;26:668–73.
3. Pollard DJ, Woodley JM. Biocatalysis for pharmaceutical intermediates: the future is now. Trends Biotechnol. 2007;25:66–73.
4. Wang SN, Xu P, Tang HZ, Meng J, Liu XL, Qing C. "Green" route to 6-hydroxy-3-succinoyl-pyridine from (S)-nicotine of tobacco waste by whole cells of a Pseudomonas sp. Environ Sci Technol. 2005;39:6877–80.
5. Schmid A, Dordick JS, Hauer B, Kiener A, Wubbolts M, Witholt B. Industrial biocatalysis today and tomorrow. Nature. 2001;409:258–68.
6. Lee SY, Kim HU. Systems strategies for developing industrial microbial strains. Nat Biotechnol. 2015;33:1061–72.
7. Choi SY, Park SJ, Kim WJ, Yang JE, Lee H, Shin J, Lee SY. One-step fermentative production of poly(lactate-co-glycolate) from carbohydrates in Escherichia coli. Nat Biotechnol. 2016;34:435–40.
8. Cheong S, Clomburg JM, Gonzalez R. Energy- and carbon-efficient synthesis of functionalized small molecules in bacteria using non-decarboxylative Claisen condensation reactions. Nat Biotechnol. 2016;34:556–61.
9. DeLoache WC, Russ ZN, Narcross L, Gonzales AM, Martin VJ, Dueber JE. An enzyme-coupled biosensor enables (S)-reticuline production in yeast from glucose. Nat Chem Biol. 2015;11:465–71.
10. Paddon CJ, Westfall PJ, Pitera DJ, Benjamin K, Fisher K, McPhee D, Leavell MD, Tai A, Main A, Eng D, et al. High-level semi-synthetic production of the potent antimalarial artemisinin. Nature. 2013;496:528–32.
11. Hara KY, Araki M, Okai N, Wakai S, Hasunuma T, Kondo A. Development of bio-based fine chemical production through synthetic bioengineering. Microb Cell Factories. 2014;13:173.
12. Junter GA, Jouenne T. Immobilized viable microbial cells: from the process to the proteome… or the cart before the horse. Biotechnol Adv. 2004;22:633–58.
13. Hartmeier W. Immobilized biocatalysts—from simple to complex systems. Trends Biotechnol. 1985;3:149–53.
14. Gong JS, Lu ZM, Li H, Shi JS, Zhou ZM, Xu ZH. Nitrilases in nitrile biocatalysis: recent progress and forthcoming research. Microb Cell Factories. 2012;11:142.
15. Qureshi N, Annous BA, Ezeji TC, Karcher P, Maddox IS. Biofilm reactors for industrial bioconversion processes: employing potential of enhanced reaction rates. Microb Cell Factories. 2005;4:24.
16. Smidsrod O, Skjakbraek G. Alginate as immobilization matrix for cells. Trends Biotechnol. 1990;8:71–8.
17. Klein J, Ziehr H. Immobilization of microbial-cells by adsorption. J Biotechnol. 1990;16:1–16.
18. Carballeira JD, Quezada MA, Hoyos P, Simeo Y, Hernaiz MJ, Alcantara AR, Sinisterra JV. Microbial cells as catalysts for stereoselective red-ox reactions. Biotechnol Adv. 2009;27:686–714.
19. Cassidy MB, Lee H, Trevors JT. Environmental applications of immobilized microbial cells: a review. J Ind Microbiol Biotechnol. 1996;16:79–101.
20. Ishikawa M, Nakatani H, Hori K. AtaA, a new member of the trimeric autotransporter adhesins from Acinetobacter sp. Tol 5 mediating high adhesiveness to various abiotic surfaces. PLoS ONE. 2012;7:e48830.

21. Hori K, Yamashita S, Ishii S, Kitagawa M, Tanji Y, Unno H. Isolation, characterization and application to off-gas treatment of toluene-degrading bacteria. J Chem Eng Japan. 2001;39:175–84.

22. Ishikawa M, Shigemori K, Hori K. Application of the adhesive bacterionanofiber AtaA to a novel microbial immobilization method for the production of indigo as a model chemical. Biotechnol Bioeng. 2014;111:16–24.

23. Linke D, Riess T, Autenrieth IB, Lupas A, Kempf VA. Trimeric autotransporter adhesins: variable structure, common function. Trends Microbiol. 2006;14:264–70.

24. Bassler J, Hernandez Alvarez B, Hartmann MD, Lupas AN. A domain dictionary of trimeric autotransporter adhesins. Int J Med Microbiol. 2015;305:265–75.

25. Hori K, Ohara Y, Ishikawa M, Nakatani H. Effectiveness of direct immobilization of bacterial cells onto material surfaces using the bacterionanofiber protein AtaA. Appl Microbiol Biotechnol. 2015;99:5025–32.

26. Yoshimoto S, Nakatani H, Iwasaki K, Hori K. An *Acinetobacter* trimeric autotransporter adhesin reaped from cells exhibits its nonspecific stickiness via a highly stable 3D structure. Sci Rep. 2016;6:28020.

27. Ishikawa M, Hori K. A new simple method for introducing an unmarked mutation into a large gene of non-competent Gram-negative bacteria by FLP/FRT recombination. BMC Microbiol. 2013;13:86.

28. Koiwai K, Hartmann MD, Linke D, Lupas AN, Hori K. Structural basis for toughness and flexibility in the C-terminal passenger domain of an *Acinetobacter* trimeric autotransporter adhesin. J Biol Chem. 2016;291:3705–24.

29. Doukyu N, Toyoda K, Aono R. Indigo production by *Escherichia coli* carrying the phenol hydroxylase gene from *Acinetobacter* sp. strain ST-550 in a water-organic solvent two-phase system. Appl Microbiol Biotechnol. 2003;60:720–5.

30. Doukyu N, Nakano T, Okuyama Y, Aono R. Isolation of an *Acinetobacter* sp. ST-550 which produces a high level of indigo in a water-organic solvent two-phase system containing high levels of indole. Appl Microbiol Biotechnol. 2002;58:543–6.

31. Mateo C, Abian O, Fernandez-Lafuente R, Guisan JM. Reversible enzyme immobilization via a very strong and nondistorting ionic adsorption on support-polyethylenimine composites. Biotechnol Bioeng. 2000;68:98–105.

32. Rueda N, Dos Santos JC, Rodriguez MD, Albuquerque TL, Barbosa O, Torres R, Ortiz C, Fernandez-Lafuente R. Reversible immobilization of lipases on octyl-glutamic agarose beads: a mixed adsorption that reinforces enzyme immobilization. J Mol Cat B. 2016;128:10–8.

33. Saylan Y, Uzun L, Denizli A. Alanine functionalized magnetic nanoparticles for reversible amyloglucosidase immobilization. Ind Eng Chem Res. 2015;54:454–61.

34. Wang J, Zhao G, Li Y, Liu X, Hou P. Reversible immobilization of glucoamylase onto magnetic chitosan nanocarriers. Appl Microbiol Biotechnol. 2013;97:681–92.

35. Chibata I, Tosa T. Industrial applications of immobilized enzymes and immobilized microbial cells. In: Wingard LB, Katchalski-Katzir E, Leon G, editors. Applied biochemistry and bioengineering. London: Academic; 1976. p. 329–57.

36. Liu D, Chen Y, Ding FY, Zhao T, Wu JL, Guo T, Ren HF, Li BB, Niu HQ, Cao Z, et al. Biobutanol production in a *Clostridium acetobutylicum* biofilm reactor integrated with simultaneous product recovery by adsorption. Biotechnol Biofuels. 2014;7:5.

37. Zhang F, Ding J, Zhang Y, Chen M, Ding ZW, van Loosdrecht MC, Zeng RJ. Fatty acids production from hydrogen and carbon dioxide by mixed culture in the membrane biofilm reactor. Water Res. 2013;47:6122–9.

38. O'Toole G, Kaplan HB, Kolter R. Biofilm formation as microbial development. Annu Rev Microbiol. 2000;54:49–79.

39. Ishii S, Unno H, Miyata S, Hori K. Effect of cell appendages on the adhesion properties of a highly adhesive bacterium, *Acinetobacter* sp. Tol 5. Biosci Biotechnol Biochem. 2006;70:2635–40.

40. Metzgar D, Bacher JM, Pezo V, Reader J, Doring V, Schimmel P, Marliere P, de Crecy-Lagard V. *Acinetobacter* sp. ADP1: an ideal model organism for genetic analysis and genome engineering. Nucleic Acids Res. 2004;32:5780–90.

10

Improving the production of 22-hydroxy-23,24-bisnorchol-4-ene-3-one from sterols in *Mycobacterium neoaurum* by increasing cell permeability and modifying multiple genes

Liang-Bin Xiong[†], Hao-Hao Liu[†], Li-Qin Xu, Wan-Ju Sun, Feng-Qing Wang[*] and Dong-Zhi Wei[*]

Abstract

Background: The strategy of modifying the sterol catabolism pathway in mycobacteria has been adopted to produce steroidal pharmaceutical intermediates, such as 22-hydroxy-23,24-bisnorchol-4-ene-3-one (4-HBC), which is used to synthesize various steroids in the industry. However, the productivity is not desirable due to some inherent problems, including the unsatisfactory uptake rate and the low metabolic efficiency of sterols. The compact cell envelope of mycobacteria is a main barrier for the uptake of sterols. In this study, a combined strategy of improving the cell envelope permeability as well as the intracellular sterol metabolism efficiency was investigated to increase the productivity of 4-HBC.

Results: *MmpL3*, encoding a transmembrane transporter of trehalose monomycolate, is an important gene influencing the assembly of mycobacterial cell envelope. The disruption of *mmpL3* in *Mycobacterium neoaurum* ATCC 25795 significantly enhanced the cell permeability by 23.4% and the consumption capacity of sterols by 15.6%. Therefore, the inactivation of *mmpL3* was performed in a 4-HBC-producing strain derived from the wild type *M. neoaurum* and the 4-HBC production in the engineered strain was increased by 24.7%. Subsequently, to enhance the metabolic efficiency of sterols, four key genes, *choM1*, *choM2*, *cyp125*, and *fadA5*, involved in the sterol conversion pathway were individually overexpressed in the engineered *mmpL3*-deficient strain. The production of 4-HBC displayed the increases of 18.5, 8.9, 14.5, and 12.1%, respectively. Then, the more efficient genes (*choM1*, *cyp125*, and *fadA5*) were co-overexpressed in the engineered *mmpL3*-deficient strain, and the productivity of 4-HBC was ultimately increased by 20.3% (0.0633 g/L/h, 7.59 g/L 4-HBC from 20 g/L phytosterol) compared with its original productivity (0.0526 g/L/h, 6.31 g/L 4-HBC from 20 g/L phytosterol) in an industrial resting cell bio-transformation system.

Conclusions: Increasing cell permeability combined with the co-overexpression of the key genes (*cyp125*, *choM1*, and *fadA5*) involved in the conversion pathway of sterol to 4-HBC was effective to enhance the productivity of 4-HBC. The strategy might also be useful for the conversion of sterol to other steroidal intermediates by mycobacteria.

Keywords: *Mycobacterium*, 22-Hydroxy-23,24-bisnorchol-4-ene-3-one (4-HBC), *mmpL3*, *choM*, *cyp125*, *fadA5*, Sterol catabolism

*Correspondence: fqwang@ecust.edu.cn; dzhwei@ecust.edu.cn
†Liang-Bin Xiong and Hao-Hao Liu contributed equally to this work
State Key Laboratory of Bioreactor Engineering, Newworld Institute
of Biotechnology, East China University of Science and Technology,
Shanghai 200237, China

Background

Mycolata actinomycetes, including *mycobacteria* [1], rhodococci [2, 3] and *gordonia* [4], are mycolic acids-rich bacteria and can catabolize natural sterols as carbon and energy sources [5]. The interruptions in the sterol metabolism pathway of mycobacteria lead to the accumulation of some important intermediates which can be used as ideal precursors to synthesize valuable steroidal pharmaceuticals [6]. For example, the C_{22} steroids, including 22-hydroxy-23,24-bisnorchol-4-ene-3-one (4-HBC), 22-hydroxy-23,24-bisnorchol-1,4-dien-3-one (1,4-HBC) and 9,22-dihydroxy-23,24-bisnorchol-4-ene-3-one (9-OHHBC), are suitable precursors for the synthesis of progestational and adrenocortical hormones [7–9].

Sterols are a kind of hydrophobic lipid compounds. In the nature, the uptake of sterols by mycobacteria mainly relies on the direct contact between the particles of sterol and the cell envelope [10]. As one of the distinctive structure of mycobacteria, the complex structure of mycolyl-arabinogalactan-peptidoglycan in the cell envelope core forms an asymmetrical and non-fluid layer outside of the cell membrane [11–14]. Moreover, the surface of this structure is decorated with a variety of non-covalently associated capsular lipids, including trehalose monomycolate (TMM), trehalose dimycolate (TDM) as well as a capsule-like coat of polysaccharide and protein [12, 14]. The polar glycolipids layer on the cell envelope surface of mycobacteria provides an optimum surface contact for material exchange [15]. However, the major frame of the complex structure in the cell envelope core guarantees a low permeability of the cells and is a main negative factor for the uptake of sterol into mycobacteria [16]. In the presence of synthesis inhibitors of the cell envelope, such as vancomycin and glycine, the uptake of sterol through the cell envelope shows significant improvements [17–20]. As a result, the sterol utilization rate and the productivity of target steroidal intermediates are remarkably increased in mycobacterial cells. However, these strategies can hardly be applied due to the high cost of the added inhibitors in large-scale production.

Mycolic acids, which accounted for 40–60% of cell dry weight, are known to play an essential role in the formation of the cell envelope surface and the cell envelope core [21]. Through a unique biosynthesis pathway in the mycobacterial cytoplasm, synthesized mycolic acids are conjugated with trehalose to form TMM, which acts as a primary mediator for adjusting the hydrophobicity of the cell envelope [14]. Then, these TMM molecules are transported from the cytoplasmic production site into the periplasm. Finally, they are transferred to the cell envelope region for the assembly of mycolic acid-related structures and other molecules, including glucose monomycolate, glycerol monomycolate, the above mentioned TDM and the critical mycolyl-arabinogalactan-peptidoglycan complex. The transmembrane transport of TMM requires some transport proteins [14]. The details of the transport process were still unclear [22], but it was revealed that the MmpL3 was probably implicated in the transport of the essential TMM from cytoplasm to periplasm in *Mycobacterium tuberculosis* [14, 21]. Interferences with the MmpL3 function in mycobacterial cells were expected to inhibit the TMM translocation to the cell envelope, thus leading to an improvement in the cell permeability. The uptake of sterols as well as the productivity of target steroidal intermediates is possibly increased accordingly. In addition, through the augmentation of key genes in engineered *M. neoaurum* strains, the production of androst-4-ene-3,17-dione (AD), androst-1,4-diene-3,17-dione (ADD) and 9α-hydroxy-4-androstene-3,17-dione (9-OHAD) was remarkably increased [23, 24]. The metabolic engineering strategy of overexpressing genes directly involved in the conversion pathway has proved to be an applicable way to increase the production of target metabolites.

In this study, the changes of cell permeability and sterol consumption after deleting of *mmpL3* in *M. neoaurum* ATCC 25795 were determined. Meanwhile, the improvement in 4-HBC production caused by the deficiency of *mmpL3* in its derived 4-HBC-producing strain was explored. Additionally, to further enhance the productivity of 4-HBC, we evaluated the effect of the individual overexpression of some key genes including *cyp125* [25, 26], *choM1*, *choM2* [23] and *fadA5* [27] in the conversion pathway of sterols on the 4-HBC production. Then, the influence of their combinatory overexpression on the 4-HBC production was assessed in an industrial resting cell bio-transformation system.

Results

Comparison of the *mmpL3* genome region in *mycobacteria*

The inhibition of MmpL3, a membrane transporter involved in the transmembrane transport of trehalose monomycolate (TMM) from cytoplasm to periplasm resulted in the accumulation of TMM in cytoplasm of *M. tuberculosis* H37Rv [14]. This modification would interfere with the normal assembly of the cell envelope [14] and possibly cause an improvement in cell envelope permeability, thus leading to a corresponding increase in the uptake of sterols by mycobacteria [17–20]. Therefore, we firstly located the possible *mmpL3* region in the genome of *M. neoaurum* ATCC 25795, which is a highly homologous strain of well-known steroidal intermediate producer strains [28, 29]. Then, the possible *mmpL3* genome region in *M. neoaurum* ATCC 25795 was compared with the homologous region in *Mycobacterium neoaurum* NRRL B-3805, *Mycobacterium neoaurum*

VKM Ac-1815D, and *M. tuberculosis* H37Rv. The parameter of sequence identity was used to preliminarily assess the functional homology between the *mmpL3* in *M. neoaurum* and its homologous genes.

A transmembrane transport protein encoded by *Mn_1721* (GeneBank: NZ_JMDW01000016.1; Region: 7540...4670, 2871-bp) in *M. neoaurum* ATCC 25795 can be annotated as a homologous protein with the MmpL3, which has been identified in the other two mycobacteria [14, 21]. The gene shows a high nucleotide sequence identity with *MyAD_02720* (2871-bp, 94%, GeneBank ID: AMO04272.1) from *M. neoaurum* NRRL B-3805, *D174_02785* (2871-bp, 94%, GeneBank ID: AHC23578.1) from *M. neoaurum* VKM Ac-1815D, and *mmpL3* (*Rv_0206*, 2835-bp, 72%, GeneBank ID: NP_214720.1) from *M. tuberculosis* H37Rv (Fig. 1; Additional file 1: Table S1). Moreover, the genomic location of *mmpL3* is highly conserved among the listed *mycobacteria*. The *mmpL3* from *M. neoaurum* ATCC 25795 is located between *Mn_1720* and *Mn_1722*, which also show the high identity (70–93%) and similar organizations with corresponding genes from the other three strains in the genus *Mycobacterium*. Comparison results of the *mmpL3* genome region suggest that the function of MmpL3 is possibly conserved in mycobacteria. Like the disruption of its homologous genes in *M. tuberculosis* H37Rv, the deletion of *mmpL3* in *M. neoaurum* ATCC 25795 is probably beneficial to the improvement in cell envelope permeability.

Deletion of *mmpL3* affects cell envelope permeability and cholesterol utilization

In order to assess the effect of MmpL3 on cell envelope permeability, the *mmpL3* was deleted from the genome of wild-type strain *M. neoaurum* by allelic replacement

(Fig. 2a). The fluorescence intensity of the Δ*mmpL3* strain labeled by fluorescein diacetate (FDA) in minimal medium (MM) was analyzed. Cell permeability of the *mmpL3*-deleted strain showed an improvement of 23.4% at 30 min compared with that of the wild-type strain (Fig. 2b). Meanwhile, the growth of the strain Δ*mmpL3* did not decline significantly, but displayed an abnormal rise compared with that of the wild-type strain was observed (Fig. 2c). The slight acceleration in the growth rate of the strain Δ*mmpL3* was inconsistent with the similar disruption study in *M. smegmatis* [21] as well as our previous phenotype analysis of the gene deletion strains of *M. neoaurum* [23, 30]. This might be interpreted as follows. The improved cell permeability increased the supplement of steroids after the deletion of *mmpL3*. To confirm this speculation, cholest-4-en-3-one was selected as a label to evaluate the steroids uptake because it was the first metabolite in the oxidation of cholesterol oxidases (Fig. 3) and could be detected by common UV detector, but the cholesterol could not be detected by common UV detector. The result indicated that the uptake of cholest-4-en-one was significantly enhanced by 33.7% after the deletion of *mmpL3* (Additional file 2: Figure S1a, 1b). This result was consistent with the enhanced cell permeability characterized by FDA. Besides, we further analyzed the cholesterol utilization of the wild-type strain *M. neoaurum*, the strain Δ*mmpL3* and the complemented strain Δ*mmpL3*+*mmpL3* (Fig. 2d).

Comparison results showed that the utilization of cholesterol in the strain Δ*mmpL3* was higher than that of the strain Δ*mmpL3*+*mmpL3* and the wild-type strain. The residual cholesterol concentration in the medium of strain Δ*mmpL3* was about 0.71 g/L at 24 h, 0.32 g/L at 48 h, and 0.11 g/L at 72 h, while the concentrations of cholesterol in the medium of the wild-type

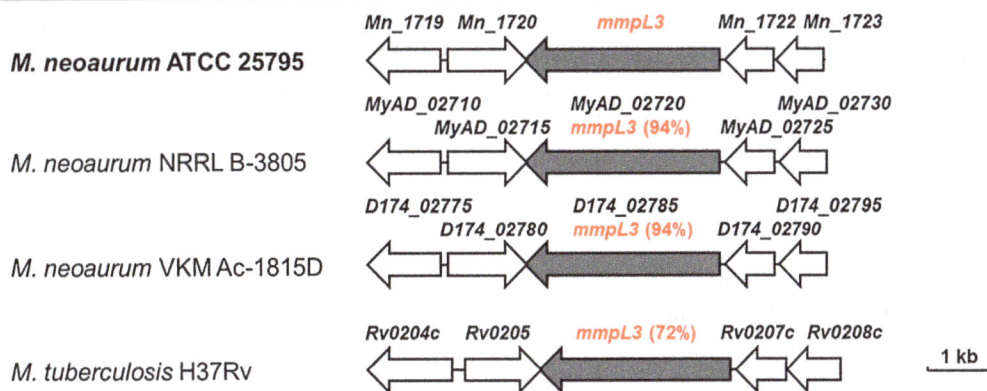

Fig. 1 Localization of the *mmpL3* homologues in the genome of *M. neoaurum* ATCC 25795 and other mycobacteria. The size and direction of genes from the predicted genome information were displayed as an arrow according to the scale. The percentages, such as 94 and 72%, indicate the sequence identity of *mmpL3* from *M. neoaurum* ATCC 25795 with the homologs in *M. neoaurum* NRRL B-3805, *M. neoaurum* VKM Ac-1815D, and *M. tuberculosis* H37Rv

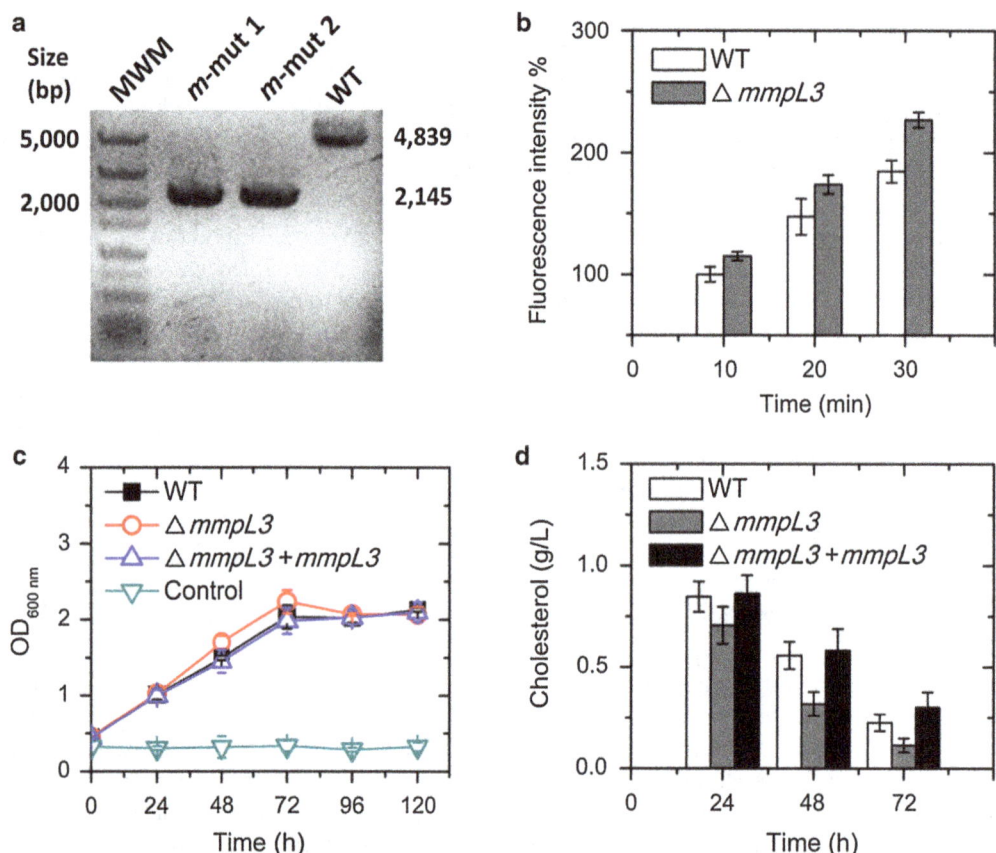

Fig. 2 Effects of deleting *mmpL3* on cell permeability and the cholesterol utilization. **a** Evidence for allelic replacement at the *mmpL3* locus of *M. neoaurum* ATCC 25795. The wild-type (WT) 4839-bp sequence was replaced by a 2145-bp fragment ligated with a 1074-bp upstream sequence and a 1071-bp downstream of the *mmpL3* (*m*) gene, thus resulting in the *mmpL3*-deficient *M. neoaurum* (*m*-mut1 and *m*-mut2). *MWM* molecular weight marker. **b** Effects of MmpL3 disruption on cell permeability. Diluted cell suspensions were stained with fluorescein diacetate (FDA) and then the mixtures were detected by a fluorescence spectrophotometer. **c** Growth characteristics of the wild-type *M. neoaurum* ATCC 25795 (WT, *squares*), the deficiency strain of *mmpL3* in the WT (Δ*mmpL3*, *open circles*) and the complementation strain of *mmpL3* in the Δ*mmpL3* (Δ*mmpL3*+*mmpL3*, *triangles*) cultured in MM with 1.0 g/L cholesterol. The control is the medium containing 1.0 g/L cholesterol without inoculum. **d** Quantitative determination of residual cholesterol from the three strains cultured in MM with 1.0 g/L cholesterol. Data represent mean ± standard deviation of three measurements

stain and the strain Δ*mmpL3*+*mmpL3* were 0.85–0.86, 0.56–0.58 and 0.23–0.30 g/L, respectively. The cholesterol utilization at 24, 48, and 72 h respectively showed the improvements of 93.3, 54.5, and 15.6%. In addition, the deletion of *mmpL3* caused no obvious inhibition on cell growth of the strain (Fig. 2c). In a word, the deficiency of *mmpL3* in the wild-type *M. neoaurum* showed an improvement in cell permeability and the cholesterol utilization.

Improvement in the 4-HBC productivity in the engineered strain by deletion of *mmpL3*

To further evaluate whether the deletion of *mmpL3* could increase the productivity of the target intermediate 4-HBC, the *mmpL3* was deleted in a typical 4-HBC-producing strain WIII (Δ*kshA*Δ*hsd4A*Δ*kstD123*) [8] to generate a strain WIIIΔ*mmpL3*.

In order to determine whether the deletion of *mmpL3* could improve cell permeability of the 4-HBC-producing strain WIII, the fluorescence intensity in cells WIII and WIIIΔ*mmpL3* in MYC/02 medium was analyzed. We found that the cell permeability of the strain WIIIΔ*mmpL3* was improved significantly compared to that of the strain WIII (Fig. 4a). The increase in the fluorescence intensity of the strain WIIIΔ*mmpL3* compared to that of the strain WIII ranged from 13.2 to 35.5% when incubation time was increased from 10 to 30 min, indicating that the cell envelope assembly in the *mmpL3*-deleted cells might be damaged to some degree.

Moreover, the growth of strain WIIIΔ*mmpL3* showed some reduction after 96 h (Fig. 4b), but the production of target intermediate 4-HBC in the *mmpL3*-deleted strain showed an improvement. The deletion of this gene increased the production of 4-HBC by 24.7% from 0.90 to 1.12 g/L at

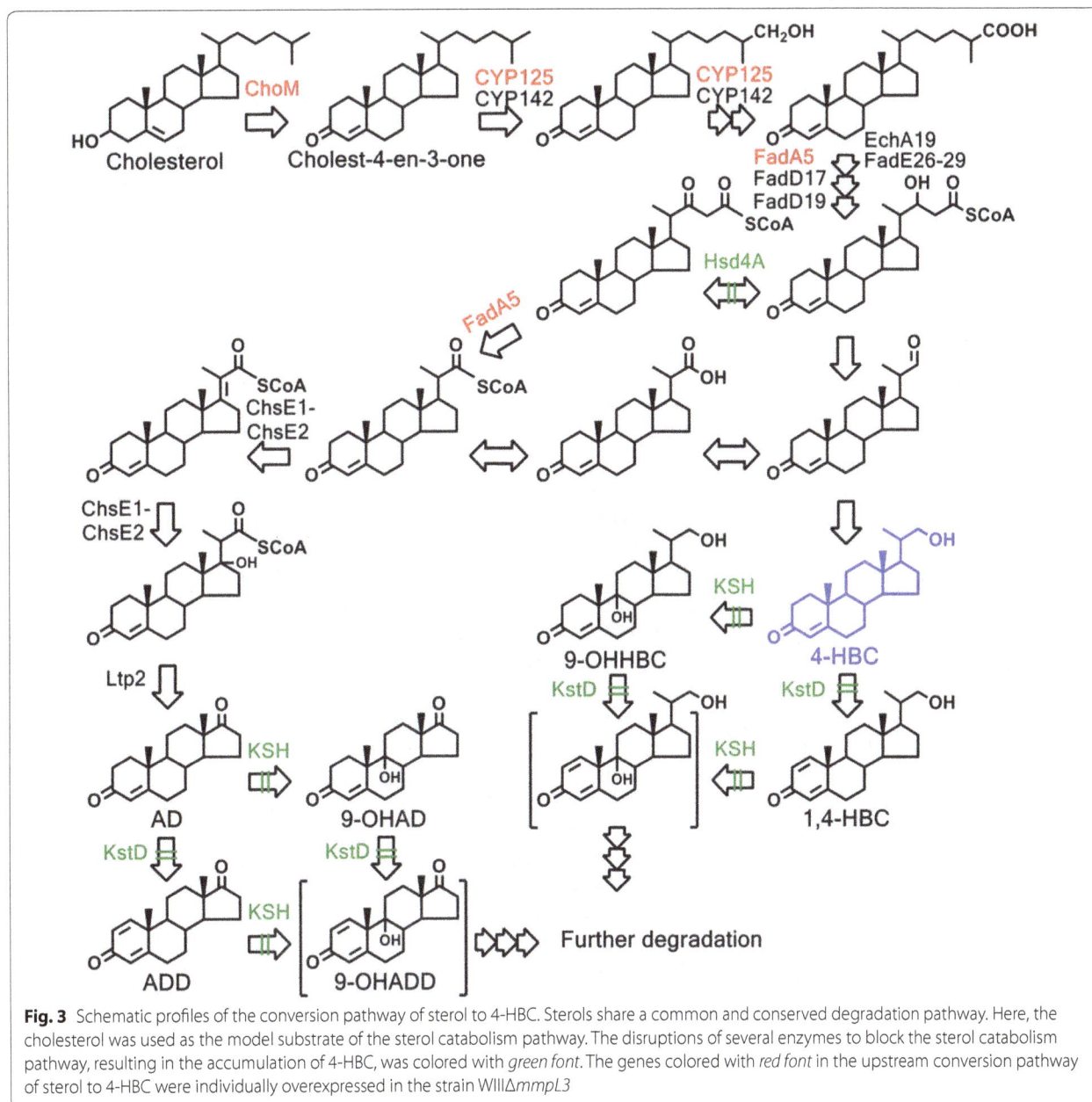

Fig. 3 Schematic profiles of the conversion pathway of sterol to 4-HBC. Sterols share a common and conserved degradation pathway. Here, the cholesterol was used as the model substrate of the sterol catabolism pathway. The disruptions of several enzymes to block the sterol catabolism pathway, resulting in the accumulation of 4-HBC, was colored with *green font*. The genes colored with *red font* in the upstream conversion pathway of sterol to 4-HBC were individually overexpressed in the strain WIIIΔ*mmpL3*

120 h (Fig. 4c). In addition, the consumption rate of phytosterol substrates in the strain WIIIΔ*mmpL3* also showed an obvious improvement after 96 h of biotransformation (Additional file 2: Figure S2). These results confirmed that the deletion of *mmpL3* did cause an improvement in cell permeability, thus leading to an increase of 4-HBC production in the engineered *mmpL3*-deficient strain.

Enhancement of 4-HBC productivity by overexpressing key genes involved in the sterol catabolic pathway of the engineered strain

We significantly increased the conversion of phytosterols to AD and ADD by overexpression of cholesterol oxidases

(ChoM1 and ChoM2) catalyzing the initial oxidation of sterol to sterone [23]. To further improve the production of the target intermediate 4-HBC, four genes, including *choM1*, *choM2*, a *cyp125* gene encoding a cytochrome P450 enzyme [25, 26] and a *fadA5* gene encoding a thiolase [27], in the conversion process of sterol to 4-HBC, were individually overexpressed in the engineered strain WIIIΔ*mmpL3* (Fig. 3).

As shown in Fig. 5, the cells overexpressing one of these genes showed higher 4-HBC productivities than their parental strain WIIIΔ*mmpL3* after 120 h of biotransformation (Fig. 5a). Among these genes, overexpressing of *choM1* showed the greatest enhancement of

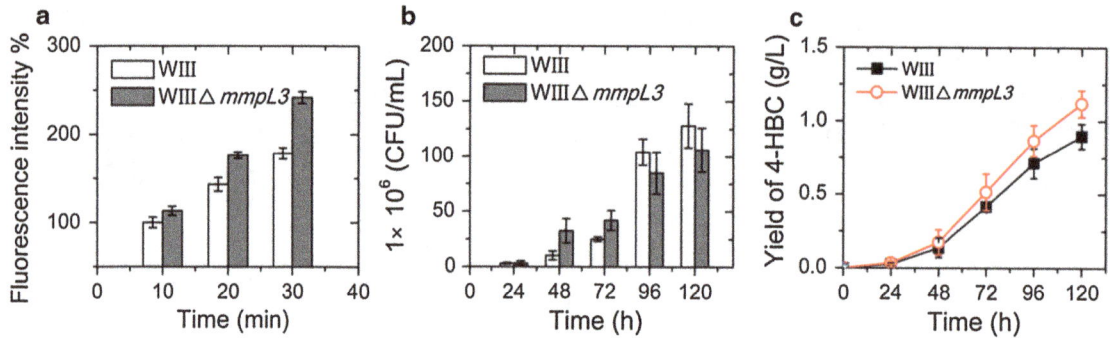

Fig. 4 Effects of deleting *mmpL3* on the improvement in 4-HBC production in the engineered mycobacterial strains. **a** Improvement in cell permeability in the *mmpL3*-deleted strain. The engineered strains were stained with FDA, incubated at 32 °C for 10 min and then the mixtures were analyzed by the fluorescence spectrophotometer. **b** Growth of the previously constructed 4-HBC-producing strain (WIII) and the *mmpL3*-deleted strain WIIIΔ*mmpL3* in MYC/02 medium with 2.0 g/L phytosterol. *CFU* colony forming unit. **c** Quantitative analyses of the 4-HBC production of the aerobic bioconversion by 4-HBC-producing strains in MYC/02 medium with 2.0 g/L phytosterol. Data represent the mean ± standard deviation of three measurements

4-HBC (18.5%), followed by the overexpression of *cyp125* (14.5%) and *fadA5* (12.1%). Overexpression of *choM2* showed the lowest improvement (8.9%) effect in the 4-HBC production. Therefore, the more efficient genes *cyp125*, *choM1* and *fadA5* were co-overexpressed to further strengthen the efficiency of the sterol catabolism in the strain WIIIΔ*mmpL3*. The growth of the engineered strain WIIIΔ*mmpL3-cyp125-choM1-fadA5* (co-overexpression of *cyp125*, *choM1* and *fadA5* in the strain WIIIΔ*mmpL3*) showed a significant decline compared with that of its ancestral strain WIII after 96 h (Fig. 5b). In order to confirm the effect of co-overexpression of the three key genes accurately, a "cyclodextrin-resting cell" system which had been applied in the industry [31] was employed here to evaluate the productivity of 4-HBC in the constructed strain. The strain WIIIΔ*mmpL3-cyp125-choM1-fadA5* yielded 7.59 g/L 4-HBC with the productivity of 0.0633 g/L/h after 120 h transformation, whereas the strain WIII yielded 6.31 g/L 4-HBC with

the productivity of 0.0526 g/L/h (Fig. 5c). The combined modifications significantly improved the productivity of the target intermediate 4-HBC by 20.3% in the industrial resting cell bio-transformation system.

Discussion

The diversity of metabolism in microorganisms, animals and plants provides us a huge database with enormous valuable natural products and metabolic intermediates [32]. Some natural products in organisms can be obtained directly using suitable extraction methods, but the concentrations of most of the metabolic intermediates are always low in organisms [33, 34].

As a traditional strategy to increase the production of target metabolites, metabolic engineering methods were used to increase the metabolism efficiency of the substrate to target metabolites [23, 24, 30]. Overexpression of key genes in the sterol catabolic pathway is effective to improve the productivity of the target 4-HBC. However,

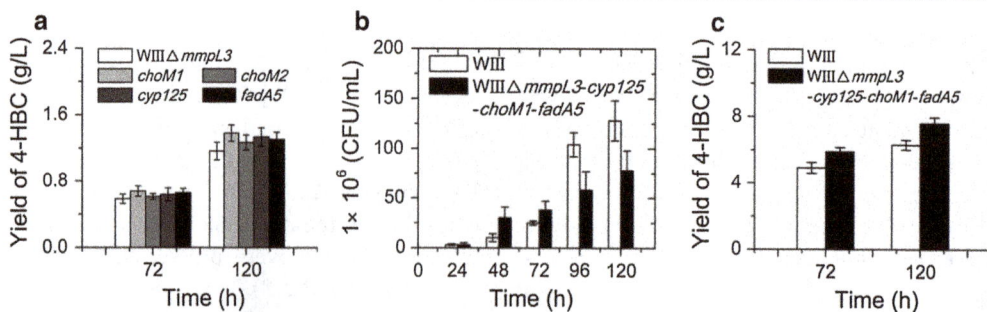

Fig. 5 Enhancement of the 4-HBC productivity by overexpressing the genes in the sterol conversion pathway. **a** Assessment of the 4-HBC production for overexpressing the genes in the upstream conversion pathway of sterol to 4-HBC. **b** Growth of the WIIIΔ*mmpL3-cyp125-choM1-fadA5* (co-overexpression of *cyp125*, *choM1*, and *fadA5* in the strain WIIIΔ*mmpL3*) and its ancestral strain WIII in MYC/02 medium with 2.0 g/L phytosterol. **c** Assessment of the 4-HBC production in the constructed 4-HBC-producing strains. A "cyclodextrin-resting cell" system with 20 g/L of phytosterol was used to determine the productivity of the engineered strains. Data represent the mean ± standard deviation of three measurements

the strategy requires an expression vector of pMV261, which is only suitable for the expression of a few genes [30]. Meanwhile, the expression plasmids in the cells may interfere with the subsequent modifications of the engineered strains. Notably, most of the key genes involved in the sterol uptake and metabolism are mainly distributed in a highly conserved gene cluster and the genes in the cluster exists as operons [35]. Therefore, in order to avoid the usage of expression plasmids for the overexpression of a large number of key genes in the early stage of modifications in engineered strains, the optimization of promoters will be adopted as the next strategy to enhance the expression levels of functional genes for further improving the productivity of target metabolites.

In order to improve cell permeability, various genes related to the transmembrane transport were tested by genetic modifications. The most effective modification was the deficiency of MmpL3. We confirmed that the assembly of mycolic acids of the cell envelope in the modified strain was partly inhibited due to the deletion of *mmpL3* (Additional file 2: Figure S3). Thus, the MmpL3 was probable not the sole transmembrane transporter for TMM in *M. neoaurum*. If it was the sole transmembrane transporter for TMM, the growth of the *mmpL3*-deficient strain would display a drastic decline due to the complete inhibition of TMM donation [14]. In addition, the deletion of *mmpL3* was proved to be a useful strategy to interfere in the normal cell envelope assembly. It is important to test whether the interference of the biosynthesis way of intracellular mycolic acids and the subsequent re-assembly of TMM, mediated by the antigen 85 proteins (FbpABC) [36] after TMM molecules were transported into the periplasm, would be beneficial to the improvement in cell permeability.

Conclusions

The deletion of *mmpL3* did elevate cell permeability without leading to an obvious growth inhibition. Both the utilization of sterol substrates and the production of 4-HBC in corresponding strains were significantly improved. Moreover, the combined modifications of CYP125, ChoM1, and FadA5, in the upstream pathway of sterol to 4-HBC, further enhanced the 4-HBC productivity.

Methods

Strains, plasmids, and culture conditions

The strains and plasmids used in this study are described in Table 1. The primers used for construction of the modified strains are described in Additional file 1: Table S2. *Escherichia coli* DH5α was used for plasmid replication. All *M. neoaurum* strains used here were derived from ATCC 25795. A typical 4-HBC-producing strain

WIII ($\Delta kshA\Delta hsd4A\Delta kstD123$) was constructed by deleting *kshAs*, *hsd4A*, *kstD1*, *kstD2*, and *kstD3* based on unmarked allelic homologous recombination [8]. The strain $\Delta mmpL3$ was constructed by deleting *mmpL3* in *M. neoaurum* ATCC 25795. The engineered strain WIII$\Delta mmpL3$ was constructed by deleting *mmpL3* in the WIII strain based on the allelic recombination.

Escherichia coli DH5α was cultured at 37 °C in 5 mL of Luria–Bertani (LB) medium (10.0 g/L tryptone, 5.0 g/L yeast extracts, 10 g/L NaCl, pH 7.0). Mycobacteria cells were firstly grew in 5 mL of LB medium ($OD_{600\,nm} = 1.0$–1.8). Then, cell suspensions were inoculated into (inoculum volume 1:10, v/v) 30 mL of MYC/01 medium (20.0 g/L glycerol, 2.0 g/L citric acid, 0.05 g/L ammonium ferric citrate, 0.5 g/L K_2HPO_4, 0.5 g/L $MgSO_4·7H_2O$, 2.0 g/L NH_4NO_3, pH7.5) in 250-mL flasks to prepare the mycobacterial strains ($OD_{600\,nm} = 1.2$–1.8).

For phenotypic identification, the cultivated mycobacterial strains were inoculated to (inoculum volume 1:10, v/v) 30 mL of MM (minimal medium, ammonium ferric citrate 0.05 g/L, K_2HPO_4 0.5 g/L, $MgSO_4·7H_2O$ 0.5 g/L, NH_4NO_3 2.0 g/L) with 1 g/L of cholesterol (purity >95.0%, Aladdin Reagents (Shanghai) Co., Ltd., Shanghai, China). For vegetative cell biotransformation, the cultivated mycobacterial strains were transferred into (inoculum volume 1:10, v/v) 30 mL of MYC/02 medium (10.0 g/L glucose, 2.0 g/L citric acid, 0.05 g/L ferric ammonium citrate, 0.5 g/L $MgSO_4·7H_2O$, 2.0 g/L NH_4NO_3, pH7.5) with 2 g/L of phytosterols (purity >95.0%, 100 g phytosterol contains 47.5 g β-sitosterol, 26.4 g campesterol, 17.7 g stigmasterol, 3.6 g brassicasterol and 4.8 g undetermined components) (Zhejiang Davi Pharmaceutical Co., Ltd., Zhejiang, China). Before steroid conversion, cholesterol or phytosterol (100.0 g/L) was firstly emulsified in Tween 80 (5% w/v) aqueous solution at 121 °C for 60 min and then added into MM or MYC/02 medium. All of the shake flask experiments were carried out at 30 °C with a shaking speed of 200 rpm in aerobic conditions for 120 h. For resting cell transformation, the cultivated mycobacterial strains were transferred into (inoculum volume 1:10, v/v) 150 mL of MYC/02 medium in 1000-mL shake flasks. After three days of growth at 30 °C under a shaking speed of 200 rpm, the cells were harvested by centrifugation at 8000×g for 15 min, washed with 20 mM KH_2PO_4, and diluted into cell suspensions (200 g/L). Resting cell transformation was performed in the system containing 100 g/L mycobacterial cells, 20 g/L phytosterols, and 80 g/L hydroxypropyl-β-cyclodextrin (HP-β-CD, RSC Chemical Industries Co. Ltd., Jiangsu, China) under non-sterile conditions in 250-mL flasks at 30 °C and 200 rpm [31]. Standard reference 4-HBC was purified and identified by ourselves [8]. Other solutions and reagents were

Table 1 Strains and plasmids used in this study

Name	Description	Source
Strains		
M. neoaurum ATCC 25795	Type strain of M. neoaurum	ATCC
ΔmmpL3	mmpL3 deleted in M. neoaurum ATCC 25795	This study
ΔmmpL3+mmpL3	mmpL3 complemented in strain ΔmmpL3	This study
WIII	kshAs, hsd4A, kstD1, kstD2 and kstD3 deleted in M. neoaurum ATCC 25795	[30]
WIIIΔmmpL3	kshAs, hsd4A, kstD1, kstD2, kstD3 and mmpL3 deleted in M. neoaurum ATCC 25795	This study
WIIIΔmmpL3-cyp125 -choM1-fadA5	cyp125, choM1and fadA5 overexpressed in strain WIIIΔmmpL3	This study
Plasmids		
p2NIL	Vector of two homologous arms for allelic recombination in mycobacteria, Kan^R	[37]
p2N-mmpL3	p2NIL carrying two homologous arms of mmpL3, Kan^R	This study
pGOAL19	Hyg, Pag85-lacZ, P_{hsp60}-sacB, PacI cassette vector, Amp^R	[37]
p19-mmpL3	p2NIL-derived with selection cassette from pGOAL19 for deletion of mmpL3 in mycobacteria	This study
pMV261	Shuttle vector of Mycobacterium and E. coli, carrying the heat shock (hsp60) promoter, Kan^R	Dr. W. R. Jacobs Jr.
p261-choM1	Recombinant pMV261, for overexpression of ChoM1 in mycobacteria	This study
p261-choM2	Recombinant pMV261, for overexpression of ChoM2 in mycobacteria	This study
p261-cyp125	Recombinant pMV261, for overexpression of CYP125 in mycobacteria	This study
p261-fadA5	Recombinant pMV261, for overexpression of FadA5 in mycobacteria	This study
p261-cyp125-choM1-fadA5	Recombinant pMV261, for overexpression of CYP125, ChoM1, and FadA5 in mycobacteria	This study
pMV306	Integration vector in Mycobacterium, without promoter, Kan^R	Dr. W. R. Jacobs Jr.
p306-mmpL3	pMV306-Phsp60-mmpL3, integrative into mycobacterial chromosomal DNA	This study

prepared according to the previously described method [24, 30].

Gene deletion, overexpression, and complementation in M. neoaurum

Target gene-deleted mutant strains were obtained via unmarked homologous recombination strategy in mycobacteria as previously described [23]. Plasmids of p2 N-mmpL3 and p19-mmpL3 were used for the knockout of mmpL3 (Table 1) [37].

To generate the target gene-overexpressed strains, the p261-gene was constructed as previously described [23]. The recombination of p261-genes was transferred into the WIIIΔmmpL3, respectively (Table 1). After PCR analysis using primer pair O-p261-F and O-p261-R, the correct monoclonal strains with sole gene-overexpressed were confirmed.

To complement the expression of target genes, the functional complementation recombinant based on pMV306 was constructed and complemented in the deletion strain according the previous method [23]. The expression cassette of mmpL3, containing a heat shock promoter hsp60, from the wild-type M. neoaurum (WT) was integrated into the double digestion sites of pMV306. Subsequently, this constructed plasmid p306-mmpL3 was transferred into the mmpL3-deleted mutant strain to complement the MmpL3 function.

Permeability analysis of the cell envelope

The permeability of the cell envelope was examined by measuring the fluorescence intensity of the cells labeled by fluorescein diacetate (FDA, Aladdin Reagents (Shanghai) Co., Ltd., Shanghai, China) according to previous procedures [38]. Cell suspensions (cell density reached 10^6 cells/mL) of 4.0 mL was mixed with 0.5 mL FDA acetone solution (2 mg/mL) and vibrated at 32 °C for 5 min before detection with a Fluoroskan Ascent Fluorescence Spectrophotometer (Thermo Labsystems Inc., PA, USA). The maximal excitation wavelength for FDA was 485 nm and the emission wavelength was 538 nm.

Analytical procedures of the mycolic acid methyl esters (MAMEs)

The MAMEs from the M. neoaurum cells were obtained according to previous procedures [36]. To avoid the possible interference from polar lipids, the cells (50 mg, in wet weight) were collected at $12,000 \times g$ for 10 min and then 0.5 mL of the mixture of methanol and chloroform (2:1, v/v) was added. The homogeneous single-phase mixture was incubated for 2 h at 60 °C and then the mixture containing the delipidated cells was centrifuged at $8000 \times g$ for 10 min. The MAMEs were prepared from the delipidated cells by incubation in 500 μL of 10% tetrabutylammonium hydroxide (Sigma-Aldrich LLC., MO, USA) overnight at 100 °C. After cooling, the mixtures were diluted with 500 μL of water, 250 μL of dichloromethane, and 62.5 μL of iodomethane (Sigma-Aldrich LLC., MO, USA), stirred for 30 min, and then centrifuged at 12,000×g for 10 min. The upper layer was removed and the lower organic layer was then washed with 1.0 mL of hydrochloric acid (1 M), followed by 1.0 mL of water. Subsequently, the reaction solution was dried under a stream of nitrogen to obtain crude MAMEs. The residue was dissolved in a mixture of toluene (0.2 mL) and acetonitrile (0.1 mL), followed by the addition of acetonitrile (0.2 mL), and then incubated for 1 h at 4 °C. The MAMEs were obtained by centrifugation and then re-suspended in dichloromethane.

The MAMEs were then analyzed by TLC on aluminum-backed silica gel 60-precoated plates F254 (Merck & Co., Inc., Hesse-Darmstadt, Germany) in a solvent system (chloroform: methanol, 90:10, v/v). The spots on the plates were observed after heating with cupric sulfate (10% w/v in an 8% v/v phosphoric acid solution).

Sterol transformation, sample extraction, and analysis

In this work, two methods including vegetative cell biotransformation and resting cell transformation were used for the assessment of steroid conversion capability [23, 31]. The conversion system was sampled (0.5 mL from the vegetative cell biotransformation; 0.1 mL from the resting cell transformation) every 24 h. The samples from the vegetative cell biotransformation were extracted with 0.5 mL of ethyl acetate and the samples from the resting cell transformation were extracted with 1.0 mL of ethyl acetate.

For gas chromatography (GC) analysis, a GC system 7820A (Agilent Technologies, CA, USA) was used in the quantitative determination of cholesterol and the mixture of phytosterols. The ethyl acetate extracts (5 μL) from the samples were injected into a DB-5 column (30 m × 0.25 μm (i.d.) × 0.25 μm film thickness, Agilent Technologies, CA, USA). The oven temperature was programmed as follows: 200 °C for 2 min, 200–280 °C within 4 min, 280 °C for 2 min, 280–305 °C within 1.5 min, and 305 °C for 10 min. Inlet and flame-ionization detector temperatures were maintained at 320 °C. Nitrogen carrier gas flow was 2 mL/min at 50 °C.

For high performance liquid chromatography (HPLC) analysis, the extracts of the samples containing 4-HBC were transferred into clean tubes, dried under vacuum, re-dissolved in methanol, and then centrifuged at 12,000×g for 20 min. The prepared samples were analyzed with a reversed-phase C18-column (250 × 4.6 mm) (Agilent Technologies, CA, USA) at 254 nm with an Agilent 1100 series HPLC (Agilent Technologies, CA, USA). The mixture of methanol and water (80:20, v/v) was used as the mobile phase.

Abbreviations

4-HBC: 22-hydroxy-23,24-bisnorchol-4-ene-3-one; TMM: trehalose monomycolate; ChoM: cholesterol oxidase; TDM: trehalose dimycolate; MAMEs: mycolic acid methyl esters; CYP125: cytochrome P450 125; FadA5: thiolase FadA5; 1,4-HBC: 22-hydroxy-23,24-bisnorchol-1,4-dien-3-one; 9-OHHBC: 9,22-dihydroxy-23,24-bisnorchol-4-ene-3-one; AD: androst-4-ene-3,17-dione; ADD: androst-1,4-diene-3,17-dione; 9-OHAD: 9α-hydroxy-4-androstene-3,17-dione; LB: Luria–Bertani; MM: minimal medium; FDA: fluorescein diacetate; HPLC: high performance liquid chromatography; GC: gas chromatography.

Author's contributions

LBX, HHL and WJS carried out the experiments. LBX and LQX analyzed the data. LBX, DZW and FQW conceived the study and reviewed the manuscript. All authors read and approved the final manuscript.

Acknowledgements

We sincerely thank T. Parish (Department of Infectious and Tropical Diseases, United Kingdom) for providing the plasmids, p2NIL and pGOAL19, and W. R. Jacobs Jr. (Howard Hughes Medical Institute) for providing the plasmids, pMV261 and pMV306.

Competing interests

The authors declare that they have no competing interests.

Funding

This work was supported by the National Natural Science Foundation of China (Grant No. 31370080) and the National Special Fund for State Key Laboratory of Bioreactor Engineering.

References

1. Wipperman MF, Sampson NS, Thomas ST. Pathogen roid rage: cholesterol utilization by Mycobacterium tuberculosis. Crit Rev Biochem Mol Biol. 2014;49:269–93.
2. Hsu FF, Soehl K, Turk J, Haas A. Characterization of mycolic acids from the pathogen Rhodococcus equi by tandem mass spectrometry with electrospray ionization. Anal Biochem. 2011;409:112–22.
3. Rosłoniec KZ, Wilbrink MH, Capyk JK, Mohn WW, Ostendorf M, Van der Geize R, Dijkhuizen L, Eltis LD. Cytochrome P450 125 (CYP125) catalyses C26-hydroxylation to initiate sterol side-chain degradation in Rhodococcus jostii RHA1. Mol Microbiol. 2009;74:1031–43.
4. Nishiuchi Y, Baba T, Yano I. Mycolic acids from Rhodococcus, Gordonia, and Dietzia. J Microbiol Methods. 2000;40:1–9.
5. Shtratnikova VY, Schelkunov MI, Fokina VV, Pekov YA, Ivashina T, Donova MV. Genome-wide bioinformatics analysis of steroid metabolism-associated genes in Nocardioides simplex VKM Ac-2033D. Curr Genet. 2016;62:643–56.
6. Wang FQ, Yao K, Wei DZ. From soybean phytosterols to steroid hormones, Soybean and Health. Croatia: Intech; 2011. p. 231–52.
7. Donova MV, Egorova OV. Microbial steroid transformations: current state and prospects. Appl Microbiol Biotechnol. 2012;94:1423–47.
8. Xu LQ, Liu YJ, Yao K, Liu HH, Tao XY, Wang FQ, Wei DZ. Unraveling and engineering the production of 23,24-bisnorcholenic steroids in sterol metabolism. Sci Rep. 2016;6:21928.
9. Toró A, Ambrus G. Oxidative decarboxylation of 17(20)-dehydro-23,24-dinorcholanoic acids. Tetrahedron Lett. 1990;31:3475–6.
10. Donova MV, Nikolayeva VM, Dovbnya DV, Gulevskaya SA, Suzina NE. Methyl-β-cyclodextrin alters growth, activity and cell envelope features of sterol-transforming mycobacteria. Microbiology. 2007;153:1981–92.
11. Daffé M, Draper P. The envelope layers of mycobacteria with reference to their pathogenicity. Adv Microb Physiol. 1998;39:131–203.
12. Draper P. The outer parts of the mycobacterial envelope as permeability barriers. Front Biosci. 1998;3:D1253–61.

13. Kaur D, Guerin ME, Škovierová H, Brennan PJ, Jackson M. Chapter 2: Biogenesis of the cell wall and other glycoconjugates of *Mycobacterium tuberculosis*. Adv Appl Microbiol. 2009;69:23–78.

14. Tahlan K, Wilson R, Kastrinsky DB, Arora K, Nair V, Fischer E, Barnes SW, Walker JR, Alland D, Barry CE III, Boshoff HI. SQ109 targets MmpL3, a membrane transporter of trehalose monomycolate involved in mycolic acid donation to the cell wall core of *Mycobacterium tuberculosis*. Antimicrob Agents Chemother. 2012;56:1797–809.

15. Atrat P, Hösel P, Richter W, Meyer HW, Hörhold C. Interactions of *Mycobacterium fortuitum* with solid sterol substrate particles. J Basic Microb. 1991;31:413–22.

16. Jankute M, Grover S, Rana AK, Besra GS. Arabinogalactan and lipoarabinomannan biosynthesis: structure, biogenesis and their potential as drug targets. Future Microbiol. 2012;7:129–47.

17. Fernandes P, Cruz A, Angelova B, Pinheiro HM, Cabral JMS. Microbial conversion of steroid compounds: recent developments. Enzyme Microb Tech. 2003;32:688–705.

18. Lisowska K, Korycka M, Hadław-Klimaszewska O, Ziółkowski A, Sedlaczek L. Permeability of mycobacterial cell envelopes to sterols: peptidoglycan as the diffusion barrier. J Basic Microb. 1996;36:407–19.

19. Sedlaczek L, Lisowska K, Korycka M, Rumijowska A, Ziółkowski A, Długoński J. The effect of cell wall components on glycine-enhanced sterol side chain degradation to androstene derivatives by mycobacteria. Appl Microbiol Biotechnol. 1999;52:563–71.

20. Rumijowska A, Lisowska K, Ziółkowski A, Sedlaczek L. Transformation of sterols by *Mycobacterium vaccae*: effect of lecithin on the permeability of cell envelopes to sterols. World J Microb Biotechnol. 1997;13:89–95.

21. Grzegorzewicz AE, Pham H, Gundi VA, Scherman MS, North EJ, Hess T, Jones V, Gruppo V, Born SE, Kordulakova J, Chavadi SS, Morisseau C, Lenaerts AJ, Lee RE, McNeil MR, Jackson M. Inhibition of mycolic acid transport across the *Mycobacterium tuberculosis* plasma membrane. Nat Chem Biol. 2012;8:334–41.

22. Takayama K, Wang C, Besra GS. Pathway to synthesis and processing of mycolic acids in *Mycobacterium tuberculosis*. Clin Microbiol Rev. 2005;18:81–101.

23. Yao K, Wang FQ, Zhang HC, Wei DZ. Identification and engineering of cholesterol oxidases involved in the initial step of sterols catabolism in *Mycobacterium neoaurum*. Metab Eng. 2013;15:75–87.

24. Xiong LB, Liu HH, Xu LQ, Wei DZ, Wang FQ. Role identification and application of SigD in the transformation of soybean phytosterol to 9α-hydroxy-4-androstene-3,17-dione in *Mycobacterium neoaurum*. J Agric Food Chem. 2017;65:626–31.

25. McLean KJ, Lafite P, Levy C, Cheesman MR, Mast N, Pikuleva IA, Leys D, Munro AW. The structure of *Mycobacterium tuberculosis* CYP125. J Biol Chem. 2009;284:35524–33.

26. Capyk JK, Kalscheuer R, Stewart GR, Liu J, Kwon H, Zhao R, Okamoto S, Jacobs WR Jr, Eltis LD, Mohn WW. *Mycobacterial cytochrome* P450 125 (Cyp125) catalyzes the terminal hydroxylation of C27 steroids. J Biol Chem. 2009;284:35534–42.

27. Nesbitt NM, Yang XX, Fontán P, Kolesnikova I, Smith I, Sampson NS, Dubnau E. A thiolase of *Mycobacterium tuberculosis* is required for virulence and production of androstenedione and androstadienedione from cholesterol. Infec Immun. 2010;78:275–82.

28. Shtratnikova VY, Bragin EY, Dovbnya DV, Pekov YA, Schelkunov MI, Strizhov N, Ivashina TV, Ashapkin VV, Donova MV. Complete genome sequence of sterol-transforming *Mycobacterium neoaurum* strain VKM Ac-1815D. Genome Announc. 2014;2:e01177–13.

29. Rodríguez-García A, Fernández-Alegre E, Morales A, Sola-Landa A, Lorraine J, Macdonald S, Dovbnya D, Smith MC, Donova M, Barreiro C. Complete genome sequence of '*Mycobacterium neoaurum*' NRRL B-3805, an androstenedione (AD) producer for industrial biotransformation of sterols. J Biotechnol. 2016;224:64–5.

30. Yao K, Xu LQ, Wang FQ, Wei DZ. Characterization and engineering of 3-ketosteroid-Δ^1-dehydrogenase and 3-ketosteroid-9α-hydroxylase in *Mycobacterium neoaurum* ATCC 25795 to produce 9α-hydroxy-4-androstene-3,17-dione through the catabolism of sterols. Metab Eng. 2014;24:181–91.

31. Gao XQ, Feng JX, Hua Q, Wei DZ, Wang XD. Investigation of factors affecting biotransformation of phytosterols to 9-hydroxyandrost-4-ene-3,-17-dione based on the HP-β-CD-resting cells reaction system. Biocatal Biotransform. 2014;32:343–7.

32. Lin YH, Sun XX, Yuan QP, Yan YJ. Extending shikimate pathway for the production of muconic acid and its precursor salicylic acid in *Escherichia coli*. Metab Eng. 2014;23:62–9.

33. Zhou YJ, Gao W, Rong QX, Jin GJ, Chu HY, Liu WJ, Yang W, Zhu ZW, Li GH, Zhu GF. Modular pathway engineering of diterpenoid synthases and the mevalonic acid pathway for miltiradiene production. J Am Chem Soc. 2012;134:3234–41.

34. Dai ZB, Liu Y, Zhang XN, Shi MY, Wang BB, Wang D, Huang LQ, Zhang XL. Metabolic engineering of *Saccharomyces cerevisiae* for production of ginsenosides. Metab Eng. 2013;20:146–56.

35. Van der Geize R, Yam K, Heuser T, Wilbrink MH, Hara H, Anderton MC, Sim E, Dijkhuizen L, Davies JE, Mohn WW, Eltis LD. A gene cluster encoding cholesterol catabolism in a soil actinomycete provides insight into *Mycobacterium tuberculosis* survival in macrophages. Proc Natl Acad Sci USA. 2007;104:1947–52.

36. Nguyen L, Chinnapapagari S, Thompson CJ. FbpA-dependent biosynthesis of trehalose dimycolate is required for the intrinsic multidrug resistance, cell wall structure, and colonial morphology of *Mycobacterium smegmatis*. J Bacteriol. 2005;187:6603–11.

37. Gordhan BG, Parish T. Gene replacement using pretreated DNA. Methods Molecular Medicine. *Mycobacterium tuberculosis* Protocols. Clifton: Humana Press; 2001. p. 77–92.

38. Shen YB, Wang M, Zhang LT, Ma YH, Ma B, Zheng Y, Liu H, Luo JM. Effects of hydroxypropyl-β-cyclodextrin on cell growth, activity, and integrity of steroid-transforming *Arthrobacter* simplex and *Mycobacterium* sp. Appl Microbiol Biotechnol. 2011;90:1995–2003.

11

Molecular and catalytic properties of fungal extracellular cellobiose dehydrogenase produced in prokaryotic and eukaryotic expression systems

Su Ma[1], Marita Preims[1], François Piumi[2], Lisa Kappel[3], Bernhard Seiboth[3], Eric Record[4], Daniel Kracher[1] and Roland Ludwig[1]* [ID]

Abstract

Background: Cellobiose dehydrogenase (CDH) is an extracellular enzyme produced by lignocellulolytic fungi. *cdh* gene expression is high in cellulose containing media, but relatively low CDH concentrations are found in the supernatant of fungal cultures due to strong binding to cellulose. Therefore, heterologous expression of CDH in *Pichia pastoris* was employed in the last 15 years, but the obtained enzymes were over glycosylated and had a reduced specific activity.

Results: We compare the well-established CDH expression host *P. pastoris* with the less frequently used hosts *Escherichia coli*, *Aspergillus niger*, and *Trichoderma reesei*. The study evaluates the produced quantity and protein homogeneity of *Corynascus thermophilus* CDH in the culture supernatants, the purification, and finally compares the enzymes in regard to cofactor loading, glycosylation, catalytic constants and thermostability.

Conclusions: Whereas *E. coli* could only express the catalytic dehydrogenase domain of CDH, all eukaryotic hosts could express full length CDH including the cytochrome domain. The CDH produced by *T. reesei* was most similar to the CDH originally isolated from the fungus *C. thermophilus* in regard to glycosylation, cofactor loading and catalytic constants. Under the tested experimental conditions the fungal expression hosts produce CDH of superior quality and uniformity compared to *P. pastoris*.

Keywords: *Aspergillus niger*, Cellobiose dehydrogenase, Cofactor loading, *Escherichia coli*, Glycoforms, Heterologous expression, *Pichia pastoris*, *Trichoderma reesei*

Background

Cellobiose dehydrogenase (CDH, EC 1.1.99.18, CAZy AA 3.1) is an extracellular flavocytochrome produced by a number of wood degrading fungi, when cellulosic materials are utilized as carbon source [1]. Recent studies showed that the physiological function of CDH is the donation of electrons to copper-dependent lytic polysaccharide monooxygenase (LPMO) [2, 3]. This oxidative CDH/LPMO system enhances the degradation rate of crystalline cellulose, and is widespread throughout the fungal kingdom together with the well-known hydrolytic cellulases [4, 5]. CDH is a monomeric enzyme that belongs to the glucose-methanol-choline (GMC) family of oxidoreductases [6]. It is composed of a large catalytic flavodehydrogenase domain (DH) containing one non-covalently bound flavin adenine dinucleotide (FAD). DH is connected to an electron transferring haem *b*-containing cytochrome domain (CYT) by a long, flexible linker enriched in hydroxy amino acids [7]. Recently reported crystal and solution structures of CDH demonstrated a

*Correspondence: roland.ludwig@boku.ac.at
[1] Department of Food Sciences and Technology, Vienna Institute of Biotechnology, BOKU-University of Natural Resources and Life Sciences, Vienna, Austria
Full list of author information is available at the end of the article

dynamic interaction between DH and CYT. CYT acts as a mobile domain that reduces the active site copper of LPMO [7]. This intricate structure makes CDH a difficult enzyme to produce. A phylogenetic analysis of CDH sequences from various fungal sources showed a division of the enzymes into three distinct classes: class I represents only basidiomycetous CDHs, which are shorter in sequence and have a highly conserved linker sequence [8]; class II exclusively comprises the more complex ascomycetous CDHs, either with or without a type-1 carbohydrate-binding module, corresponding to classes IIA and IIB, respectively. Class III contains a different branch of so far uncharacterized CDHs [9]. Due to its electron transferring properties, CDH has been widely recognized as a versatile biorecognition element in electrochemical biosensors, which is capable of detecting a wide variety of carbohydrates (cellobiose, cellodextrins, lactose, maltose, glucose) as well as quinones and catecholamines [10, 11]. This ability is also exploited for the development of CDH-based anodes in enzymatic biofuel cells [12]. Because of the high interest in this oxidoreductase, high quality preparations of CDHs are requested in large quantities.

The production of CDHs by lignocellulose-degrading fungi results in reasonable amounts, especially in media containing pure cellulose [13]. However, the cellulose-binding ability of CDH results in a relatively low enzyme concentration in the supernatant, which makes the protein purification from fungal cultures difficult and time-consuming. Therefore, several *cdh* genes have been cloned and recombinantly expressed [14–16]. Recombinant protein production allows a fast, reliable and efficient enzyme production and the possibility to generate genetically engineered enzymes. *Escherichia coli* is one of the most commonly used industrial microorganisms, but the recombinant production of intact CDH has not been achieved for this host so far. Like other secreted eukaryotic proteins, CDH is subjected to posttranslational modifications that affect the properties of the mature protein significantly [17]. These posttranslational modifications, such as O- and N-linked glycosylation, are not introduced by *E. coli*. Only the DH domain of *Phanerochaete chrysosporium* was reported to be functionally expressed in *E. coli* [18]. The produced activity after cell lysis was 733 U L^{-1} and the enzyme was purified to a specific activity of 16.7 U mg^{-1} with a yield of 7.3%. *Pichia pastoris* has been used as a heterologous expression host for several basidiomycetous and ascomycetous CDHs [8, 14, 15, 19–21]. It typically achieves high expression levels, and well established strategies for large-scale production and easy genetic manipulation are available. The yeast performs basic eukaryotic post-translational modifications and typically secretes high amounts of functional enzymes. High expression levels were achieved for most CDHs produced by *P. pastoris*. In case of *Pycnoporus cinnabarinus* CDH, 7800 U L^{-1} (351 mg L^{-1}) were measured in the culture supernatant [14]. It was, however, observed that CDHs produced by their natural hosts have a higher specific activity than CDHs recombinantly produced in *P. pastoris*. The specific activity, and consequently the turnover number, of recombinant CDH from *Corynascus thermophilus* (*Ct*CDH) was 5-times lower than observed for the enzyme isolated from the fungus [15]. This discrepancy was caused by a sub-stoichiometric occupation of catalytic sites with the FAD cofactor. Furthermore, recombinant CDHs produced by *P. pastoris* are typically over- or hyper-glycosylated, which might affect the essential intramolecular electron transfer reaction. To avoid these shortcomings, an expression host more closely related to fungal CDH producers would be of advantage. Nowadays, filamentous fungi are also well developed as expression hosts that are able to secrete large amounts of target proteins [22]. The development of new molecular genetic tools facilitates the usage of fungi in protein production [23], although only a limited number of fungal species have been explored so far, such as *Aspergillus niger*, *A. oryzae* or *Trichoderma reesei*. To date, two CDHs have been recombinantly produced in *A. oryzae* [24] and two in *A. niger* [16]. Maximal CDH activity reached 7620 U L^{-1} for the basidiomycetous *Coprinopsis cinerea* CDH, whereas only 126 U L^{-1} were measured for the ascomycetous *Podospora anserina* CDH [16]. However, the quality of the recombinant enzyme in comparison to *P. pastoris* derived CDH cannot be judged due to the limited information. Wang and Lu heterologously expressed the *cdh* gene from *P. chrysosporium* in *T. reesei* to study the synergism between CDH and cellulases [25]. But there is no further data of enzyme production or enzyme characterization. In order to identify the best expression host for high quality recombinant CDH, a detailed comparison of enzymes produced by the different expression hosts is required.

It was the objective of this study to compare four expression systems in regard to their recombinantly produced DH/CDH from *Corynascus thermophilus* (*Ct*CDH). The prokaryotic expression system *E. coli* was explored to recombinantly produce the catalytic dehydrogenase domain of *Ct*CDH (*Ct*DH). The intact, full-length *Ct*CDH was expressed in *P. pastoris* and in the two filamentous fungi *A. niger* and *T. reesei*. In order to achieve high productivity, all cultivations were carried out in a scalable bioreactor system. Finally, the recombinant enzymes were purified and their spectral properties, uniformity of glycosylation, kinetic constants and thermostability were compared.

Results

Production of the CtDH domain in E. coli and chromatographic purification

Production of CtDH was carried out in a 7-L bioreactor. In order to increase protein solubility, the operating temperature was reduced from 30 to 25 °C at an OD_{600} of 0.65. The fermentation was stopped after 28 h when the specific activity started to decline. The highest volumetric activity obtained was 648 U L^{-1} (Fig. 1a) with a corresponding specific activity of 0.42 U mg^{-1}. Approx. 1.5% (w/w) of the total proteins in the cell lysate accounted for CtDH.

The fermentation broth was harvested at a wet biomass concentration of approx. 4800 g L^{-1} and was disrupted using a homogenizer. After a one-step chromatographic purification, electrophoretic analysis showed that three major bands at approximately 53, 25 and 12 kDa were present (Additional file 1: Figure S1). Separation of these proteins was achieved by an ultrafiltration step. CtDH with a molecular mass of 53 kDa was purified to homogeneity at a yield of 60% and had a specific activity of 27.5 U mg^{-1} (Table 1). A summary of the purification procedures is presented in Additional file 1: Table S1. After purification, a bright-yellow protein solution typical for flavoproteins was obtained.

Production of CtCDH in P. pastoris and purification

Production of CtCDH in P. pastoris was carried out in a 7-L bioreactor according to a previously published protocol [15]. The volumetric CDH activity in the culture supernatant reached 376 and 320 U L^{-1} when measured with the DCIP assay and the cyt c assay, respectively. The recombinant CDH constituted up to 14% of the total proteins in the fermentation broth. CtCDH with an average molecular mass of 94 kDa was purified to homogeneity resulting in a specific activity of 9.4 U mg^{-1} (DCIP assay) and 3.5 U mg^{-1} (cyt c assay) at a final yield of 71%.

Production of CtCDH in A. niger and purification

Transformants were selected for their ability to grow on minimal medium plates without uridine. Approximately 100 uridine prototrophic transformants were obtained per microgram of expression vector and 24% of the obtained transformants showed CDH activity in the extracellular media after 10 days of cultivation.

The transformant with the highest CtCDH production was selected for further experiments. Production was carried out in a 7-L bioreactor with 4 L of A. niger culture medium inoculated with 8×10^8 asexual spores. The glucose concentration dropped below the measurable concentration after the first 3 days and fast growth of A. niger mycelium was observed before day 3. CDH activity increased gradually from day 3 to day 9, and then

reached its peak activity. The culture supernatant showed a volumetric activity of 49 U L^{-1} with the DCIP assay and 21 U L^{-1} with the cyt c assay at day 10 (Fig. 1b). The specific activity of the fermentation broth prior to purification was 0.44 U mg^{-1} with the DCIP assay and 0.19 U mg^{-1} with the cyt c assay. The recombinant CDH made up 3.5% of the total protein mass in the fermentation broth.

The recombinant enzyme was purified to homogeneity using a two-step purification protocol employing hydrophobic interaction chromatography and anion exchange chromatography. The purity of CtCDH increased 52-fold and the final purification yield was 54%. The specific activity of the homogeneous enzyme was 14.1 U mg^{-1} with the DCIP assay and 11.1 U mg^{-1} with the cyt c assay.

Production of CtCDH in T. reesei and purification

Transformants were selected based on hygromycin resistance and purified to uninuclear clones through a single-spore culture. Twenty transformants that grew well on hygromycin plates were selected for CDH expression in shaking flask cultures. To scale-up enzyme production, the highest producing clone of T. reesei was cultivated in a 2-L bioreactor. The glucose concentration decreased by 80% after the first day of incubation. The CDH activity increased rapidly from day 4 to 6, when full depletion of glucose was observed. The peak activity reached 715 U L^{-1} with the DCIP assay and 362 U L^{-1} with the cyt c assay (Fig. 1c). No further increase of CDH activity was observed from day 6 to 12, which suggests that the expression stopped at day 6 due to lack of glucose. The CtCDH concentration in the crude supernatant reached 29 mg L^{-1} based on the specific activity of the purified enzyme.

The recombinant CtCDH expressed in T. reesei was purified to apparent homogeneity as described for A. niger, and resulted in a 3.4-fold purification with a yield of 58% (Table 1). The already high specific activity of the culture supernatant (3.7 U mg^{-1}) indicates that the recombinant CtCDH is the main secreted protein (30%) in the T. reesei fermentation broth. The specific activity of the homogeneous enzyme was 14.3 U mg^{-1} with the DCIP assay and 12.5 U mg^{-1} with the cyt c assay.

Characterization of recombinant CtDH and CtCDHs
Molecular properties

Molecular masses of all recombinant enzymes were determined by SDS-PAGE (Fig. 2). The CtCDH expressed in P. pastoris showed a broad and diffuse band between 80 and 98 kDa, which was likely caused by heterogeneous glycosylation and the smearing of glycoproteins on SDS-PAGE gels, whereas the CtCDH expressed in A. niger showed a smaller and sharper band (82 ± 2 kDa).

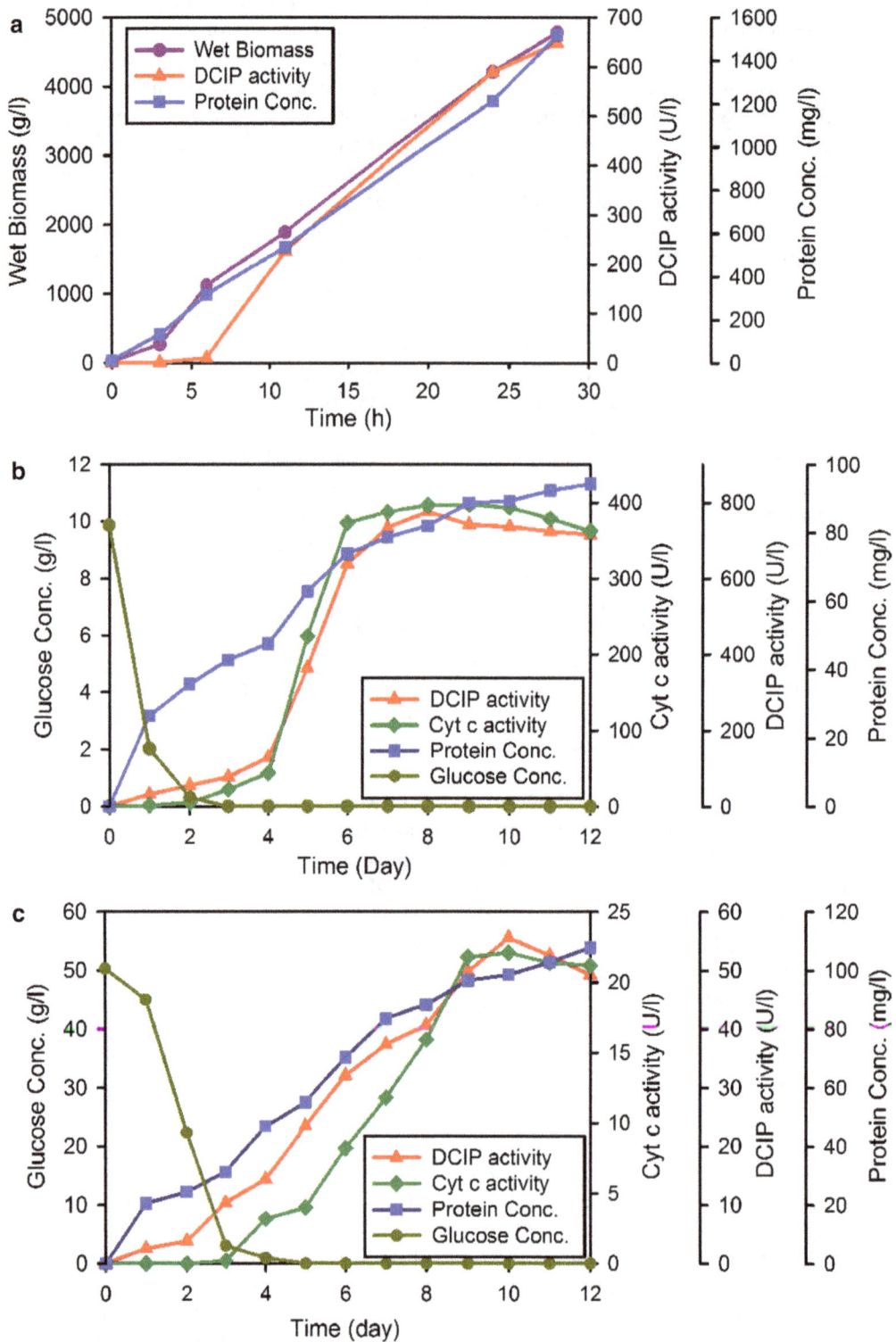

Fig. 1 Production of *CtDH/CtCDHs* expressed in *E. coli* (**a**), *A. niger* (**b**) and *T. reesei* (**c**). *Purple circles* wet biomass; *red triangles* volumetric activity (DCIP assay at acetate buffer pH 5.5), *green diamonds* volumetric activity (cyt c assay in Tris/HCl buffer, pH 7.5), *blue squares* protein concentration (per broth volume), *dark yellow circles* glucose concentration of fermentation broth. The measurements were done in triplicates; the difference between the values was less than 5%

Table 1 Comparison of enzyme production and purification for the four expression hosts

DH/CDH expressed by	E. coli	P. pastoris [15]	A. niger	T. reesei
Enzyme production				
Cultivation time (h)	28	111	288	288
Wet cell mass (g L^{-1})	4785 ± 160	275	n.d.	n.d.
Soluble protein conc. (mg L^{-1})	1517 ± 44[a]	633	108 ± 4	94 ± 3
DCIP activity (U L^{-1})	648 ± 19	376	49 ± 2	715 ± 14
Cyt c activity (U L^{-1})	–	320	21 ± 1	362 ± 11
DH/CDH expressed by	**E. coli**	**P. pastoris**	**A. niger**	**T. reesei**
Enzyme purification				
Purification steps	IMAC	Phenyl Sepharose and QSource	Phenyl Sepharose and QSource	Phenyl Sepharose and QSource
Purification yield (%)	60	71	54	58
Purification (fold)	65.5	7.7	51.8	3.4
Specific activity for DCIP (U mg^{-1})	27.5 ± 0.3	9.4 ± 0.2	14.1 ± 0.1	14.3 ± 0.1
Specific activity for cyt c (U mg^{-1})	–	3.5 ± 0.1	11.1 ± 0.2	12.5 ± 0.2
FAD loading (%)	52 ± 2	44 ± 1	56 ± 1	68 ± 3
Glycosylation (%)	0	18.9 ± 12.1	9.4 ± 2.7	1.4 ± 1.3

The measurements were done in triplicates

n.d. not determined, – the DH domain has no cyt c activity

[a] Intracellular protein

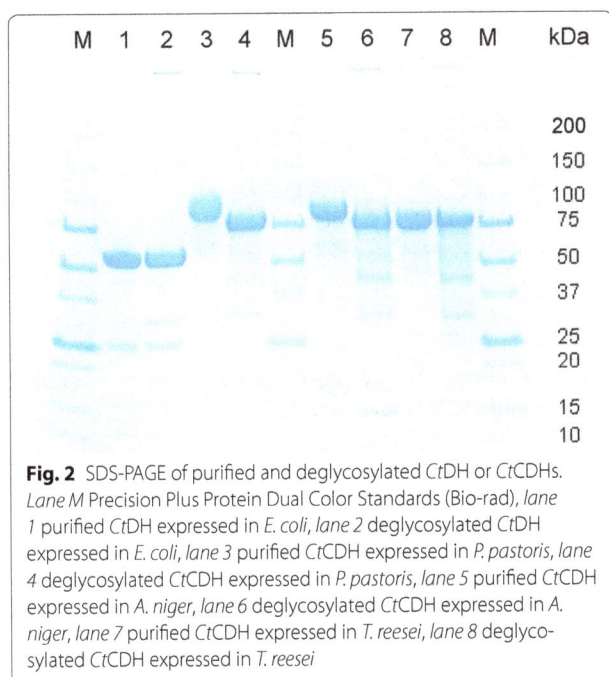

Fig. 2 SDS-PAGE of purified and deglycosylated CtDH or CtCDHs. Lane M Precision Plus Protein Dual Color Standards (Bio-rad), lane 1 purified CtDH expressed in E. coli, lane 2 deglycosylated CtDH expressed in E. coli, lane 3 purified CtCDH expressed in P. pastoris, lane 4 deglycosylated CtCDH expressed in P. pastoris, lane 5 purified CtCDH expressed in A. niger, lane 6 deglycosylated CtCDH expressed in A. niger, lane 7 purified CtCDH expressed in T. reesei, lane 8 deglycosylated CtCDH expressed in T. reesei

The CtCDH expressed in T. reesei showed the sharpest band with a molecular mass of 76 ± 1 kDa, indicating little glycosylation and a low heterogeneity of the protein. After deglycosylation under denaturing conditions with PNGase F, single sharp bands with an identical molecular mass of 75 kDa were found for all CDHs. The CtDH expressed in E. coli, before and after deglycosylation, showed identical bands with a molecular weight of 53 kDa. The additional bands at 35 kDa in the deglycosylated samples originate from PNGase F.

The UV/Vis spectrum of the purified CtDH (Fig. 3a) is characteristic for a flavoprotein. The FAD cofactor has an absorption spectrum with a maximum at 450 nm and a wide shoulder in the region of 360 nm. The FAD spectra of full-length CDHs are partially overlaid by the haem b absorbance. The spectra of all purified CtCDHs (Fig. 3b–d) show the typical characteristics of a flavocytochrome. The major peak of the oxidised spectrum at 420 nm is the Soret peak of the haem cofactor, whereas the broad shoulder between 450 and 500 nm is mainly attributed to the FAD cofactor. Upon reduction of CtCDH by its native substrate cellobiose, peaks appeared at 429, 533, and 564 nm, representing the Soret-, β- and α- peaks of the reduced haem. In this state, the absorption in the region between 450 and 500 nm decreased due to reduction of the FAD. The R_Z (A_{420}/A_{280}) values for CtCDHs expressed in P. pastoris, A. niger and T. reesei were 0.60, 0.61 and 0.61, respectively, indicating the same, high purity. After protein precipitation with trichloroacetic acid [26], the amount of the released FAD was determined. We found that 52% of the active sites of the DH domain expressed in E. coli contained FAD and that the FAD loading of CtCDH expressed in P. pastoris, A. niger and T. reesei was 44, 56 and 68%, respectively.

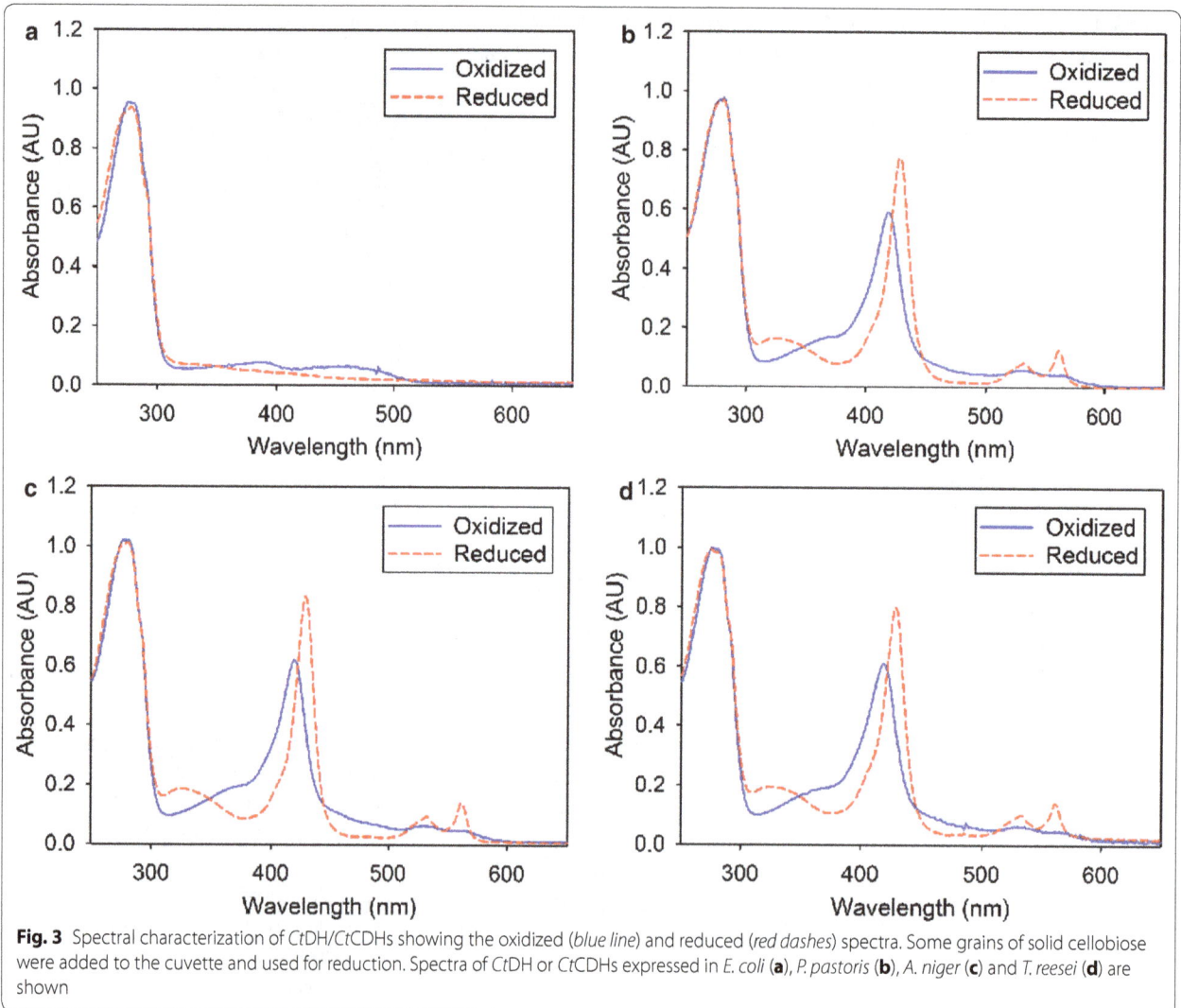

Fig. 3 Spectral characterization of *CtDH/CtCDHs* showing the oxidized (*blue line*) and reduced (*red dashes*) spectra. Some grains of solid cellobiose were added to the cuvette and used for reduction. Spectra of *CtDH* or *CtCDHs* expressed in *E. coli* (**a**), *P. pastoris* (**b**), *A. niger* (**c**) and *T. reesei* (**d**) are shown

Catalytic properties

The catalytic constants for the reduction of the two-electron acceptor DCIP were determined at pH 5.5, whereas reduction of the one-electron acceptor cyt *c* was determined at pH 7.5 (Table 2). The K_M values for cellobiose were similar for all enzymes, and were found to be around 362 µM measured with DCIP as electron acceptor and 100 µM using cyt *c* as electron acceptor. In contrast to the almost identical K_M values, a systematic difference of the k_{cat} values among the *CtDH* and the three *CtCDHs* was observed. The *CtDH* expressed in *E. coli* had the highest k_{cat} value (15.5 s^{-1}) for DCIP. *CtCDH* expressed by the eukaryotic hosts *T. reesei* and *A. niger* showed also high k_{cat} values of 15.2 and 13.3 s^{-1},

Table 2 Comparison of steady-state kinetic constants of *CtDH* and *CtCDHs* for cellobiose

Enzyme	K_M (µM)	k_{cat} (s^{-1})	k_{cat}/K_M (mM^{-1} s^{-1})
DCIP assay			
E. coli (DH)	367 ± 11	15.5 ± 0.2	42
P. pastoris (CDH)	373 ± 14	8.1 ± 0.3	22
A. niger (CDH)	357 ± 18	13.3 ± 0.3	37
T. reesei (CDH)	349 ± 21	15.2 ± 0.4	44
Cyt *c* assay			
P. pastoris (CDH)	81 ± 7	2.6 ± 0.1	32
A. niger (CDH)	102 ± 5	7.6 ± 0.2	74
T. reesei (CDH)	118 ± 11	7.9 ± 0.2	67

The measurements were done in triplicates

respectively, but CtCDH expressed in P. pastoris had a twofold lower k_{cat}. The cyt c activity for both filamentous fungal expression systems was equally good, resulting in k_{cat} values of 7.6 and 7.9 s^{-1}. For P. pastoris the turnover number was 3 times lower.

Thermostability

The thermostability of CtDH and CtCDHs was determined using the ThermoFAD method [27] and the intrinsic tryptophan fluorescence upon protein unfolding. All enzymes showed transition midpoint temperatures (T_m) in a narrow temperature range of 58.1 ± 1.5 °C. T_m values obtained with the ThermoFAD method were generally higher by 1–1.5 °C than T_m values obtained from tryptophan fluorescence (Table 3; Additional file 1: Figure S2). The T_m values for CtCDH expressed in P. pastoris and A. niger were slightly higher than those for CtCDH expressed in E. coli and T. reesei. This was observed with both methods. Protein precipitation occurred at the end of the unfolding experiments in case of the nonglycosylated CtDH and the less glycosylated CtCDHs expressed by A. niger and T. reesei.

Discussion

This study compared the suitability of four different expression systems for the production of CDH. We evaluated the titer of secreted CDH, the purification procedure and yield together with the molecular and kinetic properties of the purified, recombinant enzymes. The prokaryotic expression host E. coli can express only the DH domain of CtCDH, but produced a high amount of it in a much shorter time (28 h) than the tested eukaryotic expression systems. Per liter of fermentation medium 652 U of CDH activity were obtained, which is 1.7-fold higher than the previously reported amount of the DH domain from Phanerochaete chrysosporium (PcDH) expressed in E. coli [18].

The full-length CDH, comprising both CYT and DH, could so far only be functionally expressed in eukaryotic expression hosts. Several CDHs have been successfully expressed in the methylotrophic yeast P. pastoris,

including CtCDH [15]. The reported volumetric expression levels vary from 376 to 7800 U L^{-1}, which can be recalculated to CDH concentrations between 79 and 351 mg L^{-1}, respectively (Additional file 1: Table S2). Four CDHs have been expressed in the fungal expression hosts A. oryzae and A. niger [16, 24], with different expression levels. Additional file 1: Table S2 shows that basidiomycetous CDHs are generally expressed with higher volumetric activities than ascomycetous CDHs, which typically have lower specific activities. In terms of produced CDH concentration, the expression levels between basidiomycetous and ascomycetous CDHs are comparable, but differences arising from diverse fermentation protocols are obvious. In this work, the production level of CtCDH in A. niger was very low (55 U L^{-1}), which is in accordance with the reported data for other ascomycetous CDHs. This study reports also the first successful expression of CtCDH in T. reesei. The expression level of 800 U L^{-1} was the highest among the tested eukaryotic expression hosts in this study and compares well with other expression hosts in terms of secreted CDH concentration (32 mg L^{-1}). Since T. reesei is known to produce high concentrations of recombinant proteins, an optimization of the fermentation protocol should further increase the CDH yield [29].

The specific CDH activity in the cell lysate or the culture supernatant differed among the expression hosts and influenced the purification strategy. The weight percentage of CtDH in the E. coli cell lysate was only 1.5% and a high amount of intracellular proteins was present. For easy purification of the DH domain it was necessary to fuse a C-terminal His-tag to the enzyme, which allowed a one-step purification via immobilized metal affinity chromatography and resulted in a much higher yield of 60% compared to only 2% for recombinant PcDH without His-tag [18]. Although E. coli is not able to introduce eukaryotic posttranslational modifications, the homogeneous CtDH exhibited a high specific activity of 27.5 U mg^{-1} using the DCIP assay.

Pichia pastoris is a well-established expression system which allows for a relatively high CtCDH production (15% of total protein). However, the low cofactor occupancy and overglycosylation resulted in a CtCDH with a limited quality. The specific activity was 9.4 U mg^{-1} for DCIP and 5.3 U mg^{-1} for cyt c. Fungal hosts are thus preferable for the expression of high quality CtCDH. Homogenous CtCDH has a specific activity around 14 U mg^{-1} using the DCIP assay and 12 U mg^{-1} using the cyt c assay, which is comparable to the enzyme expressed by the native host [15]. The newly employed expression host T. reesei showed the capability of secreting the mature CtCDH into the fermentation broth, which made up 30% of all proteins in the supernatant. A. niger

Table 3 Transition midpoint temperatures (T_m) measured with the ThermoFAD method and the intrinsic tryptophan fluorescence

Enzyme	ThermoFAD	Trp unfolding
E. coli (DH)	57.7 ± 0.3	56.9 ± 0.4
P. pastoris (CDH)	59.2 ± 0.1	57.4 ± 0.6
A. niger (CDH)	59.7 ± 0.1	57.5 ± 0.5
T. reesei (CDH)	57.8 ± 0.2	56.6 ± 0.3

The measurements were done in triplicates

gave a low expression level (3.5%) for this ascomycetous CDH. The purification strategy worked similarly efficient for all three eukaryotic expression systems, giving yields between 54 and 60%.

Glycosylation is a common post-translational modification in extracellular fungal proteins. Presumably, all native CDHs contain glycan structures, although the degree of glycosylation varies significantly, from 2% [28] up to 15% [29] of the total CDH protein mass. However, the most commonly used expression host *P. pastoris* usually attaches high mannose structures as N-glycosides, which can make up for 10–48% of the total mass (Additional file 1: Table S2). Six predicted glycosylation sites are found for the intact *Ct*CDH using NetNGlyc 1.0, of which five are surface exposed and located on the DH domain. Interestingly, three glycosylation sites are close to the linker connecting DH and CYT. In addition, there are eleven putative O-glycosylation sites located on both domains (NetOGlyc 4.0). However the extent of O-glycosylation in native CDHs is uncertain. We conclude that the demand for highly active and uniform recombinant CDH preparations used for electrochemical and biocatalytic applications might be best served by CDH produced by *T. reesei*. SDS-PAGE indicated a very uniform glycosylation that makes up approx. 1.3% of the total protein mass. A higher percentage of glycosyl residues was observed for *Ct*CDH expressed in *A. niger* (8%) and the overglycosylation and heterogeneity of *Ct*CDH produced in *P. pastoris* is least favorable in this respect (7–30%).

Several sequence and structural factors have been proposed to contribute toward a greater stability of thermophilic proteins, such as the presence of prolines in loop regions or the stability of alpha-helix and surface salt bridges [30–32]. Glycosylation is another feature of eukaryotic proteins that frequently contributes to stability [33]. We compared the melting temperature of *Ct*DH and all *Ct*CDHs expressed in the study. Interestingly, neither the different FAD occupancy of the enzyme preparations nor their different degree of glycosylation influenced the overall thermostability. Using two methods, we showed that the T_m values of all enzyme preparations were within 1.5 °C, including the least flavinated *Ct*CDH expressed in *P. pastoris*.

The presence of the FAD cofactor in the DH domain of CDH is crucial for the enzymatic activity and was reported as a limiting factor in *P. pastoris* expressed *Ct*CDH [15]. In this work, the FAD loading of all four enzymes was experimentally measured and analyzed together with the catalytic turnover. For eukaryotic expression hosts, *Ct*CDH expressed in *T. reesei* had the highest FAD loading (65%) and gave the highest turnover number. In contrast, *Ct*CDH expressed in *P. pastoris*

showed the lowest specific activity and the lowest FAD content. The re-calculated k_{cat} values (normalized by the FAD loading) are very similar (21.5 ± 2.8 s^{-1}), demonstrating that the FAD loading is the key factor for the catalytic capability of the DH domain. However, the re-calculated k_{cat} value of *Ct*DH is higher (29.8 s^{-1}), which could indicate that the substrate channel is more accessible in absence of the CYT domain.

The DCIP reduction rate depends only on the catalytic reaction in the DH domain, whereas the reduction of cyt *c* is limited by the intramolecular electron transfer (IET) between DH and CYT. The cyt *c* assay has been reported to give a good estimation of the IET rate, because the electron transfer between the cofactors is, compared to the catalytic turnover at the FAD, rate limiting. It is interesting to observe that overglycosylation of the CDH molecule reduces the IET rate. A decrease of the IET is visible from the 3–5 times lower k_{cat} values of cyt *c* reduction for the highly glycosylated *Ct*CDH expressed by *P. pastoris* and *A. niger* (Table 2). The k_{cat} value of *Ct*CDH expressed in *A. niger* for cellobiose measured with the DCIP assay is about two times higher than for *Ct*CDH expressed in *P. pastoris* whereas the k_{cat} value for cellobiose measured with the cyt *c* assay is two times lower. This indicates that the *Ct*CDH expressed in *A. niger* has a reduced IET rate. A comparison of the specific cyt *c* activity of the recombinant *Ct*CDH expressed by *T. reesei* with *Ct*CDH produced by *C. thermophilus* show a similar value, but the specific activity for DCIP is only 40% compared to the published data [15]. These results suggest that for recombinant CDH expressed by fungal hosts, the FAD cofactor loading is a more important issue than glycosylation. An optimization of CDH expression in *T. reesei* seems to be the best starting point for future studies.

Conclusions

Escherichia coli is a good expression system to express the flavodehydrogenase domain of *Ct*CDH. It is easy to manipulate and fast producing. The *P. pastoris* expression system allows high protein yields, however, the low FAD loading and overglycosylation of CDH cause low specific activities. The fungal expression systems produced *Ct*CDH of superior quality and uniformity. Under the tested experimental conditions, the production of *Ct*CDH by *A. niger* results in lower amounts than *Ct*CDH production by *T. reesei*, but both fungal expression systems could be further optimized towards increased productivity by optimizing promoters and expression conditions. In conclusion, *T. reesei* is the best expression system for recombinant *Ct*CDH production. The produced *Ct*CDH has a high cofactor loading and the glycosylation and specific activity are closest to the CDH isolated from *C. thermophilus*.

Methods

Strains and media

The chemically competent *E. coli* strains NEB 5α and T7 Express were purchased from New England Biolabs (New England BioLabs, Frankfurt, Germany). *E. coli* NEB 5α was used for vector construction and propagation. *E. coli* cells were grown at 37 °C in lysogeny broth (LB) or on LB agar supplemented with ampicillin (100 mg mL^{-1}). MagicMedia *E. coli* expression medium (Thermo Fisher Scientific) was used for expression studies.

Pichia pastoris strain X-33 containing the published plasmid pPICctcdh was used for *Ct*CDH expression [15]. *P. pastoris* cells were grown in yeast peptone dextrose (YPD) broth or on YPD plates with zeocin (100 mg L^{-1}) at 30 °C and the Basal Salts Medium was used for fermentation.

Aspergillus niger strain D15#26 (*pyrG-*) [34] was used for heterologous expression of the recombinant *Ct*CDH. After cotransformation with vectors containing the *pyrG* gene and the expression cassette (Fig. 4c), respectively, transformants of *A. niger* were selected at 30 °C on solid minimal medium (without uridine) containing 70 mM NaNO$_3$, 7 mM KCl, 11 mM KH$_2$HPO$_4$, 2 mM MgSO4, and 1% (w/v) glucose and trace elements (1000× stock; 76 mM ZnSO$_4$, 178 mM H$_3$BO$_3$, 25 mM MnCl$_2$, 18 mM FeSO$_4$, 7.1 mM CoCl$_2$, 6.4 mM CuSO$_4$, 6.2 mM Na$_2$MoO$_4$, and 174 mM EDTA). The

best producing transformant for CDH expression was screened on culture medium containing 70 mM NaNO$_3$, 7 mM KCl, 200 mM Na$_2$HPO$_4$, 2 mM MgSO$_4$, 5% (w/v) glucose and trace elements at pH 5.5 inoculating it with 2×10^6 spores mL^{-1}. The same medium was used for fermentation.

Trichoderma reesei Δ*xyr1* strain [35] was used for *Ct*CDH expression. This strain derived from *T. reesei* QM9414 (ATCC 26921) is deleted in the major cellulase and xylanase regulator which provides a (hemi)cellulase free background. It was maintained on potato dextrose agar (PDA) plates at 28 °C. Transformants of *T. reesei* were grown for selection on PDA with hygromycin B (50 mg L^{-1}). For *Ct*CDH expression, strains were grown at 28 °C in a modified MA-Medium (Mandels–Andreotti) [36] containing 10 g L^{-1} glucose, 1.4 g L^{-1} (NH$_4$)$_2$SO$_4$, 2.0 g L^{-1} KH$_2$PO$_4$, 0.3 g L^{-1} MgSO$_4$·7H$_2$O, 0.4 g L^{-1} CaCl$_2$·2H$_2$O, 1 g L^{-1} peptone, and 1/50 (v/v) of the trace element solution (0.25 g L^{-1} FeSO$_4$·7H$_2$O, 0.08 g L^{-1} MnSO$_4$·H$_2$O, 0.07 g L^{-1} ZnSO$_4$·7H$_2$O, 0.1 g L^{-1} CoCl$_2$·2H$_2$O) at pH 5.0 inoculated with 10^6 spores mL^{-1}. The same medium was used for fermentation.

Chemicals and vectors

All chemicals were purchased from Sigma, Fluka, Roth or VWR and were of the highest purity available. Primers were synthesized by Microsynth and nucleotide

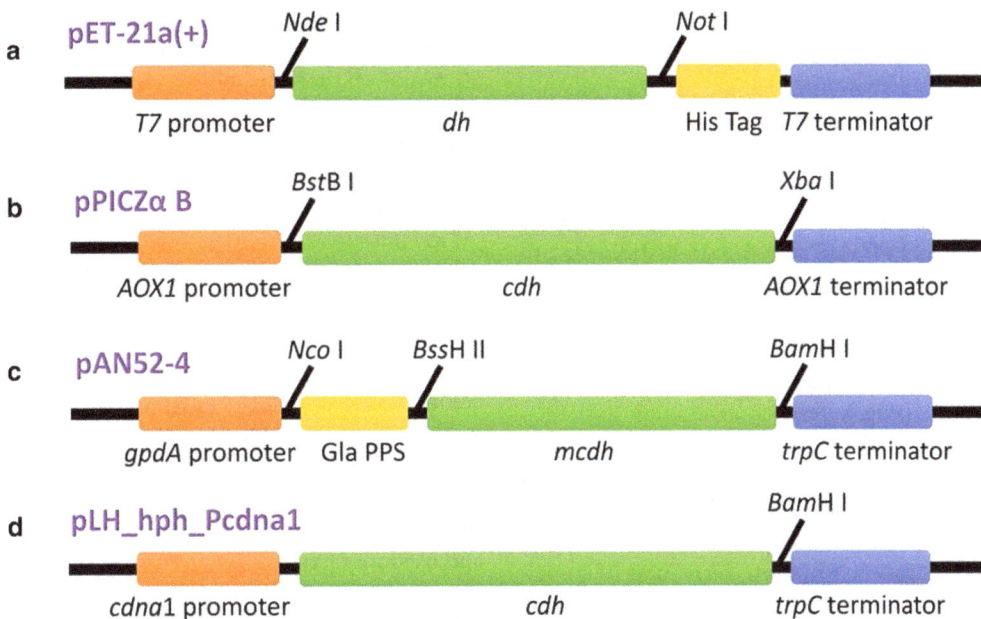

Fig. 4 Expression plasmids for *Ct*DH/*Ct*CDHs production. Schematic presentation of the *Ct*DH/*Ct*CDHs expression cassettes used in this study. *dh* the encoding gene of the flavin domain of *Ct*CDH, *cdh* the encoding gene of the full-length of *Ct*CDH with its own signal peptide, *mcdh* the encoding gene of the mature *Ct*CDH without its own signal peptide, *Gla PPS* the gene of the glucoamylase (GLA) prepro sequence. See experimental procedures for more information

Table 4 Nucleotide sequences of primers used in this study

Primer name	Sequence (from 5′ to 3′)
Pecoli-F1	GGAATTC CATATG GACACGTATGATTACATCGT
Pecoli-R1	ATAAGAAT GCGGCCGC ATAACGCAGGGACAGGATGC
Paniger-F1	TTG GCGCGC TCAGATGACCGAAGGGACGTA
Paniger-R1	CG GGATCC CTAATACCGCAGGGACAGGA
Ptreesei-F1	TCATCGATGTCGACCATGAAGCTTCTCAGCCGCGTTG
Ptreesei-R1	CG GGATCC CTAATACCGCAGGGACAGGATG
Ptrpc-F1	CG GGATCC GAAGCTTGAGATCCAC
Ptrpc-R1	CCAAGCTTGCATGCCAAGAAGGATTACCTCTAAACAAG

The underlined characters indicate the restriction site

sequences are shown in Table 4. Restriction enzymes, dNTP mix and T4 DNA ligase were from Fermentas and the Phusion polymerase from New England Biolabs. The plasmid pET-21a (+) from Novagen was used for expression in *E. coli* under control of the T7 promoter. For *Ct*CDH expression in *A. niger*, pAN52-4 [37] and pAB4-1 [38] containing the *pyrG* selection marker were used. In expression vector pAN52-4, the *A. niger* constitutive *gpdA* promoter and the *A. nidulans trpC* terminator were used to drive the expression of recombinant *Ct*CDH. Vector pLH_hph_Pcdna1 [39] containing an hygromycin B expression cassette as fungal selection marker followed by the constitutive *cdna1* promoter region was used for *Ct*CDH expression in *T. reesei*.

Construction of the expression vectors
The gene of the CDH flavin domain was PCR-amplified from pPIC*ctcdh* (Fig. 4b), a vector previously created to express the entire CDH in *P. pastoris* under the control of the AOX1 promoter [15]. Primers Pecoli-F1 and Pecoli-R1 (Table 4) were used to introduce the *Nde*I and *Not*I restriction sites, respectively. The PCR product was purified, digested and ligated into the 5′-*Nde*I and 3′-*Not*I sites of pET-21a(+) in-frame with the six-histidine tag to generate plasmid pET21-CtDH (Fig. 4a).

The DNA fragment encoding the mature *Ct*CDH was amplified using primers Paniger-F1 and Paniger-R1 (Table 4). The amplicon was integrated into the pAN52-4 expression vector using a restriction cloning approach with *Bss*HII and *Bam*HI enzymes. In addition, the synthesized oligonucleotides of the glucoamylase (GLA) prepro sequence [40] was digested with *Nco*I and *Bss*HII and ligated into the respective sites of pAN52-4 using the Rapid DNA Ligation Kit from Fermentas (Fig. 4c). The final construct pAN52-CtCDH was used to express *Ct*CDH in *A. niger*.

The gene encoding the full-length CDH with its own signal peptide was PCR-amplified from pPIC*ctcdh* using primers Ptreesei-F1 and Ptreesei-R1 (Table 4). A

PCR-amplicon of the *A. nidulans trpC* terminator was digested by *Bam*HI and ligated with the same predigested PCR product of *Ct*CDH. The vector pLH_hph_Pcdna1 was linearized by *Sbf*I restriction enzyme and ligated with DNA fragment of *Ct*CDH and TrpC terminator using In-Fusion HD Cloning Kits following the manufacturer's protocol (Takara Bio Europe, Saint-Germain-en-Laye, France) (Fig. 4d). Correct insertion of the gene was checked by DNA sequencing.

Selection of the production clones
Correct insertion of the *Ct*DH gene was checked by DNA sequencing and verified plasmid was transformed into *E. coli* T7 Express. The cells harboring pET21-CtDH were cultivated in small-scale with the auto-inducing MagicMedia to optimize the expression conditions. Baffled shaken flasks (250 mL) were filled with 78 mL MagicMedia and 2 mL of overnight seed culture (1:40 dilution) and were incubated at 30 °C in a shaking incubator (130 rpm) until an $OD_{600} > 6.0$ was reached. Then the temperature was reduced to 25 °C for 18 h more. Cells were harvested by centrifugation, resuspended in Tris–HCl buffer (50 mM, pH 7.5; 0.5 M NaCl), and disrupted by ultrasonication. After centrifugation, the enzymatic activity of the supernatant was compared using the DCIP assay.

Aspergillus niger cotransformation was carried out as previously described by Punt and van den Hondel [41], using the pAN52-CtCDH and pAB4-1 in a 10:1 ratio. Transformants were selected for uridine prototrophy on selective solid minimum medium (without uridine) and incubated for 10 days. In order to screen the best transformant for enzyme production in liquid medium, 50 mL of culture medium was inoculated with 2×10^6 spores mL^{-1} in a 250-mL baffled flask. The culture was monitored for 14 days at 30 °C and 130 rpm. The pH was adjusted to 5.5 with 1 M citric acid and activity was checked daily.

Trichoderma reesei transformation was performed as described [42] using uncut plasmid DNA. Transformants were streaked twice onto PDA plates that contained 50 µg mL^{-1} of hygromycin B, and purified by plating conidiospores onto PDA plates with 0.1% Triton X-100 as colony restrictor. Candidate transformants were screened in shaking flasks (160 rpm, 28 °C) in a modified MA-Medium (50 mL in a 250-mL shaken flask) for 10 days. The pH was adjusted daily to 5.0 with 2 M KOH. Culture supernatants were collected by filtration with Miracloth and used for enzymatic analysis.

Production of recombinant *Ct*DH/*Ct*CDHs in different systems
An overnight seed culture of the *E. coli* transformant (selected from a LB plate with 100 µg mL^{-1} ampicillin)

was inoculated at a 1:40 dilution into 6 L MagicMedia (which starts the expression of CtDH by autoinduction) with 100 µg mL^{-1} ampicillin in a 7-L glass vessel fermenter (MBR Bioreactor, Wetzikon, Switzerland). The initial cultivation temperature was 30 °C, the variable airflow rate was around 6 L min^{-1}, and the agitation was set to 500 rpm. After 6 h, the cultivation temperature was changed to 25 °C. Samples were taken initially and after 3, 6, 11, 24 and 28 h. Cells were collected by centrifugation and used to determine wet biomass. The pellet was resuspended in 50 mM Tris–HCl buffer and disrupted by ultrasonication. The intracellular CDH activity and protein concentration were assayed after centrifugation. The corresponding specific activity of the crude extract was calculated from the volumetric activity of the crude extract divided by the specific activity of the purified enzyme.

Aspergillus niger fermentation was carried out in 7-L MBR fermenter, inoculating 8×10^9 spores in 4 L culture medium. The following set points were used: pH 5.5, T = 30 °C, Airflow 6 L min^{-1}. The pH was maintained using 2 M citrate acid. Initial agitation speed was 350 rpm. At day 5, agitation was increased to 600 rpm and 0.5 L water was added to submerge the mycelium. Samples were taken daily and the CDH activity, protein and glucose concentration of the supernatant were measured.

Cultivation of *T. reesei* was performed in a Sixfors bioreactor (Infors HT, Bottmingen, Switzerland) with a working volume of 0.4 L. Inoculum culture was pre-grown in 250 mL Erlenmeyer flasks on a rotary shaker (160 rpm) at 28 °C, containing 50 mL modified MA-medium. Pre-culture grown for 24 h was filtered, the biomass washed with sterile water and transferred into the fermenter. Operating conditions were: pH 5.0, 28 °C, 400 rpm and 0.4 vvm (volumes of air per volume of liquid per minute). Samples were taken daily and activity was measured.

Enzyme production with each expression system was repeated twice (biological duplicates). The expression of volumetric activity and protein concentration differed by less than 10% (data not shown). We only report the results of the fermentation from which we purified the enzyme. All measurements of volumetric activities and protein concentrations were performed three times (technical triplicates). Protein concentrations were determined by the Bradford method using a prefabricated assay from Bio-Rad Laboratories and bovine serum albumin (BSA) as the calibration standard [43]. Glucose concentrations were measured by D-glucose Assay Kit (Megazyme, Wicklow, Ireland).

Purification of recombinant CtDH/CtCDHs in different systems
The *E. coli* fermentation broth was centrifuged at 4000×g for 10 min at 4 °C, the pellets were suspended in 1.5 L

Tris–HCl buffer (50 mM, pH 7.5; 0.5 M NaCl) and disrupted using a APV Rannie und Gaulin homogenizer. The crude extract obtained by centrifugation (6000×g, 30 min, 4 °C) was loaded onto a HiTrap HP column (65 mL, GE Healthcare). The His-tagged CtDH was eluted by a linear gradient of imidazole (5–500 mM) in Tris–HCl buffer (50 mM, pH 7.5; 0.5 M NaCl). Active fractions (~60 mL) were pooled and desalted using a HiTrap Desalting Column. The enzyme solution was further purified and concentrated 10-fold using an ultrafiltration module (30-kDa cutoff, Sartorius Stedim, Germany).

The purification strategy for the supernatants of *A. niger* and *T. reesei* fermentations was similar as previously described [15]. The culture broth was clarified by filtration with Miracloth and centrifuged at 6000×g for 20 min. Saturated ammonium sulfate solution was gently added to give a 20% saturation and particles were removed by centrifugation (6000×g, 30 min, 4 °C). The clear supernatant was applied on a Phenyl FF hydrophobic column (70 mL, GE Healthcare) equilibrated with 50 mM sodium acetate buffer pH 5.5 containing 20% (saturation) ammonium sulfate. The proteins were then eluted by a linear gradient from 20 to 0% $(NH_4)_2SO_4$ in the same buffer in 5 column volumes at a flow rate of 1 mL min^{-1}. Absorbances at 280, 420 (haem b), and 450 nm (FAD) were measured online along with the conductivity values. Fractions were collected automatically and tested by the cyt c activity assay. Active fractions were pooled and then diafiltered with sodium acetate buffer using an ultrafiltration module with a 10-kDa cutoff until a conductivity of 2–3 mS cm^{-1} was obtained. The partially deionized enzyme solution was loaded on a Q15-source anion exchange column (19 mL, GE Healthcare), previously equilibrated with 50 mM sodium acetate buffer pH 5.5. Proteins were eluted by increasing the amount of elution buffer (50 mM sodium acetate buffer, pH 5.5, containing 500 mM NaCl) linearly from 0 to 100% in 10 column volumes at a flow rate of 0.5 mL min^{-1}. Fractions containing CDH activity were pooled and concentrated by ultrafiltration through a polyether–sulfone membrane with a 10 kDa molecular mass cutoff (Vivaflow crossflow cassette, Sartorius, Les Ulis, France). The purity and molecular weight of the recombinant enzymes were determined using SDS-PAGE.

Electrophoretic analysis
SDS-PAGE was carried out by using Mini-Protean TGX precast gels (Bio-Rad Laboratories, Austria) with a gradient of 4–15% polyacrylamide. Proteins were visualized by Coomassie brilliant blue staining. The molecular mass under denaturing conditions was determined with a Precision Plus Protein Dual Color Standard (Bio-Rad

Laboratories, Austria). All procedures were done according to the manufacturer's recommendations.

To estimate the degree of glycosylation, homogenous enzymes were treated with N-glycosidase F (PNGase F) (New England BioLabs, Frankfurt, Germany), which cleaves the bond between N-Acetyl-glucosamine and Asparagine in N-linked glycoproteins under denaturing conditions, according to the manufacturer's instructions. Reaction mixtures were incubated for 1 h at 37 °C and analysed on SDS-PAGE gels.

UV/Vis spectra and determination of FAD loading

Spectra of homogeneously purified proteins were recorded from 700 to 250 nm at room temperature in both the oxidized and reduced states using a Hitachi U-3000 spectrophotometer (Hitachi, Tokyo, Japan). The purified proteins, which were in the oxidized state after purification, were diluted in 50 mM sodium acetate buffer, pH 5.5, to an absorbance at 280 nm of ~1 before recording a spectrum. The spectrum of the reduced enzyme was measured immediately after addition of a 100-fold molar excess of cellobiose to the cuvette. The molar absorption coefficients at 280 nm for all proteins were calculated by using the mature amino acid sequence and the ProtParam program (http://web.expasy.org/protparam/). The calculated molar absorption coefficients of $CtDH$ at 280 nm ($\varepsilon_{280} = 109$ mM^{-1} cm^{-1}) and $CtCDH$ ($\varepsilon_{280} = 149$ mM^{-1} cm^{-1}) were used for the determination of the protein concentration. To assess the purity of the various CDH preparations, the R_Z values, defined by the ratio A_{420}/A_{280}, were calculated for each CDH in its oxidized state.

The FAD loading was determined according to a published protocol using trichloroacetic acid to release the non-covalently bound FAD cofactor from the protein by precipitation [26]. After precipitation, the pH of the solution was carefully titrated with grains of solid sodium carbonate to a pH of 6 7. Then a spectrum was taken immediately to quantify the amount of FAD in the solution. The amount FAD was calculated using the molar absorption coefficient for free FAD ($\varepsilon_{450} = 11.3$ mM^{-1} cm^{-1}).

Enzyme activity assays

Enzyme activity was assayed spectrophotometrically using cyt c ($\varepsilon_{550} = 19.6$ mM^{-1} cm^{-1}) as electron acceptor for intact CDH or DCIP ($\varepsilon_{550} = 6.8$ mM^{-1} cm^{-1}) as an electron acceptor for both the intact holoenzyme and the dehydrogenase domain. The activities were determined by monitoring the reduction of 300 µM 2,6-dichlorophenol indophenol (DCIP) in 50 mM sodium acetate buffer (pH 5.5) containing 30 mM lactose and 4 mM of sodium fluoride (sodium fluoride was used as a laccase inhibitor). The cyt c-based assay contained 50 mM Tris–HCl buffer, pH 7.5, 20 µM cyt c and 30 mM lactose. The reaction was monitored for 180 s at 30 °C in a Lambda 35 UV–Visible spectrophotometer featuring a temperature-controlled 8-cell changer (Perkin Elmer, Massachusetts, USA). Enzyme activity was defined as the amount of enzyme that oxidizes 1 µmol of the electron acceptor per minute under the assay conditions.

Steady-state-kinetic measurements

The kinetic parameters were determined for cellobiose oxidation measured at 30 °C in 100 mM McIlvaine buffer pH 5.5 using DCIP or pH 7.5 using cyt c. The concentration of cellobiose ranged from 10 µM to 5 mM with both electron acceptors (DCIP and cyt c). Triplicates were run to ensure reliable kinetic parameter determination. Catalytic constants were calculated using nonlinear least squares regression by fitting the observed data to the Michaelis–Menten equation (Sigma Plot 12.0, Systat Software). Protein concentration was determined by measuring absorbance at 280 nm (A280).

Thermal stability measurements

$Thermo$FAD assays were performed according to Forneris et al. [27] and Reich et al. [44]. This method is based on the intrinsic fluorescence of the flavin cofactor of the flavoproteins and the fluorescence of the flavin cofactor is quenched by the protein environment. 25 µL of enzyme solution (15, 7.5 and 1.5 µM) was heated from 30 to 85 °C in increments of 1 °C per min using a MyiQ Real-Time PCR cycler (Bio-Rad Laboratories, California, USA) using an excitation wavelength range between 470 and 500 nm and an SYBR Green fluorescence emission filter (523–543), which falls within the same fluorescence range as the isoalloxazine ring of the FAD (470–570). The T_m values were obtained as the maximum of the first derivative of the sigmoid curves and the reported T_m values are the mean value of 12 independent experiments (triplicate for each enzyme concentration).

Alternatively, unfolding of the intact CDH was monitored by following the intrinsic tryptophan fluorescence upon heating [45]. Tryptophan is located on both CYT and DH (from 21 Trp residues 5 are located in CYT and 16 are located in DH). Experiments were performed in a temperature controlled Cary Eclipse fluorescence spectrophotometer (Agilent Technologies, California, USA) at a total volume of 2 mL and CDH or DH concentration of 1.5 µM. Cuvettes were stirred throughout the experiment. A heat ramp from 30 to 85 °C was applied at a rate of 1 °C per min. The excitation wavelength was 279 nm; the emission was recorded at 320 and 360 nm. The T_m was determined by fitting the tryptophan fluorescence emission ratio of 360–320 nm to a sigmoidal function. All experiments were performed in triplicates.

Abbreviations

CDH: cellobiose dehydrogenase; CtCDH: CDH from Corynascus thermophilus; CtDH: dehydrogenase domain of CtCDH; CYT: cytochrome domain of CDH; cyt c: cytochrome c; DCIP: 2,6-dichloroindophenol; DET: direct electron transfer; MET: mediated electron transfer; DH: dehydrogenase domain of CDH; FAD: flavin adenine dinucleotide; IET: intramolecular electron transfer; LPMO: lytic polysaccharide monooxygenase.

Authors' contributions

SM planned and performed the majority of experiments, analyzed the data and wrote the manuscript. RL coordinated the study and revised the manuscript. MP performed part of the screening and production of CtCDH in A. niger. DK designed the thermostability experiments, performed part of the experiments and data processing. FP and ER performed the transformation and screening of CtCDH in A. niger. LK and BS performed part of the transformation of CtCDH in T. reesei. All authors read and approved the final manuscript.

Author details

[1] Department of Food Sciences and Technology, Vienna Institute of Biotechnology, BOKU-University of Natural Resources and Life Sciences, Vienna, Austria. [2] UMR BDR, INRA, ENVA, Université Paris Saclay, 78350 Jouy en Josas, France. [3] Research Area Biochemical Technology, Institute of Chemical Engineering, TU Wien, Gumpendorferstrasse 1a, Vienna, Austria. [4] Aix Marseille Université, INRA, BBF (Biodiversité et Biotechnologie Fongiques), Marseille, France.

Acknowledgements

We thank Nenad Mardetko, Qingying Yu, Tanja Lesic, Stephen Nagle and Diarmaid de Barra for technical assistance.

Competing interests

The authors declare that they have no competing interests.

Funding

Financial support from the European Commission is gratefully acknowledged (project BIOENERGY FP7-PEOPLE-2013-ITN-607793).

References

1. Cameron MD, Aust SD. Cellobiose dehydrogenase-an extracellular fungal flavocytochrome. Enzyme Microb Technol. 2001;28:129–38.
2. Correa TLR, dos Santos LV, Pereira GAG. AA9 and AA10: from enigmatic to essential enzymes. Appl Microbiol Biotechnol. 2016;100:9–16.
3. Hemsworth GR, Johnston EM, Davies GJ, Walton PH. Lytic polysaccharide monooxygenases in biomass conversion. Trends Biotechnol. 2015;33:747–61.
4. Phillips CM, Beeson WT IV, Cate JH, Marletta MA. Cellobiose dehydrogenase and a copper-dependent polysaccharide monooxygenase potentiate cellulose degradation by Neurospora crassa. ACS Chem Biol. 2011;6:1399–406.
5. Beeson WT, Vu VV, Span EA, Phillips CM, Marletta MA. Cellulose degradation by polysaccharide monooxygenases. Annu Rev Biochem. 2015;84:923–46.
6. Zamocky M, Hallberg M, Ludwig R, Divne C, Haltrich D. Ancestral gene fusion in cellobiose dehydrogenases reflects a specific evolution of GMC oxidoreductases in fungi. Gene. 2004;338:1–14.
7. Tan TC, Kracher D, Gandini R, Sygmund C, Kittl R, Haltrich D, Hallberg BM, Ludwig R, Divne C. Structural basis for cellobiose dehydrogenase action during oxidative cellulose degradation. Nat Commun. 2015;6:7542.
8. Zhang R, Fan Z, Kasuga T. Expression of cellobiose dehydrogenase from Neurospora crassa in Pichia pastoris and its purification and characterization. Protein Expr Purif. 2011;75:63–9.
9. Harreither W, Sygmund C, Augustin M, Narciso M, Rabinovich ML, Gorton L, Haltrich D, Ludwig R. Catalytic properties and classification of cellobiose dehydrogenases from ascomycetes. Appl Environ Microbiol. 2011;77:1804–15.
10. Felice AK, Sygmund C, Harreither W, Kittl R, Gorton L, Ludwig R. Substrate specificity and interferences of a direct-electron-transfer-based glucose biosensor. J Diabetes Sci Technol. 2013;7:669–77.
11. Stoica L, Lindgren-Sjolander A, Ruzgas T, Gorton L. Biosensor based on cellobiose dehydrogenase for detection of catecholamines. Anal Chem. 2004;76:4690–6.
12. Ludwig R, Ortiz R, Schulz C, Harreither W, Sygmund C, Gorton L. Cellobiose dehydrogenase modified electrodes: advances by materials science and biochemical engineering. Anal Bioanal Chem. 2013;405:3637–58.
13. Ludwig R, Salamon A, Varga J, Zamocky M, Peterbauer CK, Kulbe KD, Haltrich D. Characterisation of cellobiose dehydrogenases from the white-rot fungi Trametes pubescens and Trametes villosa. Appl Microbiol Biotechnol. 2004;64:213–22.
14. Bey M, Berrin JG, Poidevin L, Sigoillot JC. Heterologous expression of Pycnoporus cinnabarinus cellobiose dehydrogenase in Pichia pastoris and involvement in saccharification processes. Microb Cell Fact. 2011;10:113.
15. Harreither W, Felice AK, Paukner R, Gorton L, Ludwig R, Sygmund C. Recombinantly produced cellobiose dehydrogenase from Corynascus thermophilus for glucose biosensors and biofuel cells. Biotechnol J. 2012;7:1359–66.
16. Turbe-Doan A, Arfi Y, Record E, Estrada-Alvarado I, Levasseur A. Heterologous production of cellobiose dehydrogenases from the basidiomycete Coprinopsis cinerea and the ascomycete Podospora anserina and their effect on saccharification of wheat straw. Appl Microbiol Biotechnol. 2013;97:4873–85.
17. Zamocky M, Ludwig R, Peterbauer C, Hallberg BM, Divne C, Nicholls P, Haltrich D. Cellobiose dehydrogenase—a flavocytochrome from wood-degrading, phytopathogenic and saprotropic fungi. Curr Protein Pept Sci. 2006;7:255–80.
18. Ferri S, Sode K. Functional expression of Phanerochaete chrysosporium cellobiose dehydrogenase flavin domain in Escherichia coli. Biotechnol Lett. 2010;32:855–9.
19. Yoshida M, Ohira T, Igarashi K, Nagasawa H, Aida K, Hallberg BM, Divne C, Nishino T, Samejima M. Production and characterization of recombinant Phanerochaete chrysosporium cellobiose dehydrogenase in the methylotrophic yeast Pichia pastoris. Biosci Biotechnol Biochem. 2001;65:2050–7.
20. Zamocky M, Schumann C, Sygmund C, O'Callaghan J, Dobson AD, Ludwig R, Ludwig R, Haltrich D, Peterbauer CK. Cloning, sequence analysis and heterologous expression in Pichia pastoris of a gene encoding a thermostable cellobiose dehydrogenase from Myriococcum thermophilum. Protein Expr Purif. 2008;59:258–65.
21. Sygmund C, Kracher D, Scheiblbrandner S, Zahma K, Felice AK, Harreither W, Kittl R, Ludwig R. Characterization of the two Neurospora Crassa cellobiose dehydrogenases and their connection to oxidative cellulose degradation. Appl Environ Microbiol. 2012;78:6161–71.
22. Liu L, Yang H, Shin H, Chen R, Li J, Du G, Chen J. How to achieve high-level expression of microbial enzymes: strategies and perspectives. Bioengineered. 2013;4:212–2.
23. Punt PJ, van Biezen N, Conesa A, Albers A, Mangnus J, van den Hondel C. Filamentous fungi as cell factories for heterologous protein production. Trends Biotechnol. 2002;20(5):200–6.
24. Langston JA, Brown K, Xu F, Borch K, Garner A, Sweeney MD. Cloning, expression, and characterization of a cellobiose dehydrogenase from

Thielavia terrestris induced under cellulose growth conditions. Biochim Biophys Acta. 2012;1824:802–12.

25. Wang M, Lu X. Exploring the Synergy between cellobiose dehydrogenase from *Phanerochaete chrysosporium* and cellulase from *Trichoderma reesei*. Front Microbiol. 2016;7:620.

26. Macheroux P. UV-visible spectroscopy as a tool to study flavoproteins. Methods Mol Biol. 1999;131:1–7.

27. Forneris F, Orru R, Bonivento D, Chiarelli LR, Mattevi A. *Thermo*FAD, a *Thermo*fluor-adapted flavin ad hoc detection system for protein folding and ligand binding. FEBS J. 2009;276:2833–40.

28. Schou C, Christensen MH, Schulein M. Characterization of a cellobiose dehydrogenase from *Humicola insolens*. Biochem J. 1998;330:565–71.

29. Baminger U, Subramaniam SS, Renganathan V, Haltrich D. Purification and characterization of cellobiose dehydrogenase from the plant pathogen *Sclerotium (Athelia) rolfsii*. Appl Environ Microbiol. 2001;67:1766–74.

30. Lee CW, Wang HJ, Hwang JK, Tseng CP. Protein thermal stability enhancement by designing salt bridges: a combined computational and experimental study. PLoS ONE. 2014;9:e112751.

31. Kumar S, Tsai CJ, Nussinov R. Factors enhancing protein thermostability. Protein Eng. 2000;13:179–91.

32. Li W, Zhou X, Lu P. Structural features of thermozymes. Biotechnol Adv. 2005;23:271–81.

33. Shirke AN, Su A, Jones JA, Butterfoss GL, Koffas MA, Kim JR, Gross RA. Comparative thermal inactivation analysis of *Aspergillus oryzae* and *Thielavia* terrestris cutinase: role of glycosylation. Biotechnol Bioeng. 2016. doi:10.1002/bit.26052.

34. Gordon CL, Khalaj V, Ram AF, Archer DB, Brookman JL, Trinci AP, Jeenes DJ, Doonan JH, Wells B, Punt PJ, van den Hondel CA, Robson GD. Glucoamylase:green fluorescent protein fusions to monitor protein secretion in *Aspergillus niger*. Microbiology. 2000;146(Pt 2):415–26.

35. Stricker AR, Grosstessner-Hain K, Wurleitner E, Mach RL. Xyr1 (xylanase regulator 1) regulates both the hydrolytic enzyme system and D-xylose metabolism in *Hypocrea jecorina*. Eukaryot Cell. 2006;5:2128–37.

36. Vaheri M, Leisola M, Kauppinen V. Transglycosylation products of cellulase system of *Trichoderma reesei*. Biotechnol Lett. 1979;1:41–6.

37. Conesa A, van den Hondel CA, Punt PJ. Studies on the production of fungal peroxidases in *Aspergillus niger*. Appl Environ Microbiol. 2000;66:3016–23.

38. van Hartingsveldt W, Mattern IE, van Zeijl CM, Pouwels PH, van den Hondel CA. Development of a homologous transformation system for *Aspergillus niger* based on the *pyr*G gene. Mol Gen Genet. 1987;206:71–5.

39. Uzbas F, Sezerman U, Hartl L, Kubicek CP, Seiboth B. A homologous production system for *Trichoderma reesei* secreted proteins in a cellulase-free background. Appl Microbiol Biotechnol. 2012;93:1601–8.

40. Piumi F, Levasseur A, Navarro D, Zhou S, Mathieu Y, Ropartz D, Ludwig R, Faulds CB, Record E. A novel glucose dehydrogenase from the white-rot fungus *Pycnoporus cinnabarinus*: production in *Aspergillus niger* and physicochemical characterization of the recombinant enzyme. Appl Microbiol Biotechnol. 2014;98:10105–18.

41. Punt PJ, van den Hondel CA. Transformation of filamentous fungi based on hygromycin B and phleomycin resistance markers. Methods Enzymol. 1992;216:447–57.

42. Gruber F, Visser J, Kubicek C, De Graaff L. The development of a heterologous transformation system for the cellulolytic fungus *Trichoderma reesei* based on a *pyr*G-negative mutant strain. Curr Genom. 1990;18:71–6.

43. Bradford MM. A rapid and sensitive method for the quantitation of microgram quantities of protein utilizing the principle of protein-dye binding. Anal Biochem. 1976;72:248–54.

44. Reich S, Kress N, Nestl BM, Hauer B. Variations in the stability of NCR ene reductase by rational enzyme loop modulation. J Struct Biol. 2014;185:228–33.

45. Permyakov EA, Burstein EA. Some aspects of studies of thermal transitions in protein by means of their intrinsic fluorescence. Biophys Chem. 1984;19:265–71.

12

A versatile one-step CRISPR-Cas9 based approach to plasmid-curing

Ida Lauritsen[†], Andreas Porse[†], Morten O. A. Sommer and Morten H. H. Nørholm[*][iD]

Abstract

Background: Plasmids are widely used and essential tools in molecular biology. However, plasmids often impose a metabolic burden and are only temporarily useful for genetic engineering, bio-sensing and characterization purposes. While numerous techniques for genetic manipulation exist, a universal tool enabling rapid removal of plasmids from bacterial cells is lacking.

Results: Based on replicon abundance and sequence conservation analysis, we show that the vast majority of bacterial cloning and expression vectors share sequence similarities that allow for broad CRISPR-Cas9 targeting. We have constructed a universal plasmid-curing system (pFREE) and developed a one-step protocol and PCR procedure that allow for identification of plasmid-free clones within 24 h. While the context of the targeted replicons affects efficiency, we obtained curing efficiencies between 40 and 100% for the plasmids most widely used for expression and engineering purposes. By virtue of the CRISPR-Cas9 targeting, our platform is highly expandable and can be applied in a broad host context. We exemplify the wide applicability of our system in Gram-negative bacteria by demonstrating the successful application in both *Escherichia coli* and the promising cell factory chassis *Pseudomonas putida*.

Conclusion: As a fast and freely available plasmid-curing system, targeting virtually all vectors used for cloning and expression purposes, we believe that pFREE has the potential to eliminate the need for individualized vector suicide solutions in molecular biology. We envision the application of pFREE to be especially useful in methodologies involving multiple plasmids, used sequentially or simultaneously, which are becoming increasingly popular for genome editing or combinatorial pathway engineering.

Keywords: CRISPR-Cas9, Plasmid-curing, pFREE, Replicon analysis, *Pseudomonas putida*, Genome engineering

Background

Since their discovery in the early 1950s, plasmids have played a pivotal role in the advancement of molecular biology, and form the basis for DNA cloning and gene expression in modern biotechnology [1]. While the diversity and applications of cloning vectors have grown dramatically, the vector backbones used today are, for historical reasons, build upon a limited set of parts [2–6].

A central property of a plasmid is its replication machinery that determines the copy-number and ability of plasmids to co-exist [7]. One group of cloning vectors that display a relatively high copy-number is based on the

ColE1-like replication machinery, including the pMB1 replicon of pBR322 and its high-copy derivatives found in e.g. pUC18/19, pBluescript® and pJET1.2® [5, 8]. All of the ColE1-derived replicons function via anti-sense RNA for replication control but are able to co-reside to some degree. This group of RNA-controlled ColE-like replicons also contains the widely used p15A replicon that can stably exist together with ColE1-like plasmids and is maintained in fewer copies per cell [3, 9]. A large proportion of naturally occurring plasmids replicate through the use of replication (Rep) proteins that act in a self-inhibitory fashion to control plasmid copy-number [10]. These include the replicons of pBBR1, RK2 and RSF1010 that are found in cloning vectors and function in a broad host context [11]. Similarly, the Rep protein based pSC101 vector was the first to be used for recombinant gene

*Correspondence: morno@biosustain.dtu.dk
[†]Ida Lauritsen and Andreas Porse contributed equally to this work
Novo Nordisk Foundation Center for Biosustainability, Technical University of Denmark, 2800 Kongens Lyngby, Denmark

expression, and is popular due to its relatively high stability in spite of a low copy-number (<8 copies per cell) along with the ability to co-exist with ColE1-like and p15A replicons [1, 12].

While techniques for transfer of plasmid DNA into many bacterial hosts are well established, obtaining plasmid-free cells still poses a significant challenge [13, 14]. In genome and metabolic engineering, the introduction of one or more plasmid-based genetic tools is often required, although a plasmid-free strain is eventually desired [15–18]. For example, sequential steps of plasmid-based genome editing, and the use of screening and characterization tools for strain engineering might involve multiple vectors that need removal prior to final application of the strain [17, 19].

Due to the high copy-number and intrinsic stability of modern cloning vectors, plasmid-curing is often tedious. Traditional methods for plasmid-curing are based on prolonged growth under stressful conditions, such as elevated temperature or the addition of DNA intercalating agents, to interfere with plasmid replication [14]. Other methods based on replicon-incompatibility exploit competition between identical replicons but require precise knowledge of the replication machinery of the target plasmid, as well as subsequent curing of the interfering plasmid [13, 20]. A considerable downside of the existing methods is the variable efficiency, time consumption, and the risk of accumulating unwanted mutations due to prolonged growth regimes and the use of mutagenic curing agents [17, 21].

To accommodate the need for efficient removal of cloning vectors when needed, temperature sensitive plasmid-replicons have been developed [22]. However, the relatively large size, temperature restrictions, low copy-number and little variety of these vectors, complicates cloning procedures and limits their application for multi-plasmid and broad-host purposes. Another way to facilitate the selection of plasmid free clones is by incorporating a counter-selectable marker into the plasmid backbone [23]. Although this strategy allows for rapid identification of cells lacking the marker gene, these do not actively remove the plasmid and negative selection markers are prone to mutational escape and often have stringent requirements to the growth media and host background [23–25].

With the advent of CRISPR-Cas9 technology, mimicking the natural bacterial defense against plasmid and phage intruders, a powerful and flexible approach to precise DNA targeting is now available for a wide range of organisms [26, 27]. Although CRISPR-Cas9 has been applied for specific targeting of certain plasmid features, a generally applicable platform for quick and efficient curing of cloning vectors will constitute a highly useful tool in molecular biology [28].

Here we exploit the common origin of modern plasmid vectors to develop a broadly applicable CRISPR-Cas9-based curing platform. We show that our system enables fast and efficient curing of all major plasmid replicons used in modern molecular biology laboratories and can be applied in a broad phylogenetic context.

Results

We first explored the distribution of cloning vector replicons by performing a BLAST search of selected replicons against all bacterial plasmids with full nucleotide sequences available in the Addgene plasmid repository [29] (Fig. 1). The ColE1-like (including p15A) and pSC101 replicons accounted for 91% of the plasmids in the Addgene database. The vast majority of these plasmids belonged to the ColE1 family (86.4%), underlining the popularity of these vectors in molecular biology (Fig. 1). Surprisingly, a considerable fraction of vectors annotated with the pBBR1 and RK2 broad host-range replicons also contained full-sized ColE1-like replicon sequences. Including these redundant replicons in our calculations, a plasmid-curing system targeting the ColE1 and pSC101 plasmid groups will cover 93.3% of the (at present 4657) bacterial vectors deposited in Addgene (Additional file 1: Figure S1).

Through sequence alignments of representative replicons from each replicon-group, we identified highly conserved regions between all ColE1-like replicons that were also shared with p15A (Fig. 2a). These regions were used to design CRISPR-Cas9-compatible guide RNA (gRNA) that, upon recognition by Cas9, target all ColE1-like and p15A vectors. Because the protein-based mechanism of

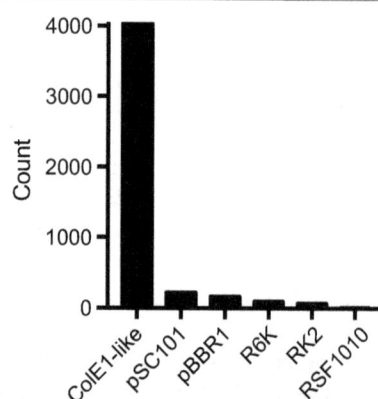

Fig. 1 Frequency of major replicons in bacterial cloning and expression vectors deposited to Addgene. A BLAST search was performed against all complete bacterial vector sequences (4657) in the Addgene database (Feb. 2017). ColE1-like plasmids include the closely related RNA-based replicons of ColE1, pBR322/pMB1, pUC18/19, pJET1.2®, colA and p15A

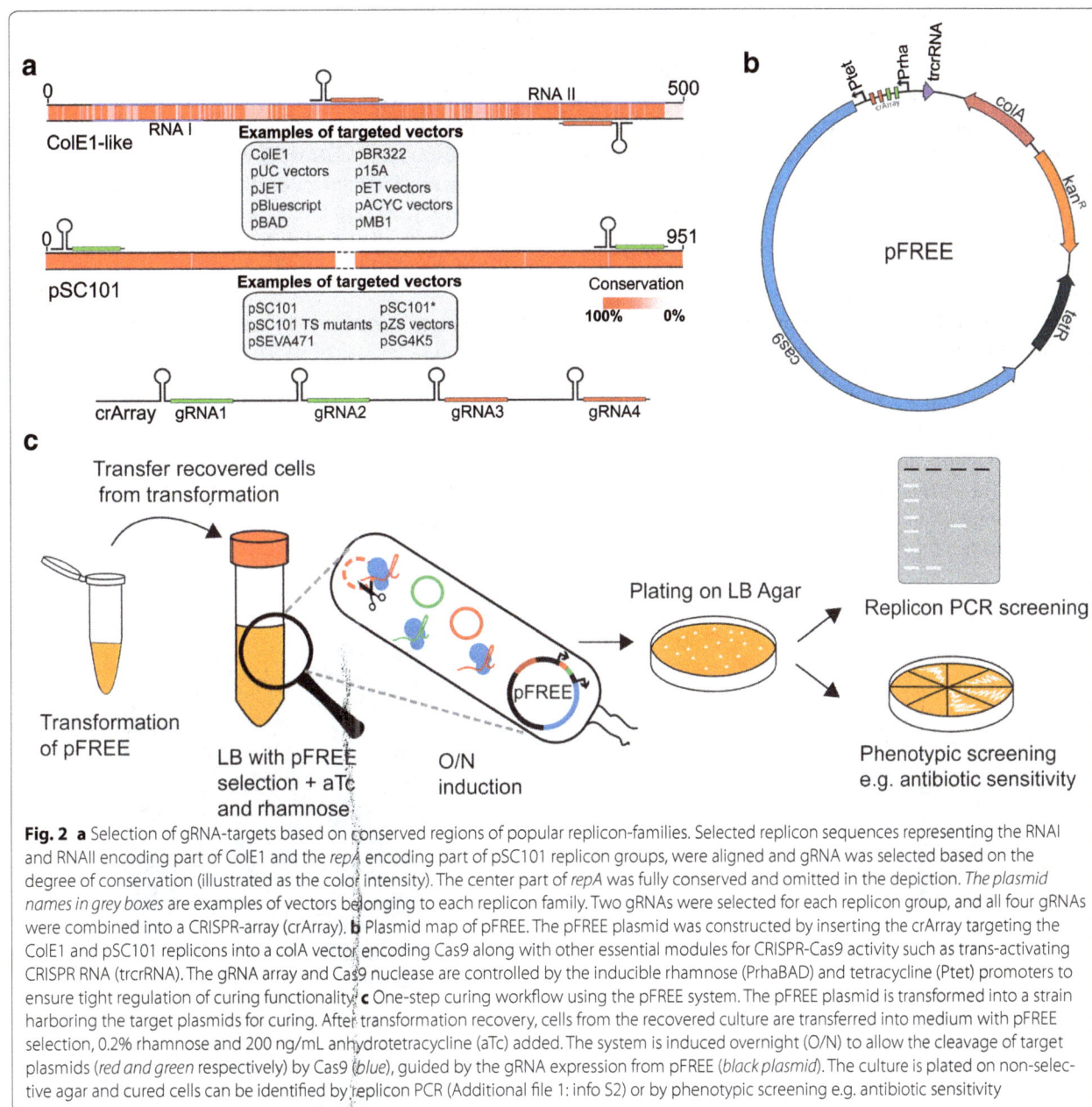

Fig. 2 a Selection of gRNA-targets based on conserved regions of popular replicon-families. Selected replicon sequences representing the RNAI and RNAII encoding part of ColE1 and the *repA* encoding part of pSC101 replicon groups, were aligned and gRNA was selected based on the degree of conservation (illustrated as the color intensity). The center part of *repA* was fully conserved and omitted in the depiction. *The plasmid names in grey boxes* are examples of vectors belonging to each replicon family. Two gRNAs were selected for each replicon group, and all four gRNAs were combined into a CRISPR-array (crArray). **b** Plasmid map of pFREE. The pFREE plasmid was constructed by inserting the crArray targeting the ColE1 and pSC101 replicons into a colA vector encoding Cas9 along with other essential modules for CRISPR-Cas9 activity such as trans-activating CRISPR RNA (trcrRNA). The gRNA array and Cas9 nuclease are controlled by the inducible rhamnose (PrhaBAD) and tetracycline (Ptet) promoters to ensure tight regulation of curing functionality. **c** One-step curing workflow using the pFREE system. The pFREE plasmid is transformed into a strain harboring the target plasmids for curing. After transformation recovery, cells from the recovered culture are transferred into medium with pFREE selection, 0.2% rhamnose and 200 ng/mL anhydrotetracycline (aTc) added. The system is induced overnight (O/N) to allow the cleavage of target plasmids (*red and green* respectively) by Cas9 (*blue*), guided by the gRNA expression from pFREE (*black plasmid*). The culture is plated on non-selective agar and cured cells can be identified by replicon PCR (Additional file 1: info S2) or by phenotypic screening e.g. antibiotic sensitivity

pSC101 replication is fundamentally different from that of the ColE1-like replicons, we designed separate gRNA to facilitate curing of the pSC101-based vectors. To increase curing efficiency and counteract the potential for mutational escape, we included two gRNA targets for each replicon group (Table 1).

The four gRNAs were implemented as a CRISPR-array, along with the tracrRNA and Cas9-components and incorporated into a single vector containing all parts necessary to form the fully functional curing system designated "pFREE" (Fig. 2b). An important feature of

Table 1 Selected gRNAs and their target replicons

gRNA	Sequence (5′ to 3′)	Targeted replicon group
gRNA1	ATGAACTAGCGATTA GTCGCTATGACTTAA	*pSC101*
gRNA2	AACCACACTAGAGAA CATACTGGCTAAATA	*pSC101*
gRNA3	GGTTGGACTCAAGAC GATAGTTACCGGATA	*ColE1-like except colA*
gRNA4	GGCGAAACCCGACAG GACTATAAAGATACC	*ColE1-like including colA (self-curing of pFREE)*

a plasmid-curing system is a suicide functionality that renders the resulting cells completely plasmid-free without any additional incubation steps. The pFREE vector is based on the colA replicon that resembles ColE1-like replicons to some degree but colA is only recognized by one of the ColE1-targeting gRNAs. Due to the self-curing feature of pFREE, plasmid-curing can be done in a one-step workflow directly after transformation of pFREE as outlined in Fig. 2c.

Quantification of curing efficiency

In order to test the efficiency of our plasmid-curing system, we constructed three target plasmids by inserting *gfp* under control of a constitutive promoter into similar backbones of the pZ vector system [30]. These three plasmids differ only by their ColE1, p15A or pSC101 replicons and are designated pZE-GFP, pZA-GFP and pZS-GFP. The curing efficiency was quantified at different time points, and the loss of fluorescence reflected plasmid-curing of the *gfp* expressing vectors. The curing

rates were comparable between the target plasmids and after 24 h the vast majority of all three populations were cured with 80–90% of the plated cells being plasmid-free (Fig. 3). Non-fluorescent cells were assessed for self-curing of the pFREE plasmid by kanamycin sensitivity, and no pFREE-carrying cells were detected after 24 h. These results clearly demonstrate effective plasmid-curing of vectors with ColE1, p15A and pSC101 replicons, targeted by the crArray of the pFREE system, and efficient self-curing of the pFREE plasmid.

pFREE cures major cloning vector-systems used in *E. coli*

Seven representatives of widely used cloning and expression vector systems were selected to demonstrate the general applicability of the pFREE system to cure commonly used vectors with similar replicons but variable backbone content. The majority of these plasmids contained variations of ColE1 replicons including the pJET1.2®, pUC19 and pBluescript® high copy-number variants as well as a pET-vector most commonly used for

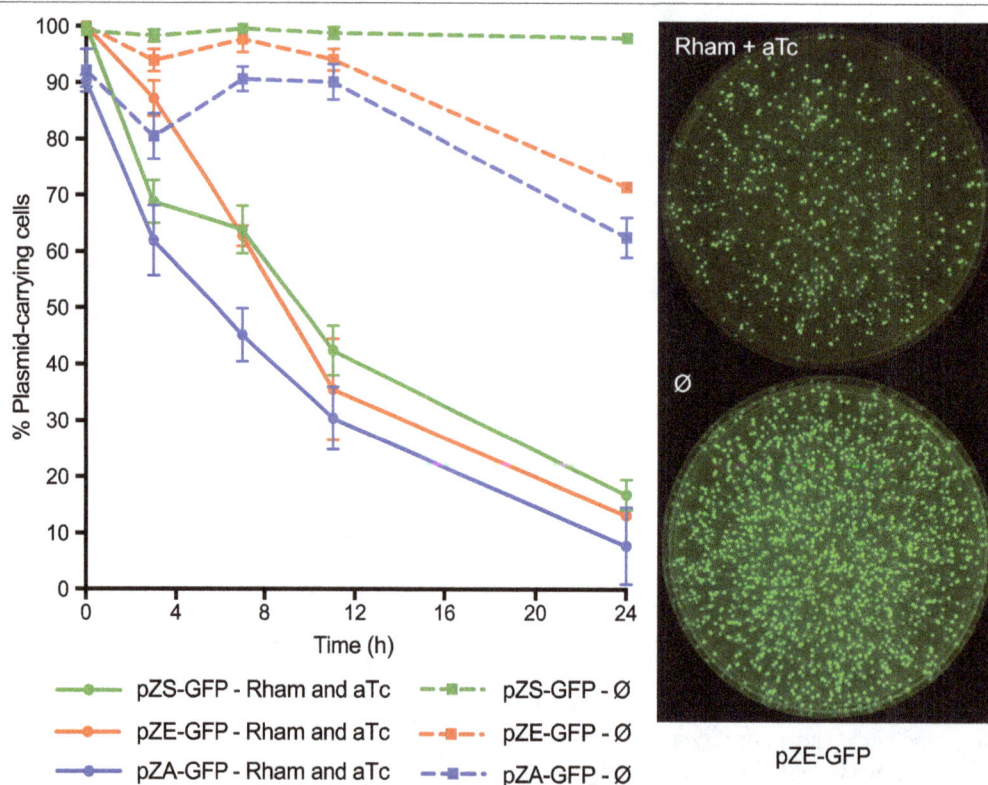

Fig. 3 Time course characterization of the pFREE plasmid-curing system. Curing of pZ-plasmids expressing GFP with either pSC101 (pZS-GFP, *green*), ColE1 (pZE-GFP, *red*) or p15A (pZA-GFP, *blue*) replicon. *The solid lines* indicate induced cultures with rhamnose (Rham) and anhydrotetracycline (aTc), whereas the *dashed lines* refer to non-induced (Ø). Plating was performed at induction time (0) and 3, 7, 11 and 24 h after induction. Between 100 and 150 colony forming units (CFUs) were counted from each replicate and the ratio between fluorescent and non-fluorescent cells were determined. The percentage of plasmid-carrying cells is depicted. Of the non-fluorescent and tested cells, all had lost the pFREE plasmid after 24 h. Data points represent mean value of three biological replicates with *error-bars* showing standard deviation. Representative LB agar plates for pZE-GFP with equal number of cells plated with cultures induced with rhamnose and aTc (*top*) and non-induced (Ø) (*bottom*) of the pFREE system after 24 h

protein production. In addition, the low copy-number pSEVA471 [31] plasmid harboring a pSC101 replicon and the p15A-based pACYC-Duet-1 medium-copy plasmid was also included.

While the pSEVA471 and pACYC-Duet-1 plasmids were cured with similar efficiency to what was observed for the pZ plasmids (Figs. 3, 4), the ColE1-like replicons were cured with efficiencies ranging from 40 to 100%. These results exemplify that, although replicon context does play a role, the pFREE system can be used for efficient curing of the most common commercial plasmid vectors with varying copy-numbers and auxiliary content.

One-step curing of multiple plasmids

To improve the practical application of the pFREE system as a fast and simple curing system, we developed a one-step workflow as displayed in Fig. 2c. Plasmid-curing is induced directly after pFREE-transformation and completely plasmid-free clones (without target and pFREE plasmids) can easily be detected either by phenotypic screening (e.g. antibiotic sensitivity) or faster by the set of universal replicon amplifying PCR oligonucleotides that we developed (Additional file 1: info S2). To test the one-step protocol and to evaluate the performance of the pFREE system for curing multiple plasmids simultaneously, we prepared a strain containing three compatible target plasmids. After transformation of pFREE into this

strain, the target plasmids were cured directly from the transformation mix and plated on non-selective LB agar after overnight induction. From the tested cells, 80% were completely cured whereas 10% or less contained one or more plasmids and all cells had lost pFREE (Fig. 5).

Self-curing dynamics of pFREE

To investigate the dynamics of the pFREE self-targeting feature, we quantified the self-curing efficiency of the pFREE plasmid over time. In the absence of plasmid selection, 90% of the cells were cured of pFREE after 7 h of induction whereas 65% of the cells were cured in the presence of plasmid selection (Additional file 1: Figure S2). After 10 h of pFREE induction, all cells were cured for pFREE regardless of the plasmid selection.

pFREE-RK2: a temperature sensitive and broad host-range version of pFREE

The curing efficiency of the pFREE system was between 40 and 100% as depicted in Figs. 3 and 4. Due to the highly efficient self-curing of pFREE, we speculated that over-efficient self-targeting could be a bottleneck preventing complete curing of the target plasmids. In that case, a system allowing self-curing to take place only after the target plasmid-curing has occurred, might increase duration of CRISPR expression and consequently the

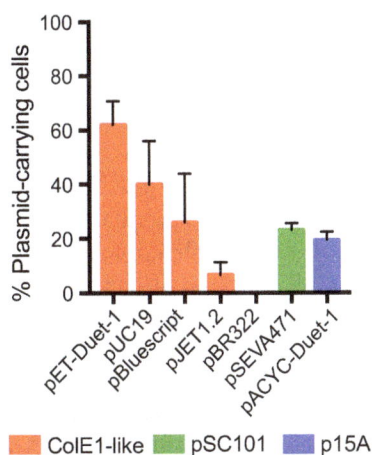

Fig. 4 pFREE-mediated curing of selected widely used cloning vectors with either ColE1-like (*red*), pSC101 (*green*) or p15A (*blue*) replicons. 50 CFUs from each replicate of each target plasmid was checked for antibiotic sensitivity after 24 h of induction of the pFREE system. The percentage of plasmid-carrying cells is depicted. The pFREE plasmid was cured in all colonies tested. Curing of the target plasmids was verified by replicon PCR (Additional file 1: info S2). The *bars* represent mean value of three biological replicates with *error-bars* showing standard deviation

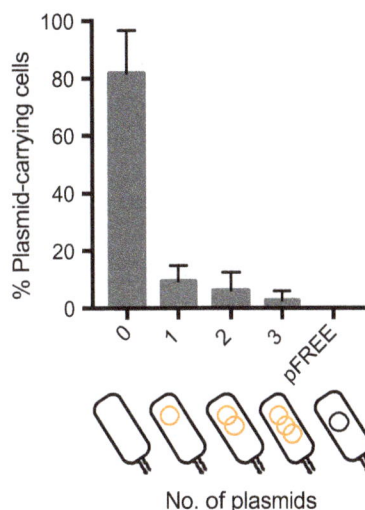

Fig. 5 Curing of multiple co-residing plasmids using the one-step transformation protocol depicted in Fig. 2c. The pFREE plasmid was transformed into a strain harboring the same three pZ plasmids as used in Fig. 3. After recovery, plasmid-curing with the pFREE system was induced and cultures were plated after 24 h of induction. 50 CFUs from each replicate were checked for antibiotic sensitivity on LB agar plates. The percentage of cells carrying 0, 1, 2, 3 (*orange plasmids*) or pFREE (*black plasmid*) is depicted. The *bars* represent mean value of three biological replicates with *error-bars* showing standard deviation

curing efficiency. To compare the effect of self-curing mechanism and copy-number on curing outcomes, we designed a pFREE version with the temperature-sensitive, low-copy RK2 replicon that replicates in a broad representation of Gram-negative bacteria [32] designated pFREE-RK2 (Additional file 1: Figure S3). The RK2 replicon is not targeted by the pFREE crArray, thus omitting the CRISPR-Cas9-based self-curing feature of the pFREE system but carries a *trfA* mutant that allows curing at elevated temperatures instead [33]. The curing efficiency of the pFREE-RK2 system was quantified in the same way as for pFREE and exhibited comparable curing efficiencies of 35–100% (Additional file 1: Figures S4, S5). The highly similar curing efficiencies observed for pFREE and pFREE-RK2, indicates that simultaneous self-curing and curing of target plasmid does not significantly affect the overall curing efficiency of the system. Multi-plasmid curing by the one-step protocol was also tested for pFREE-RK2 and showed comparable efficiency (Additional file 1: Figure S6), demonstrating that a lower copy-number and fundamentally different mechanism of self-curing does not alter CRISPR-Cas9 targeting efficiency.

Curing in *Pseudomonas putida*

To demonstrate the versatility of our plasmid-curing system in a broader phylogenetic context, we set out to test the pFREE-RK2 in an alternative host bacterium supported by the RK2 replicon. We chose the Gram-negative soil bacterium *Pseudomonas putida* as our model host due to its promise as a new and powerful chassis for metabolic engineering and production of fine chemicals [34]. Using the *P. putida* strain KT2440 harboring the *gfp*-expressing pSEVA441-GFP plasmid, we targeted the ColE1-based pRO1600/ColE1 fusion replicon without the need to change any components of the pFREE-RK2 curing plasmid. After overnight induction of the curing system, approximately half of the *P. putida* population (53% SD ± 5.1%, three biological replicates) was cured for pSEVA441-GFP, whereas no detectable curing was observed without pFREE-RK2. These results demonstrate that our CRISPR-based curing system can be applied in a broader host context and that CRISPR-Cas9 technology can be successfully applied in *P. putida*.

pFREE enables precise curing without off-target effects

The curing functionality of pFREE is tightly regulated and the gRNAs were selected to avoid potential off-target effects of CRISPR-Cas9 expression [35]. However, to ensure that the curing activity of pFREE did not induce off-target effects, we whole-genome sequenced three individual isolates of *E. coli* and *P. putida* harboring pFREE or pFREE-RK2 respectively before and after the curing procedure. The sequencing results showed that 24 h of induction with the pFREE system did not cause mutations (SNPs and small INDELS) or larger rearrangements in the host genomes; confirming the orthogonality of pFREE in these hosts.

Discussion

Plasmids are fundamental in all aspects of molecular biology due to their role as genetic scaffolds that are easy to modify and transfer between hosts. However, when plasmids carry functions that are only temporarily necessary, or a clean strain background is needed, limited options are currently available for efficient plasmid-curing of the most widely used cloning vectors in bacteria.

Existing methods for plasmid-curing are based on curing agents or incompatibility mediated plasmid displacement [13, 20]. However, these methods require sequential rounds of growth in stressful or non-selective conditions to promote the appearance of plasmid-free segregants. Such methods increase the chance of accumulating unwanted mutations and are time-consuming. Prior work has demonstrated that plasmids, traditionally considered incompatible can co-exist stably for multiple growth cycles [36], which only complicates incompatibility-based plasmid-curing further; especially for plasmids maintained at high copy-numbers.

To address this methodological bottleneck, we developed the pFREE system as a fast and simple one-step plasmid curing-method based on sequence conservation within replicon groups and CRISPR-Cas9-targeted plasmid cleavage. Using this system, curing of one or multiple target plasmids can be performed directly after transformation of the pFREE plasmid and cured cells can easily be screened for specific phenotypes (e.g. antibiotic resistance) or by the diagnostic PCR developed here (Additional file 1: info S2). In the absence of prior plasmid sequence information, the PCR based replicon identification protocol is also useful for replicon profiling prior to curing (Additional file 1: info S2).

Using the pFREE system, we cured both single and multiple plasmids with an efficiency of 40–100% (Figs. 3, 4). We investigated the dynamics of the pFREE self-curing feature and observed complete curing of pFREE already after 10 h with kanamycin added for pFREE selection. Although the inclusion of pFREE selection during curing reduced the self-curing rate, and allows for the one-step transformation protocol, it also shows that cells that are actively cured during selective culturing are not necessarily killed (Additional file 1: Figure S2; Fig. 2c).

Such persistence may result from slower degradation of the resistance conferring aminoglycoside phosphotransferase enzyme compared to the rate of plasmid-curing, or could be an effect of indirect resistance were the antibiotic sensitive cells are protected by pFREE-carrying cells [37].

We first speculated that the highly efficient self-curing of pFREE was limiting the trans-curing efficiency of pFREE. However, similar curing efficiencies were observed when the self-targeted colA replicon was replaced with the temperature sensitive RK2 replicon (pFREE-RK2).

Differences in curing efficiencies were observed for the individual plasmids tested here; presumably caused by variations in copy-number, plasmid incompatibility or fitness constrains originating from other factors present in the plasmid backbones. We did not observe a clear correlation between copy-number and curing efficiency, with the extremely high copy vectors of pJET1.2®, pBluescript® and pUC19 displaying curing efficiencies similar to the low and medium-copy-number pSEVA471 and pACYC-Duet-1 (Fig. 4). Although the overall curing efficiency was similar between pFREE and pFREE-RK2, there were small differences in the relative efficiency against the different replicon families (Fig. 3; Additional file 1: Figure S4). Such differences, e.g. the higher efficiency for curing of replicons more similar to the replicon of the curing plasmid, could be a result of partial replicon incompatibility and might explain the higher loss of p15A and ColE1-like RNA-based replicons for pFREE when co-residing without induction of the curing system (Fig. 3; Additional file 1: Figure S4). Surprisingly, the pBR322 plasmid was completely cured for both versions of pFREE, whereas the pET-Duet-1, carrying the exact same replicon, displayed the lowest curing efficiency (40% cured) observed here; indicating a substantial effect of auxiliary plasmid factors on plasmid stability (Fig. 4; Additional file 1: Figure S5). Such differences can be caused by factors such as resistance markers or other genetic cargo that affects plasmid persistence at the population level. The pBR322 is known to inflict a fitness cost on E. coli hosts due to the expression of the costly tetracycline efflux pump encoded by tetA [38]. If a high fitness benefit of losing the plasmid exists, the expansion of the plasmid-free population will contribute exponentially to the observed plasmid loss and synergistically improve the curing outcome.

Mutations in target plasmids or in the CRISPR platform of pFREE along with biological stochasticity could also explain the non-perfect curing of target plasmids by our system. CRISPR-Cas9 systems are widely used for genome-editing purposes, and other applications have shown similar susceptibility to small subpopulations of "escapers" that avoid targeting [39].

We developed the crArray encoded by pFREE to target replicons belonging to the ColE1/p15A and pSC101 groups based on the distribution of replicons in bacterial

vectors deposited in the Addgene database, which agreed with historical trends in cloning vector usage [1, 5, 9]. Additionally, we discovered that the selected gRNAs in pFREE indirectly target vectors with other replicons such as pBBR1 and RK2 due to redundant replicon sequences present in a high proportion of these backbones (Additional file 1: Figure S1). Although we target the majority of replicons used for routine cloning in E. coli there are exceptions within the broad host-range vectors and R6K (Additional file 1: Figure S1). The R6K replicon is primarily used as a suicide vector and is of little relevance in a curing perspective [40]. Since the vast majority of plasmid vectors that are used belong to the ColE1-like, p15A or pSC101 replicon groups (Fig. 1), the chance of a target plasmid being covered by the pFREE system is high. Hence, less knowledge about the replicon group of target plasmids is needed prior to curing compared to incompatibility-based curing methods [13] and only a few colonies will have to be screened to identify a cured variant.

Due to the broad functionality of the CRISPR-Cas9 technology in a variety of organisms [27] the pFREE curing system has great potential as a universal plasmid-curing tool in bacteria, as shown here for both E. coli and P. putida, and can in theory be expanded to eukaryotic organisms such as yeast where plasmids are also employed [41]. With decreasing cost of nucleic acid synthesis, custom crArrays for targeting of other plasmids than the ones included here are easily implemented into the pFREE backbones. It is possible that a similar approach can be used clinically to combat the increasing medical burden of plasmid-encoded multidrug resistance in pathogenic bacteria. Although the diversity of natural plasmid replicons by far exceeds that of cloning vectors, the most endemic plasmid-families encoding virulence and antibiotic resistance factors do share conserved features within their replication, stability, resistance and conjugation modules that could be targeted for future expansion of our plasmid-curing system [42, 43].

Conclusions

We show that all major replicons used for cloning and expression purposes share sequence features that allow for universal CRISPR-Cas9 targeting and use this information to develop a fast and one-step plasmid-curing platform that allows for targeting of the major classes of vectors used in molecular biology. Using our curing protocol, we demonstrate efficient curing of major cloning and expression vectors in biotechnology and perform in-depth characterization of the curing dynamics. To facilitate subsequent identification of plasmid-cured variants, we supply a set of universal primers that allow for rapid PCR screening directly from a culturing plate. Furthermore, we construct a temperature-sensitive and broad host-range version of pFREE (pFREE-RK2) that provides an efficient curing solution for broad range of Gram-negative bacteria including the upcoming cell factory *Pseudomonas putida*.

Methods

Replicon prevalence and conservation analysis

Bioinformatic analysis was performed using the CLC Main Workbench (QIAGEN Bioinformatics) and R (version 3.3.1) software. Replicons used in multiple sequence alignments and BLAST searches were downloaded from GenBank or Addgene. Sequences with the following accession numbers were included as ColE1-like replicons: ColE1 (GenBank NC_001371), pBR322/pMB1 (GenBank J01749.1), pUC19 (Addgene plasmid #49793), pJET1.2® (GenBank EF694056.1), ColA (Addgene plasmid #73962). In addition, p15A and pSC101 replicons were included: p15A (GenBank V00309.1) and pSC101 *repA* (GenBank K00042.1), temperature sensitive pSC101 *repA* of pKD46 (GenBank AY048746) and pGRG36 (GenBank DQ460223.1). For the BLAST analysis, the R6K (GenBank KX485333.1) and broad host-range replicons of pBBR1 (GenBank U02374.1), *trfA* gene of RK2 (GenBank U05774.1) and RSF1010 (GenBank M28829.1) were also included.

Multiple alignments were used to identify conserved regions in the selected replicon sequences. gRNA was selected based on broad conservation in replicons, as well as the absence of matches to proteobacterial chromosomes in NCBIs RefSeq database; where at least four chromosomal mismatches were present for each gRNA sequence.

We downloaded all (4657) Addgene entries of bacterial plasmids where the full nucleotide sequence was accessible from the search function at https://www.addgene.org/ (accessed Feb. 2017). Replicon frequencies and positive gRNA hits were assessed using BLAST [44]. An e-value cutoff of 1e-130 was used for replicon BLAST and allowed proper classification according to database annotations. For gRNA BLAST searches, only hits with a perfect match to the query replicons were included as positive hits.

Plasmid construction

pFREE was constructed by amplification of the pMAZ-SK backbone [18] using oligonucleotides 1 and 2. See Additional file 1: Table S3 for all plasmids and references and Additional file 1: Table S4 for all oligonucleotides used in this study. The crArray encodes four different gRNAs of 30 nts, separated by direct repeats of 36 nts. The crArray was constructed by PCR using two ultramer oligonucleotides 3 and 4 (size of 200 nts

and 173 nts respectively) with an overlapping region of 72 nts mixed with the two uracil-containing oligonucleotides 5 and 6, and cloned into the pMAZ-SK amplified PCR backbone by USER cloning as described previously [45]. Insertion of the tetratracycline repressor (*tetR*) was performed by Gibson assembly as described elsewhere [46] with oligonucleotides 7 and 8 for pMAZ-SK backbone amplification and *tetR* amplification from plasmid pZS4Int-tetR with oligonucleotides 9 and 10. Oligonucleotides 11 and 12 were used to amplify the pFREE backbone and the *Cas9* gene was amplified from pMA7CR_2.0 [18] with oligonucleotides 13 and 14 and cloned into pFREE by Gibson assembly.

The temperature sensitive broad host-range version of pFREE (pFREE-RK2) was constructed by amplification of the temperature sensitive RK2 replicon from pSIM9 [47] using oligonucleotides 15 and 16, including the *trfA* gene and *oriV* regions. The backbone from pFREE was amplified with oligonucleotides 17 and 18 to insert the RK2 replicon into the pFREE backbone via USER cloning [48]. Likewise, versions of pFREE and pFREE-RK2 with ampicillin, chloramphenicol and zeocin resistance genes were constructed and all pFREE constructs are available through Addgene.

Bacterial strains, media and growth conditions

Escherichia coli Top10 (Thermo Fisher Scientific, Waltham, MA, USA) was used for cloning and curing experiments. *E. coli* cultures were grown in lysogeny broth (LB) at 30 °C with shaking at 270 rpm. The antibiotics ampicillin (100 µg/mL), chloramphenicol (34 µg/mL), zeocin (100 µg/mL) and kanamycin (50 µg/mL) were added when needed. *crArray* expression was induced with 0.2% L-rhamnose (w/v) and expression of *Cas9* endonuclease was induced with 200 ng/mL anhydrotetracycline (aTc).

Time course curing dynamics of pFREE and pFREE-RK2

Overnight cultures of *E. coli* Top10 harboring the pZA-GFP, pZE-GFP or pZS-GFP plasmid respectively and pFREE or pFREE-RK2 were diluted 2000-fold in 10 mL LB broth containing 0.2% L-rhamnose, 200 ng/mL aTc and 50 µg/mL kanamycin. Cultures were grown from three randomly picked colonies. Time course assessment of curing efficiency was done by plating on non-selective LB agar. For each time-point, the ratio between fluorescent cells and non-fluorescent cells were determined by quantification of GFP-fluorescent colonies among 100–150 CFUs from each plate. To assess self-curing of pFREE, at least 10 colonies were checked for growth on LB agar plates containing kanamycin (50 µg/mL).

Curing of widely used cloning and expression vectors

The plasmids pET-Duet-1, pBR322, pJET1.2®, pUC19, pACYC-Duet-1, pSEVA471 pBluescript® and pSEVA441-GFP were cured with the protocol described above. To test for plasmid-curing, individual colonies were checked for growth on LB agar plates containing the relevant antibiotic. Plasmid-curing was verified by replicon PCR with oligonucleotides S1, S2, S3, S4 and S5 (Additional file 1: info S2).

One-step curing of multiple plasmids

Escherichia. coli Top10 strain harboring three (pZA-GFP, pZE-GFP and pZS-GFP) plasmids was grown in 5 mL LB containing antibiotics for plasmid selection. The culture was grown to an OD_{600} of 0.3 and made electrocompetent by three steps of washing in MilliQ water. 50 µL of competent cells were transformed with 50 ng of pFREE or pFREE-RK2 by electroporation (1.65 kV, 200 Ohm, 25 µF) and recovered for 2 h in 500 µL SOC medium at 30 °C and shaking (500 rpm). After recovery, 50 µL of the recovered cells were transferred to 10 mL LB medium with 0.2% L-rhamnose, 200 ng/mL aTc and 50 µg/mL kanamycin added. The cultures were plated on non-selective LB agar after 24 h of incubation. 50 colonies of each replicate were checked on relevant antibiotics to assess the curing efficiency.

Assessment of genomic off-target effects

Two parallel cultures were initiated from each of three individual colonies of *E. coli* carrying pFREE and *P. putida* carrying pFREE-RK2. One culture was induced to activate the curing process while the other functioned as a control without induction of pFREE. Both cultures were grown at 30 °C shaking (250 rpm) for 24 h and genomic DNA was purified using the QIAGEN blood and tissue DNA isolation kit. The genomic DNA was prepared for sequencing using the KAPA HyperPlus Kit (Kapa Biosystems) and the resulting libraries were sequenced on an Illumina NextSeq platform. Fastq output files were imported into CLC Genomics Workbench software (QIAGEN) where all analysis was performed. Reads were trimmed and quality filtered before mapping of the reads, originating from the plasmid cured genomes, to the assembled control genomes. SNP and small INDEL variants were detected using quality based variant detection and larger INDELS and structural variants were assessed using the "Structural Variants and InDels" pipeline as well as by manual inspection of read mappings.

Authors' contributions
IL, AP, MHHN and MOAS designed the experiments. IL and AP performed the experiments. IL and AP wrote the manuscript with input from MHHN and MOAS. All authors read and approved the final manuscript.

Acknowledgements
The authors would like to acknowledge Pablo Ivan Nikel and Patricia Maria Calero Valdayo for providing the *P. putida* strain and plasmid.

Competing interests
The authors declare that they have no competing interests.

Funding
This work was supported by the EU H2020 ERC-20104-STG LimitMDR (638902) Grant, the Danish Council for Independent Research Sapere Aude programme DFF-4004-00213 and the Novo Nordisk Foundation.

References
1. Cohen SN. DNA cloning: a personal view after 40 years. Proc Natl Acad Sci USA. 2013;110:15521–9.
2. Cohen SN, Chang AC, Boyer HW, Helling RB. Construction of biologically functional bacterial plasmids in vitro. Proc Natl Acad Sci USA. 1973;70:3240–4.
3. Chang ACY, Cohen SN. Construction and characterization of amplifiable DNA cloning vectors derived from P15A cryptic plasmid. J Bacteriol. 1978;134:1141–56.
4. Chang ACY, Cohen SN. Genome construction between bacterial species in vitro: replication and expression of *Staphylococcus* plasmid genes in *Escherichia coli*. Proc Natl Acad Sci USA. 1974;71:1030–4.
5. Hershfield V, Boyer HW, Yanofsky C, Lovett MA, Helinski DR. Plasmid ColEl as a molecular vehicle for cloning and amplification of DNA. Proc Natl Acad Sci USA. 1974;71:3455–9.
6. Rosano GL, Ceccarelli EA. Recombinant protein expression in *Escherichia coli*: advances and challenges. Front Microbiol. 2014;5:1–17.
7. Novick RP. Plasmid incompatibility. Microbiol Rev. 1987;51:381–95.
8. Bolivar F, Rodriguez RL, Greene PJ, Betlach MC, Heyneker HL, Boyer HW, et al. Construction and characterization of new cloning vehicles. II. A multipurpose cloning system. Gene. 1977;2:95–113.
9. Selzer G, Som T, Itoh T, Tomizawa JI. The origin of replication of plasmid p15A and comparative studies on the nucleotide sequences around the origin of related plasmids. Cell. 1983;32:119–29.
10. del Solar G, Giraldo R, Ruiz-Echevarría MJ, Espinosa M, Díaz-Orejas R. Replication and control of circular bacterial plasmids. Microbiol Mol Biol Rev. 1998;62:434–64.
11. Jain A, Srivastava P. Broad host range plasmids. FEMS Microbiol Lett. 2013;348:87–96.
12. Tucker WT, Miller CA, Cohen SN. Structural and functional analysis of the par region of the pSC 101 plasmid. Cell. 1984;38:191–201.
13. Hale L, Lazos O, Haines AS, Thomas CM. An efficient stress-free strategy to displace stable bacterial plasmids. Biotechniques. 2010;48:223–8.
14. Trevors J. Plasmid curing in bacteria. FEMS Microbiol Lett. 1986;32:149–57.
15. Lee J, Saddler JN, Um Y, Woo HM. Adaptive evolution and metabolic engineering of a cellobiose- and xylose- negative *Corynebacterium glutamicum* that co-utilizes cellobiose and xylose. Microb Cell Fact. 2016;15:20.
16. Ginesy M, Belotserkovsky J, Enman J, Isaksson L, Rova U. Metabolic engineering of *Escherichia coli* for enhanced arginine biosynthesis. Microb Cell Fact. 2015;14:29.
17. Schlegel S, Genevaux P, de Gier JW. De-convoluting the genetic adaptations of *E. coli* C41 (DE3) in real time reveals how alleviating protein production stress improves yields. Cell Rep. 2015;10:1758–66.
18. Ronda C, Pedersen LE, Sommer MOA, Nielsen AT. CRMAGE: CRISPR optimized MAGE recombineering. Scientific Reports. Nature Publishing Group. 2016; 6:19452.
19. Raman S, Rogers JK, Taylor ND, Church GM. Evolution-guided optimization of biosynthetic pathways. Proc Natl Acad Sci. 2014;111:201409523.
20. Kamruzzaman M, Shoma S, Thomas CM, Partridge SR, Iredell JR. Plasmid interference for curing antibiotic resistance plasmids in vivo. PLoS ONE. 2017;12:e0172913.

21. Crameri R, Davies JE, Hütter R. Plasmid curing and generation of mutations induced with ethidium bromide in streptomycetes. J Gen Microbiol. 1986;132:819–24.

22. Hashimoto-Gotoh T, Franklin FCH, Nordheim A, Timmis KN. Specific-purpose plasmid cloning vectors I. Low copy number, temperature-sensitive, mobilization-defective pSC101-derived containment vectors. Gene. 1981;16:227–35.

23. Reyrat J-M, Pelicic V, Gicquel B. Counterselectable markers: untapped tools for bacterial genetics and pathogenesis. Infect Immun. 1998;66:4011–7.

24. Gregg CJ, Lajoie MJ, Napolitano MG, Mosberg JA, Goodman DB, Aach J, et al. Rational optimization of tolC as a powerful dual selectable marker for genome engineering. Nucleic Acids Res. 2014;42:4779–90.

25. Podolsky T, Fong S, Lee B. Direct selection of tetracycline-sensitive Escherichia coli cells using nickel salts. Plasmid. 1996;115:112–5.

26. Jinek M, Chylinski K, Fonfara I, Hauer M, Doudna JA, Charpentier E. A programmable dual-RNA—guided DNA endonuclease in adaptive bacterial immunity. Science. 2012;337:816–22.

27. Singh V, Braddick D, Dhar PK. Exploring the potential of genome editing CRISPR-Cas9 technology. Gene. 2017;599:1–18 **(Elsevier B. V.)**.

28. Kim JS, Cho DH, Park M, Chung WJ, Shin D, Ko KS, et al. CRISPR/Cas9-mediated re-sensitization of antibiotic-resistant Escherichia coli harboring extended-spectrum Beta-lactamases. J Microbiol Biotechnol. 2015;26:394–401.

29. Kamens J. The Addgene repository: an international nonprofit plasmid and data resource. Nucleic Acids Res. 2015;43:D1152–7.

30. Lutz R, Bujard H. Independent and tight regulation of transcriptional units in Escherichia coli via the LacR/O, the TetR/O and AraC/I1-I2 regulatory elements. Nucleic Acids Res. 1997;25:1203–10.

31. Silva-Rocha R, Martínez-García E, Calles B, Chavarría M, Arce-Rodríguez A, De Las Heras A, et al. The standard european vector architecture (SEVA): a coherent platform for the analysis and deployment of complex prokaryotic phenotypes. Nucleic Acids Res. 2013;41:666–75.

32. Karunakaran P, Blatny JM, Ertesvåg H, Valla S. Species-dependent phenotypes of replication-temperature-sensitive trfA mutants of plasmid RK2: a codon-neutral base substitution stimulates temperature sensitivity by leading to reduced levels of trfA expression. J Bacteriol. 1998;180:3793–8.

33. Haugan K, Karunakaran P, Blatny JM, Valla S. The phenotypes of temperature-sensitive mini-RK2 replicons carrying mutations in the replication control gene trfA are suppressed nonspecifically by intragenic cop mutations. J Bacteriol. 1992;174:7026–32.

34. Nikel PI, Chavarría M, Danchin A, de Lorenzo V. From dirt to industrial applications: Pseudomonas putida as a synthetic biology chassis for hosting harsh biochemical reactions. Curr Opin Chem Biol. 2016;34:20–9.

35. Schaefer KA, Wu W-H, Colgan DF, Tsang SH, Bassuk AG, Mahajan VB. Unexpected mutations after CRISPR–Cas9 editing in vivo. Nat Publ Gr. 2017;14:547–8.

36. Velappan N, Sblattero D, Chasteen L, Pavlik P, Bradbury ARM. Plasmid incompatibility: more compatible than previously thought? Protein Eng Des Sel. 2007;20:309–13.

37. Nicoloff HH, Andersson DI. Indirect resistance to several classes of antibiotics in cocultures with resistant bacteria expressing antibiotic-modifying or -degrading enzymes. J Antimicrob Chemother. 2016;71:100–10.

38. Lenski RE, Simpson SC, Nguyen TT. Genetic analysis of a plasmid-encoded, host genotype-specific enhancement of bacterial fitness. J Bacteriol. 1994;176:3140–7.

39. Li Y, Lin Z, Huang C, Zhang Y, Wang Z, Tang YJ, et al. Metabolic engineering of Escherichia coli using CRISPR-Cas9 mediated genome editing. Metab Eng. 2015;31:13–21 **(Elsevier)**.

40. Martínez-García E, Calles B, Arévalo-Rodríguez M, de Lorenzo V. pBAM1: an all-synthetic genetic tool for analysis and construction of complex bacterial phenotypes. BMC Microbiol. 2011;11:38.

41. Stearns T, Ma H, Botstein D. Manipulating yeast genome using plasmid vectors. Methods Enzymol. 1990;185:280–97.

42. Shintani M, Sanchez ZK, Kimbara K. Genomics of microbial plasmids: classification and identification based on replication and transfer systems and host taxonomy. Front Microbiol. 2015;6:1–16.

43. Johnson TJ, Wannemuehler YM, Johnson SJ, Logue CM, White DG, Doetkott C, et al. Plasmid replicon typing of commensal and pathogenic Escherichia coli isolates. Appl Environ Microbiol. 2007;73:1976–83.

44. Altschul SF, Gish W, Miller W, Myers EW, Lipman DJ. Basic local alignment search tool. J Mol Biol. 1990;215:403–10.

45. Cavaleiro AM, Kim SH, Seppälä S, Nielsen MT, Nørholm MHH. Accurate DNA assembly and genome engineering with optimized uracil excision cloning. ACS Synth Biol. 2015;4:1042–6.

46. Gibson DG, Young L, Chuang R-Y, Venter JC, Hutchison CA, Smith HO, et al. Enzymatic assembly of DNA molecules up to several hundred kilobases. Nat Methods. 2009;6:343–5.

47. Datta S, Costantino N, Court DL. A set of recombineering plasmids for gram-negative bacteria. Gene. 2006;379:109–15.

48. Nour-Eldin HH, Hansen BG, Nørholm MHH, Jensen JK, Halkier BA. Advancing uracil-excision based cloning towards an ideal technique for cloning PCR fragments. Nucleic Acids Res. 2006;34:e122.

13

A bacterial negative transcription regulator binding on an inverted repeat in the promoter for epothilone biosynthesis

Xin-jing Yue, Xiao-wen Cui, Zheng Zhang, Ran Peng, Peng Zhang, Zhi-feng Li and Yue-zhong Li[*] iD

Abstract

Background: Microbial secondary metabolism is regulated by a complex and mostly-unknown network of global and pathway-specific regulators. A dozen biosynthetic gene clusters for secondary metabolites have been reported in myxobacteria, but a few regulation factors have been identified.

Results: We identified a transcription regulator Esi for the biosynthesis of epothilones. Inactivation of *esi* promoted the epothilone production, while overexpression of the gene suppressed the production. The regulation was determined to be resulted from the transcriptional changes of epothilone genes. Esi was able to bind, probably via the N-terminus of the protein, to an inverted repeat sequence in the promoter of the epothilone biosynthetic gene cluster. The Esi-homologous sequences retrieved from the RefSeq database are all of the Proteobacteria. However, the Esi regulation is not universal in myxobacteria, because the *esi* gene exists only in a few myxobacterial genomes.

Conclusions: Esi binds to the epothilone promoter and down-regulates the transcriptional level of the whole gene cluster to affect the biosynthesis of epothilone. This is the first transcription regulator identified for epothilone biosynthesis.

Keywords: Negative transcription regulator, Transcription inhibition, Epothilone synthesis, Inverted repeat sequence in promoter, Myxobacteria, Proteobacteria

Background

Many kinds of microorganisms, such as actinomycetes, bacilli and myxobacteria, are excellent producers of secondary metabolites with various biological activities, potentially intriguing in anti-infection, anticancer and other pharmaceutical applications. The production of secondary metabolites in microorganisms is normally limited, probably due to the complex and mostly-unknown network of global and pathway-specific regulators, as well as the large biosynthetic gene cluster containing many gene modules [1–4]. Myxobacteria are able to produce diverse secondary metabolites [5–8]. Although a dozen gene clusters for the biosynthesis of secondary metabolites have been reported in

myxobacteria, a few regulation factors have been identified. StiR is the first regulator identified for secondary mechanism in myxobacteria, which was found involving in the positive regulation of stigmatellin production in *Cystobacter fuscus* Cb f17.1 [9]. ChiR was found to serve as a pleiotropic regulator, which not only positively regulated the chivosazol biosynthesis, but also affected the fruiting body development in *Sorangium cellulosum* So ce56 [10]. NtcA, a famous global transcriptional regulator responsive to the concentration of nitrogen, was reported to be able to regulate the biosynthesis of chivosazol and etnangien negatively in *S. cellulosum* So ce56 [11].

Epothilones are originally produced by some strains of *S. cellulosum*, and are a group of antitumor compounds mimicking paclitaxel in the polymerization of tubulin [6, 12]. Epothilones are potential anticancer agents by their greater water solubility and efficacy against paclitaxel-resistant tumors [13, 14]. Researchers tried

*Correspondence: lilab@sdu.edu.cn
State Key Laboratory of Microbial Technology, School of Life Science, Shandong University, Jinan 250100, China

many methods to improve the epothilones yield, including co-cultivation of different strains [15], mutation and high-throughput screening of high-producing strains [16], immobilization of strains [17], heterologous expressions [18–24] and manipulation of related genes [25, 26]. However, the yield is still limited. The biosynthetic gene cluster of epothilone is a large operon of approximately 56 kb in size, containing seven homodromous open reading frames (ORFs) [27, 28]. Because of the lack of efficient protocols for genetic manipulation in S. cellulosum, the regulation mechanisms of this big operon have not yet been investigated and thus remains mostly unknown.

A key regulatory step that modulates bacterial gene expressions is the promoter recognition by RNA polymerase and transcription initiation [29]. A transcription factor can increase the rate of transcription initiation (activators) or prevent RNA polymerase from initiating transcription (repressors). Many repressors prevent transcription by binding DNA at positions directly interfering the binding of RNA polymerases. Thus, in the promoters subject to repression, operator sequences for a repressor are often found to overlap or be immediately adjacent to the transcription start site [30]. For example, HrcA, a classic heat-shock regulator, binds to the controlling inverted repeat of chaperone expression (CIRCE) to regulate expressions of the downstream groEL-ES operon and dnaK gene [31–33]. Transcription regulators achieve specific binding in normally a dimer or further multimer form; and most operators contain direct or inverted repeats of a 4–5 base pair sequence [34].

Sorangium cellulosum So0157-2 is an epothilone producing strain [35]. We previously sequenced the genome of So0157-2 [36], studied in details the characteristics of epothilone operon promoter in Escherichia coli (E. coli) [37], and randomly integrated the epothilone biosynthetic gene cluster into M. xanthus genome by transposition [22]. In this study, we reported the identification of a negative transcription regulator for the biosynthesis of epothilones.

Results
Inactivation of esi promotes the epothilone production in S. cellulosum So0157-2
Using the conjugation and subsequent homologous recombination protocols as previously described [38], we inactivated genes flanking the epothilone gene cluster in S. cellulosum So0157-2 to determine whether they were associated with the synthesis, regulation or modification of epothilones. The SCE1572_25030 gene locates approximately 500 bp upstream to the epothilone promoter (Fig. 1a). We amplified an internal fragment of the gene and cloned it into the pCC11 plasmid [38] to construct the inactivation vector, which was introduced into

the genome of So0157-2. The gene was inactivated by insertional mutagenesis, which was confirmed by amplification using Tail-PCR and sequencing of the insertion region.

We fermented the mutant and the wild type strain in 50 mL of EMP medium supplemented with 2% of XAD-16 resin [16]. After 7 days of shaking cultivation, the resin was harvested, and the absorbed compounds were extracted for High Performance Liquid Chromatography (HPLC) analysis to determine the yields of epothilones. The results showed that the insertion mutation of SCE1572_25030 resulted in a remarkable increase in the yields of epothilones A and B, from 2.2 and 1.4 mg/L in the wild type strain to 4.7 and 10.1 mg/L in the mutant (Fig. 1b; t test, p value <0.01). The total yields of epothilones A and B increased approximately 4.1 times, from 3.6 to 14.8 mg/L. The results suggested that the SCE1572_25030 gene probably encoded an epothilone synthesis inhibitor (Esi). The mutant strain was termed as So0157-2-esi⁻.

The presence of esi gene suppresses the epothilone production in M. xanthus
Due to the genetic manipulation difficulty in S. cellulosum cells, we investigated influences of the esi gene on the production of epothilones in M. xanthus. We previously integrated the So0157-2 epothilone biosynthetic gene cluster randomly into M. xanthus genome by transposition; and the esi gene upstream to the epothilone gene cluster was also included in the transferred fragment [22]. We chose the ZE9 transformant derived from M. xanthus DZ2 for the assay. Bioinformatics analysis indicated that there was no homologous sequence of esi in DZ2, which was further confirmed by PCR amplification of an esi fragment against the DZ2 genome. Firstly, we deleted the esi gene from ZE9, producing the knockout mutant ZE9Δesi. The ZE9 and ZE9Δesi strains had similar growth curves in CYE medium (Fig. 2a). Methyl oleate is able to greatly increase the epothilone production ability in Myxococcus cells [39]. When fermented in the CYE medium supplemented with methyl oleate (CMO medium), the two strains showed different production abilities of epothilones. After 6 days of incubation, the yields of epothilones A and B increased 72.4 and 24.2%, and the total yields of epothilones were 9.9 mg/L in ZE9 and 13.7 mg/L in ZE9Δesi, respectively (Fig. 2b). The deletion of esi significantly increased the production of epothilones in Myxococcus cells (t test, p value <0.01).

Furthermore, we constructed an esi-overexpressing mutant strain ZE9 att::esi by using plasmid pSWU30-p630-esi to introduce an additional esi gene into the attB site of ZE9 genome. Using the same process, the empty plasmid pSWU30-p630 was also introduced into ZE9,

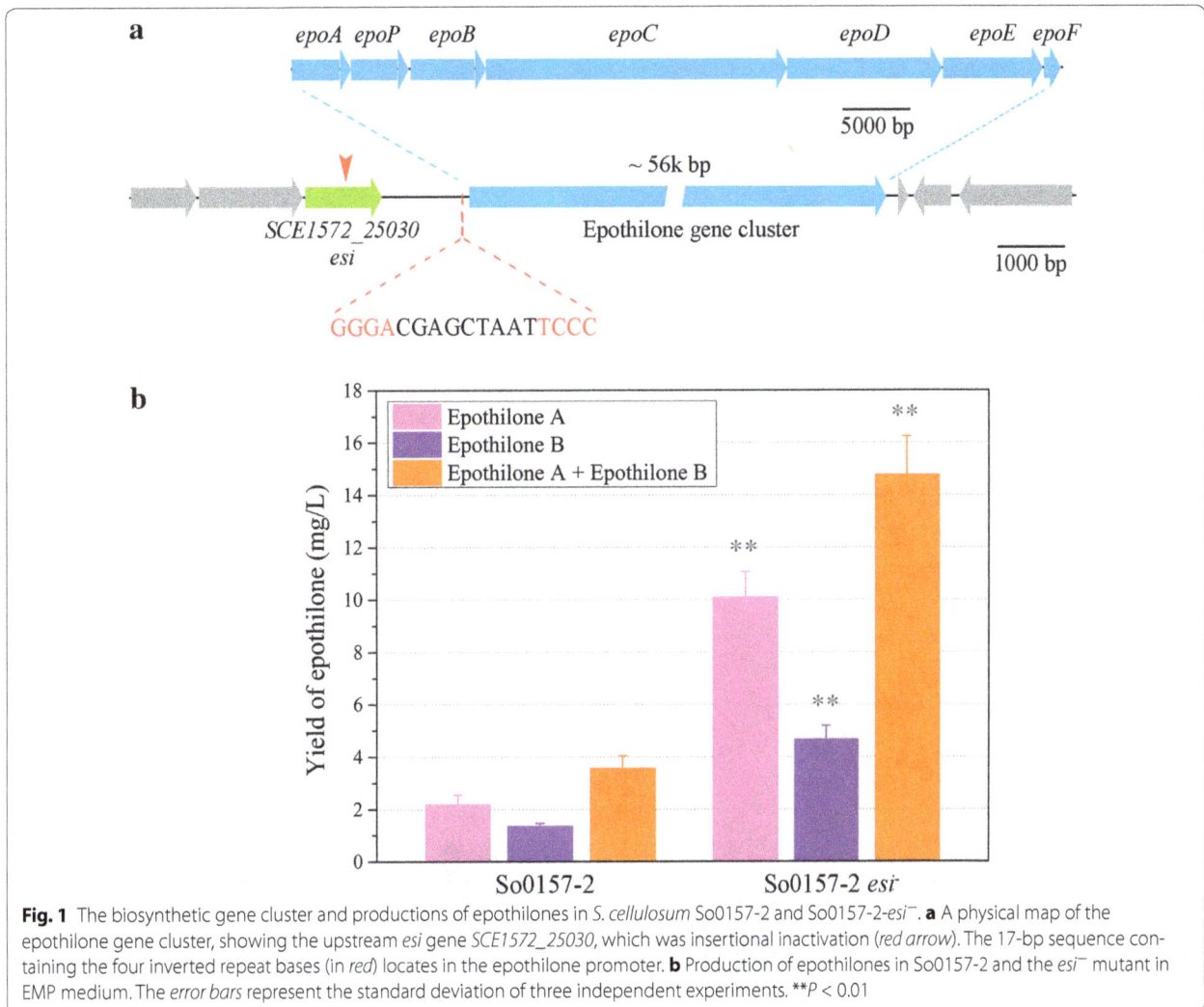

Fig. 1 The biosynthetic gene cluster and productions of epothilones in *S. cellulosum* So0157-2 and So0157-2-*esi⁻*. **a** A physical map of the epothilone gene cluster, showing the upstream *esi* gene *SCE1572_25030*, which was insertional inactivation (*red arrow*). The 17-bp sequence containing the four inverted repeat bases (*in red*) locates in the epothilone promoter. **b** Production of epothilones in So0157-2 and the *esi⁻* mutant in EMP medium. The *error bars* represent the standard deviation of three independent experiments. **$P < 0.01$

forming the ZE9 *att::Tet* strain as a control. These two strains were inoculated in CYE medium supplemented with 5 µg/mL tetracycline for 48 h, which were used as seeds to inoculate fresh CYE medium to assay their growth curves. The results showed that the ZE9 *att::esi* and ZE9 *att::Tet* strains also had a similar growth curve as the ZE9 strain (Fig. 2a). However, the introduction of an additional *esi* gene obviously suppressed the epothilone production (9.9 mg/L in ZE9 vs. 5.5 mg/L in ZE9 *att::esi*; *t* test, *p* value <0.01), whereas the ZE9 *att::Tet* had a similar epothilone yield as ZE9 (*t* test, *p* value >0.05) (Fig. 2b). Accordingly, Esi is a negative regulator for the production of epothilones, functioning not only in the producing *Sorangium* strains, but also in *Myxococcus*.

Esi negatively regulates the transcription of epothilone biosynthetic genes

To investigate the regulation mechanism of Esi on the production of epothilones, we comparatively analyzed the transcriptional levels of epothilone biosynthetic genes in the *esi*-deletion and overexpression mutants with their parent strain *M. xanthus* ZE9 using quantitative real-time polymerase chain reaction (RT-qPCR). In ZE9 strain, compared with the expression of the *epoA* gene, the *esi* gene expressed at a low level (Fig. 3). In the early growth stage (24 h of incubation), expression of the *esi* gene was approximately 21% of the *epoA* gene; then weakly decreased with the increase of incubation. After the growth, the expression level of *esi* gene was greatly decreased, remaining less than 2% of the 24 h-expression at the 60 h of incubation (the stable stage of the growth curve, referred to Fig. 2b). Comparatively, in the ZE9 *att::esi* mutant, the additional *esi* gene was under the control of the 630-bp promoter of *pilA*, which held the total expression level of *esi* at high levels. The *esi* expressions at the 24, 48 and 60 h were 4.35-, 3.16- and 2.70-fold of that at the 24 h in the ZE9 strain (Fig. 3).

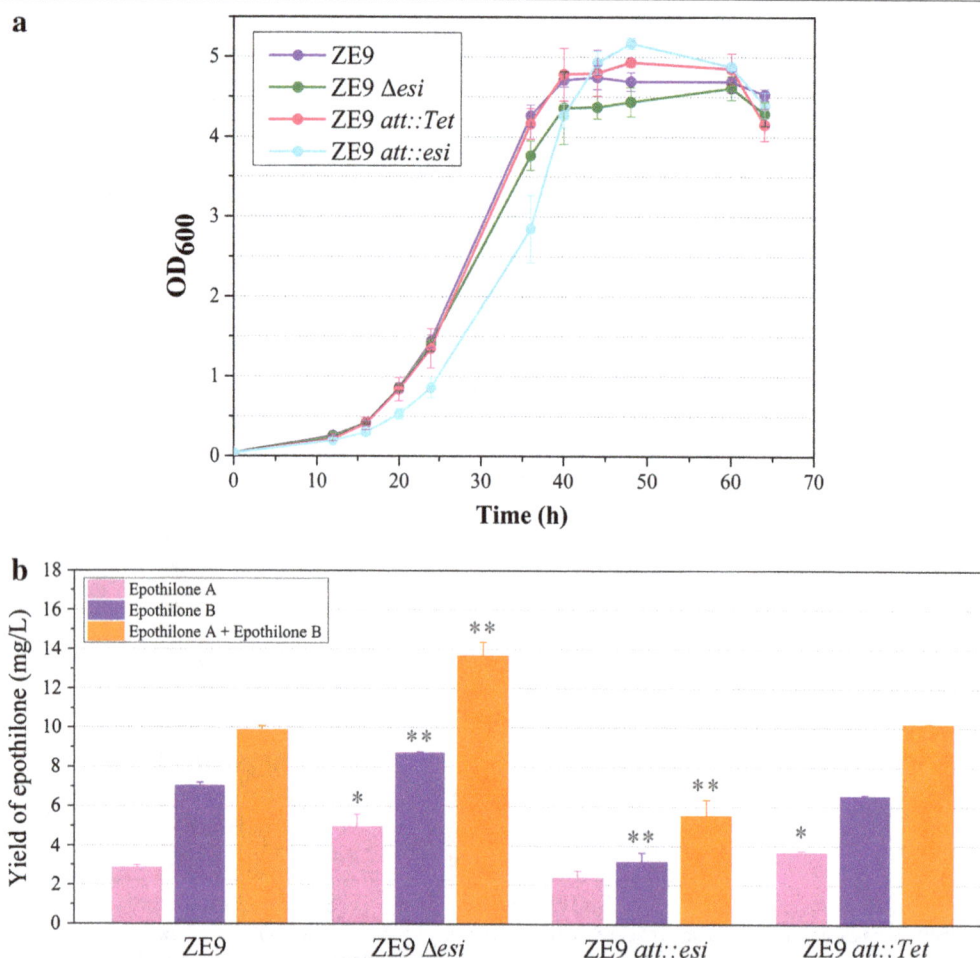

Fig. 2 The growth curves (**a**) and production abilities of epothilones (**b**) in *M. xanthus* ZE9 and its mutants. The growth curves were assayed in CYE medium. The production of epothilones were detected in the CYE medium supplemented with methyl oleate (CMO medium). The *error bars* represent the standard deviation of three independent experiments. **$P < 0.01$ and *$P < 0.05$, respectively

Similar to that in our previous report [22], transcriptional levels of the epothilone biosynthetic genes markedly varied in ZE9. With the increase of incubation time, transcriptions of the seven ORFs in the gene cluster each increased approximately ten times during the growth stage from 24 to 48 h, and then rapidly decreased in the following 12 h (Fig. 3). The transcriptional pattern was consistent with the production curves of epothilones in *M. xanthus* cells [22] or *S. cellulosum* cells [16], which was different from other microbial secondary metabolisms, usually expressed in stationary growth phase [2, 4].

After the deletion of the *esi* gene, the transcriptional level of each of the seven epothilone ORFs was markedly up-regulated at different incubation time points (*t* test, *p* value <0.05 or *p* value <0.01) (Fig. 3). The transcriptional increases ranged from 99% (*epoF*) to 200% (*epoC*) at 24 h, 88% (*epoF*) to 574% (*epoD*) at 48 h and 124% (*epoD*) to 611% (*epoB*) at 60 h of incubation. However, with the

overexpression of *esi*, the epothilone genes were correspondingly down-regulated in the ZE9 *att::esi* strain (Fig. 3). The above RT-qPCR analysis demonstrated that Esi negatively regulated the transcription of epothilone genes.

The structure model of Esi and its potential regulation mechanism

It is known that the repression of transcription initiation often occurs simply by steric hindrance [29]. From the above results, we suggested that the *esi* protein product probably functioned by directly binding on the epothilone promoter sequence to block the transcription. Esi (WP_020736929.1) is a protein with 373 amino acids in size. Domain prediction by the SMART program showed that the protein had no signal peptide or transmembrane domain. We constructed the three-dimensional (3D) structure of the Esi protein using the

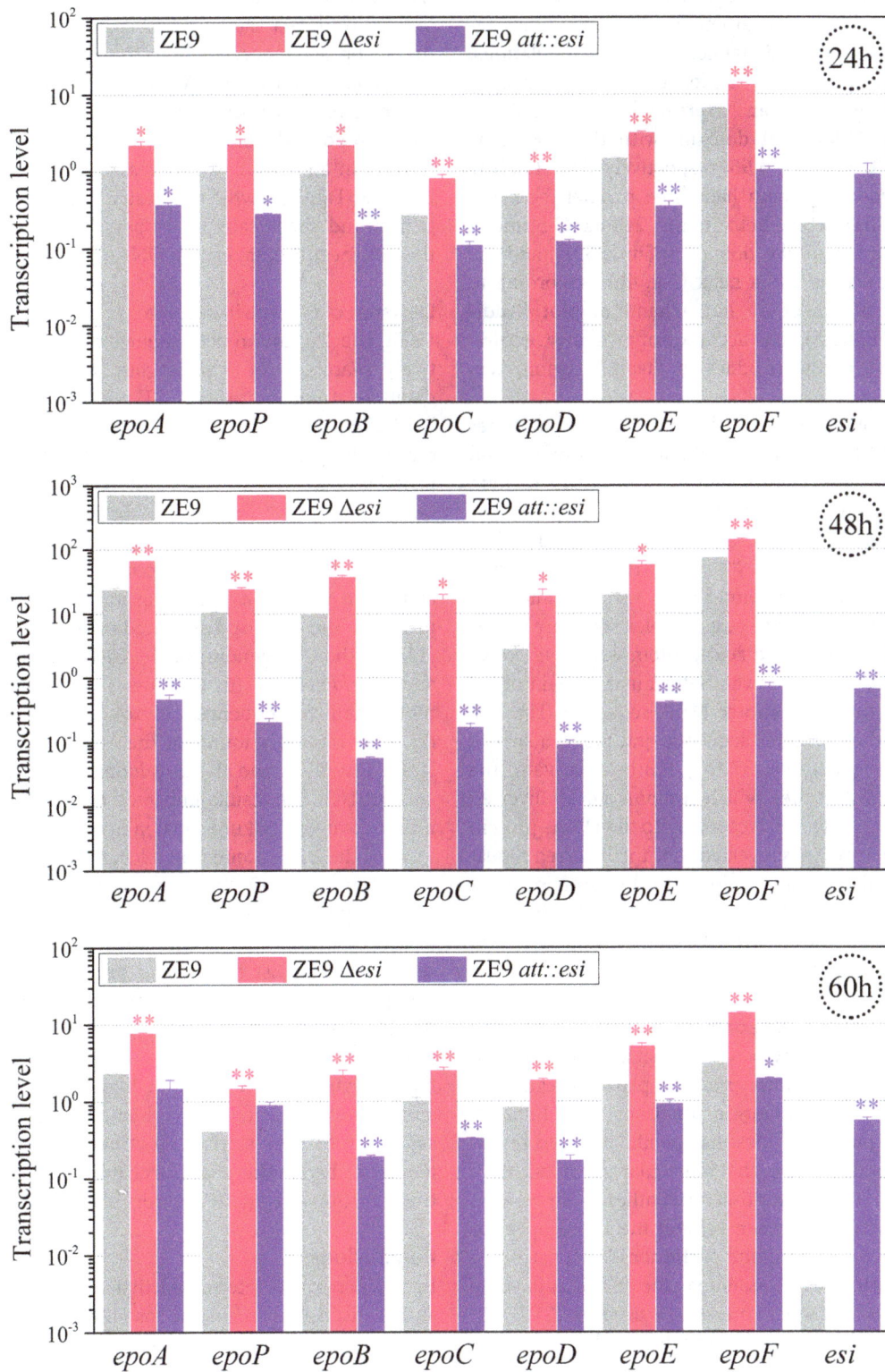

Fig. 3 RT-qPCR analysis of expressions of the *esi* gene and the seven epothilone ORFs in the ZE9, ZE9 Δ*esi* and ZE9 *att::esi* strains after 24, 48 and 60 h of incubation in CMO medium. The expression level of *epoA* in the ZE9 strain at 24 h of incubation was set to be 1. The *error bars* represent the standard deviation of three independent experiments. **P < 0.01 and *P < 0.05, respectively

I-TASSER and QUARK programs. All the structural models were refined at the atomic level by using the fragment-guided molecular dynamics (FG-MD) simulations. Based on the constructed model, the Esi protein contained three regions, i.e. an N-terminal domain, a linker region and a C-terminal domain, with the isoelectric points of 6.90, 5.87 and 5.95, respectively (Fig. 4a). The 106-aa N-terminal domain had three parallel β-pleated sheets and three alpha helixes, the 248-aa C-terminal domain was a Cupin-like domain (pfam13621), and the linker region was an alpha helix (Fig. 4b). According to the quality assessment by Ramachandran plot (Additional file 1: Figure S1), the accuracy of protein structural models (80.6% of the residues in favored region) was acceptable [40].

The repressors, usually in dimer or further multimer form, bind to the operator sequence of promoter and thus block the transcription; and most operators contain direct or inverted repeats of a 4–5 base pair sequence for the binding [34]. We further constructed potential DNA–protein complex structure. We reconstructed a 3D structure of the epothilone promoter using w3DNA, and then calculated to determine the best complex structure of Esi-promoter by using nucleic acid-protein dock (NPDock). The results suggested that the N-terminal domain of Esi connected with the promoter DNA sequence (Fig. 4c). We checked the promoter sequence and found a special 17-bp DNA sequence (−137 to −153 bp relative to the transcriptional start site), which contains a 4-bp inverted repeat sequence (Fig. 1a). According to the DNA–protein complex structure, we suggested that the inverted repeat was the best Esi-binding region. The Esi protein might work as a dimer, and the two N-terminal domains bind specifically to the inverted GGGA of the double-strand promoter DNA sequence.

Esi negatively regulates on epothilone promoter sequence

In our previous report, we revealed that a 24-bp fragment in the epothilone promoter probably played a regulation role for the RNA polymerase functions [36]. Interestingly, this fragment includes the specific inverted repeat sequence. To determine the molecular mechanism of Esi regulation on the epothilone synthesis, we assayed the binding activity between Esi and the 17-bp operator DNA sequence of epothilone promoter. We constructed the *esi* gene into the expression vector pET-28a, forming pET-28a-*esi*, which was expressed in *E. coli* BL21 (DE3) cells. The Esi protein was tagged with a 6-His sequence. After induction with 0.1 mM IPTG, the Esi proteins were extracted from the cultivated cells, and purified based on the fused His-tag (Fig. 5a).

The double-strand 17-bp sequence was obtained by PCR annealing with the primers labeled with biotin.

Electrophoretic Mobility Shift Assay (EMSA) was performed to detect the specific binding activity of Esi and the 17-bp sequence. Each of the reaction system contained biotin-labeled DNA and LightShift™ Poly (dI–dC). The results clearly showed that the His-Esi protein was able to bind to the biotin-labeled DNA. If the molar ratio of protein and DNA was increased, the more DNA was blocked (Fig. 5b), which indicated that the His-Esi protein bound specifically with operator DNA, rather than the nonspecific competition DNA (Poly dI–dC).

Existence of *esi* gene in bacteria

Using the Esi amino acid sequence as a seed, we iteratively searched the homologous sequences in RefSeq database using the psi-BLAST program, and found 156 Esi-homologues with more than 90% of alignment coverage. The genes encoding for the 156 Esi homologous proteins distributed in 203 different bacterial genomes (Additional file 2: Table S1), all belonging to the Proteobacteria. These genomes contained single *esi* gene copy, except *Pseudomonas fluorescens* FW300-N2C3, which contained two genes encoding for the Esi homologous proteins. The *esi* gene distributed mainly in *Pseudomonas* (143 of the 203 genomes), *Acinetobacter* (41 genomes) and *Burkholderia* (6 genomes). Surprisingly, among more than 20 sequenced myxobacterial genomes, the *esi* gene was only found in the *Stigmatella aurantiaca* strain DW4/3-1 and the *S. cellulosum* strains So0157-2 and So0163. The results indicated that the Esi regulation mechanism was not universal in myxobacteria.

All of the Esi proteins had a length ranging from 356aa to 419aa, and their functions have not been reported, indicating that the *esi* gene codes a novel negative regulator. The evolutionary conservation of each amino acid residue in the Esi homologue proteins was calculated using ConSurf [41], which showed that the C-terminal domain had the highest conservation, followed by the N-terminal domain and the linker region (Additional file 3: Figure S2). Because of high similarities of their amino acid sequences, these homologous proteins were suggested to have similar 3D structures as the Esi protein described in this paper, probably playing functions as transcriptional regulation factors like the Esi protein.

Conclusions

In this paper, we determined that the Esi protein played as a transcription regulator for epothilone biosynthesis. Inactivation of *esi* gene markedly promotes epothilone production. The Esi binds, probably via the N-terminus of the protein, to the inverted repeat of epothilone promoter and down-regulates the transcription of each gene in the gene cluster to suppress epothilones synthesis. Esi regulation is not universal in myxobacteria. This is the

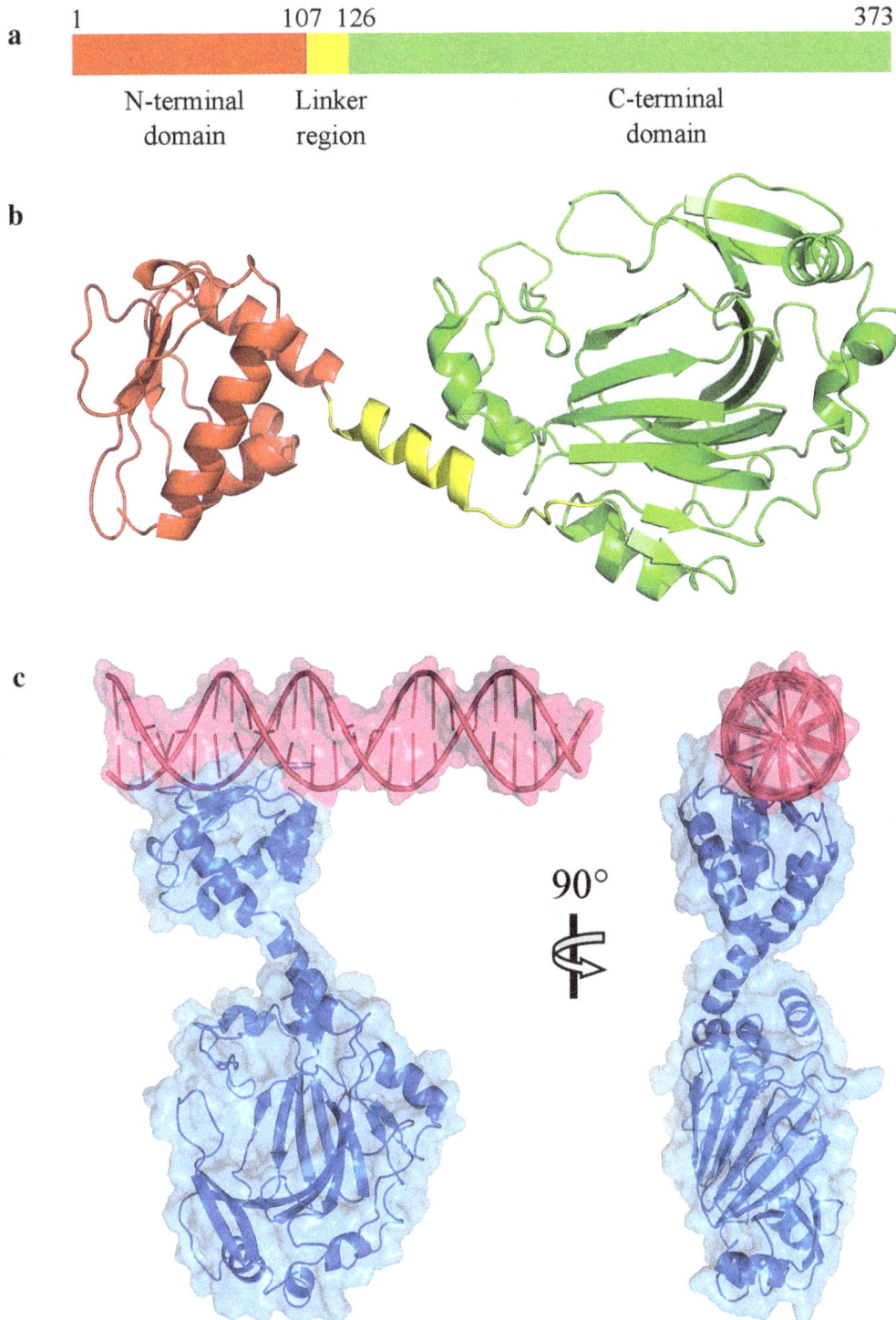

Fig. 4 The structure modeling of Esi protein and the DNA-Esi binding complex. **a** Positions of N-terminal domain, linker region and C-terminal domain. **b** The structure model of Esi protein. The protein backbone is shown in Cartoon mode. There were three parallel β-pleated sheets and three alpha helixes in the N-terminal domain and an alpha helix in the linker region. The C-terminal region is a cupin-like domain. **c** The DNA–protein complex structure of Esi protein (*blue*) and epothilone promoter (*red*). The *top* of the N-terminal domain of Esi binds with the promoter DNA sequence

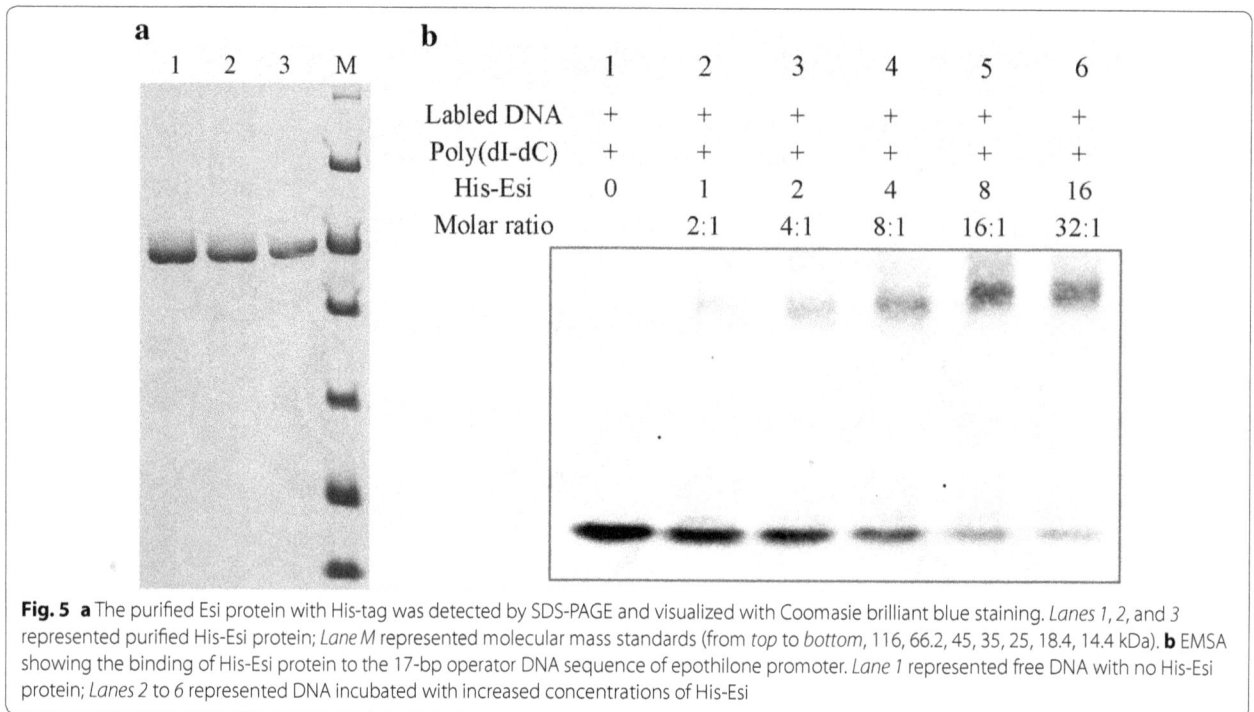

Fig. 5 a The purified Esi protein with His-tag was detected by SDS-PAGE and visualized with Coomasie brilliant blue staining. *Lanes 1, 2*, and *3* represented purified His-Esi protein; *Lane M* represented molecular mass standards (from *top* to *bottom*, 116, 66.2, 45, 35, 25, 18.4, 14.4 kDa). **b** EMSA showing the binding of His-Esi protein to the 17-bp operator DNA sequence of epothilone promoter. *Lane 1* represented free DNA with no His-Esi protein; *Lanes 2 to 6* represented DNA incubated with increased concentrations of His-Esi

first transcription regulator identified for the biosynthesis of epothilones.

Discussion

The Esi protein contained three regions, i.e. an N-terminal domain, a linker region and a C-terminal domain. The Esi protein connected with an operator of the promoter for the biosynthesis of epothilones, and the operator sequence contained a 4-bp inverted repeat in a 17-bp DNA fragment. We speculated that Esi worked in a similar way as HrcA [32, 33]: two molecules of Esi protein interacted to each other through the C-terminal domain, forming a mirrored dimer and the two N-terminal domains bind to the 4-bp inverted repeat of each DNA strand to repress transcription initiation. Although bacterial Esi proteins have not yet been studied, many of the homologous proteins were annotated to be cupin, a diverse superfamily of proteins named after its conserved barrel domain [42, 43] or transcription factor jumonji, a protein having a DNA binding domain and two conserved *jmj* domains [44]. The JmjC domains have been identified to play transcription factors in numerous eukaryotic proteins, such as PHD, C2H2, ARID/BRIGHT and zinc fingers, functioning by a histone demethylation mechanism that is conserved from yeast to human [44, 45]. However, since there is no histone in prokaryotes, the functioning mechanism of Esi was not like the eukaryotic JmjC proteins, and detailed molecular functioning pattern requires further investigations.

Methods
Bacterial strains, plasmids and culture conditions

Bacterial strains and plasmids used in this study are listed in Additional file 4: Table S2. *E. coli* strains XL1-Blue, DH5α λ pir and DH5α were used for cloning of pBJ113 plasmids, pSWU30 plasmids [46] and pET-28a plasmids, respectively. *E. coli* BL21 (DE3) was used to express the Esi protein. *E. coli* strains were grown at 37 °C in Luria-Broth (LB) medium (10 g/L peptone, 5 g/L yeast extract and 5 g/L NaCl, pH 7.2), and *M. xanthus* strains were grown at 30 °C in CYE medium [10 g/L casitone, 5 g/L yeast extract, 10 mM 3-(N-morpholino) propanesulfonic acid (MOPS) and 4 mM $MgSO_4$, pH 7.6] or CMO medium [10 g/L casitone, 5 g/L yeast extract, 10 mM MOPS, 4 mM $MgSO_4$ and 7 mL/L methyl oleate, pH 7.6]. The medium were supplemented with the following different antibiotics if required: ampicillin [Amp], 100 µg/mL; kanamycin [Km], 40 µg/mL; Apramycin, [Apra], 30 µg/mL, tetracycline [Tet], 5 µg/mL, and Gentamycin [Gm]. The primers used in constructing vectors and mutant strains were list in Additional file 5: Table S3.

Inactivation of *esi* gene in So0157-2

Inactivation of *esi* was performed by insertion mutation. Using genomic DNA of So0157-2 as template, the homologous arm was amplified by PCR with primers *esi* F1 and *esi* R1. The PCR fragment was phosphorylated with T4 Polynucleotide Kinase (Fermentas, Canada) and then cloned into the SmaI restriction site of pCC11 [38], leading to the insertion plasmid pCC-*esi*. After verification by DNA sequence analysis, the pCC-*esi* was transformed into *E. coli* DH5α (λ *pir*) harbouring the pRK2003(Km^r) plasmid, the resulting strain containing double plasmids was designed as ET-*esi*. The transformation of the plasmid pCC-*esi* into the genome of the So0157-2 was performed by biparental mating and conjugation, as described previously [38]. The *E.coli* ET-*esi* and So0157-2 were mixed with a ratio of 1:50, inoculated on the nitrocellulose membrane on VY/2 plate (7 µg/mL of Gm, 5 µg/mL of Cm), incubated for 40 h and then transferred to the selection VY/2 plate (15 µg/mL of Gm, 3 µg/mL of Cm). The clones appeared after two weeks of incubation at 30 °C were verified by Tail-PCR (data not shown) as described previously [22] and purified on CNST plate (15 µg/mL of Gm, 10 µg/mL of Cm). A Genome Walking kit (TaKaRa, Japan) was used to amplified the flanking sequence of insertion sites with three specific primers, Cm1, Cm2 and Cm3, designed in this work and a random primer AP1 provided by the kit. The PCR products with appropriate size from the third PCR round were withdrawn by the DNA Extraction kit (Promega, USA) and sequenced after cloning to the pMD19-T vector (Takara, Japan).

Knockout and overexpression of *Esi* gene in ZE9

Myxococcus xanthus ZE9 was constructed with inserting the epothilone biosynthetic gene cluster and some flanking sequence, including the *esi* gene, into the genome of DZ2 by transposition in previous work [22]. The deletion of *esi* gene was performed by construction of knockout vector pBJ-*esi*. The arms for homologous recombination were amplified from the genome of So0157-2 by PCR using the primers *esi*-up F and *esi*-up R for upstream arm and *esi*-down F and esi-down R for downstream arm. The 1.6-Kb fragment of *esi*-ud was amplified by overlap PCR with *esi*-up mixed with *esi*-down as template and the primers *esi*-up F and *esi*-down R, digested with EcoRI and KpnI, and then cloned into pBJ113, constructing the knockout vectors pBJ-*esi*. The pBJ-*esi* was introduced into ZE9 by electroporation as described previously [22]. Resistant colonies that appeared after 6 days of incubation on CYE plates with 40 µg/mL of Km were checked by colony PCR with primers KG-test F and KG-trst R. The Km-resistant colonies were resuspended in 2.5 mL-tube with CYE liquid medium and diluted in gradient for 10^2–10^5 times, mixed with 2.5 mL of 0.5% soft agar and the mixture were spread on CYE selection agar plates containing 0.1% galactose to select the second recombination event. The galactose-resistant but Km-sensitive colonies were screened and checked by colony PCR with primers *Esi*-up F and *Esi*-down R. The knockout mutant, designated as ZE9Δ*Esi*, was confirmed by sequence analysis of the PCR products.

To construct the overexpression mutant, the 630-bp promoter of the *pilA* gene (MXAN_RS28035) was amplified from the genomic DNA of DK1622 by PCR using the primers p630 F and p630 R. The PCR fragment was digested with HindIII and XbaI and then ligated into the plasmid pSWU30 [46], generating the vector of pSWU30-p630. The 1.2-Kb *esi* gene was amplified from the genomic DNA of So0157-2 by PCR with the primers *esi* F2 and *esi* R2 and ligated into XbaI and KpnI sites of pSWU30-p630 to produce the overexpression vector pSWU30-p630-*esi*. The pSWU30-p630-*esi* was electroporated into ZE9 and Tet-resistant colonies were screened by colony PCR using the primer p630 F and esi R2. The mutant strain ZE9 *att::Esi* was purified by being gradient diluted, mixed with soft agar (0.5%), and spread on CYE plates with 5 µg/mL of Tet. As a control, the empty plasmid pSWU30-p630 was also introduced into ZE9, producing ZE9 *att::Tet*.

Transcriptional analysis of *esi* gene and epothilone genes

The mutant strains and wild type strains were cultured in 50 mL of CYE liquid medium for 20 h (OD600 was about 1.5), then transplanted into fresh CMO medium with a start OD600 of 0.04, and harvested at the early stage, middle stage and stable stage of exponential growth successively. Total RNA were extracted according to the protocol provided by the BIOZOL Toal RNA Extraction Regent (BioFast, China), and transcribed reversely into cDNA with PrimeScript™ Regent Kit with DNAase (Takara, Japan). We analyzed the relative transcriptional level of the *esi* gene and epothilone gene cluster by RT-qPCR on LightCycler® 480 (Switzerland) with SYBR® Premix Ex Taq™ GC Dye (Takara, Japan). The *gapA* gene (MXAN_RS13645), encoding glyceraldehyde-3-phosphate dehydrogenase, was used as the reference gene. The primers used in RT-qPCR were list in Additional file 6: Table S4.

Extraction and detection of epothilones

So0157-2 and So0157-2-Esi$^−$ were cultured in M26 medium (8 g/L potato starch, 2 g/L soy peptone, 2 g/L glucose, 2 g/L yeast powder, 1 g/L MgSO$_4$, 1 g/L CaCl$_2$, 1 mL/L EDTA-Fe^{3+}, 1 mL/L microelement, pH 7.2) for 3 days and inoculated into 50 mL of EPM medium (4.7 g/L dextrin, 1.7 g/L soybean cake powder, 0.8 g/L glucose, 0.5 g/L saccharose, 2.2 g/L MgSO$_4$·7H$_2$O, 1 g/L CaCl$_2$, 2 mL/L EDTA-Fe3$^+$, 1 mL/L corn steep liquor, 1 mL/L microelement, pH 7.5) supplemented with 2% of resin XAD 16 for the absorbtion of epothilone products. ZE9 and *esi* mutants were grown overnight in 50 mL of CYE medium supplemented with Apra (30 µg/mL). The cultures were inoculated at a ratio of 2:100 into 50 mL of CMO medium containing 2% of the XAD-16 resin.

The resin were harvested with strainer after 9 days (for *S. cellulosum*) and 6 days (for *M. xanthus*) and extracted with 3 mL of methanol by shaking at room temperature overnight [16]. The supernatant was centrifuged for 10 min at 12,000 rpm and filtered with 0.22 μm filter to remove the impurities. 20 μL of the sample was injected into High Performance Liquid Chromatography (HPLC, SHIMADZU, Japan) and analyzed on a Shim-pack MRC-ODS RP C18 column (4.6 mm × 250 mm, 4.60 μm; Shimadzu, Japan) and monitored at 250 nm, with a mobile phase of 60% of methanol (HPLC grade) and 40% of H_2O at a flow rate of 1.0 mL/min. The yield of epothilone was quantified from the peak area in the UV chromatogram, by reference against a calibration standard.

Bioinformatics analysis of Esi

The signal peptides and transmembrane regions were predicted with SignalP [47] and TMHMM [48], respectively. Esi protein domain was annotated by SMART [49] and PFAM [50]. We constructed the N-terminal domain of Esi protein by ab initio protein folding and protein structure prediction with QUARK program [51], and then modeled the three-dimensional structures model of the whole Esi protein using the I-TASSER program based on a threading approach [52]. All the structural models were refined in the atomic-level by the fragment-guided molecular dynamics (FG-MD) simulations [53]. According to quality assessment by Ramachandran plot [40], the accuracy of the protein structural models were acceptable. The three-dimensional nucleic-acid structures of the epothilone promoter sequence was reconstructed with w3DNA server [54]. The NPDock (Nucleic acid-Protein Dock) was used to model the complex structures of Esi protein and epothilone promoter [55].

The position-specific iterative basic local alignment search tool (PSI-BLAST) [56] was used to search for homologous proteins of Esi in the US National Center for Biotechnology Information (NCBI) reference sequence (RefSeq) database [57]. The amino acid sequence of Esi protein was used as query sequence. A multiple sequence alignment of the Esi protein and its homologous proteins was established using the MAFFT program [58]. The evolutionary conservation of amino acid positions in the Esi protein and its homologous sequences was estimated by using ConSurf algorithm [41]. The JTT substitution matrix was used and the computation was based on the empirical Bayesian paradigm. The conservation scale was defined from the most variable amino acid positions (grade 1), which were considered to be evolved rapidly, to the most conservative positions (grade 9), which were considered to be evolved slowly. The sequence and modeled structure of the Esi protein were shown in nine-color conservation grades. The statistical analysis was conducted using IBM SPSS Statistics.

Expression and purification of Esi protein in *E. coli*

Expression vector was constructed by amplifying the *esi* gene fragment, fused with a His-tag at N-terminus, with PCR using the primers of *esi* F3/*esi* R3 and ligating it into the XbaI and KpnI sites of the plasmid pET-28a, generating pET-28a-*esi*. *E. coli* BL21 (DE3) cells were transformed with plasmid pET-28a-*esi*. For the heterologous expression of Esi, 500 mL LB medium containing Km was inoculated with 5 mL of the overnight culture, prepared from a single colony, and grown at 37 °C for about 2 h. 0.1 mM IPTG was added at an optical density (OD600 nm) of 0.8. After 24 h of induced-expression at 16 °C, cells were harvested by centrifugation at 12,000 rpm for 5 min, resuspended with 1/10 volume of Lysis Buffer (25 mM Tris–HCl, 250 mM NaCl, pH 8.0) and disrupted by sonication. After centrifugation at 12,000 rpm for 30 min, the supernatant was incubated overnight with Ni–nitrilotriacetic acid HiszBind resin (Novagen), and then eluated with Lysis Buffer containing gradient imidazole ranging from 20 to 250 Mm to collect soluble purified Esi protein.

Electrophoretic mobility shift assay

The biotin-labeled DNA fragment of operator DNA sequence was generated by PCR annealing (95 °C, 3 min; 90 °C, 1 min; 83 °C, 1 min; 76 °C, 1 min; 70 °C, 1 min; 65 °C, 1 min; 60 °C, 30 min) using the following oligonucleotides: 5′-GGGACGAGCTAATTCCC-3′ and 5′-GGGAATTAGCTCGTCCC-3′. The binding reactions were performed in 25 μL final volume containing 2.5 μL of reaction buffer (120 mM HEPES, 40 mM Tris, 600 mM KCl, 50 mM $MgCl_2$, 1 mM EDTA, pH 8.0), 1 ul of glycerol buffer (20% glycerol in 20 Mm HEPES), 1 μL of biotin-labeled DNA fragment (12 pmol), 1 μL of Poly (dI–dC), N μL of protein and (17.5-N) μL of 50 mM NaH_2PO_4. The reaction samples were incubated at 30 °C for 30 min and then loaded into a 5% non-denaturing polyacrylamide gel to conduct electrophoresis with TGE buffer (14.3 g/L glycine, 3 g/L Tris, 1 mM EDTA, pH 8.0) for 2.5 h at 10 mA and 4 °C. The samples were electro-blotted onto an Amersham Hybond-N + membrane (GE, UK) by use of a mini transblot electrophoretic transfer cell (Bio-Rad). Hybridization and blotting was performed according to the protocol described by Chemiluminescent Nucleic Acid Detection Module (Thermo, USA). After completion of hybridization, the membrane was dried, autoradiographed, and the bands were detected by use of a Chemi Doc™ XRS + with Image Lab™ software (BIO-RAD).

Additional files

Additional file 1: Figure S1. The ramachandran plot of the Esi structure model, which showed 80.6% residues in favored region, 13.5% residues in allowed region, 5.9% residues in outlier region.

Additional file 2: Table S1. Homologous amino acid sequences of Esi protein.

Additional file 3: Figure S2. Evolution conservative analysis of amino acid residues, calculated from the sequences of 156 Esi homologues. (A) Conservation scale is defined from the most variable residue sites (grade 1, color represented by turquoise; in rapid evolution) to conservative residue sites (grade 9, color represented by maroon; in slow evolution). (B) Conservation analysis of the Esi sequence sites in ConSurf grades.

Additional file 4: Table S2. Strains and plasmids used in this study.

Additional file 5: Table S3. Primers used in construction of vectors and mutant strains.

Additional file 6: Table S4. Primers used in RT-qPCR.

Authors' contributions
XJY, ZFL and YZL designed the experiments. XJY, RP, PZ and XWC performed the experiments. XJY, ZZ, and YZL analyzed the data. XJY and YZL wrote the manuscript. All authors read and approved the final manuscript.

Acknowledgements
Not applicable.

Competing interests
The authors declare that they have no competing interests.

Funding
This work was financially supported by the National Natural Science Foundation of China (NSFC) (31471183 and 31670076) to YZL and (31370123) to ZFL.

References
1. Macheleidt J, Mattern DJ, Fischer J, Netzker T, Weber J, Schroeckh V, Valiante V, Brakhage AA. Regulation and Role of fungal secondary metabolites. Annu Rev Genet. 2016;50:371–92.
2. Bibb MJ. Regulation of secondary metabolism in streptomycetes. Curr Opin Microbiol. 2005;8:208–15.
3. Bode HB, Muller R. Analysis of myxobacterial secondary metabolism goes molecular. J Ind Microbiol Biotechnol. 2006;33:577–88.
4. Brakhage AA. Regulation of fungal secondary metabolism. Nat Rev Microbiol. 2013;11:21–32.
5. Gerth K, Irschik H, Reichenbach H, Trowitzsch W. The myxovirescins, a family of antibiotics from Myxococcus virescens (Myxobacterales). J Antibiot (Tokyo). 1982;35:1454–9.
6. Gerth K, Bedorf N, Hofle G, Irschik H, Reichenbach H. Epothilons A and B: antifungal and cytotoxic compounds from *Sorangium cellulosum* (Myxobacteria). Production, physico-chemical and biological properties. J Antibiot (Tokyo). 1996;49:560–3.
7. Sasse F, Steinmetz H, Heil J, Hofle G, Reichenbach H. Tubulysins, new cytostatic peptides from myxobacteria acting on microtubuli. Production, isolation, physico-chemical and biological properties. J Antibiot (Tokyo). 2000;53:879–85.
8. Gerth K, Pradella S, Perlova O, Beyer S, Muller R. Myxobacteria: proficient producers of novel natural products with various biological activities—past and future biotechnological aspects with the focus on the genus Sorangium. J Biotechnol. 2003;106:233–53.
9. Rachid S, Sasse F, Beyer S, Muller R. Identification of StiR, the first regulator of secondary metabolite formation in the myxobacterium Cystobacter fuscus Cb f17.1. J Biotechnol. 2006;121:429–41.
10. Rachid S, Gerth K, Kochems I, Muller R. Deciphering regulatory mechanisms for secondary metabolite production in the myxobacterium *Sorangium cellulosum* So ce56. Mol Microbiol. 2007;63:1783–96.
11. Rachid S, Gerth K, Muller R. NtcA: a negative regulator of secondary metabolite biosynthesis in *Sorangium cellulosum*. J Biotechnol. 2009;140:135–42.
12. Bollag DM, McQueney PA, Zhu J, Hensens O, Koupal L, Liesch J, Goetz M, Lazarides E, Woods CM. Epothilones, a new class of microtubule-stabilizing agents with a taxol-like mechanism of action. Cancer Res. 1995;55:2325–33.
13. Kowalski RJ, Giannakakou P, Hamel E. Activities of the microtubule-stabilizing agents epothilones A and B with purified tubulin and in cells resistant to paclitaxel (Taxol(R)). J Biol Chem. 1997;272:2534–41.
14. Su DS, Balog A, Meng D, Bertinato P, Danishefsky SJ, Zheng YH, Chou TC, He L, Horwitz SB. Structure–activity relationship of the epothilones and the first in vivo comparison with paclitaxel. Angewandte Chemie Int Ed Engl. 1997;36:2093–6.
15. Li PF, Li SG, Li ZF, Zhao L, Wang T, Pan HW, Liu H, Wu ZH, Li YZ. Co-cultivation of *Sorangium cellulosum* strains affects cellular growth and biosynthesis of secondary metabolite epothilones. FEMS Microbiol Ecol. 2013;85:358–68.
16. Gong GL, Sun X, Liu XL, Hu W, Cao WR, Liu H, Liu WF, Li YZ. Mutation and a high-throughput screening method for improving the production of epothilones of Sorangium. J Ind Microbiol Biotechnol. 2007;34:615–23.
17. Gong GL, Huang YY, Liu LL, Chen XF, Liu H. Enhanced production of epothilone by immobilized *Sorangium cellulosum* in porous ceramics. J Microbiol Biotechnol. 2015;25:1653–9.
18. Fu J, Wenzel SC, Perlova O, Wang J, Gross F, Tang Z, Yin Y, Stewart AF, Muller R, Zhang Y. Efficient transfer of two large secondary metabolite pathway gene clusters into heterologous hosts by transposition. Nucleic Acids Res. 2008;36:e113.
19. Julien B, Shah S. Heterologous expression of epothilone biosynthetic genes in *Myxococcus xanthus*. Antimicrobial Agents Chemother. 2002;46:2772–8.
20. Park SR, Park JW, Jung WS, Han AR, Ban YH, Kim EJ, Sohng JK, Sim SJ, Yoon YJ. Heterologous production of epothilones B and D in *Streptomyces venezuelae*. Appl Microbiol Biotechnol. 2008;81:109–17.
21. Osswald C, Zipf G, Schmidt G, Maier J, Bernauer HS, Muller R, Wenzel SC. Modular construction of a functional artificial epothilone polyketide pathway. ACS Synth Biol. 2014;3:759–72.
22. Zhu LP, Yue XJ, Han K, Li ZF, Zheng LS, Yi XN, Wang HL, Zhang YM, Li YZ. Allopatric integrations selectively change host transcriptomes, leading to varied expression efficiencies of exotic genes in *Myxococcus xanthus*. Microb Cell Fact. 2015;14:105.
23. Mutka SC, Carney JR, Liu Y, Kennedy J. Heterologous production of epothilone C and D in *Escherichia coli*. Biochemistry. 2006;45:1321–30.
24. Tang L, Shah S, Chung L, Carney J, Katz L, Khosla C, Julien B. Cloning and heterologous expression of the epothilone gene cluster. Science. 2000;287:640–2.
25. Ye W, Zhang W, Chen Y, Li H, Li S, Pan Q, Tan G, Liu T. A new approach for improving epothilone B yield in *Sorangium cellulosum* by the introduction of vgb epoF genes. J Ind Microbiol Biotechnol. 2016;43:641–50.
26. Han SJ, Park SW, Park BW, Sim SJ. Selective production of epothilone B by heterologous expression of propionyl-CoA synthetase in *Sorangium cellulosum*. J Microbiol Biotechnol. 2008;18:135–7.
27. Molnar I, Schupp T, Ono M, Zirkle R, Milnamow M, Nowak-Thompson B, Engel N, Toupet C, Stratmann A, Cyr DD, et al. The biosynthetic gene cluster for the microtubule-stabilizing agents epothilones A and B from *Sorangium cellulosum* So ce90. Chem Biol. 2000;7:97–109.

28. Julien B, Shah S, Ziermann R, Goldman R, Katz L, Khosla C. Isolation and characterization of the epothilone biosynthetic gene cluster from *Sorangium cellulosum*. Gene. 2000;249:153–60.

29. Browning DF, Busby SJ. The regulation of bacterial transcription initiation. Nat Rev Microbiol. 2004;2:57–65.

30. Gralla JD. Activation and repression of *E. coli* promoters. Curr Opin Genet Dev. 1996;6:526–30.

31. Zuber U, Schumann W. CIRCE, a novel heat shock element involved in regulation of heat shock operon dnaK of *Bacillus subtilis*. J Bacteriol. 1994;176:1359–63.

32. Wiegert T, Schumann W. Analysis of a DNA-binding motif of the *Bacillus subtilis* HrcA repressor protein. FEMS Microbiol Lett. 2003;223:101–6.

33. Liu J, Huang C, Shin DH, Yokota H, Jancarik J, Kim JS, Adams PD, Kim R, Kim SH. Crystal structure of a heat-inducible transcriptional repressor HrcA from *Thermotoga maritima*: structural insight into DNA binding and dimerization. J Mol Biol. 2005;350:987–96.

34. Browning DF, Busby SJ. Local and global regulation of transcription initiation in bacteria. Nat Rev Microbiol. 2016;14:638–50.

35. Li ZF, Zhao JY, Xia ZJ, Shi J, Liu H, Wu ZH, Hu W, Liu WF, Li YZ. Evolutionary diversity of ketoacyl synthases in cellulolytic myxobacterium Sorangium. Syst Appl Microbiol. 2007;30:189–96.

36. Han K, Li ZF, Peng R, Zhu LP, Zhou T, Wang LG, Li SG, Zhang XB, Hu W, Wu ZH, et al. Extraordinary expansion of a *Sorangium cellulosum* genome from an alkaline milieu. Sci Rep. 2013;3:2101.

37. Zhu LP, Li ZF, Sun X, Li SG, Li YZ. Characteristics and activity analysis of epothilone operon promoters from *Sorangium cellulosum* strains in *Escherichia coli*. Appl Microbiol Biotechnol. 2013;97:6857–66.

38. Xia ZJ, Wang J, Hu W, Liu H, Gao XZ, Wu ZH, Zhang PY, Li YZ. Improving conjugation efficacy of *Sorangium cellulosum* by the addition of dual selection antibiotics. J Ind Microbiol Biotechnol. 2008;35:1157–63.

39. Lau J, Frykman S, Regentin R, Ou S, Tsuruta H, Licari P. Optimizing the heterologous production of epothilone D in *Myxococcus xanthus*. Biotechnol Bioeng. 2002;78:280–8.

40. Lovell SC, Davis IW, Arendall WB 3rd, de Bakker PI, Word JM, Prisant MG, Richardson JS, Richardson DC. Structure validation by Calpha geometry: phi, psi and Cbeta deviation. Proteins. 2003;50:437–50.

41. Ashkenazy H, Erez E, Martz E, Pupko T, Ben-Tal N. ConSurf 2010: calculating evolutionary conservation in sequence and structure of proteins and nucleic acids. Nucleic Acids Res. 2010;38:W529–33.

42. Dunwell JM. Cupins: a new superfamily of functionally diverse proteins that include germins and plant storage proteins. Biotechnol Genet Eng Rev. 1998;15:1–32.

43. Dunwell JM, Purvis A, Khuri S. Cupins: the most functionally diverse protein superfamily? Phytochemistry. 2004;65:7–17.

44. Tsukada Y, Fang J, Erdjument-Bromage H, Warren ME, Borchers CH, Tempst P, Zhang Y. Histone demethylation by a family of JmjC domain-containing proteins. Nature. 2006;439:811–6.

45. Clissold PM, Ponting CP. JmjC: cupin metalloenzyme-like domains in jumonji, hairless and phospholipase A2beta. Trends Biochem Sci. 2001;26:7–9.

46. Wu SS, Kaiser D. Genetic and functional evidence that type IV pili are required for social gliding motility in *Myxococcus xanthus*. Mol Microbiol. 1995;18:547–58.

47. Petersen TN, Brunak S, von Heijne G, von Nielsen H. SignalP 4.0: discriminating signal peptides from transmembrane regions. Nat Methods. 2011;8:785–6.

48. Krogh A, Larsson B, von Heijne G, Sonnhammer EL. Predicting transmembrane protein topology with a hidden Markov model: application to complete genomes. J Mol Biol. 2001;305:567–80.

49. Letunic I, Doerks T, Bork P. SMART: recent updates, new developments and status in 2015. Nucleic Acids Res. 2015;43:D257–60.

50. Finn RD, Coggill P, Eberhardt RY, Eddy SR, Mistry J, Mitchell AL, Potter SC, Punta M, Qureshi M, Sangrador-Vegas A, et al. The Pfam protein families database: towards a more sustainable future. Nucleic Acids Res. 2016;44:D279–85.

51. Xu D, Zhang Y. Ab initio protein structure assembly using continuous structure fragments and optimized knowledge-based force field. Proteins. 2012;80:1715–35.

52. Yang J, Yan R, Roy A, Xu D, Poisson J, Zhang Y. The I-TASSER Suite: protein structure and function prediction. Nat Methods. 2015;12:7–8.

53. Zhang J, Liang Y, Zhang Y. Atomic-level protein structure refinement using fragment-guided molecular dynamics conformation sampling. Structure. 2011;19:1784–95.

54. Zheng G, Lu XJ, Olson WK. Web 3DNA–a web server for the analysis, reconstruction, and visualization of three-dimensional nucleic-acid structures. Nucleic Acids Res. 2009;37:W240–6.

55. Tuszynska I, Magnus M, Jonak K, Dawson W, Bujnicki JM. NPDock: a web server for protein-nucleic acid docking. Nucleic Acids Res. 2015;43:W425–30.

56. Johnson M, Zaretskaya I, Raytselis Y, Merezhuk Y, McGinnis S, Madden TL. NCBI BLAST: a better web interface. Nucleic Acids Res. 2008;36:W5–9.

57. O'Leary NA, Wright MW, Brister JR, Ciufo S, Haddad D, McVeigh R, Rajput B, Robbertse B, Smith-White B, Ako-Adjei D, et al. Reference sequence (RefSeq) database at NCBI: current status, taxonomic expansion, and functional annotation. Nucleic Acids Res. 2016;44:D733–45.

58. Katoh K, Standley DM. MAFFT multiple sequence alignment software version 7: improvements in performance and usability. Mol Biol Evol. 2013;30:772–80.

14

Overexpression of a C_4-dicarboxylate transporter is the key for rerouting citric acid to C_4-dicarboxylic acid production in *Aspergillus carbonarius*

Lei Yang[1], Eleni Christakou[2], Jesper Vang[1], Mette Lübeck[1] and Peter Stephensen Lübeck[1*]

Abstract

Background: C_4-dicarboxylic acids, including malic acid, fumaric acid and succinic acid, are valuable organic acids that can be produced and secreted by a number of microorganisms. Previous studies on organic acid production by *Aspergillus carbonarius,* which is capable of producing high amounts of citric acid from varieties carbon sources, have revealed its potential as a fungal cell factory. Earlier attempts to reroute citric acid production into C_4-dicarboxylic acids have been with limited success.

Results: In this study, a glucose oxidase deficient strain of *A. carbonarius* was used as the parental strain to overexpress a native C_4-dicarboxylate transporter and the gene *frd* encoding fumarate reductase from *Trypanosoma brucei* individually and in combination. Impacts of the introduced genetic modifications on organic acid production were investigated in a defined medium and in a hydrolysate of wheat straw containing high concentrations of glucose and xylose. In the defined medium, overexpression of the C_4-dicarboxylate transporter alone and in combination with the *frd* gene significantly increased the production of C_4-dicarboxylic acids and reduced the accumulation of citric acid, whereas expression of the *frd* gene alone did not result in any significant change of organic acid production profile. In the wheat straw hydrolysate after 9 days of cultivation, similar results were obtained as in the defined medium. High amounts of malic acid and succinic acid were produced by the same strains.

Conclusions: This study demonstrates that the key to change the citric acid production into production of C_4-dicarboxylic acids in *A. carbonarius* is the C_4-dicarboxylate transporter. Furthermore it shows that the C_4-dicarboxylic acid production by *A. carbonarius* can be further increased via metabolic engineering and also shows the potential of *A. carbonarius* to utilize lignocellulosic biomass as substrates for C_4-dicarboxylic acid production.

Keywords: *Aspergillus carbonarius*, Citric acid, C_4-dicarboxylate transporter, Lignocellulosic biomass, Malic acid, Metabolic engineering, Succinic acid

Background

C_4-dicarboxylic acids, including malic acid, fumaric acid and succinic acid, are amongst top value added chemicals with their large and growing markets due to wide spectra of applications [1]. In addition to their traditional uses such as food additives, chelators and acidulants, C_4-dicarboxylic acids have been for the past decades extensively exploited to be key building blocks for deriving varieties of commodity and specialty chemicals [2]. To reduce the dependence of the global economy on petroleum industry, bio-refinery of renewable biomass is considered to be an alternative approach to support industrial manufacture [3]. The fact that C_4-dicarboxylic acids are present as key intermediates in primary metabolism of living cells indicates the potential of using microbial systems

*Correspondence: psl@bio.aau.dk
[1] Section for Sustainable Biotechnology, Department of Chemistry and Bioscience, Aalborg University Copenhagen, A. C. Meyers Vænge 15, 2450 Copenhagen SV, Denmark
Full list of author information is available at the end of the article

to produce them from fermentable sugars derived from renewable biomass and the feasibilities of improving the production strains via metabolic engineering [4–6]. In recent years, bio-based production of C_4-dicarboxylic acids has received increasing research attention and achieved an important status in bio-economy. So far, bio-based succinic acid production has succeeded with a number of commercialized processes using bacterial strains (*Escherichia coli* and *Actinobacillus succinogenes*) and yeast strains (*Saccharomyces cerevisiae*) [7], and bio-technological processes for malic acid and fumaric acid are under research development [8, 9]. The bottlenecks in the current biotechnologies for production of C_4-dicarboxylic acids are relatively low productivity in production processes and high production cost due to the choice of substrates (glucose) and downstream product purification [10]. Although the research effort to address those technical constraints now mainly focus on the industrial candidate strains, exploiting new cell factories with their special genetic and physiological traits may open the window of opportunity for future technical breakthrough in the bio-based production of C_4-dicarboxylic acids.

Application of fungal technology in industrial production of organic acids has been demonstrated with several *Aspergillus* species, such as citric acid production by *Aspergillus niger* and itaconic acid production by *Aspergillus terreus* [11–13]. *Aspergillus carbonarius*, a member of black aspergilli, possesses several valuable virtues to be a competent cell factory for organic acid production. It can produce different types of organic acids (citric acid, gluconic acid and malic acid) from varieties of substrates ranging from mono-sugars to polysaccharides such as glucose, xylose, cellulose and starch, and tolerate stress conditions, especially low pH, during organic acid production [14, 15]. Organic acid profiling of *A. carbonarius* has shown its capability of producing high amounts of citric acid and gluconic acid under different pH conditions [16]. To further strengthen its abilities for organic acid production, a series of genetic modifications has been introduced targeted to the primary metabolic pathways and the regulatory system in *A. carbonarius* [14, 17, 18]. However, significant impacts have only been obtained on the production of citric acid rather than other organic acids e.g. C_4-dicarboxylic acids. Deletion of glucose oxidase that converts glucose to gluconic acid in pH buffered conditions that are suitable for C_4-dicarboxylic acid production, completely eliminated accumulation of gluconic acid and increased citric acid production, but only improved malic acid production at a very limited level [18]. Overexpression of enzymes carrying out the carboxylation of phosphoenolpyruvate in *A. carbonarius* supposed to increase the carbon flux towards the rTCA branch from which C_4-dicarboxylic acids are produced as

key intermediates [19], gave no significant impact on the production of C_4-dicarboxylic acids, and the increased carbon flux seemed to flow towards citric acid production [16]. This phenomenon was also observed in another well-known citric acid producing species, *A. niger*, when three genes involved in the rTCA branch were overexpressed [20]. In addition to central carbon metabolic pathways, export of C_4-dicarboxylic acids is an essential step to consider in metabolic engineering of microbial strains for production of C_4-dicarboxylic acids. In *Aspergillus oryzae* and *S. cerevisiae*, synergistic impacts on malic acid production were obtained when C_4-dicarboxylate transporters were overexpressed in combination with other genetic modifications [9, 21]. In *A. carbonarius*, there is not yet any report regarding the C_4-dicarboxylate transporter.

In this study, we identified a gene *dct* encoding a putative C_4-dicarboxylate transporter (DCT) from the genome of *A. carbonarius* and overexpressed it in a glucose oxidase deficient strain to examine its effect on C_4-dicarboxylic acid production. The Δgox strain is used as a parental strain as it provides an ideal platform to evaluate the impact of introduced genetic modifications in glucose containing media under pH buffered conditions without the interference of extracellular conversion of glucose to gluconic acid [18]. Furthermore, we expressed the *frd* gene encoding a NADH dependent fumarate reductase from *Trypanosoma brucei* in combination with the *dct* gene to increase succinic acid production (Fig. 1).

Methods
Strains and cultivation conditions
A glucose oxidase deficient Δgox strain, which is not able to produce gluconic acid [18], was selected as the parental strain to construct the derived strains in this study. *A. carbonarius* wild type ITEM5010 was used as a donor strain for obtaining the *dct* gene. The cultivation of the strains for spore production was carried out in potato dextrose agar medium at 30 °C for 4–6 days. For purification of total RNA, strains were inoculated into yeast extract peptone dextrose (YEPD) medium and incubated stationary at 30 °C for 2 days.

Identification of a C_4-dicarboxylate transporter in *A. carbonarius*
The amino acid sequences of the C_4-dicarboxylate transporter (C4T318) in *A. oryzae* and the malic acid transporter (MAE1) in *Schizosaccharomyces pombe* (accession no. BAE58879.1 and NP594777.1) were selected to identify an orthologous gene in *A. carbonarius*. The amino acid sequence with high identity was identified as the putative C_4-dicarboxylate transporter and the encoding

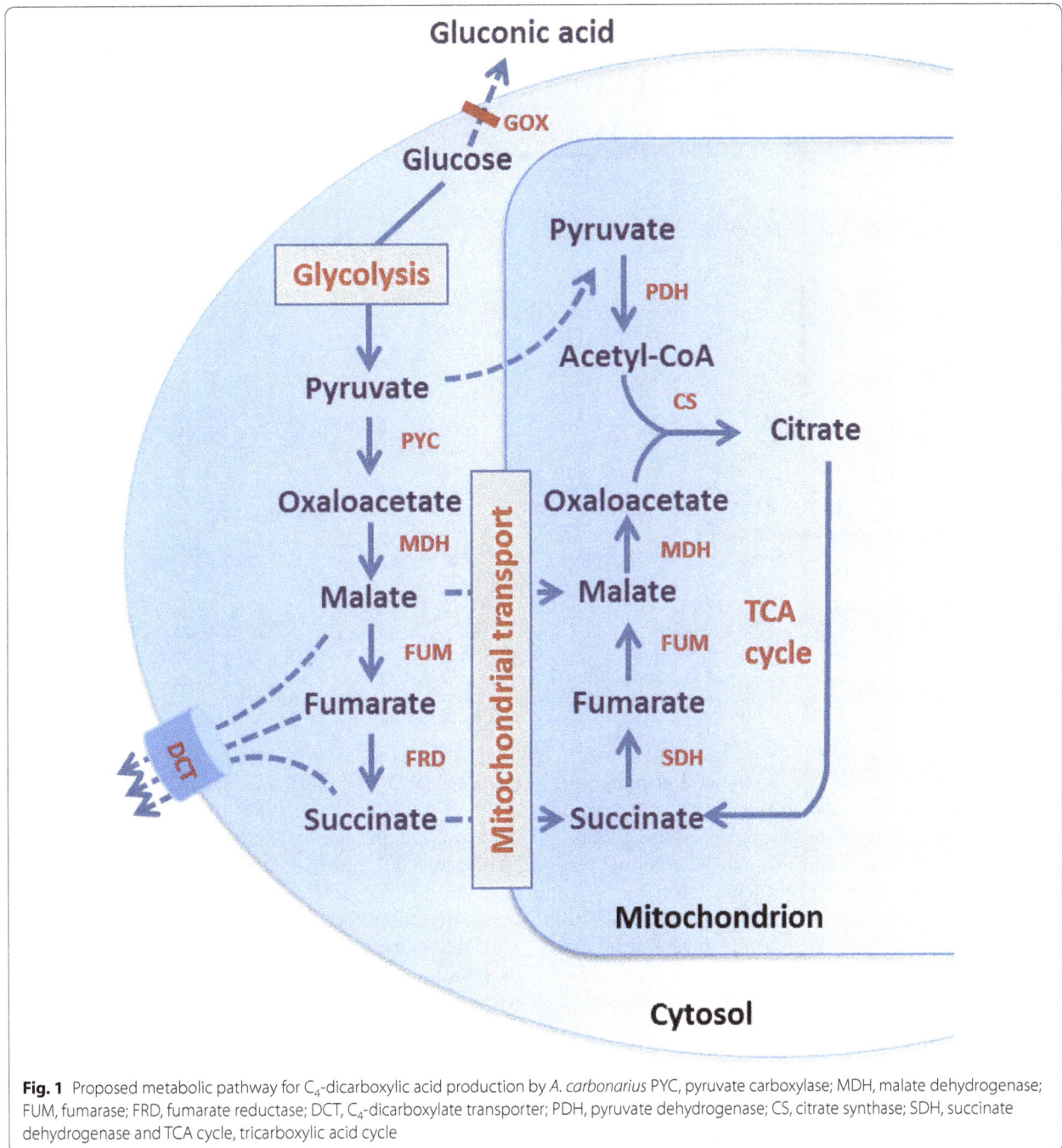

Fig. 1 Proposed metabolic pathway for C$_4$-dicarboxylic acid production by *A. carbonarius* PYC, pyruvate carboxylase; MDH, malate dehydrogenase; FUM, fumarase; FRD, fumarate reductase; DCT, C$_4$-dicarboxylate transporter; PDH, pyruvate dehydrogenase; CS, citrate synthase; SDH, succinate dehydrogenase and TCA cycle, tricarboxylic acid cycle

gene was termed *dct* (accession no. KY178298) in this study.

Plasmid construction and fungal transformation
The *dct* gene encoding the putative C$_4$-dicarboxylate transporter was amplified with primers DctFw and DctRv from the cDNA that was synthesized from total RNA of the wild type as previously described [16]. The *frd* gene

encoding a NADH dependent fumarate reductase in *Trypanosoma brucei* was codon optimized, synthesized and cloned as previously described [22]. Both genes were inserted individually via Simple USER cloning [23] into a fungal expression cassette consisting of a constitutive promoter gpdA and a terminator TrpC in plasmid pSBe3 that carries the phleomycin resistance gene *bleo* (Fig. 2). For co-transformation, the *dct* gene flanked with

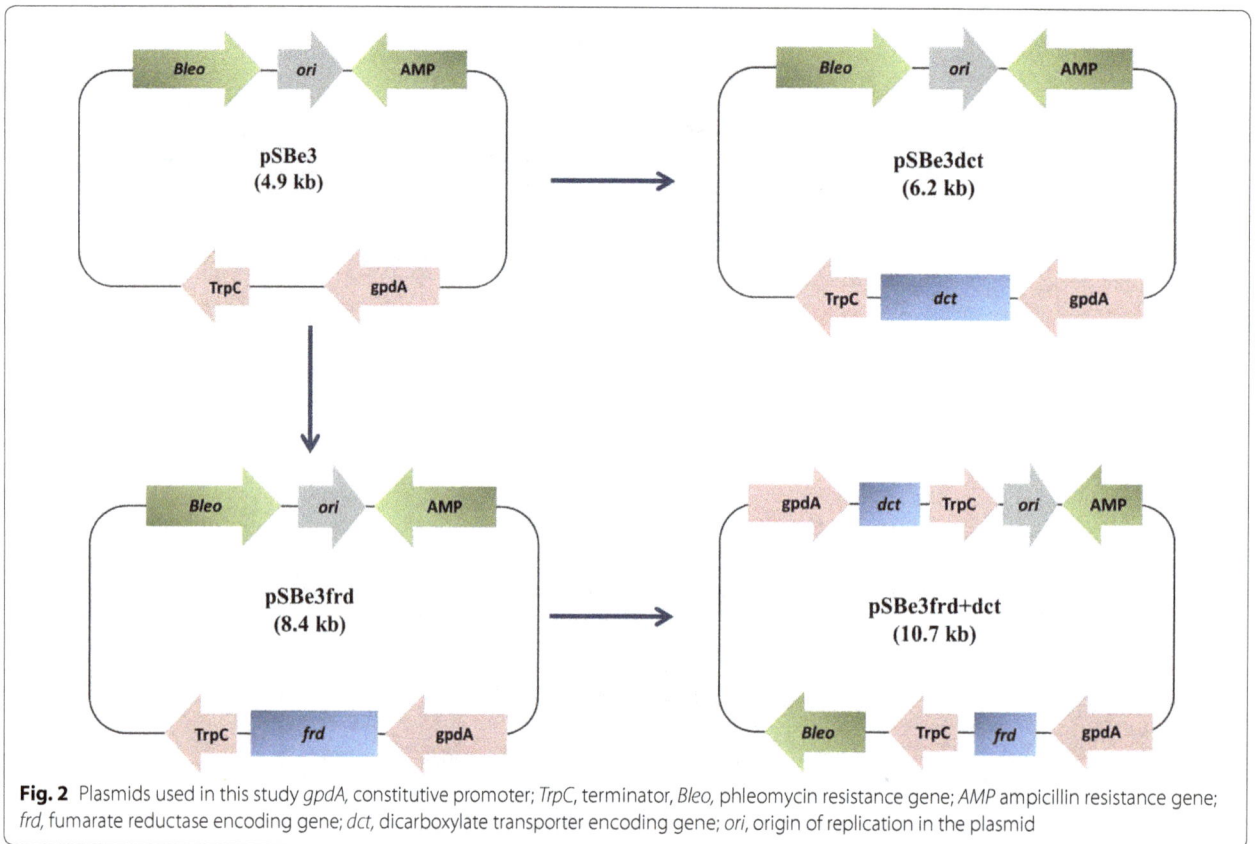

Fig. 2 Plasmids used in this study *gpdA*, constitutive promoter; *TrpC*, terminator, *Bleo*, phleomycin resistance gene; *AMP* ampicillin resistance gene; *frd*, fumarate reductase encoding gene; *dct*, dicarboxylate transporter encoding gene; *ori*, origin of replication in the plasmid

the gpdA promoter and TrpC terminator regions were amplified with primers GpdFw and TrpRv from plasmid pSBe3dct and inserted into plasmid pSBe3frd (Table 1 and Fig. 2) to construct pSBe3frd-dct. All resulting plasmids were verified by DNA sequencing (StarSEQ®).

Protoplast transformation was carried out with the above mentioned plasmids as previously described [18]. Minimal medium with agar (MMA) consisted of glucose, 10 g/L; sorbitol, 182 g/L; NaNO$_3$, 6 g/L; KCl,

0.5 g/L; MgSO$_4$·7H$_2$O, 0.5 g/L; KH$_2$PO$_4$, 1.5 g/L; ZnSO$_4$, 0.005 g/L; FeSO$_4$·7H$_2$O, 0.003 g/L; CuSO$_4$, 0.001 g/L; MnCl$_2$, 0.002 g/L; biotin, 0.001 g/L; thiamine, 0.001 g/L; riboflavin, 0.001 g/L; para-aminobenzoic acid, 0.001 g/L; agar, 18 g/L and 100 µg/mL phleomycin. To stabilize the derived transformants from protoplast transformation, the spores from transformants were streaked out on potato dextrose agar (PDA) plate and pieces of agar with single colonies were transferred back to MMA. This

Table 1 Primers used in this study

Name	Sequence (5′ → 3′)	Annotation
DctFw	AGAGCGAUATGCATGTCCACGACACC	SimpleUSER cloning of *dct* gene
DctRv	TCTGCGAUTCATTCAGACACATCCTCGTC	SimpleUSER cloning of *dct* gene
FrdFw	AGAGCGAUATGGTTGATGGTCGGTCGT	SimpleUSER cloning of *frd* gene
FrdRv	TCTGCGAUTTAGCTACCCGACGGTTCAGTT	SimpleUSER cloning of *frd* gene
GpdFw	GGCATTAAUTCGTGGACCTAGCTGATTCTG	PCR amplification of expression cassette
TrpRv	GGTCTTAAUTCGAGTGGAGATGTGGAGTG	PCR amplification of expression cassette
qDctFw	TTCCTTCCACCTCAACTGGT	qPCR *dct* gene
qDctRv	GCTGAGCAGGACAAAGATGA	qPCR *dct* gene
qActinFw	AGAGCGGTGGTATCCATGAG	qPCR beta-actin gene
qActinRv	TGGAAGAGGGAGCAAGAGCG	qPCR beta-actin gene

stabilization procedure was repeated three times before the obtained transformants were verified for the transformed genes by PCR.

cDNA synthesis and transcription analysis

For cDNA synthesis, total RNA was extracted from fresh fungal mycelia with a total RNA isolation kit (A&A Biotech®) and treated with DNaseI (Thermo Scientific®) according to the manufacturer's manuals. The cDNA was synthesized from the treated RNA samples with a reverse transcription kit (Bio-Rad®) as previously described [16]. For transcription analysis of *dct* gene, quantitative real-time PCR (qPCR) was set up by mixing Ultra-Fast SYBR green qPCR mix, the synthesized cDNA and corresponding primers according to the manufacturer's manual. The qPCR reaction was carried out in a Rotor-Gene 6000 RT-PCR machine using the following program: initial incubation at 95 °C for 5 min; 40 cycles of at 95 °C for 20 s. and 60 °C for 35 s. The threshold cycle (Ct), baseline and efficiency of amplification were determined in RG6000 application software. The relative expression level of the *dct* gene was calculated by normalizing the gene expression level of the *dct* gene to the reference gene beta-actin based on Ct values obtained from biological triplicates, and changes of the *dct* gene expression level between the selected transformants and parental strain were calculated using the $2^{-\Delta\Delta Ct}$ method [24].

Enzyme assay of fumarate reductase

The expression of fumarate reductase in the selected transformants was verified with measurement of fumarate reductase activity. The *frd* overexpression strain and the parental strain were cultivated in the YPED medium for 2 days and the fresh grown mycelia was used for preparation of cell extract. The cell extract was prepared as previously described [22], and the cytosolic faction was used immediately for measurement of enzyme activity. Fumarate reductase was assayed spectrophotometrically in 1.5 mL cuvettes (1.0 cm light path) at 30 °C. The assay mixture was composed of 50 mM phosphate buffer (pH 6.5), 20 mM fumarate, 0.2 mM NADH and the enzyme activities were determined by monitoring the absorbance due to the oxidation of NADH at 340 nm for 10 min [25]. One unit fumarate reductase activity was defined as the oxidation of 1 μmol NADH based on the amount of protein per minute at 30 °C and pH 6.5

Organic acid production

All the strains in this study were firstly cultivated in a pre-culture medium containing 3.6 g/L yeast extract and 10 g/L peptone. The freshly harvested spores were inoculated in pre-culture medium at final concentration of 10^5/mL. The pre-cultivation was carried out in 50 mL falcon tubes containing 10 mL pre-culture medium at 30 °C for 2 days with agitation speed of 180 rpm. After pre-cultivation, the pre-culture was filtered with Mira-cloth and the pre-grown fungal cells were transferred into the production medium. The cultivation was carried out in 100 mL Erlenmeyer flasks containing 20 mL acid production (AP) medium at 30 °C with agitation speed of 180 rpm. The defined medium consisted of glucose, 80 g/L; NH_4NO_3, 1.5 g/L; KH_2PO_4, 0.15 g/L; $MgSO_4{\cdot}7H_2O$, 0.8 g/L; $CaCl_2{\cdot}2H_2O$, 0.2 g/L; NaCl, 0.15 g/L; $ZnSO_4$, 0.0015 g/L; $FeSO_4{\cdot}7H_2O$, 0.03 g/L; biotin, 1×10^{-5} g/L and $CaCO_3$, 80 g/L [26]. The liquid fraction of the wheat straw hydrolysate was prepared as previously described and supplemented with the same amounts of nutrients (except glucose and xylose) and calcium carbonate as mentioned in the defined medium [15]. The pH of hydrolysate was adjusted to 6.5 using 10 M NaOH followed by sterilization with 0.2 μm sterile filter (Nalgene®). The initial concentration of glucose and xylose in the sterilized hydrolysate were 66 and 55 g/L respectively.

Analysis of sugars and organic acids

All samples were acidified and filtered before HPLC analysis. Acidification of samples was achieved by adding 50 μL 50% sulfuric acid into 1 mL samples. The acidified samples were then incubated at 80 °C for 15 min and centrifuged at 14,000 rpm for 1 min. The supernatant was filtered with 0.45 μM filter before sampling for HPLC analysis. Analysis of sugars and organic acids were carried out with an Aminex 87H column (Biorad®) at 60 °C by using a HPLC mobile phase (5 mM H_2SO_4) at a flow rate of 0.6 mL/min. Malic acid in the samples was analyzed with a L-malate assay kit (Megazyme®).

Fungal biomass measurement

For measurement of fungal biomass, 30 mL culture was acidified with 1 N·HCl to dissolve insoluble calcium carbonate and filtered with filter paper followed by thoroughly washing with distilled water. The washed fungal biomass was dried on filter paper at 100 °C for 48 h before weighing. All the filter papers were dried at 100 °C for 24 h before use.

Results

Identification of C4-dicarboxylate transporter and fungal transformation

The *dct* gene was identified using two amino acid sequences of reported C4-dicarboxylate transporters in fungi. As seen in Fig. 3, the amino acid sequence of the C4-dicarboxylate transporter in *A. carbonarius* shows high identity (68%) to the reported C4-dicarboxylate transporter from *A. oryzae*, especially in the predicted transmembrane domains whereas the MAE1 sequence

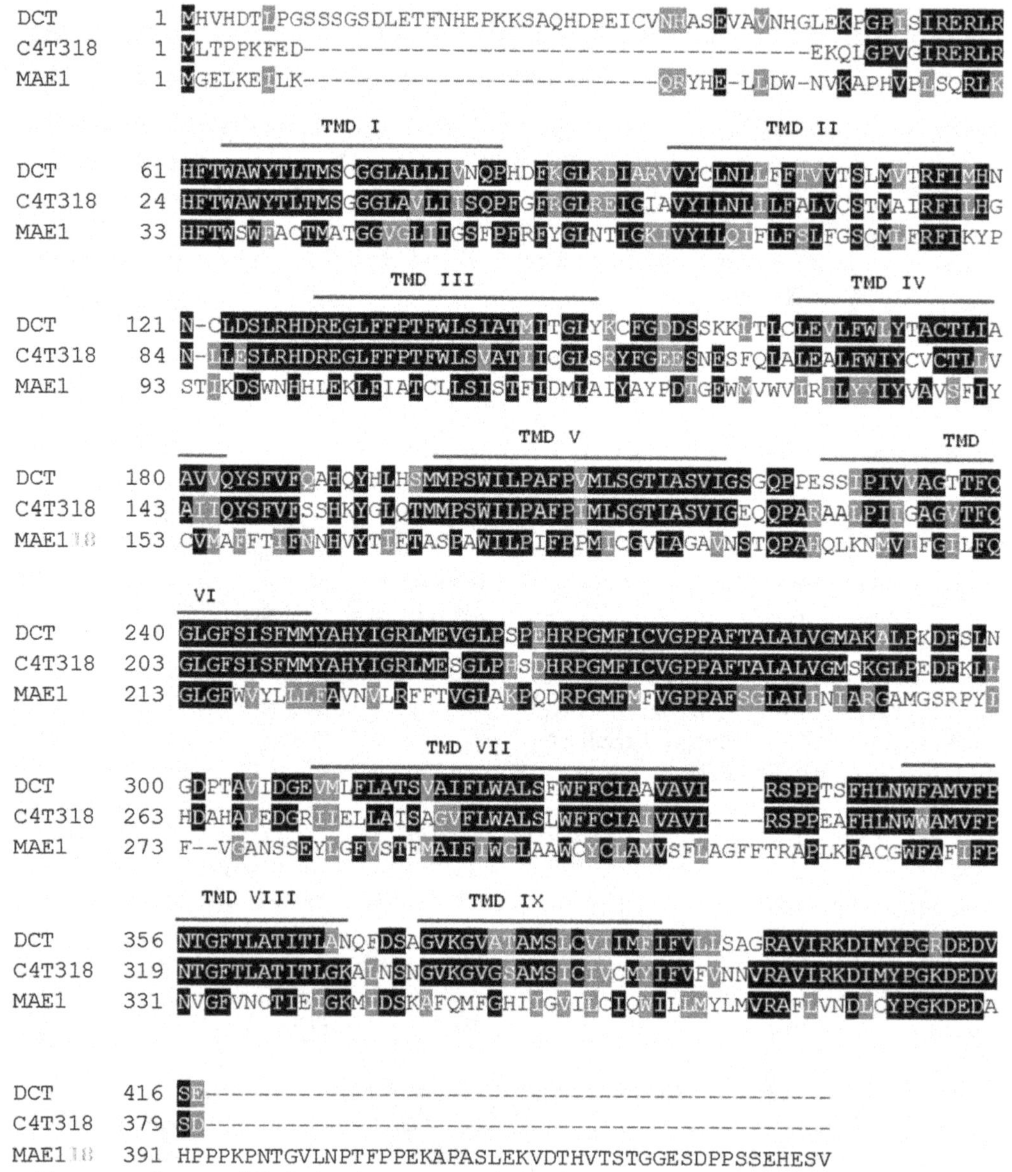

Fig. 3 Amino acid sequence alignment of C4-dicarboxylate transporters in *A. carboanrius* (DCT), *A. oryzae* (C4T318) and *S. pombe* (MAE1) (Please note that the positions of transmembrane domains (TMD) were predicted and annotated based on the amino acid sequence of C4T318 from *A. oryzae*)

from *S. pompe* has much lower (35%) identity to the DCT in *A. carbonarius*.

Protoplast transformation of the parental strain (*Δgox*) with three plasmids (pSBe3frd, pSBe3dct and pSBe-3frd + dct) resulted in 8 *frd* transformants, 12 *dct* transformants and 6 *frd-dct* transformants, respectively. The integration of the transformed expression cassettes containing target genes was verified with PCR amplification of the intact expression cassette. The transformants that produced the highest titers of C4-dicarboxylic acids in the first screening were selected for the following study and comparison.

Overexpression of the C$_4$-dicarboxylate transporter and expression of the fumarate reductase

Overexpression of the *dct* gene was verified by comparing relative expression level of the *dct* gene in the *dct* and *frd-dct* strains with the parental strain. The relative expression level of the *dct* gene in the *dct* and *frd-dct* strains increased 33- and 39-folds respectively compared with the parental strain (Table 2). Heterologous expression of fumarate reductase was confirmed in the *frd* strain and the *frd-dct* strain by measuring the activities of fumarate reductase. Significant activities of fumarate reductase were detected in both of the *frd* and *frd-dct* strains (0.013

Table 2 Relative expression level of the *dct* gene in over-expressing strains

	Fold change
Parental strain (*Δgox*)	Set to 1
dct strain	33
frd-dct strain	39

and 0.018 U/mg) but there was no detectable activity in the parental strain (Table 3).

Impacts of overexpressing the C4-dicarboxylate transporter and fumarate reductase on organic acid production

Comparison of organic acid production by the selected derived strains and the parental strain were at first made in a defined medium containing 80 g/L glucose under pH buffered condition. Overexpression of the *dct* gene substantially increased C4-dicarboxylic acid production in the *dct* and *frd-dct* strains (Fig. 4b, c; Table 4). From day 2, the two strains (*dct* and *frd-dct*) began to secrete malic acid and succinic acid simultaneously with production of citric acid, whereas the *frd* strain and the parental strain only produced citric acid (Fig. 4b–d). However, it seems that increased production of malic acid and succinic acid in the *dct* and *frd-dct* strains did not have any significant impact on citric acid production in the early phase of organic acid production. After day 4 a deceleration of citric acid production by these two strains was observed whereas the parental strain and the *frd* strain continued producing citric acid. The *dct* and the *frd-dct* strains continued producing malic acid and succinic acid as the glucose was consumed during the cultivation. In addition, low amounts of fumaric acid were detected in the *dct* and the *frd-dct* strains after 9 days (Table 4). Expression of fumarate reductase in the *frd* strain did not lead to an overproduction of succinic acid (Fig. 4c), only a slight elevation was found compared to the parental strain. Citric acid was still produced in the *frd* strain as the only major organic acid at high quantity similar to the organic acid profile obtained in the parental strain (Fig. 4b–d). When the C4-dicarboxylate transporter was overexpressed in

Table 3 The specific activity of fumarate reductase in the parental strain *Δgox* and the FRD overexpressing strains

	Fumarate reductase
Parental strain (*Δgox*)	n.d
frd strain	0.013
frd-dct strain	0.018

The enzyme activity (U/mg protein) was measured in the cells after 40 h of incubation in the YPD medium

combination with the fumarate reductase in the *frd-dct* strain, succinic acid production increased dramatically. After 9 days, the *frd-dct* strain produced 16 g/L succinic acid which was significantly higher than titers obtained from the *dct* strain (7.4 g/L) and the *frd* strain (0.13 g/L) (Table 4). Production of malic acid and fumaric acid was lower in the *frd-dct* strain compared with the *dct* strain.

From day 4 efficient production of C4-dicarboxylic acids also increased glucose consumption by the *dct* and *frd-dct* strains compared with the parental strain and the *frd* strain (Fig. 4a). After 9 days, glucose was almost depleted by the *dct* and *frd-dct* strain but there was still over 10 g/L glucose left in the *frd* strain and the parental strain. There were also significant differences in fungal biomass among the strains. The *dct* and *frd-dct* strains produced lower fungal biomass (352 mg/g glucose and 364 mg/g glucose) than parental strain and *frd* strain (461 mg/g glucose and 459 mg/g glucose) after 9 days (Table 4).

The ability of the developed strains to produce C4-dicarboxylic acids in the wheat straw hydrolysate buffered with calcium carbonate was investigated. All the tested strains were able to grow in the hydrolysate and utilize glucose and xylose simultaneously for organic acid production (Fig. 5). The concentration of glucose and xylose decreased significantly after day 3, but the sugar consumption varied among the tested strains. After day 3 the *dct* and *frd-dct* strains started consuming sugars more rapidly than the parental strain and the *frd* strain, and in total, the *dct* and *frd-dct* strains consumed 111 and 105 g/L sugar respectively after 9 days, which was higher than the amounts of sugars consumed by the parental strain (77 g/L) and the *frd* strain (79 g/L). The production of organic acids began in compliance with the sugar consumption in all the tested strains (Fig. 5). The *frd* strain and the parental strain produced higher amounts of citric acid than the *dct* and *frd-dct* strains after 9 days. High quantities of malic acid and succinic acid were obtained only in the *dct* and *frd-dct* strains. The *dct* strain produced 20 g/L malic acid and 2.1 g/L succinic acid from the hydrolysate, and the *frd-dct* strain produced less malic acid (17 g/L) but more succinic acid 10 g/L. Low amount of fumaric acid was also produced by the *dct* and *frd-dct* strains (Table 5). For the parental strain and the *frd* strain, low amounts of malic acid and succinic acid but no fumaric acid could be detected in the early phase of cultivation and remained at this level until day 9. Citric acid was produced from day 2 as one of the major organic acids by all the tested strains. The parental strain and *frd* strain produced 19 and 17 g/L citric acid respectively, which were higher than the amounts obtained from the *dct* (8.5 g/L) and *frd-dct* strains (7.5 g/L). The fungal biomass of all the tested

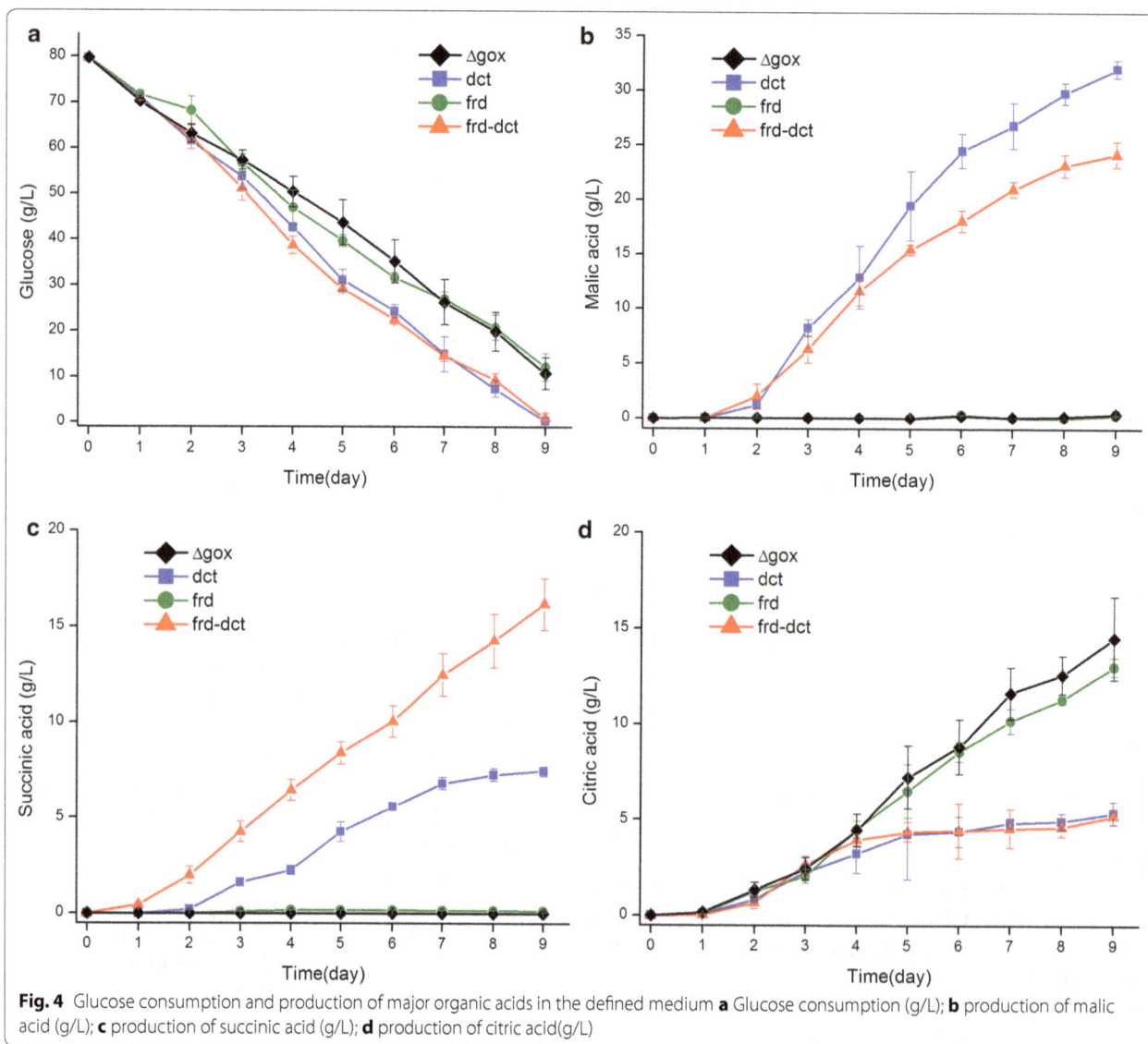

Fig. 4 Glucose consumption and production of major organic acids in the defined medium **a** Glucose consumption (g/L); **b** production of malic acid (g/L); **c** production of succinic acid (g/L); **d** production of citric acid(g/L)

strains, as well as the organic acid yield, showed the same pattern as obtained in the defined medium. The *dct* and *frd-dct* strains had lower biomass yield but much higher yield of C_4-dicarboxylic acids than the parental strains and the *frd* strain (Table 5).

Discussion

Aspergillus carbonarius is known as an efficient citric acid producing species. It has shown a potential to be a cell factory for production of various organic acids, but the main barrier for C_4-dicarboxylic acid production is that the intracellular carbon flux primarily flows towards citric acid rather than other acids such as C_4-dicarboxylic acids when genetic modifications targeting to primary metabolic pathways are introduced.

In this study, a C_4-dicarboxylate transporter in *A. carbonarius* was identified and overexpressed in order to facilitate the export of C_4-dicarboxylic acids (malic acid, fumaric acid and succinic acid). Fundamental studies on C_4-dicarboxylate transporters in microorganisms have focused mainly on bacteria and yeasts e.g. *E. coli* and *S. pombe* [27–29]. The known C_4-dicarboxylate transporters are normally responsible for transferring several C_4-dicarboxylic acids rather than a specific one. For instance, the C_4-dicarboxylate transporters from *E. coli* and *Pseudomonas aeruginosa* can transport malic acid, fumaric acid, oxaloacetic acid, and succinic acid [30, 31]. Although a number of fungal strains are naturally capable of producing high amounts of C_4-dicarboxylic acids, e.g. *Aspergillus flavus* and *Rhizopus oryzae* [32, 33], there are

Table 4 Comparison of sugar consumption, production of organic acids and fungal biomass in the defined medium after 9 days

Strain	Glucose consumption	Concentration (g/L)				Yield (mg/g glucose)				Fungal biomass
		Malic acid	Fumaric acid	Succinic acid	Citric acid	Malic acid	Fumaric acid	Succinic acid	Citric acid	
Δgox	69	0.4	n.d	n.d	14	5.8	n.d	n.d	203	461
dct	79	32	0.96	7.4	5.3	404	12	93	66	352
frd	67	0.31	n.d	0.13	13	4.6	n.d	1.9	192	459
frd-dct	79	24	0.8	16	5.2	307	10	205	66	364

C_4-dicarboxylic acid production from wheat straw hydrolysate

still very limited amount of information regarding their C_4-dicarboxylate transporters. The identified C_4-dicarboxylate transporter from *A. carbonarius* belongs to the same dicarboxylate transporter family as the two reference transporters, C4T318 and MAE1. In *S. pombe*, MAE1 functions as a malate permease transporting C_4-dicarboxylates via proton symport [27]. Expression of MAE1 in *S. cerevisiae* could increase both uptake and export of C_4-dicarboxylates depending on the growth conditions [21, 27]. Based on the sequence similarity between MAE1 and the C_4-dicarboxylate transporter of *A. carbonarius*, the transport of C_4-dicarboxylates may be achieved via the same mechanism in *A. carbonarius*, and the function of C_4-dicarboxylate transporter, when it is naturally expressed at a low level, is probably to mediate the uptake of C_4-dicarboxylates as alternative carbon sources. The recent successes in metabolic engineering for malic acid production have indicated an essential role of C_4-dicarboxylate transporters for production of C_4-dicarboxylic acids also in filamentous fungi. Overexpression of a dicarboxylate transporter, C4T318, significantly increased production of malic acid in combination with other genetic modifications in *Aspergillus oryzae* [9]. *A. carbonarius*, compared with *A. oryzae*, is unable to naturally excrete any of the C_4-dicarboxylic acids. Previous efforts on increasing the carbon flux towards the rTCA branch in *A. carbonarius* led to enhanced production of citric acid but only had a limited impact on malic acid production [16]. This result implies that there exist some limiting steps for C_4-dicarboxylic acid production in *A. carbonarius*. When the identified putative C_4-dicarboxylate transporter was overexpressed in the *dct* strain, malic acid production increased dramatically and the titer of malic acid reached 32 g/L in a defined medium after 9 days (Fig. 4b). This indicates that the carbon flux has been partially shunt into malic acid production in the *dct* strain after a more efficient export of malic acid was achieved through overexpression of the C_4-dicarboxylate transporter. Moreover, the *dct* strain is also able to produce fumaric acid and succinic acid compared

with the parental strain, which indicates that the overexpressed C_4-dicarboxylate transporter, as shown from other known C_4-dicarboxylate transporters, can transport different C_4-dicarboxylic acids instead. Accordingly, the *dct* strain produced lower amounts of citric acid (5.3 g/L) than the parental strain (14 g/L), which implies that export of C_4-dicarboxylate reduced the carbon flux towards biosynthesis of citrate in *A. carbonarius* in buffered conditions. Although there is not yet any study on metabolic flux in pathways related to citric acid production in *A. carbonarius*, the correlation between intracellular concentration of C_4-dicarboxylates and citric acid production has been illustrated in a well-known citric acid producing strain of *A. niger*, where the increased concentration of cytosolic C_4-dicarboxylates triggers the citric acid production in the early phase and leads to enhanced citric acid production probably via anti-port of C_4-dicarboxylates and citrate across mitochondrial membrane [20]. Therefore, the export of C_4-dicarboxylic acids from the cytosol can theoretically reduce the amounts of C_4-dicarboxylic acids that are transported into mitochondria for biosynthesis of citric acid and in turn decrease citric acid production in *A. carbonarius*. In addition, the expression of the C_4-dicarboxylate transporter also influenced the sugar consumption and biomass yield. Compared with the parental strain, the sugar consumption rate increased in the *dct* overexpressing strains after the production of malic acid and succinic acid began at day 3. It seems that the export of C_4-dicarboxylic acids creates the extra outlet of intracellular carbon flux, which improves the sugar utilization in the *dct* and *frd-dct* strains. On the other hand, the fungal biomass decreased in the *dct* and *frd-dct* strains compared with the parental strain. This indicates that a re-programming of the carbon metabolism might cause a slow-down of the biomass growth in the derived strains due to the overexpression of the *dct* gene.

While expression of the fumarate reductase from *Trypanosoma brucei* in the natural malic acid producer *Aspergillus saccharolyticus* significantly increased production

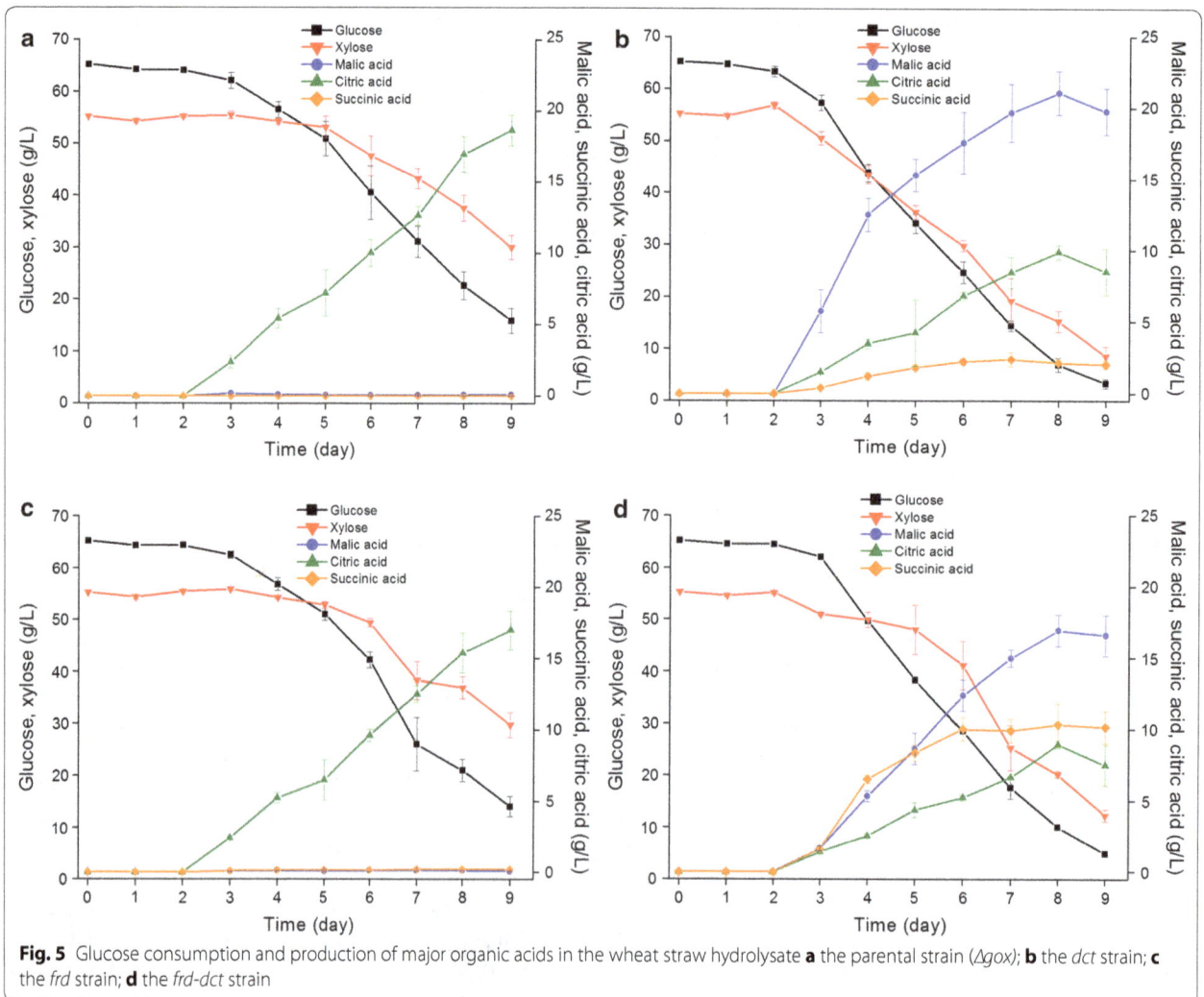

Fig. 5 Glucose consumption and production of major organic acids in the wheat straw hydrolysate **a** the parental strain (Δgox); **b** the dct strain; **c** the frd strain; **d** the frd-dct strain

of succinic acid [22], expression of the fumarate reductase in *A. carbonarius* did not change the production pattern. However, when expressing the fumarate reductase in combination with the C_4-carboxylate transporter the production of succinic acid was significantly increased. This again indicates the essential role of an efficient succinate export system for enhanced succinic acid production by *A. carbonarius*. As the *dct* strain showed an elevated production of all three C_4-dicarboxylic acids, it was assumed that the intracellular fumarate can be used as substrate for fumarate reductase to produce succinate in the cytosol and that the overexpressed C_4-dicarboxylate transporter facilitates the export of succinate across the plasma membrane. As expected, the succinic acid production in the *frd-dct* strain increased over twofolds compared with the *dct* strain. This demonstrates the feasibility of improving succinic acid production in *A. carbonarius* by converting fumarate to succinate in the cytosolic reductive pathway.

Currently, the industries for bio-based production of C_4-dicarboxylic acids are seeking feasible solutions to lower the production cost [10]. Carbohydrates existing in lignocellulosic biomass are considered as cheap alternative substrates for organic acid production [6]. *A. carbonarius* has been reported for its efficient co-utilization of glucose and xylose during the cultivation and its ability to produce different types of organic acids from the hydrolysate such as citric acid and gluconic acid [15]. In this study, we have further demonstrated its ability to produce C_4-dicarboxylic acids from lignocellulosic biomass with the developed strains. In the hydrolysate, the sugar consumption by all the strains began later than that observed in the defined medium. The inhibitory effects on spore germination and fungal growth in the hydrolysate which delayed sugar consumption in the first two days have been reported in our previous study [15]. However, inoculating with fungal mycelia from pre-culture in this study did not improve

Table 5 Comparison of sugar consumption, production of organic acids and fungal biomass in the wheat straw hydrolysate after 9 days

Strain	Sugar (glucose/xylose) consumption	Concentration (g/L)				Yield (mg/g sugar)				Fungal biomass
		Malic acid	Fumaric acid	Succinic acid	Citric acid	Malic acid	Fumaric acid	Succinic acid	Citric acid	
Δgox	77 (51/26)	0.1	n.d	n.d	19	1.4	n.d	n.d	247	515
dct	111 (62/49)	20	0.9	2.1	8.5	179	7.9	18	76	342
frd	79 (52/27)	0.1	n.d	0.2	17	1.0	n.d	2.8	216	505
frd-dct	105 (60/45)	17	0.5	10	7.5	157	5.3	96	71.2	345

the inhibitor tolerance or accelerate the sugar consumption in the early phase of cultivation. Due to the lagged sugar consumption, the organic acid production was also delayed in the hydrolysate. The patterns of organic acids produced by all the tested strains remained the same as that obtained in the defined medium, but the yields of organic acids were lower in the hydrolysate. Fungal organic acid production is significantly affected by a number of factors in the medium, including cultivation pH, carbon sources, nitrogen and metal ions [33–36]. Although the cultivation pH in the defined medium and the hydrolysate was maintained steadily at 6.5 by adding the calcium carbonate, the concentration of nutrients in the hydrolysate was not kept at the same level as in the defined medium due to the complex composition of the wheat straw hydrolysate. This may result in the different yields of C_4-dicarboxylic acids between these two types of media. For future perspective, the nutrients supplement needs to be optimized based on compositional analysis of hydrolysate to improve C_4-dicarboxylic acid production.

Conclusions

This study shows that the key to change the citric acid production of a non-natural C_4-dicarboxylic acid producing strain into production of C_4-dicarboxylic acids is the C_4-dicarboxylate transporter and that the C_4-dicarboxylic acid production can be further increased via metabolic engineering. Finally, it reveals the potential of *A. carbonarius* to utilize lignocellulosic biomass as substrate for C_4-dicarboxylic acid production.

Authors' contributions
LY has made substantial contributions to experimental design, performed identification of transporter gene, construction of plasmids and transformants, biochemical characterization (fermentation and enzyme assays), data analysis and drafted the manuscript. EC and JV performed cDNA synthesis, gene cloning and plasmid construction. ML and PSL planned the project (conception and experimental design) and edited the manuscript. All authors read and approved the final manuscript.

Author details
[1] Section for Sustainable Biotechnology, Department of Chemistry and Bioscience, Aalborg University Copenhagen, A. C. Meyers Vænge 15, 2450 Copenhagen SV, Denmark. [2] Section for Biotechnology, Aalborg University Copenhagen, Fredrik Bajers Vej 7H, 9220 Aalborg, Denmark.

Acknowledgements
We thank laboratory technician Gitte Hinz-Berg for HPLC analysis.

Competing interests
The authors declare that they have no competing interests.

Funding
Financial support from the Danish Strategic Research Program MycoFuelChem (DSF Grant No. 11-116803) is acknowledged.

References
1. Werpy T, Petersen G. Top value added chemicals from biomass. Results of Screening for Potential Candidates from Sugars and Synthesis Gas. 2004; 1.
2. Sauer M, Porro D, Mattanovich D, Branduardi P. Microbial production of organic acids: expanding the markets. Trends Biotechnol. 2008;26(2):100–8.
3. Kamm B, Gruber PR, Kamm M. Biorefineries–industrial processes and products. Hoboken: Wiley Online Library; 2006.
4. Tan Z, Chen J, Zhang X. Systematic engineering of pentose phosphate pathway improves *Escherichia coli* succinate production. Biotechnol Biofuels. 2016;9(1):262.
5. Bradfield MFA, Mohagheghi A, Salvachúa D, Smith H, Black BA, Dowe N, Beckham GT, Nicol W. Continuous succinic acid production by *Actinobacillus succinogenes* on xylose-enriched hydrolysate. Biotechnol Biofuels. 2015;8(1):181.
6. Mondala AH. Direct fungal fermentation of lignocellulosic biomass into itaconic, fumaric, and malic acids: current and future prospects. J Ind Microbiol Biotechnol. 2015;42(4):487–506.
7. Ahn JH, Jang Y, Lee SY. Production of succinic acid by metabolically engineered microorganisms. Curr Opin Biotechnol. 2016;42:54–66.
8. Zhang B, Skory CD, Yang S. Metabolic engineering of *Rhizopus oryzae*: effects of overexpressing pyc and pepc genes on fumaric acid biosynthesis from glucose. Metab Eng. 2012;14(5):512–20.
9. Brown SH, Bashkirova L, Berka R, Chandler T, Doty T, McCall K, McCulloch M, McFarland S, Thompson S, Yaver D, Berry A. Metabolic engineering

of *Aspergillus oryzae* NRRL 3488 for increased production of l-malic acid. Appl Microbiol Biotechnol. 2013;97(20):8903–12.

10. Debabov V. Prospects for biosuccinic acid production. Appl Biochem Microbiol. 2015;51(8):787–91.

11. Show PL, Oladele KO, Siew QY, Aziz Zakry FA, Lan JC, Ling TC. Overview of citric acid production from *Aspergillus niger*. Front Life Sci. 2015;8(3):271–83.

12. Huang X, Lu X, Li Y, Li X, Li J. Improving itaconic acid production through genetic engineering of an industrial *Aspergillus terreus* strain. Microb Cell Fact. 2014;13(1):119.

13. Yang L, Lübeck M, Lübeck PS. *Aspergillus* as a versatile cell factory for organic acid production. Fungal Biol Rev. 2016;31(1):33–49.

14. Weyda I, Lübeck M, Ahring BK, Lübeck PS. Point mutation of the xylose reductase (XR) gene reduces xylitol accumulation and increases citric acid production in *Aspergillus carbonarius*. J Ind Microbiol Biotechnol. 2014;41(4):733–9.

15. Yang L, Lübeck M, Souroullas K, Lübeck PS. Co-consumption of glucose and xylose for organic acid production by *Aspergillus carbonarius* cultivated in wheat straw hydrolysate. World J Microbiol Biotechnol. 2016;32(4):57.

16. Yang L, Lübeck M, Lübeck PS. Effects of heterologous expression of phosphoenolpyruvate carboxykinase and phosphoenolpyruvate carboxylase on organic acid production in *Aspergillus carbonarius*. J Ind Microbiol Biotechnol. 2015;42(11):1533–45.

17. Linde T, Zoglowek M, Lübeck M, Frisvad JC, Lübeck PS. The global regulator LaeA controls production of citric acid and endoglucanases in *Aspergillus carbonarius*. J Ind Microbiol Biotechnol. 2016;43(8):1139–47.

18. Yang L, Lübeck M, Lübeck PS. Deletion of glucose oxidase changes the pattern of organic acid production in *Aspergillus carbonarius*. AMB Expr. 2014;4(1):54.

19. Zhang T, Ge C, Deng L, Tan T, Wang F. C4-dicarboxylic acid production by overexpressing the reductive TCA pathway. FEMS Microbiol Lett. 2015;. doi:10.1093/femsle/fnv052 **(Epub 2015 Apr 9)**.

20. de Jongh WA, Nielsen J. Enhanced citrate production through gene insertion in *Aspergillus niger*. Metab Eng. 2008;10(2):87–96.

21. Zelle RM, De Hulster E, Van Winden WA, De Waard P, Dijkema C, Winkler AA, Geertman J-A, Van Dijken JP, Pronk JT, Van Maris AJA. Malic acid production by *Saccharomyces cerevisiae*: engineering of pyruvate carboxylation, oxaloacetate reduction, and malate export. Appl Environ Microbiol. 2008;74(9):2766–77.

22. Yang L, Lübeck M, Ahring BK, Lübeck PS. Enhanced succinic acid production in *Aspergillus saccharolyticus* by heterologous expression of fumarate reductase from *Trypanosoma brucei*. Appl Microbiol Biotechnol. 2016;100(4):1799–809.

23. Hansen NB, Lübeck M, Lübeck PS. Advancing USER cloning into simpleUSER and nicking cloning. J Microbiol Methods. 2014;96(1):42–9.

24. Livak KJ, Schmittgen TD. Analysis of relative gene expression data using real-time quantitative PCR and the 2-$\Delta\Delta$CT method. Methods. 2001;25(4):402–8.

25. Coustou V, Besteiro S, Rivière L, Biran M, Biteau N, Franconi J, Boshart M, Baltz T, Bringaud F. A mitochondrial NADH-dependent fumarate reductase involved in the production of succinate excreted by procyclic *Trypanosoma brucei*. J Biol Chem. 2005;280(17):16559–70.

26. Goldberg I, Rokem JS, Pines O. Organic acids: old metabolites, new themes. J Chem Technol Biotechnol. 2006;81(10):1601–11.

27. Camarasa C, Bidard F, Bony M, Barre P, Dequin S. Characterization of *Schizosaccharomyces pombe* malate permease by expression in *Saccharomyces cerevisiae*. Appl Environ Microbiol. 2001;67(9):4144–51.

28. Six S, Andrews SC, Unden G, Guest JR. Escherichia coli possesses two homologous anaerobic C4-dicarboxylate membrane transporters (DcuA and DcuB) distinct from the aerobic dicarboxylate transport system (Dct). J Bacteriol. 1994;176(21):6470–8.

29. Sousa MJ, Mota M, Leão C. Transport of malic acid in the yeast *Schizosaccharomyces pombe*: evidence for proton-dicarboxylate symport. Yeast. 1992;8(12):1025–31.

30. Janausch IG, Zientz E, Tran QH, Kröger A, Unden G. C4-dicarboxylate carriers and sensors in bacteria, Biochimica et Biophysica Acta (BBA)—Bioenergetics. 2002;1553(1–2):39–56.

31. Valentini M, Storelli N, Lapouge K. Identification of C(4)-dicarboxylate transport systems in *Pseudomonas aeruginosa* PAO1. J Bacteriol. 2011;193(17):4307–16.

32. He H, Li S, Xu Q, Zhang K, Huang H. Effect of cycloheximide on regulation of metabolic pathway for L-malic acid accumulation by *Rhizopus oryzae*. Guocheng Gongcheng Xuebao Chin J Process Eng. 2009;9(1):153–6.

33. Battat E, Peleg Y, Bercovitz A, Rokem JS, Goldberg I. Optimization of L-malic acid production by *Aspergillus flavus* in a stirred fermentor. Biotechnol Bioeng. 1991;37(11):1108–16.

34. Papagianni M. Advances in citric acid fermentation by *Aspergillus niger*: biochemical aspects, membrane transport and modeling. Biotechnol Adv. 2007;25(3):244–63.

35. Peleg Y, Stieglitz B, Goldberg I. Malic acid accumulation by *Aspergillus flavus* - I. Biochemical aspects of acid biosynthesis. Appl Microbiol Biotechnol. 1988;28(1):69–75.

36. Rhodes RA, Moyer AJ, Smith ML, Kelley SE. Production of fumaric acid by *Rhizopus arrhizus*. Appl Microbiol. 1959;7(2):74–80.

15

Development of a counterselectable seamless mutagenesis system in lactic acid bacteria

Yongping Xin, Tingting Guo, Yingli Mu and Jian Kong*⊙

Abstract

Background: Lactic acid bacteria (LAB) are receiving more attention to act as cell factories for the production of high-value metabolites. However, the molecular tools for genetic modifying these strains are mainly vector-based double-crossover strategies, which are laborious and inefficient. To address this problem, several counterselectable markers have been developed, while few of them could be used in the wild-type host cells without pretreatment.

Results: The *pheS* gene encoding phenylalanyl-tRNA synthetase alpha subunit was identified in *Lactococcus lactis* NZ9000 genome. When mutant *pheS* gene (*pheS**) under the control of the *Lc. lactis* NZ9000 L-lactate dehydrogenase promoter (P_{ldh}) was expressed from a plasmid, the resulted PheS* with an A312G substitution rendered cells sensitive to the phenylalanine analog *p*-chloro-phenylalanine (*p*-Cl-Phe). This result suggested *pheS** was suitable to be used as a counterselectable marker in *Lc. lactis*. However, the expression level of *pheS** from a chromosomal copy was too low to confer *p*-Cl-Phe sensitivity. Therefore, a strategy of cascading promoters was attempted for strengthening the expression level of *pheS**. Expectedly, a cassette 5Pldh-*pheS** with five tandem repetitive promoters P_{ldh} resulted in a sensitivity to 15 mM *p*-Cl-Phe. Subsequently, a counterselectable seamless mutagenesis system PheS*/pG⁺host9 based on a temperature-sensitive plasmid pG⁺host9 harboring a 5Pldh-*pheS** cassette was developed in *Lc. lactis*. We also demonstrated the possibility of applying *pheS** to be a counterselectable marker in *Lactobacillus casei* BL23.

Conclusions: As reported in *E. coli*, *pheS** as a counterselectable marker has been demonstrated to be functional in targeted gene(s) deletion in *Lc. lactis* as well as in *L. casei*. Moreover, the efficiency and timesaving counterselectable seamless mutagenesis system PheS*/pG⁺host9 could be used in the wild-type host cells without pretreatment.

Keywords: Lactic acid bacteria, Temperature-sensitive plasmid, Seamless mutagenesis, Counterselectable marker, *pheS*

Background

Lactic acid bacteria (LAB) are important microorganisms used as starter cultures in the dairy fermented processes [1, 2]. Due to their generally recognized as safe status, some LAB strains have been used as cell factories or vaccine delivery vehicles for the heterogeneous production of specific compounds [3, 4] or pharmaceutical molecules [5–8]. Also, since the wealth of genomic data being delivered by massively parallel sequencing, interests in development of high-efficiency genome engineering tools for rerouting natural metabolic pathways to produce high valuable end products were increasing [9, 10]. Considering of the importance of the recombineering system in LAB, significant efforts have been recently concentrated on the exploitation of gene targeting techniques in LAB to accelerate genome engineering or gene functional analysis, as the recently reported single-stranded DNA recombineering (SSDR) system and double-stranded DNA recombineering (DSDR) system [11–13].

The SSDR system was established in *Lactobacillus reuteri* and *Lactococcus lactis* and could generate precision genome mutations without leaving any other foreign

*Correspondence: kongjian@sdu.edu.cn
State Key Laboratory of Microbial Technology, Shandong University, 27 Shanda Nanlu, Jinan 250100, People's Republic of China

DNA [11]. The major limitation is the ability to achieve efficiencies that would allow the modification of any sites in the genome and easily recover the mutants without selection [14]. To address this problem, enhanced SSDR has been achieved with the assistant of the type-II clustered regularly interspaced short palindromic repeats locus from *Streptococcus pyogenes* in *L. reuteri* and >99.99% of non-recombinants could be eliminated without antibiotic selection [12]. But it could not be used for gene(s) deletion or insertion in other LAB because it has not proved the functional application of type-II CRISPR–Cas system in LAB except for *L. reuteri*. The DSDR technique was established in *L. plantarum* which was involved in the efficiently generation of gene(s) deletion or insertion [13]. However, this genetic system was not functional in other LAB and still left a *lox72* site, a heterologous DNA sequence after the antibiotic selection marker excised from the genome.

Seamless mutagenesis refers to targeted mutagenesis without any other micro-change, such as the presence of the selectable marker used to screen mutants or a *loxP* site after excising the selectable marker [15]. The seamless mutagenesis strategy usually appropriate for mutating the protein coding region in which any extraneous sequence introduced could interfere with protein expression. So far, several seamless mutagenesis methods based on homologous double-crossover have been successfully achieved in LAB, but the most widely used seamless mutagenesis strategy was based on a temperature-sensitive plasmid such as pG+host9 [16] or pG+host5 [17]. The merit of these plasmids is that both the non-replicate temperature at 37 °C and the replicate temperature at 28 °C are the adaptive temperature for the growth of most LAB. With plasmid pG+host9 [16], several chromosomal deletion derivatives of *Lc. lactis* and *Streptococcus thermophilus* were obtained in our laboratory [18–20]. However, these vectors, while powerful, suffer from a relatively low rate of recombination events and require labor-intensive screening procedures to distinguish clones with the desired seamless mutants. Therefore, improving the efficiency of this seamless mutagenesis system for fast analysis the function of gene(s) in LAB is very instant.

In recent years, a two-step selection/counterselection strategy has been demonstrated to be functional in improving the efficiency of method for fast generating seamless mutagenesis in the genome, which is normally consist of a positive selectable marker (usually an antibiotic resistance gene) and a counterselectable cassette. Counterselectable markers, including the genes *upp* [21–23] and *oroP* [24], have been characterized and functionally analyzed. However, the counterselectable marker *upp* could not be made in wild-type LAB strains without

pre-treatment while *oroP* has not been widely used for other LAB strains [21–24].

Recently, gene *pheS* encoding phenylalanyl-tRNA synthetase alpha subunit has been demonstrated to function as a host strain-independent counterselectable marker in *Thermus thermophilus*, *Bacteroides* sp., *Escherichia coli* and *Streptococcus mutans* [25–28], but has yet not been used in the model strain *Lc. lactis*. In *E. coli*, only an A294G substitution in the protein PheS altered the specificity of the phenylalanyl-tRNA synthetase which resulted in the sensitivity to phenylalanine analogs such as *p*-chloro-phenylalanine (*p*-Cl-Phe) [27]. In this study, we identified a conserved alanine residue in the PheS protein, and demonstrated that the dominant-negative mutant protein PheS* with an A312G amino acid substitution rendered cells sensitive to 15 mM *p*-Cl-Phe in *Lc. lactis* NZ9000 and 10 mM *p*-Cl-Phe in *L. casei* BL23. To employ this conditional lethal gene *pheS** as a negative selectable marker, a high-efficiency seamless mutagenesis system PheS*/pG+host9 based on a temperature-sensitive plasmid pG+host9 carrying a 5P$_{ldh}$-*pheS** cassette was constructed in *Lc. lactis* NZ9000. The aim of this study is to explore the potential of using *pheS** as a counterselectable marker for rapidly screening mutants for targeted gene analysis or genome engineering in LAB.

Methods

Plasmids, bacterial strains, and growth conditions

The plasmids and bacterial strains used in this study are shown in Table 1. *E. coli* DH5α was used as the host for cloning procedures and grown aerobically in Luria–Bertani (LB) medium at 37 °C. Unless otherwise specified, *Lc. lactis* and *L. casei* were grown statically at 30 °C in M17 (Oxoid) broth supplemented with 0.5% glucose (GM17) and at 37 °C in MRS (Oxiod) broth, respectively. For counterselection, semi-defined M9 plates [29] supplemented with 0.4% glucose, namely GM9 plates, were added with 15 mM *p*-Cl-Phe (Sigma) for *Lc. lactis* and 10 mM *p*-Cl-Phe (Sigma) for *L. casei*. If required, antibiotics were added as follows: 10 μg/ml erythromycin or 5 μg/ml chloramphenicol for *Lc. lactis*, 5 μg/ml erythromycin for *L. casei*, 10 μg/ml chloramphenicol, 100 μg/ml ampicillin and 300 μg/ml erythromycin for *E. coli* DH5α.

Reagents and enzymes

All enzymes used in this study were purchased from TaKaRa. Restriction enzymes and T4 DNA ligases were used according to standard procedures. PCR amplicons for cloning purposes were generated by 2× PrimeSTAR max premix, and PCR reactions for screening purposes were performed with rTaq DNA polymerase. All oligonucleotides used in this study are listed in Table 2.

Table 1 Plasmids and bacterial strains used in this study

Strain or plasmid	Characteristic(s)	Source
Strains		
Escherichia coli DH5α	F⁻ *supE44 ΔlacU169* Φ80*lacZ ΔM15 hsdR17 recA1 endA1 gyrA96 thi-1 relA1*	Novagen
Lactococcus lactis strains		
NZ9000	Derivative of MG1363, *pepN::nisRK*	[33]
lGn	Derivative of NZ9000, *galK*::pG⁺UD-nP-pheS*	This work
dA	Derivative of NZ9000, *ΔaldB*	This work
Lactobacillus casei BL23	Derivative of *L. casei* ATCC 393 (pLZ15⁻)	[36]
Plasmids		
pG⁺host9	Erm'; temperature-sensitive vector	[16]
pG⁺UD	pG⁺host9 derivative with up and down homology arms of the part of galactose operon	This work
pG⁺UD-nP-pheS*	pG⁺UD derivative with *pheS** driven by n P_{ldh}	This work
pG⁺-nP-pheS*	pG⁺host9 derivative with *pheS** driven by n P_{ldh}	This work
pG⁺UD2-5P-pheS*	pG⁺-5P-pheS* derivative with upstream and downstream sequences of the *aldB* gene	This work
pUC19	Amp'; cloing vector	This work
pSec:Leiss:Nuc	pWV01 replicon; expresses Nuc under PnisA control; Cm'	[34]
pleiss-P-pheS*	pSec:Leiss:Nuc derivative with *pheS** driven by a P_{ldh}	This work
pTRKH2	Erm'; expressing vector	[37]
pOgfp	Source of *gfp* gene	[18]

Bioinformatic analysis

A multiple-sequence alignment was performed using Clustal X, version 2.0 [30] and ESPript 3.0 [31]. The amino acid sequences of PheS proteins from six LAB strains were aligned with the amino acid sequences of PheS proteins from *E. coli* [27] and *E. faecalis* [32].

Construction of the counterselectable system PheS*/pG⁺host9 in *Lc. lactis*

The counterselectable P_{ldh}-*pheS** cassette was constructed using an overlap extension PCR strategy. The constitutive promoter region of the L-lactate dehydrogenase gene (*ldh*) (accession number: NC_017949) in *Lc. lactis* NZ9000 [33] was amplified by PCR using primer pair ldhF1 and ldhR1. The *pheS** gene was generated as two fragments by PCR using the *Lc. lactis* NZ9000 chromosomal DNA as a template with the primer pairs pheSF and siteR, siteF and pheSR, respectively. The point mutation responsible for *p*-Cl-Phe sensitivity was introduced by the primers siteF and siteR annealing internal to the wild-type *pheS* gene. There are overlapping regions among the three amplicons, which allowed an overlap extension PCR step using primers ldhF1 and pheSR to create P_{ldh}-*pheS** cassette. The generated 1270 bp P_{ldh}-*pheS** cassette was digested with *Eco*RI and *Bgl*II and ligated to the compatible sites of *Lc. lactis*/*E. coli* shuttle vector pSec:Leiss:Nuc [34], creating pleiss-P-pheS*.

To investigate the feasibility of the counterselectable P_{ldh}-*pheS** cassette, the plasmid pleiss-P-pheS* was introduced into the competent cells of *Lc. lactis* NZ9000

by electroporation [35]. The recombinant strain *Lc. lactis* NZ9000/pleiss-P-pheS* was incubated in GM17 with 5 µg/ml chloramphenicol. Overnight cultures were tenfold serially diluted, and 5 µl of diluted solution were pipetted onto air dried GM9 plates containing 0, 5, 10, 15 mM *p*-Cl-Phe, the cell survival was measured.

Construction of plasmids pleiss-nP-gfp

To demonstrate whether cascading promoters could increase the *gfp* gene expression, a series of plasmids pleiss-nP-gfp carrying promoter clusters nP_{ldh}-*gfp* were constructed as follows: the *gfp* gene was PCR amplified using primers gfpF and gfpR from plasmid pOgfp [18]. the promoter P_{ldh} and *gfp* gene were fused by an overlap extension PCR using primers ldhF2 and gfpR. The resulting product P_{ldh}-*gfp* was digested with *Bgl*II and *Eco*RI and ligated into the corresponding sites of pSec:Leiss:Nuc [34] to create plasmid pleiss-P-gfp. The promoter P_{ldh} was generated by PCR with primers ldhF2 and ldhR2, and the PCR product was digested with *Bgl*II and *Bam*HI and inserted into the *Bgl*II site of pleiss-P-gfp to generate pleiss-2P-gfp. The same procedure was carried out to construct the plasmid pleiss-nP-gfp (n: the copy number of P_{ldh} in the promoter clusters nP_{ldh}-*gfp*). Then, the above plasmids pleiss-nP-gfp were introduced into *Lc. lactis* NZ9000.

Fluorescence assay

Recombinant strains harboring the pleiss-nP-gfp were grown aerobically in 5 ml GM17 broth containing 5 µg/

Table 2 Oligonucleotide primers used in this study

Primer	Sequence (5′–3′)[a]	Restriction site
aldB-uF	AGGGTACCGGCGAAAGTCATGTAACAATCC	KpnI
aldB-uR	CTGACATGATATTTCTCTTTTCTAT	
aldB-dR	CCGCTCGAGTGCTGACAGATGGCTGGCTGTG	XhoI
aldB-dF	GAAAAGAGAAATATCATGTCAGTAATTGCTTAAATTTCTTTAGC	
aldB-testF	ATATTTCTGCCACAATTTTCATGCC	
aldB-testR	CCAATCCTGTACCAATAACAGCAAT	
pheSF	ATAAAAAATCGAAAAGGAGATAAAAATGAACTTACAAGAAAAAATTGAAG	
pheSR	AAAGATCTTCAGTCGAATTGTTCTAAGAATC	BglII
pheSR2	AAGAATTCTCAGTCGAATTGTTCTAAGAATC	EcoRI
siteR	ACCAAAACCAGAATAAACAGAAG	
siteF	CTGTTTATTCTGGTTTTGGTTTTGGACTCGGTCAAGAACG	
ldhF1	CCGAATTCATTCATTTTACACATTGTA	EcoRI
ldhF2	GACAGATCTATTCATTTTACACATTGTA	BglII
ldhF3	AGACGTCGACATTCATTTTACACATTGTA	SalI
ldhR1	TTTTATCTCCTTTTCGATTTTTAT	
ldhR2	CGGGATCCTTTTATCTCCTTTTCGATTTTTTAT	BamHI
ldhR3	CCGCTCGAGTTTTATCTCCTTTTCGATTTTTTAT	XhoI
upF	AGGGTACCATGTCAATAGTTGTCGAAAA	KpnI
upR	CAGTTTCTGCTAAGGTATCA	
downF	TGATACCTTAGCAGAAACTGATGAATTAGCACAGCAAGTG	
downR	CCGCTCGAGCTCTAGTAAAATGTTCCTCA	XhoI
testF	TTAAGGAAATGAATTTAGAGGAGAG	
testR	AAACCTTCATGTCCTTCTTGAGT	
BL-pheSF	AAAGATCTATGGATCTTCAAACCAAACTTG	BglII
BL-siteR	ACCAAAACCGCCGTAAACGTC	
BL-siteF	GACGTTTACGGCGGTTTTGGTTTTGGCCTTGGTCCTGATCG	
BL-pheSR	GATCTGCAGTTAACCCTCCTGGCTGAATTGC	PstI
gfpF	ATAAAAAATCGAAAAGGAGATAAAAAGATATGAGCAAAGGAG	
gfpR	CGCGAATTCTTAGTAGAGCTCATC	EcoRI

[a] The restriction sites in the primer sequences are underlined

---

ml chloramphenicol at 30 °C. Samples for measurement were taken out after 12 h and harvested by centrifugation at $10,000 \times g$ for 3 min. After being resuspended twice with PBS buffer (137 mM NaCl, 2.7 mM KCl, 10 mM Na_2HPO_4, 2 mM KH_2PO_4, pH7.4), 200 µl of bacterial suspension was transferred into a 96-well plate in which OD_{600} and fluorescence were read with excitation at 485 nm and emission at 528 nm using a Multi-Detection Microplate Reader, Synergy HT (BioTek). For each sample, three repetitions were performed with PBS buffer as a blank.

Construction of counterselectable cassettes nP$_{ldh}$-*pheS** and plasmids pG⁺-nP-pheS*

To increase the expression of PheS* protein, a series of counterselectable cassettes, namely nP$_{ldh}$-*pheS** were constructed as follows: the P$_{ldh}$-*pheS** cassette was PCR amplified using primer pair ldhF3 and pheSR2 from

plasmid pleiss-P-pheS*. The resulting DNA fragment was digested with SalI and EcoRI and ligated into the corresponding sites of pUC19 to create plasmid pUC-P-pheS*. The promoter P$_{ldh}$ was generated by PCR with primers ldhF3 and ldhR3, and the PCR product was digested with XhoI and SalI and inserted into the SalI site of pUC-P-pheS* to generate pUC-2P-pheS. The same procedure was carried out to construct the plasmid pUC-nP-pheS (n: the copy number of P$_{ldh}$ in the nP$_{ldh}$-*pheS** cassettes). To develop a counterselectable system in *Lc. lactis*, the above plasmids pUC-nP-*pheS** were digested with SalI and EcoRI, and the generating nP$_{ldh}$-*pheS** cassettes (Fig. 4a) were ligated to the SalI and EcoRI sites of pG⁺host9 to yield plasmid pG⁺-nP-pheS*.

Functional analysis of the pG⁺-nP-pheS* in *Lc. lactis*

To knockout the 709 bp fragment of the galactose operon in *Lc. lactis* NZ9000, pG⁺UD-nP-pheS* was constructed as follows. Upstream with 1000 bp in size (amplified with primers upF and upR) and downstream with 1011 bp (amplified with primers downF and downR) homology arms were PCR amplified from the genomic DNA of *Lc. lactis* NZ9000 and spliced by an overlap extension PCR using primers upF and downR; Subsequently, the fused fragment was digested by *Kpn*I and *Xho*I and inserted into the corresponding sites of the temperature-sensitive vectors pG⁺host9 [16] and pG⁺-nP-pheS*, resulting the plasmids pG⁺UD and pG⁺UD-nP-pheS*, respectively. The plasmid pG⁺UD and pG⁺UD-nP-pheS* were introduced into *Lc. lactis* NZ9000 to perform double-crossover homologous recombination as described previously [16]. Briefly, the recombinants were grown at 28 °C until OD$_{600}$ 0.8–1.0, then transferred to 37 °C for 2 h to allow the single-crossover integrants growth. Appropriate cultures were plated onto GM17 medium with 5 μg/ml erythromycin at 37 °C. Subsequently, the single-crossover integrants were cultured in GM17 medium without erythromycin at 28 °C for excision of the vector by a second crossover process. The cultures were then plated onto GM9 plates containing 15 mM *p*-Cl-Phe at 37 °C. The single-crossover integrants and double-crossover mutants were both verified utilizing primer pair testF and testR.

To further confirm the function of pG⁺-5P-pheS*, the *aldB* gene encoding α-acetolactate decarboxylase was knocked out from the *Lc. lactis* NZ9000 genome using the above protocols. Primer pairs of aldB-uF/aldB-uR and aldB-dF/aldB-dR were utilized for amplifying the upstream and downstream homology arms, and fused by an overlap extension PCR. The resultant ~2.0 kb fragment was digested and inserted into the *Kpn*I and *Xho*I sites of the vector pG⁺-5P-pheS*. Subsequently, the yielding pG⁺UD2-5P-pheS* was transferred into *Lc. lactis* NZ9000 to perform the double-crossover homologous recombination as described above. The mutant *Lc. lactis* dA was verified by PCR with the primer pair aldB-testF and aldB-testR, and the mutant genotype was also confirmed by sequence analysis (Biosune Company, Shanghai, China).

Extending this counterselectable marker *pheS** to other LAB

To extend this counterselectable marker *pheS** to other LAB, *L. casei* was selected as a host. The mutant gene *pheS** was generated as two fragments by an overlap extension PCR using the *L. casei* BL23 [36] genomic DNA as a template with the primers BL-pheSF and BL-siteR, BL-pheSR and BL-siteF, respectively. The point mutation responsible for *p*-Cl-Phe sensitivity was PCR amplified by the primer pair BL-siteF and BL-siteR annealing internal

to *pheS* gene. The generated *pheS** was digested with *Pst*I and *Bgl*II and ligated to the compatible sites of pTRKH2 [37], creating pTRKH2-pheS*.

To investigate the feasibility of the counterselectable marker *pheS**, the plasmid pTRKH2-pheS* was introduced into *L. casei* BL23 by electroporation [13]. The recombinant *L. casei* BL23/pTRKH2-pheS* was incubated into MRS with 5 μg/ml erythromycin. Overnight cultures were streaked onto a GM9 plate containing 10 mM *p*-Cl-Phe, the cell survival was measured.

Results
Bioinformatic analysis of PheS protein in selected LAB species

Previously, it was reported that only a point mutant *pheS** gene encoding an A294G substitution in *E. coli* PheS [27] or an A312G substitution in *Enterococcus faecalis* PheS [32] resulted in the obviously sensitivity to the phenylalanine analog *p*-Cl-Phe. Hence, to identify the amino acid residue for site-directed mutagenesis, the amino acid sequences of PheS from *Lc. lactis*, *L. casei*, *L. plantarum*, *L. brevis*, *L. rhamnosus* and *S. thermophilus* were aligned with the amino acid sequences from *E. coli* and *E. faecalis* by Clustal X version 2.0 [30] and ESPript 3.0 [31]. As shown in Fig. 1, the amino acid residues A312 in *Lc. lactis* NZ9000, A312 in *L. casei*, A312 in *L. plantarum*, A312 in *L. brevis*, A312 in *L. rhamnosus* and A314 in *S. thermophilus* are strictly conserved compared to the residue A294 in *E. coli* [27] and A312 in *E. faecalis* [32], indicating that this alanine residue of PheS protein was highly conserved in LAB.

To verify the above putative result, we chose the amino acid residues A312 in *Lc. lactis* NZ9000 PheS for site-directed mutagenesis. The *pheS* gene (accession number: NC_009004) encoding phenylalanyl-tRNA synthetase alpha subunit was identified from the genome of *Lc. lactis* NZ9000. It was 1038 bp in size and composed of 345 amino acid residues. After precision mutation GCT to GGT by an overlap extension PCR strategy in codon 312 of the *pheS* gene, an A312G point mutation was introduced into the PheS protein, resulting a dominant-negative mutant protein PheS*. Moreover, we also tested the potential of A312G point mutation of PheS for counterselection in *L. casei* BL23.

Functional analysis of the counterselectable marker P$_{ldh}$-*pheS** in *Lc. lactis*

To test the feasibility of the gene *pheS** as a counterselectable marker in *Lc. lactis*, it was driven constitutively by a strong promoter of L-lactate dehydrogenase gene (*ldh*) in *Lc. lactis* NZ9000. Subsequently, this resultant P$_{ldh}$-*pheS** cassette was inserted into the *Lc. lactis*/*E. coli*

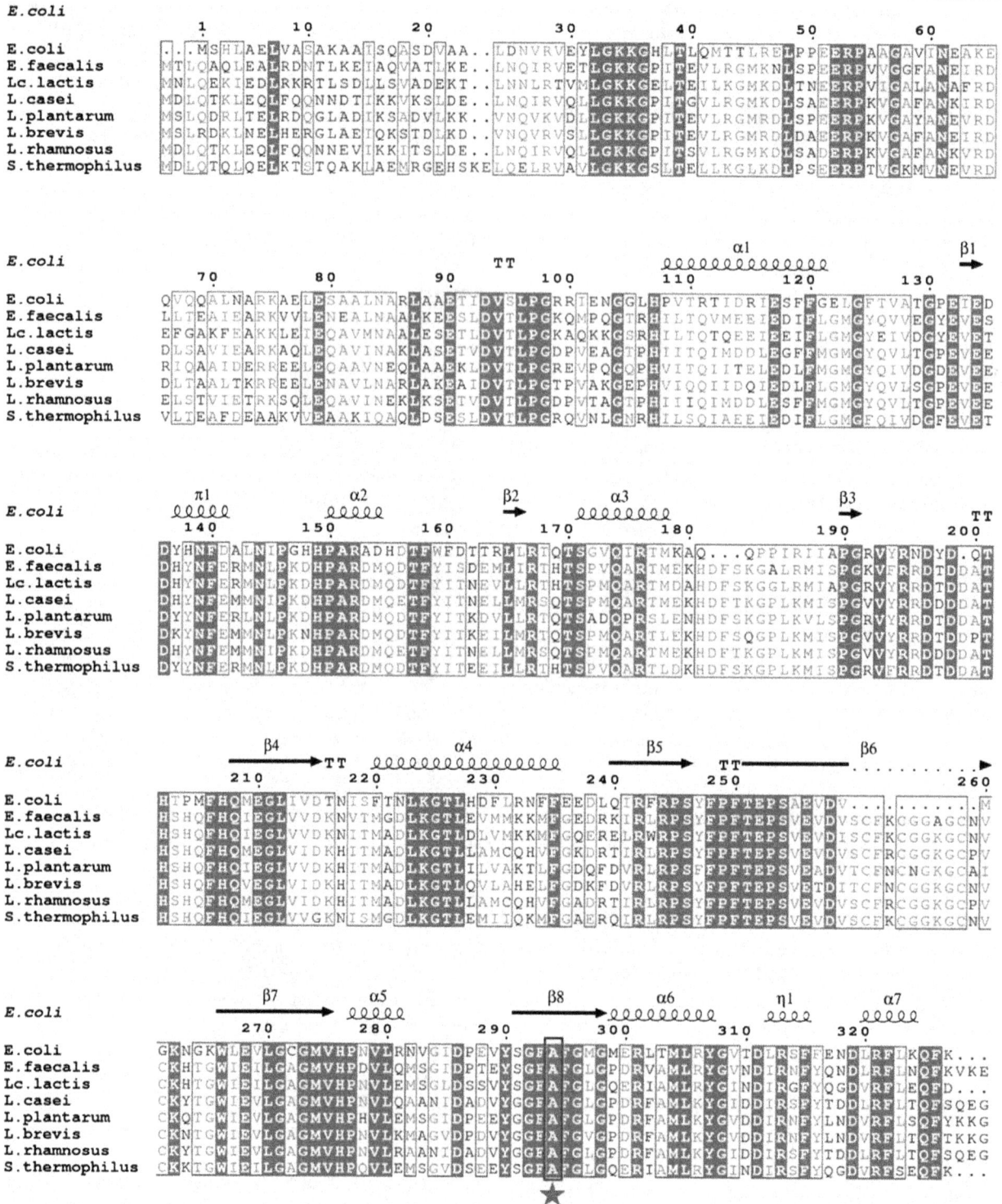

Fig. 1 A multiple-sequence alignment of PheS from a variety of distantly related species. Full length sequences of PheS were determined using Clustal X. The secondary structure of PheS in *E. coli* (PDB code: 3PCO) is shown at the *top* of each set of sequence. The conserved alanine residues mutated to generate *p*-Cl-Phe sensitivity were boxed with a *thick line* and indicated with a pentagram. *α* α-helix, *β* β-sheet, *π* π-helix, *η* 3_{10}-helix, *TT* β-turn

shuttle vector pSec:Leiss:Nuc, yielding the recombinant plasmid pleiss-P-pheS*. The schedule of construction of pleiss-P-pheS* was shown in Fig. 2.

After introduction of the plasmid pleiss-P-pheS* into *Lc. lactis* NZ9000, the sensitivity of the recombinant *Lc. lactis* NZ9000/pleiss-P-pheS* was measured on the GM9 plates containing 0, 5, 10, 15 mM phenylalanine analog *p*-Cl-Phe, respectively. As shown in Fig. 3, *Lc. lactis* NZ9000/pleiss-P-pheS* grew well on the GM9 plate without *p*-Cl-Phe, while the growth was completely inhibited in the presence of 15 mM of *p*-Cl-Phe. But, unlike the *Lc. lactis* NZ9000/pleiss-P-pheS*, the growth of *Lc. lactis* NZ9000 equipped with the plasmid pSec:Leiss:Nuc as a control was not inhibited under the equivalent concentrations of *p*-Cl-Phe, indicating that the dominant-negative mutant protein PheS* is functional as a stringent counterselectable marker in the presence of 15 mM *p*-Cl-Phe.

Creation of a counterselectable cassette 5P$_{ldh}$-*pheS** in *Lc. lactis*

The temperature sensitive plasmid pG$^+$host9 was widely used for gene(s) deletion and insertion in LAB. Here, a counterselectable system PheS*/pG$^+$host9 was constructed based on a pG$^+$host9 carrying the fragment P$_{ldh}$-*pheS** from the plasmid pleiss-P-pheS*, and yielding pG$^+$-P-pheS*. To investigate the feasibility of this vector for gene deletion, upstream and downstream homology arms of the 709 bp fragment of galactose operon were spliced and inserted into pG$^+$host9 and pG$^+$-P-pheS*, resulting pG$^+$UD and pG$^+$UD-P-pheS*. The single-crossover integrants *Lc. lactis* IG0 and *Lc. lactis* IG1 were both pipetted onto GM9 plates containing 15 mM *p*-Cl-Phe, respectively. Unfortunately, *Lc. lactis* IG1 was not completely inhibited by *p*-Cl-Phe. We supposed that this unexpected phenomenon might resulted from the low expression of PheS* protein from a chromosomal copy.

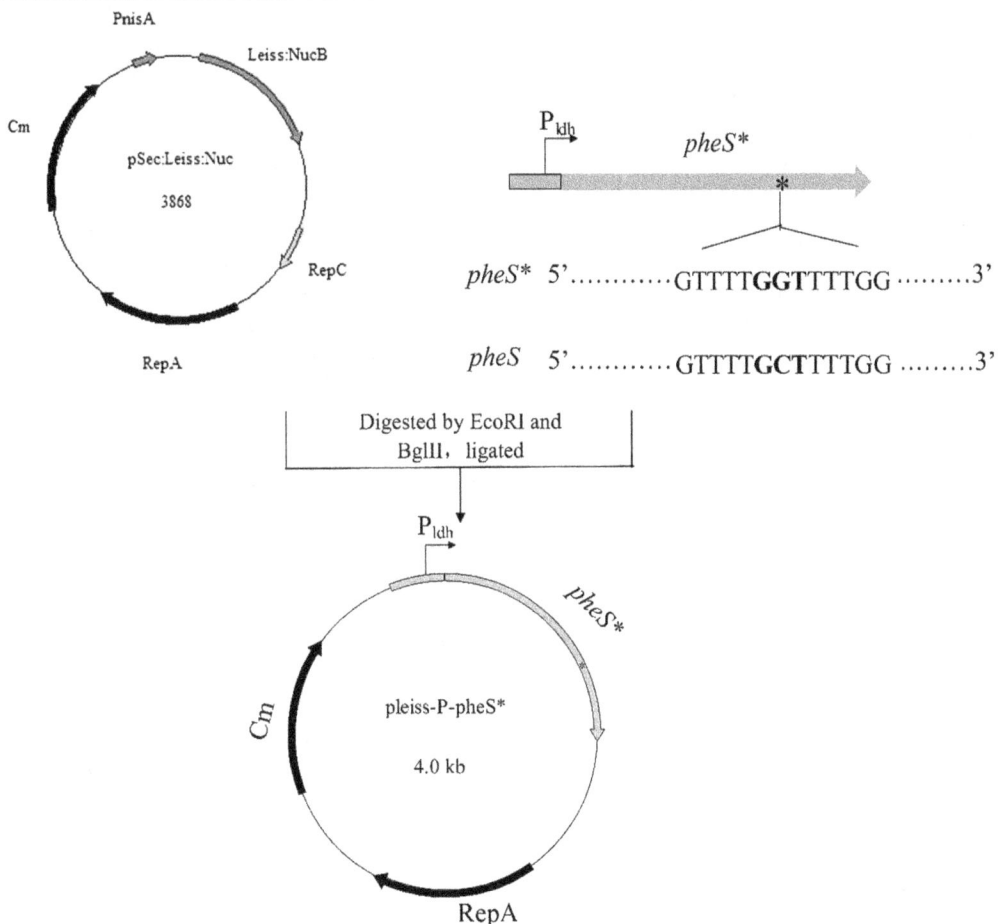

Fig. 2 Construction of a vector for detecting the PheS*/pG$^+$host9 counterselectable system. The wild-type *pheS* was changed to *pheS** using an overlap extension PCR to introduce a point mutation of GCT to GGT. P$_{ldh}$: the promoter region of the L-lactate dehydrogenase gene (*ldh*) (accession number: NC_017949) from *Lc. lactis* NZ9000

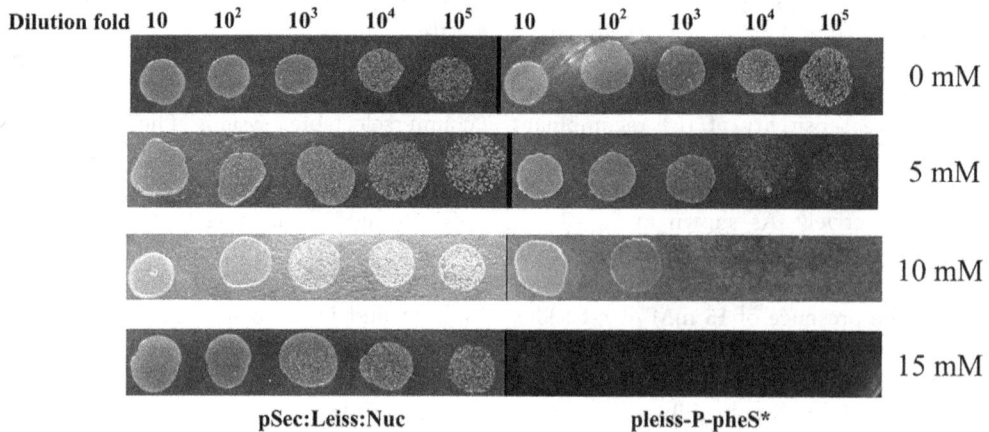

Fig. 3 Detection of the sensitivity of *Lc. lactis* NZ9000/pleiss-P-pheS* to *p*-Cl-Phe. Wild-type *Lc. lactis* NZ900 carrying either an empty vector (pSec:Leiss:Nuc) or a *Lc. lactis* derivative carrying the P_{ldh}-*pheS** cassette (pleiss-P-pheS*) were cultivated overnight in GM9 plates containing the indicated concentration of *p*-Cl-Phe. This experiment was performed in triplicate with the similar results

To increase the expression level of *pheS**, we firstly tested whether cascading promoters could be functional in *Lc. lactis*. A series of plasmids pleiss-nP-gfp carrying promoter clusters nP_{ldh}-*gfp* were constructed and introduced into the *Lc. lactis* NZ9000. To optimize the constructed nP_{ldh}-*gfp* promoter clusters, fluorescence intensity of each construct radiated from the green fluorescence protein after 12 h of aerobic incubation was determined. By analyzing the cell growth and relevant fluorescence of each recombinant strain, we found that the more copies of the P_{ldh} promoter were present in the expression cassette, the higher the specific fluorescence was (Fig. 4a). This result indicated that cascading promoters could improve the expression level of *gfp* gene.

Subsequently, various copies of the P_{ldh} were driven the expression of dominant-negative mutant protein PheS* (Fig. 4b). As shown in Fig. 4c, the increase of the P_{ldh} copies in the nP_{ldh}-*pheS** cassettes inserted into the chromosomal locus of the integrants resulted in the enhanced sensitivity to 15 mM *p*-Cl-Phe. When the *pheS** gene expressed from five copies of the P_{ldh}, the growth of *Lc. lactis* IG5 was substantially inhibited. Thus, we chose the $5P_{ldh}$-*pheS** cassette as a negative selectable marker for development of a counterselectable seamless mutagenesis system PheS*/pG+host9 in *Lc. lactis*.

Functional analysis of the counterselectable system PheS*/pG+host9 in *Lc. lactis*

To verify the potential of the counterselectable system PheS*/pG+host9 in *Lc. lactis* NZ9000, a 709 bp fragment of the galactose operon was selected as a targeting region for deletion through the vector pG+UD-5P-pheS*. After *p*-Cl-Phe counterselection, 24 resistance colonies picked randomly on a GM9 plate containing 15 mM *p*-Cl-Phe

were all double-crossover occurred and 10 out of them (approximately 42%) were shown to possess the expected mutant genotype by PCR determination and sequencing. The result was closed to the theoretical value (50%) since the target region of the galactose operon was not essential for the growth of *Lc. lactis* NZ9000 (Fig. 5a, b). Hence, we supposed that the $5P_{ldh}$-*pheS** cassette based counterselectable system was functional in *Lc. lactis* NZ9000 to perform seamless gene deletion.

To further determine whether the counterselectable system PheS*/pG+host9 would be feasible for genome engineering, *aldB* gene which encodes for α-acetolactate decarboxylase catalyzing α-acetolactate to acetoin in the diacetyl biosynthesis in *Lc. lactis* was deleted by this system (Fig. 5a). Twenty-one colonies were selected randomly and detected by PCR amplification. The double-crossover events also occurred in 100%, and six out of them were the expected mutants (Fig. 6). This result indicated that the efficiency of screening double-crossover mutants was significantly improved compared with using pG+host9 alone in our laboratory previously [18].

Potential of the counterselectable marker *pheS** in other LAB

To test the feasibility of the gene *pheS** as a counterselection marker in other LAB, the strain *L. casei* BL23 was selected as a host. After insertion of *pheS** into pTRKH2 [37], the obtained plasmid pTRKH2-pheS* was introduced into *L. casei* BL23. Subsequently the sensitivity of the recombinant *L. casei* BL23/pTRKH2-pheS* to *p*-Cl-Phe was measured on the GM9 plates containing 10 mM *p*-Cl-Phe. As shown in Fig. 7, the recombinant *L. casei* BL23/pTRKH2 grew well on the GM9 plate containing 10 mM *p*-Cl-Phe, while the growth of *L. casei* BL23/

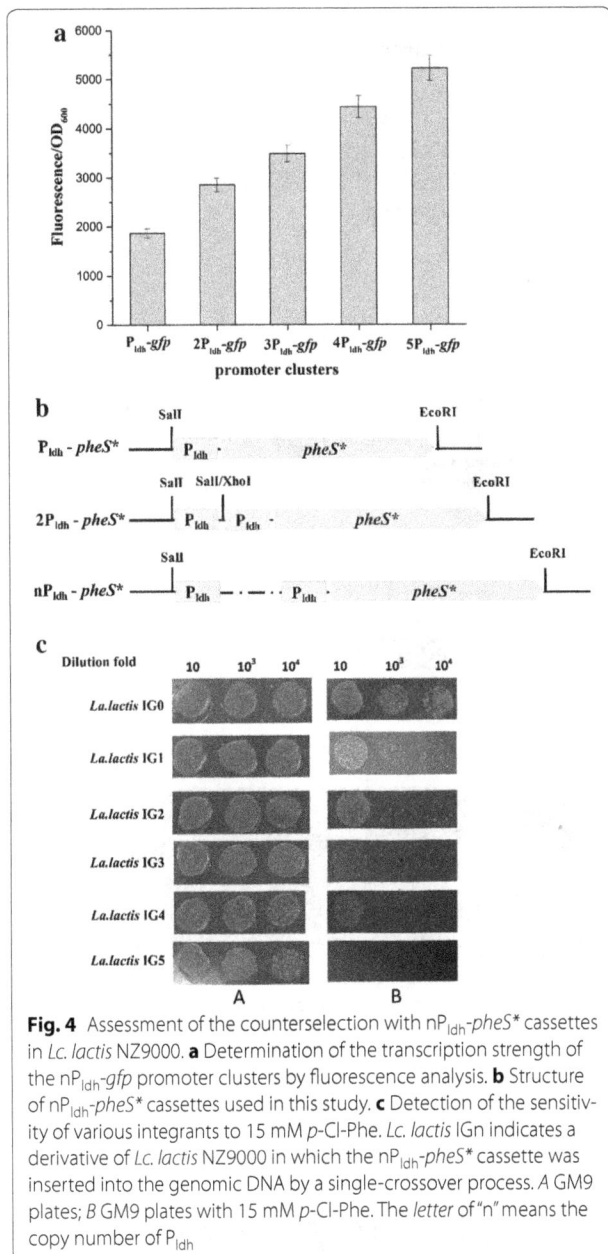

Fig. 4 Assessment of the counterselection with nP_{ldh}-*pheS** cassettes in *Lc. lactis* NZ9000. **a** Determination of the transcription strength of the nP_{ldh}-*gfp* promoter clusters by fluorescence analysis. **b** Structure of nP_{ldh}-*pheS** cassettes used in this study. **c** Detection of the sensitivity of various integrants to 15 mM *p*-Cl-Phe. *Lc. lactis* IGn indicates a derivative of *Lc. lactis* NZ9000 in which the nP_{ldh}-*pheS** cassette was inserted into the genomic DNA by a single-crossover process. *A* GM9 plates; *B* GM9 plates with 15 mM *p*-Cl-Phe. The *letter of* "n" means the copy number of P_{ldh}

the metabolic flux [38]. In this study, a seamless negative selectable mutagenesis system $PheS*/pG^+host9$ was developed. We also demonstrated its feasibility by constructing strains bearing the targeting seamless deletion of a 709 bp fragment in lactococcal galactose operon and *aldB* gene. Expectedly, the ratio of the double-crossover event was 100% after counterselection by *p*-Cl-Phe.

To our knowledge, this is the first report that the mutated *pheS* allele can be used as a counterselection marker for efficient and rapid genomic engineering in *Lc. lactis*. Previously, the development of a *pheS* based counterselection system in *Streptococcus mutans*, which is a close relative to *Lc. lactis*, has been reported [28]. However, *S. mutans* is a pathogenic bacterium distributed in the dental caries and could not be applied in the food field and used as a cell factory [39]. We expected that combining the counterselectable marker *pheS** with the traditional genetic tool pG^+host9 [16] would overcome the bottleneck of laboriously screening of the double-crossover recombinants, and this system has greatly potential in genome engineering in LAB.

Protein sequence analysis suggested that the alanine residue of PheS protein is highly conserved in LAB (Fig. 1). Here we have demonstrated the feasibility of *pheS** as a counterselectable marker in *Lc. lactic* and *L. casei*, these results were consistent with the previously results in *S. mutans* and *Enterococcus faecalis* [28, 32]. Therefore, we speculated that the dominant-negative mutant gene *pheS** might be widely used as a counterselectable marker in a variety of lactic acid bacterial species. However, the sensitivity of the cells to *p*-Cl-Phe was depended on strain specific manner, such as 15 mM *p*-Cl-Phe for *Lc. lactis* NZ9000, 20 mM *p*-Cl-Phe for *S. mutans* [28]. Hence, optimization of the PheS* expression is needed when employing *pheS** as a counterselectable marker in other LAB strains [25, 28].

In this study, the PheS* protein under the control of a promoter P_{ldh} has the ability of completely inhibiting the growth of *Lc. lactis* NZ9000 at 15 mM *p*-Cl-Phe, suggesting it is possible to use P_{ldh}-*pheS**cassette as a counterselectable marker in *Lc. lactis*. However, the growth of the recombinants with P_{ldh}-*pheS** inserted into the chromosomal locus was not completely inhibited by even higher concentration of *p*-Cl-Phe. This unexpected result means that the ratio of screening double-crossover recombinants would not be 100% after *p*-Cl-Phe counterselection. We speculated that this phenomenon was caused by low expression level of PheS* [28], because the copy number of P_{ldh}-*pheS** from the chromosomal locus was lower than that in the plasmid pleiss-P-pheS*. Lower yield of PheS* was insufficient to compete with the background expression of wild-type PheS to form complexes with phenylalanyl-tRNA synthetase beta subunit (PheT)

pTRKH2-pheS* was obviously inhibited, indicating that the conditional-lethal mutant gene *pheS** has the potential as a counterselectable marker in *L. casei* and other LAB.

Discussion

In consideration of the increasing use in industrial and medical area, LAB are intensively studied on their genetics and metabolism [9, 10]. Therefore, efficient genome engineering tools are necessary for target gene(s) deletion or insertion for functional analysis or rerouting

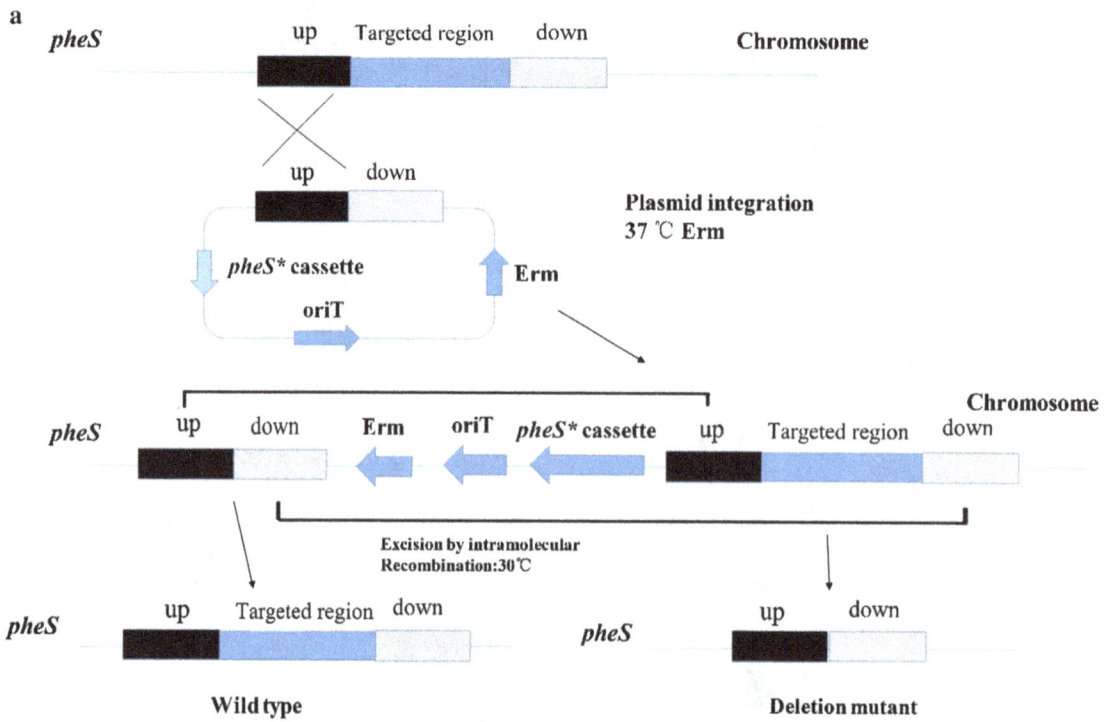

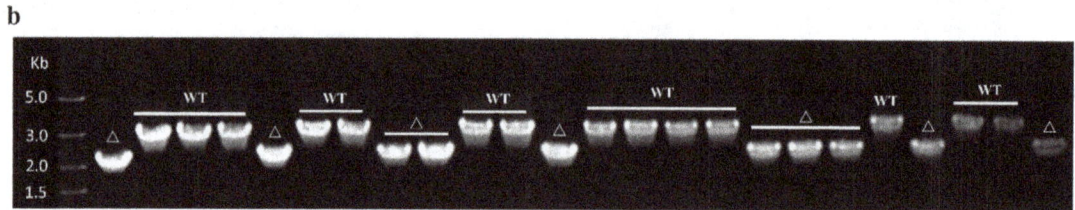

Fig. 5 Construction of the PheS*/pG+host9 counterselectable system in *Lc. lactis* NZ9000. **a** An effcient counterselectable system PheS*/pG+host9 used to create gene deletions in *Lc. lactis* NZ9000. The "*pheS*" cassette" indicates the *pheS*" gene under the control of five cascading P$_{\text{ldh}}$. "*up*" and "*down*" indicate the upstream and downstream homology arms of the targeted region. "Erm" indicates the erythromycin resistant gene. "OriT" indicates the temperature sensitive origin of replication. **b** Twenty-four *p*-Cl-Phe-resistant colonies were amplified by PCR to screen for the deletion of 709 bp fragment of the galactose operon. The expected PCR fragment from the mutant type (Δ) is approximately 2.0 kb, while the band from the wild-type (WT) is about 2.7 kb

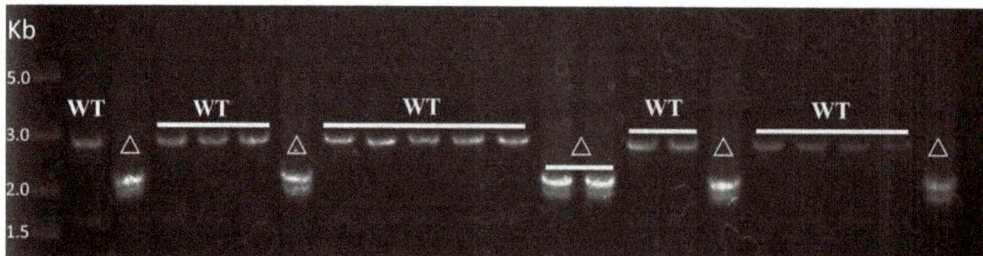

Fig. 6 Generation of a seamless in-frame *aldB* deletion mutant. Twenty-one *p*-Cl-Phe-resistant colonies were amplified by PCR to screen for the deletion of the *aldB* gene (711 bp). The expected PCR fragment from the mutant type (Δ) is approximately 2.0 kb, while the band from the wild-type (WT) is about 2.7 kb

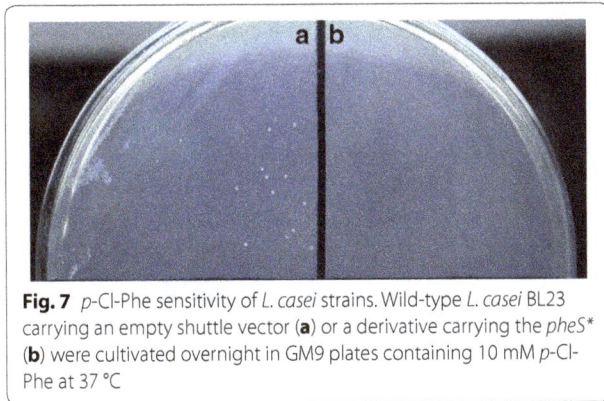

Fig. 7 *p*-Cl-Phe sensitivity of *L. casei* strains. Wild-type *L. casei* BL23 carrying an empty shuttle vector (**a**) or a derivative carrying the *pheS** (**b**) were cultivated overnight in GM9 plates containing 10 mM *p*-Cl-Phe at 37 °C

[40]. In these cases, a strategy of cascading promoters [41] was employed to improve the expression level of protein PheS*. Surprisingly, when protein PheS* was driven simultaneously by five copies of the P_{ldh}, the generating integrant *Lc. lactis* IG5 was substantially inhibited in the presence of 15 mM *p*-Cl-Phe and the ratio of screening double-crossover recombinants was 100%, suggesting recombination among the promoters was not occurred and the use of repeated P_{ldh} promoters could not confer genetic instability [41]. This strategy provides a new idea to address the issue of the low expression of the exogenous protein(s) in LAB.

Several strategies have been employed to fulfill the genome engineering in LAB by homologous double-crossover using a solely conditional replication plasmid [38] or combining with other counterselectable system, such as *upp* [22] or *oroP* [24] based cassettes. Compared with those methods, the negative selectable system PheS*/pG+host9 has several advantages. (1) It greatly simplifies the procedure for screening double-crossover recombinants. For example, taking only 2 days to screen double-crossover variants after the single-crossover integrants were subcultured at 28 °C. The ratio of the double-crossover recombinants was 100% after *p*-Cl-Phe counterselection. However, the ratio between the deletion and wild-type strains may not be the theoretical value (1:1), it can vary considerably depending on the function of gene(s) to be deleted. (2) To our knowledge, among all the reported counterselectable markers, only *pheS** has the greatly potential to be widely utilized in wild-type LAB without pretreatment. In contrast to other counterselectable system, the variants required the counterselectable marker deficient strains, as in the case of *upp* [21–23] and *oroP* [24]. Recently, a new counterselection method for wild-type *Lc. lactis* genome engineering based on class IIa bacteriocin sensitivity was reported [42]. However, the li006Dlitation of this method to be widely used in LAB was the sensitivity to bacteriocins

which would depend on the interaction between the listerial MpnC and the native PtnD [42]. (3) Strains without *pheS** can naturally grow on GM9 medium with 15 mM *p*-Cl-Phe. This means 15 mM *p*-Cl-Phe has no side-effect on the growth of the expected mutants.

Moreover, this mutagenesis system PheS*/pG+host9 allowed gene deletion without any genomic scarring [15] in *Lc. lactis*, as the case of the *aldB* gene. The generating genetically modified microorganisms (GMOs) [14] were seamless mutagenesis which means only leaving self-DNA in its native genome location [15]. Therefore, this system is useful in seamless gene deletions in industrial strains. However, this seamless mutagenesis system PheS*/pG+host9 remains challenging in large DNA fragment deletions or insertions. In this study, the limited length of the targeted DNA fragment was mostly from the low efficient homologous recombination mediated by RecA [17]. In consideration of the ratio of the double-crossover recombinants was 100% after *p*-Cl-Phe counterselection, the ideal goal for deletion or insertion of large DNA fragment is the new genome engineering tools responsible for targeted fragments replacement by selection and the $5P_{ldh}$-*pheS** cassette responsible for selectable marker excision by counterselection [15].

Conclusions

A seamless mutagenesis system PheS*/pG+host9 based on a counterselectable marker *pheS** and a temperature sensitive plasmid pG+host9 was developed in *Lc. lactis*. Moreover, this system can be used for rapidly constructing a seamless mutagenesis (deleted or inserted) strain. We also extended the counterselectable marker *pheS** to *L. casei*. Although the feasibility of *pheS** as a counterselectable marker used in other LAB remains to be demonstrated, we speculated that this conterselectable marker will accelerate the analysis of genes with unknown function and metabolic engineering research in LAB.

Abbreviations
LAB: lactic acid bacteria; *pheS*: gene of phenylalanyl-tRNA synthetase alpha subunit; *pheS**: the mutant *pheS* gene; PheS*: phenylalanyl-tRNA synthetase alpha subunit with an A312G substitution; P_{ldh}: promoter of the *Lc. lactis* NZ9000 L-lactate dehydrogenase; *p*-Cl-Phe: *p*-chloro-phenylalanine; SSDR: single-stranded DNA recombineering; DSDR: double-stranded DNA recombineering.

Authors' contributions
YPX, TTG and JK conceived and designed the experiments, YPX carried out the experimental work. YPX, TTG, YLM and JK wrote and revised the manuscript. All authors read and approved the final manuscript.

Acknowledgements
We thank N. Galleron for kindly providing the plasmid pSec:leiss:Nuc and *Lc. lactis* NZ9000. We also thank I. Biswas and E. Maguin for their generous gift of the plasmid pG+host9.

Competing interests

The authors declare that they have no competing interests.

Funding

This work was supported by the National Natural Science Foundation of China (Grants 31400077 and 31471715) and the Public Service Sectors (Agriculture) Special and Scientific Research Projects Hi-Tech Research and Development Program of China (Grant 201503134).

References

1. Cavanagh D, Fitzgerald GF, McAuliffe O. From field to fermentation: the origins of Lactococcus lactis and its domestication to the dairy environment. Food Microbiol. 2015;47:45–61.
2. Leroy F, De Vuyst L. Lactic acid bacteria as functional starter cultures for the food fermentation industry. Trends Food Sci Technol. 2004;15:67–78.
3. Kleerebezem M, Hugenholtz J. Metabolic pathway engineering in lactic acid bacteria. Curr Opin Biotech. 2003;14:232–7.
4. Liu J, Dantoft SH, Wurtz A, Jensen PR, Solem C. A novel cell factory for efficient production of ethanol from dairy waste. Biotechnol Biofuels. 2016;9:33.
5. Guimaraes VD, Innocentin S, Lefevre F, Azevedo V, Wal JM, Langella P, Chatel JM. Use of native lactococci as vehicles for delivery of DNA into mammalian epithelial cells. Appl Environ Microbiol. 2006;72:7091–7.
6. Kruger C, Hu Y, Pan Q, Marcotte H, Hultberg A, Delwar D, van Dalen PJ, Pouwels PH, Leer RJ, Kelly CG, et al. In situ delivery of passive immunity by lactobacilli producing single-chain antibodies. Nat Biotechnol. 2002;20:702–6.
7. Steidler L, Hans W, Schotte L, Neirynck S, Obermeier F, Falk W, Fiers W, Remaut E. Treatment of murine colitis by Lactococcus lactis secreting interleukin-10. Science. 2000;289:1352–5.
8. Wyszynska A, Kobierecka P, Bardowski J, Jagusztyn-Krynicka EK. Lactic acid bacteria–20 years exploring their potential as live vectors for mucosal vaccination. Appl Microbiol Biotechnol. 2015;99:2967–77.
9. Rossi M, Amaretti A, Raimondi S. Folate production by probiotic bacteria. Nutrients. 2011;3:118–34.
10. Thakur K, Tomar SK, De S. Lactic acid bacteria as a cell factory for riboflavin production. Microb Biotechnol. 2016;9:441–51.
11. van Pijkeren JP, Britton RA. High efficiency recombineering in lactic acid bacteria. Nucleic Acids Res. 2012;40:e76.
12. Oh JH, van Pijkeren JP. CRISPR-Cas9-assisted recombineering in Lactobacillus reuteri. Nucleic Acids Res. 2014;42:e131.
13. Yang P, Wang J, Qi Q. Prophage recombinases-mediated genome engineering in Lactobacillus plantarum. Microb Cell Fact. 2015;14:154.
14. van Pijkeren JP, Britton RA. Precision genome engineering in lactic acid bacteria. Microb Cell Fact. 2014;13(Suppl 1):S10.
15. Wang H, Bian X, Xia L, Ding X, Muller R, Zhang Y, Fu J, Stewart AF. Improved seamless mutagenesis by recombineering using ccdB for counterselection. Nucleic Acids Res. 2014;42:e37.
16. Maguin E, Duwat P, Hege T, Ehrlich D, Gruss A. New thermosensitive plasmid for gram-positive bacteria. J Bacteriol. 1992;174:5633–8.
17. Biswas I, Gruss A, Ehrlich SD, Maguin E. High-efficiency gene inactivation and replacement system for gram-positive bacteria. J Bacteriol. 1993;175:3628–35.
18. Guo T, Kong J, Zhang L, Zhang C, Hu S. Fine tuning of the lactate and diacetyl production through promoter engineering in Lactococcus lactis. PLoS ONE. 2012;7:e36296.
19. Lu W, Kong J, Kong W. Construction and application of a food-grade expression system for Lactococcus lactis. Mol Biotechnol. 2013;54:170–6.
20. Wang T, Lu W, Lu S, Kong J. Protective role of glutathione against oxidative stress in Streptococcus thermophilus. Int Dairy J. 2015;45:41–7.
21. Goh YJ, Azcarate-Peril MA, O'Flaherty S, Durmaz E, Valence F, Jardin J, Lortal S, Klaenhammer TR. Development and application of a upp-based counterselective gene replacement system for the study of the S-layer protein SlpX of Lactobacillus acidophilus NCFM. Appl Environ Microbiol. 2009;75:3093–105.
22. Martinussen J, Hammer K. Cloning and characterization of upp, a gene encoding uracil phosphoribosyltransferase from Lactococcus lactis. J Bacteriol. 1994;176:6457–63.
23. Martinussen J, Hammer K. Powerful methods to establish chromosomal markers in Lactococcus lactis: an analysis of pyrimidine salvage pathway mutants obtained by positive selections. Microbiology. 1995;141(Pt 8):1883–90.
24. Solem C, Defoor E, Jensen PR, Martinussen J. Plasmid pCS1966, a new selection/counterselection tool for lactic acid bacterium strain construction based on the oroP gene, encoding an orotate transporter from Lactococcus lactis. Appl Environ Microbiol. 2008;74:4772–5.
25. Carr JF, Danziger ME, Huang AL, Dahlberg AE, Gregory ST. Engineering the genome of Thermus thermophilus using a counterselectable marker. J Bacteriol. 2015;197:1135–44.
26. Kino Y, Nakayama-Imaohji H, Fujita M, Tada A, Yoneda S, Murakami K, Hashimoto M, Hayashi T, Okazaki K, Kuwahara T. Counterselection employing mutated pheS for markerless genetic deletion in Bacteroides species. Anaerobe. 2016;42:81–8.
27. Kast P, Hennecke H. Amino acid substrate specificity of Escherichia coli phenylalanyl-tRNA synthetase altered by distinct mutations. J Mol Biol. 1991;222:99–124.
28. Xie Z, Okinaga T, Qi F, Zhang Z, Merritt J. Cloning-independent and counterselectable markerless mutagenesis system in Streptococcus mutans. Appl Environ Microbiol. 2011;77:8025–33.
29. Chu LL, Pandey RP, Jung N, Jung HJ, Kim EH, Sohng JK. Hydroxylation of diverse flavonoids by CYP450 BM3 variants: biosynthesis of eriodictyol from naringenin in whole cells and its biological activities. Microb Cell Fact. 2016;15:135.
30. Larkin MA, Blackshields G, Brown NP, Chenna R, McGettigan PA, McWilliam H, Valentin F, Wallace IM, Wilm A, Lopez R, et al. Clustal W and Clustal X version 2.0. Bioinformatics. 2007;23:2947–8.
31. Robert X, Gouet P. Deciphering key features in protein structures with the new ENDscript server. Nucleic Acids Res. 2014;42:W320–4.
32. Kristich CJ, Chandler JR, Dunny GM. Development of a host-genotype-independent counterselectable marker and a high-frequency conjugative delivery system and their use in genetic analysis of Enterococcus faecalis. Plasmid. 2007;57:131–44.
33. Kuipers OP, de Ruyter PGGA, Kleerebezem M, de Vos WM. Quorum sensing-controlled gene expression in lactic acid bacteria. J Biotechnol. 1998;64:15–21.
34. Le Loir Y, Gruss A, Ehrlich SD, Langella P. A nine-residue synthetic propeptide enhances secretion efficiency of heterologous proteins in Lactococcus lactis. J Bacteriol. 1998;180:1895–903.
35. Holo H, Nes IF. High-frequency transformation, by electroporation, of Lactococcus lactis subsp. cremoris grown with glycine in osmotically stabilized media. Appl Environ Microbiol. 1989;55:3119–23.
36. Hazebrouck S, Pothelune L, Azevedo V, Corthier G, Wal JM, Langella P. Efficient production and secretion of bovine beta-lactoglobulin by Lactobacillus casei. Microb Cell Fact. 2007;6:12.
37. O'Sullivan DJ, Klaenhammer TR. High- and low-copy-number Lactococcus shuttle cloning vectors with features for clone screening. Gene. 1993;137:227–31.
38. Fang F, O'Toole PW. Genetic tools for investigating the biology of commensal lactobacilli. Front Biosci (Landmark Ed). 2009;14:3111–27.
39. Legenova K, Bujdakova H. The role of Streptococcus mutans in the oral biofilm. Epidemiol Mikrobiol Imunol. 2015;64:179–87.
40. Mermershtain I, Finarov I, Klipcan L, Kessler N, Rozenberg H, Safro MG. Idiosyncrasy and identity in the prokaryotic Phe-system: crystal structure of E. coli phenylalanyl-tRNA synthetase complexed with phenylalanine and AMP. Protein Sci. 2011;20:160–7.
41. Li M, Wang J, Geng Y, Li Y, Wang Q, Liang Q, Qi Q. A strategy of gene overexpression based on tandem repetitive promoters in Escherichia coli. Microb Cell Fact. 2012;11:19.
42. Wan X, Usvalampi AM, Saris PE, Takala TM. A counterselection method for Lactococcus lactis genome editing based on class IIa bacteriocin sensitivity. Appl Environ Microbiol. 2016;100:9661–9.

16

Heterologous expression of abaecin peptide from *Apis mellifera* in *Pichia pastoris*

Denis Prudencio Luiz[1*], Juliana Franco Almeida[2], Luiz Ricardo Goulart[2], Nilson Nicolau-Junior[1] and Carlos Ueira-Vieira[1]

Abstract

Background: Antimicrobial peptides (AMPs) are the first line of host immune defense against pathogens. Among AMPs from the honeybee *Apis mellifera*, abaecin is a major broad-spectrum antibacterial proline-enriched cationic peptide.

Results: For heterologous expression of abaecin in *Pichia pastoris*, we designed an ORF with HisTag, and the codon usage was optimized. The gene was chemically synthetized and cloned in the pUC57 vector. The new ORF was subcloned in the pPIC9 expression vector and transformed into *P. pastoris*. After selection of positive clones, the expression was induced by methanol. The supernatant was analyzed at different times to determine the optimal time for the recombinant peptide expression. As a proof-of-concept, *Escherichia coli* was co-incubated with the recombinant peptide to verify its antimicrobial potential.

Discussion: Briefly, the recombinant Abaecin (rAbaecin) has efficiently decreased *E. coli* growth ($P < 0.05$) through an in vitro assay, and may be considered as a novel therapeutic agent that may complement other conventional antibiotic therapies.

Keywords: Abaecin, Proline-rich antimicrobial peptide, Heterologous expression, *Pichia pastoris*, *Apis mellifera*

Background

Insects have cellular and humoral defenses in the innate immunological system. The circulating hemocytes are responsible for the first, as in phagocytosis cases and pathogens nodulation [1, 2] and the production of peptides by the fat bodies related to the second [3].

The antimicrobial peptides (AMPs) are molecules present in the immune system of multinuclear organisms, acting in the defense against invaders such as gram-positive, gram-negative bacteria and fungi [4].

The abaecin peptide, found in *Apis mellifera*, is one of the largest proline-rich antimicrobial peptide, with 34 amino acids containing 10 prolines (29%) and no cysteine residues. The total charge is 4^+, grouped within positions 12, 13, 27, and 29. Prolines are uniformly distributed through the peptide length, preventing the α-helical conformation [5, 6].

The yeast *Pichia pastoris* is methylotrophic, which means a capacity of using the methanol as its only carbon source. The yeast oxidizes the methanol producing formaldehyde and hydrogen peroxide, using oxygen molecules and alcohol oxidase enzymes. The yeast produces large quantity of this enzyme, due to its low affinity to oxygen, being regulated by the alcohol oxidase 1 promoter (AOX1). The AOX1 promoter within the vector, induced by methanol, leads to expression of the gene of interest at high levels. The protein expression using the pPIC9 vector occurs in an extracellular manner, decreasing the steps between expression and obtaining the peptide of interest [7, 8].

Antimicrobial peptides were isolated from many kind of organism like vertebrates and invertebrates animals, plants, bacteria, fungi, viruses and artificially synthesized in laboratory for experiments [9]. However, our aim was to construct a synthetic gene to express the abaecin

*Correspondence: prudenis@yahoo.com.br
[1] Genetics Laboratory, Institute of Genetics and Biochemistry, Federal University of Uberlândia, 1720 Pará, Uberlândia, MG 38400902, Brazil
Full list of author information is available at the end of the article

peptide from *A. mellifera* in *P. pastoris*. Our purpose was to optimize the expression of this recombinant antimicrobial peptide with biological effects, leading to a possible new drug development.

Methods
Gene design and synthesis
The recombinant peptide coding sequence was designed based on the deposited sequence in the National Center for Biotechnology Information (NCBI) NP_001011617.1, of *A. mellifera*'s abaecin peptide. Some nucleotides sequences were added to facilitate cloning and optimization of this protein expression in yeast *P. pastoris*. Two restriction sites were added: in the 5′ position for EcoRI enzyme, and another one in 3′ position for NotI enzyme. It was also added hexahistidine-tag (Histag) for expression confirmation process, and a stop codon was included as well. The peptide signal of the entire abaecin sequence was not used.

The sequence was optimized with the following parameters: length: 129, GC%: 45.08, GAATTC-TACGTTCC ATTGCCTAACGTTCCACAACCTGGTAGAAG ACCATTTCCTACTTTCCCAGGTCAAGGACC TTTTAACCCTAAGATTAAATGGCCTCAGGGAT ATCGTCGACATCACCATCACCATCACTAA-GCGG CCGC, therefore, the new abaecin sequence was deduced: YVPLPNVPQPGRRPFPTFPGQGPFNPKIK-WPQGYHHHHHH. Then, to optimize the sequence for *P. pastoris*, the codon adaptation index (CAI) was used. It measures the extension level of differential usage of codons in highly expressed genes [10]. The measurements in this technique use the OptimumGeneTM (GenScript® Corporation) software.

After analysis and deduction of the optimized DNA sequence, the chemical synthesis of the minigene was performed, cloned in pUC57, and sequenced for confirmation (GenScript® Corporation).

Sub-cloning of the synthetic gene in the expression vector pPIC9
The insert was removed from vector pUC57 using the NotI and EcoRI enzymes. The vector pPic9 was also digested with the same enzymes, quantified and dephosphorylated with SAP enzyme. It was used 16.7 ng of the insert in the ligation reaction with 50 ng of pPIC9 vector and 1U of the DNA ligase I (Invitrogen) at 14 °C overnight. The product of this reaction was named pPic9abacin, confirmed by electrophorese at 0.8% of agarose. The *Escherichia coli* Top10 bacteria were transformed with the pPic9abaecin by electroporation in 0.2 cm cuvette with lysogeny broth (LB), following the parameters: 2.5 kV, 200 Ω and 25 µF, electroporated in Bio-Rad Gene Pulser (Bio-Rad). These clones were cultivated in LB

medium containing 50 µg/mL of Ampicillin. The plasmid extraction was performed using QIAprep® (Qiagen®), following the kit protocol.

Transformation and electroporation of *Pichia*
The recombinant plasmid was linearized with SacI enzyme (FastDigest, Fermentas) 1 µL/µg (enzyme/DNA), purified with phenol/chloroform/125:24:1 and used to transform the strain GS115 *P. pastoris* yeast. The GS115 strain was cultivated in 50 mL of yeast extract peptone dextrose (YPD, 1% yeast extract, 2% peptone and 2% dextrose) at 30 °C overnight, then inoculated 0.5 mL of this culture in 500 mL of fresh YPD at 30 °C overnight until $OD_{600} = 1.3$–1.5. The cells were centrifuged at $1500 \times g$ for 5 min at 4 °C and resuspended, for each of the followed steps: 500 and 250 mL sterilized water, then in 20 and 1 mL of sorbitol 1 M, all processes realized in ice-cold. The transformation system of 10 µg of linearized DNA in 10 µL TE Buffer (10 mM Tris–HCl, 1 mM EDTA, pH 8.0) and 80 µL of competent *P. pastoris* cells was incubated in ice-cold electroporation cuvette of 0.2 cm for 5 min in ice. The electroporation followed the parameters of 1.5 kV, 200 Ω and 25 µF, cells were immediately incubated with 1 mL of 1 M sorbitol and spread 200 µL of aliquots on RDB plates (regeneration dextrose base, 9.3:1 sorbitol/agar, w/w) without histidine. The plates were incubated at 30 °C until colonies appear [8]. The transformation was confirmed by PCR and electrophorese in agarose gel 1.5% of the clones.

Expression of the recombinant peptide
A recombinant colony was inoculated in buffered minimal glycerol (BMG) medium containing 100 mM potassium phosphate, pH 6.0, 1.34% (w/v) of yeast nitrogen base (YNB), 4×10^{-5}% (w/v) of biotin, and 1% glycerol, for two days at 28 °C. Then, the cells were centrifuged and re-suspended in buffered minimal methanol (BMM) medium, composed of 100 mM of potassium phosphate, pH 6.0, 1.34% (w/v) YNB, 4×10^{-5}% (w/v) biotin, and methanol 0.5%, in order to avoid the action of proteases [11]. at the optical density $OD_{600} = 1.0$, under the temperature of 28 °C, at 250 rpm.

To better verify the induction time of expression rates, supernatants were collected at times 0, 6, 12, 24, 36, 48, 60, 72 and 96 (h) with the addition of 0.5% (w/v) of methanol at times 0, 24, 48 and 72 (h). The expression confirmation was analyzed by Tricine-SDS-PAGE electrophoresis [12], and silver-stained.

Growth inhibition test
The inhibition test was done in 96-well microtiter plate containing, in the first row, 200 µL of LB, as negative control. A mix containing 200 µL of LB with *E. coli* DH5α, in

an $OD_{595} = 0.3$, was added in the second row as first positive control (only *E. coli*). In the third, fourth and fifth rows was added 200 μL of LB with *E. coli* and new lyophilized BMM medium in quantities of 1, 10 and 25 μg, respectively, control for supernatant without recombinant peptide (named BMM). In the sixth, seventh and eighth rows were added 200 μL of LB with *E. coli* and BMM medium with abaecin in quantities of 1, 10 and 25 μg, respectively. The test was run in quadruplicated and the average of OD values from the negative control were subtracted from all wells.

Abaecin peptide modeling

The abaecin peptide was modeled by ab initio method, which is used when there is little or no initial information about the molecule structure in the data banks. The Rosetta 3.5 software [13], specifically the protocol of the AbinitioRelax software [14–19] was used to this goal. Approximately 2000 models were generated, one being selected after evaluation and validation using the dDFIRE [20]. The dDFIRE generates a score based on the free energy of the structure, therefore we select, among the thousands of models, one who has minor free energy. The peptide was visualized using the Chimera 1.1 software [21], where views of surface of the Coulombic electrostatic potential and amino acids hydrophobicity were generated.

Statistical analysis

The data obtained was analyzed using the Prism 4.0 software (GraphPad, San Diego, CA). After verification of data distribution, it was used analysis of variance for repeated variables, and Bonferroni tests to compare the obtained p value ($P < 0.05$).

Results

Gene design and synthesis and codon optimization

After the deduction and optimization of the new gene sequence, the abaecin of *A. mellifera* and the modified abaecin sequences were aligned by Clustal W software to check the non-modified regions and the optimized codons. The chemically synthesized gene of abaecin, with the size of 143 bp, was cloned into pUC57 clone vector.

Expression of recombinant peptide

The colonies of *P. pastoris* transformed with pPic9abaecin were grown in BMG medium for development and then in BMM for expression. Supernatants were collected from the expression medium in different times (Fig. 1). At times T3 (72 h) and T4 (96 h), it is possible to visualize the 5.2 kDa band showing that the selected colony produced the isolated peptide, not excluding the possibility of also containing peptides in the truncated form.

Inhibition growth test

Due to non-purification of the peptide in the present study, we adapted the antibiogram assay using the crude expression supernatant. After the confirmation of the production of recombinant peptide, the supernatant containing peptide was lyophilized and tested in *E. coli* culture in 96-well microtiter plate for antimicrobial activity confirmation.

The test showed that the quantities of 10 and 25 μg of lyophilized medium, containing the recombinant peptide, was sufficient to significantly inhibit the *E. coli* growth after 24 h treatment (Fig. 2b, c). There was no interference in the abaecin peptide inhibition by the

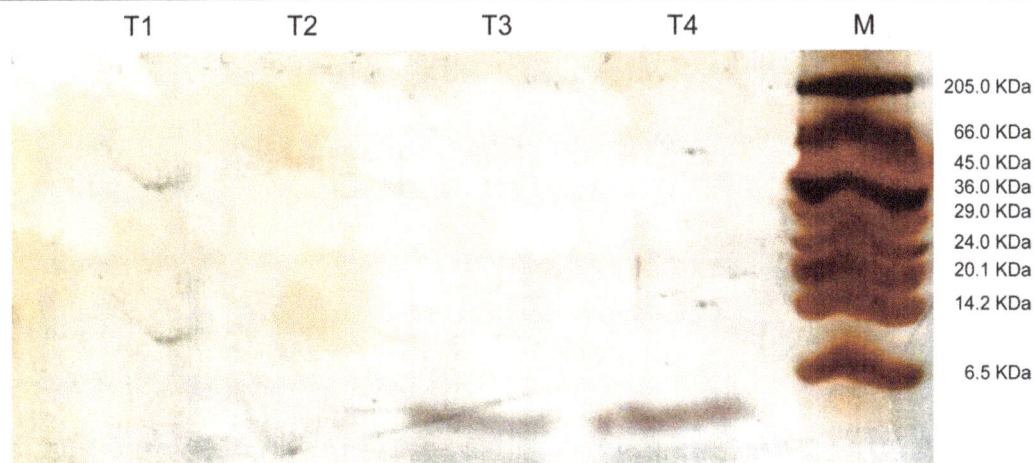

Fig. 1 Electrophoresis Tricine-SDS-PAGE, silver-stained. Expression time: T1 (24 h), T2 (48 h), T3 (72 h), T4 (96 h) with addition of 0.5% of methanol at all times. It shows the production of the peptide abaecin, of approximately 5.2 kDa, migrating lower than the last band of the marker (M) of 6.5 kDa. The beginning of the peptide's expression happened at the 72 h of incubation under 30 °C

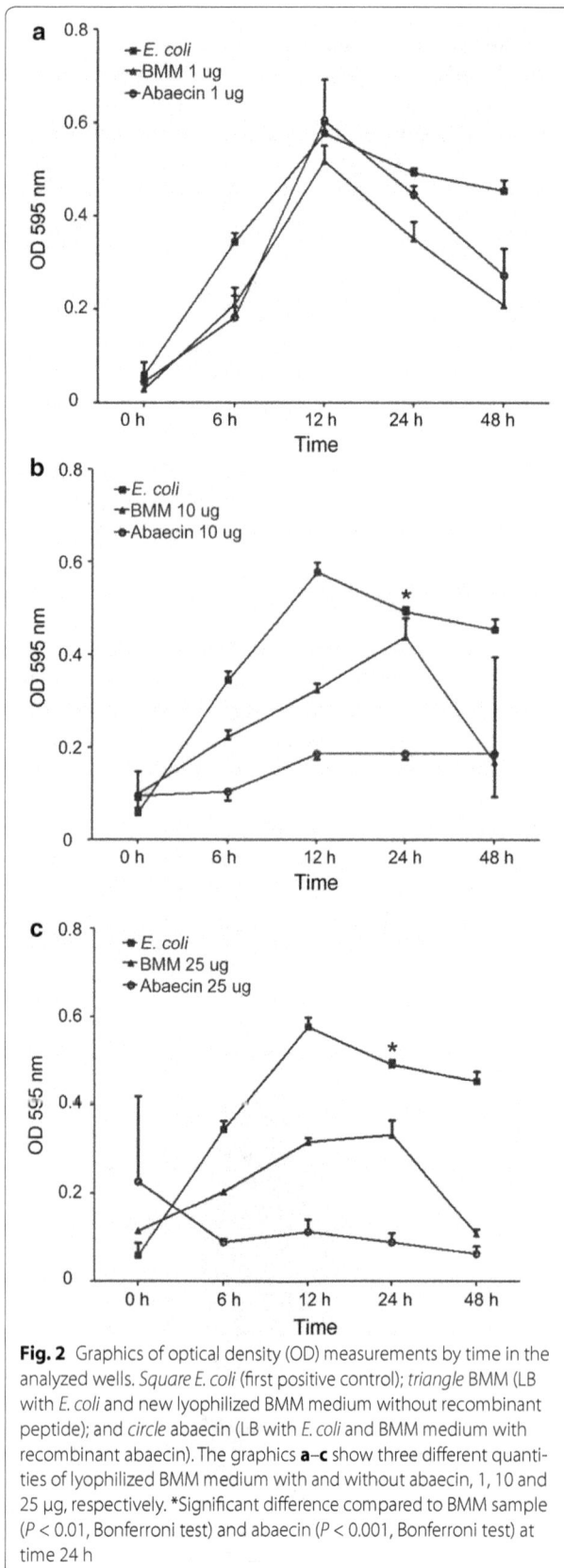

Fig. 2 Graphics of optical density (OD) measurements by time in the analyzed wells. *Square E. coli* (first positive control); *triangle* BMM (LB with *E. coli* and new lyophilized BMM medium without recombinant peptide); and *circle* abaecin (LB with *E. coli* and BMM medium with recombinant abaecin). The graphics **a–c** show three different quantities of lyophilized BMM medium with and without abaecin, 1, 10 and 25 µg, respectively. *Significant difference compared to BMM sample (*P* < 0.01, Bonferroni test) and abaecin (*P* < 0.001, Bonferroni test) at time 24 h

lyophilized BMM medium. The quantity of 1 µg did not inhibit bacterial growth (Fig. 2a).

The statistical test showed significant difference in rAbaecin activity in relation to controls (*E. coli* and BMM only). It presented antibacterial activity in concentrations of 10 and 25 µg, respectively, after time 12, accentuating after 24 h.

Abaecin structure

The peptide sequence optimized abaecin (43 aas + histag) showed a α-helix structure in C-terminal portion composed by (KWPQGYHHHHHH) residues in red on structural analyses. The views of surface of the Coulombic electrostatic potential and amino acids hydrophobicity showed in (Fig. 3). This conformation suggests an interaction with the bacterial membrane.

Discussion

The antimicrobial effect of peptides, and their production, has been studied in the immunological system of animals, such as bees. A study on the bee specie *Bombus pascuorum* identified four types of antimicrobial peptides: defensin, hymenoptaecin, apidaecin and abaecin. These peptides were directly extracted from the animal's hemolymph, which were purified and tested in bacterial culture, presenting antimicrobial activity [22].

In wasps from the specie *Pteromalus puparum* an abaecin cDNA was identified. The protein was chemically synthetized and tested in *E. coli* culture using this cDNA sequence, showing inhibitory activity, and no hemolytic activity was shown [23].

Previous DNA recombination methods were heavily used to high levels of peptide expression [24, 25]. For a long time, the system of proteins expression by *E. coli* was used due to its easy maintenance and fast culture growth [26]. It was not possible to use *E. coli* as an expression system because the recombining product is toxic to the bacteria. Because of that, the *P. pastoris* expression system was used.

The *P. pastoris* yeast, being a eukaryote, has advantages in the protein expression and processing, being able to make post-translational modifications. Another great advantage is the fact that, it is easier to manipulate when compared to both *E. coli* and *Saccharomyces cerevisiae* expression. The methodology used is also faster and cheaper compared to other expression systems like baculovirus or other culture tissues, allowing a high expression index [8, 27, 28].

The use of *P. pastoris* yeast has been used as an important protein expression method. Studies of antimicrobial peptides based on shrimp [29], butterfly, frog [30], *Drosophila melanogaster*, spine soldier bug [31] and human [32] genes have proven the effectiveness of this system for the production of these peptides without affecting

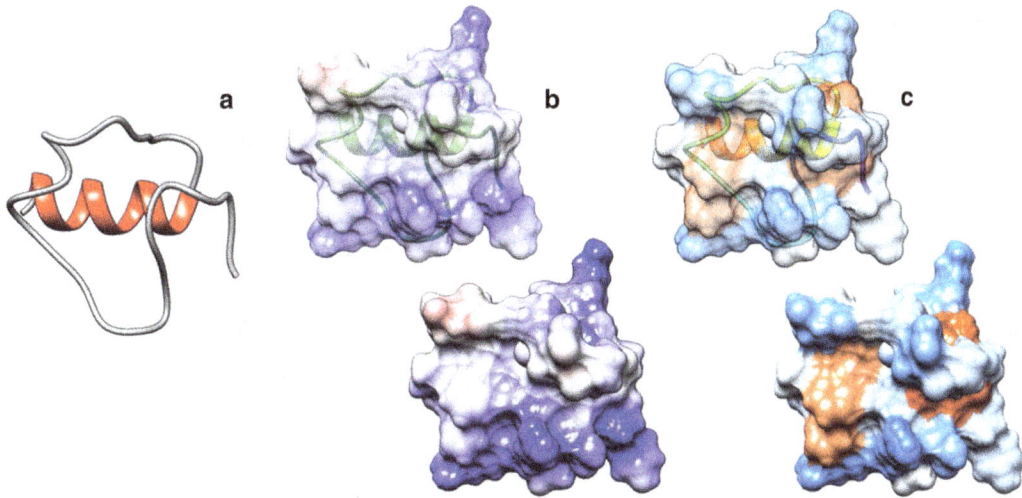

Fig. 3 Structural analysis of the heterologous peptide. In **a** structure abaecin with α-helix in red, in **b** analysis of electrostatic surface: positive charges in blue and negative charges in red, and **c** hydrophobicity surface analysis, warmer colors indicate hydrophobic regions

their inhibitory activity. Here we describe a production of recombinant abaecin (from *A. mellifera*) in *P. pastoris* host.

The chemical synthesis of the abaecin gene from *A. mellifera*, to express the peptide in *P. pastoris* yeast, allowed the production of the antimicrobial peptide. This system, which has shown capacity of a fast production and easy maintenance, can be used to deliver the expected results. The use of pPIC9 vector of expression allowed the peptide to be expressed in an extracellular way, reducing the number of steps in the manipulation methodology. By using the expression optimization, it was possible to estimate the necessary time for the production.

In order to the abaecin peptide be tested in *E. coli* samples, the optimum incubation period was 24 h. After this time the OD of the analyzed samples decreases, supposedly because bacterial growth reaches a high concentration, causing bacterial death due to the lack of supply and low environmental conditions in the culture mediums [33]. The treated samples with the abaecin in the lyophilized BMM medium peptide at 10 μg, showed an antibacterial activity after 12 h keeping an OD lower than 0.2. On the other hand, the sample treated with 25 μg showed antibacterial activity profile between 6 and 12 h with an OD close to 0.1 (Fig. 2).

The antibacterial activity depend on the hydrophobicity and the structural ability to assume an amphipathic helical conformation [34, 35]. Other patterns affect the potency and spectrum of the AMPs α-helix, such as sequence, charge and amphipathicity. These features are interrelated and they are the key for the development of more potent AMPs with a straight action [36, 37].

Antimicrobial peptides with more than 20 aas proline rich, like abaecin, showed effectiveness against gram-positive bacteria and fungi [38–40]. In another study, the chemically synthetized abaecin peptide, based on *B. pascuorum*, didn't show any effect without the aid of another peptide. When combined with other peptide, it exhibited action against *E. coli* growth [41]. Our heterologous abaecin, expressed in the *P. pastoris* system, showed antibacterial activity without the use of another AMP.

The charge of the proline-rich AMPs peptides (verified by in silico modeling) may provide the crossing of the membrane cell in non-lytic manner, interacting with proteins from the cytoplasm of the bacterium hampering protein synthesis [42]. This process shortens the quantity of the total bacteria in the samples, showed by the decreasing levels of the OD in the reading 595 nm in the test (Fig. 2).

It was shown that the methodology used facilitated the results of the proposed study. The chemical synthesis based in the sequence of the abaecin gene of *A. mellifera*, and optimized to *P. pastoris* (pPic9abaecin) proved to be efficient inside the expression system. The pPIC9 vector was of great importance to simplify the steps of the process, enabling it to obtain the peptide abaecin in the supernatant expression medium. The rAbaecin expressed in a heterologous manner at time 72 h, showing that there is an inhibition variation of microbial growth of *E. coli* DH5α bacteria within both time 6, 12 and 24 h.

Further studies are needed to the large-scale production and purification of this compound, in order to produce an antibiotic with such elements.

This results also shows that the heterologous abaecin peptide has antimicrobial activity against *E. coli*, and have biotechnological potential for the production of new antimicrobial drug, which acts against bacteria resistance to current drugs.

Abbreviations
AMP: antimicrobial peptides; AOX 1: alcohol oxidase 1 promoter; pPic9aba-cin: pPIC9 vector-insert; rAbaecin: recombinant abaecin; YPD: yeast extract peptone dextrose; RDB: regeneration dextrose base; BMG: buffered minimal glycerol; YNB: yeast nitrogen base; BMM: buffered minimal methanol.

Authors' contributions
CUV and JFA designed the experiments, DPL and JFA conducted most of the experiments. NNJ modeled the peptide structure. DPL assayed the recombinant protein expression and growth inhibition test. DPL wrote this manuscript and CUV, LRG and JFA helped to revise. All authors read and approved the final manuscript.

Author details
[1] Genetics Laboratory, Institute of Genetics and Biochemistry, Federal University of Uberlândia, 1720 Pará, Uberlândia, MG 38400902, Brazil. [2] Nanobiotechnology Laboratory, Institute of Genetics and Biochemistry, Federal University of Uberlândia, 1720 Pará, Uberlândia, MG 38400902, Brazil.

Acknowledgements
We would like to express our gratitude to all staff from the Laboratory of Biochemistry and Animal Toxins (LaBiTox) and Nanobiotechnology Laboratory at UFU for the support and availability of the equipment used.

Competing interests
The authors declare that they have no competing interests.

Funding
This work was supported by the Coordination of Improvement of Higher Education Personnel (CAPES); National Counsel of Technological and Scientific Development (CNPq, Grant Number 445679/2014-0) and Minas Gerais State Research Foundation (Fapemig).

References
1. Jiravanichpaisal P, Lee BL, Söderhäll K. Cell-mediated immunity in arthropods: hematopoiesis, coagulation, melanization and opsonization. Immunobiology. 2006;211:213–36.
2. Strand MR. The insect cellular immune response. Insect Sci. 2008;15:1–14.
3. Arrese EL, Soulages JL. Insect fat body: energy, metabolism, and regulation. Annu Rev Entomol. 2010;55:207–25.
4. Izadpanah A, Gallo RL, Diego S. Antimicrobial peptides. J Am Acad Dermatol. 2005;52:381–90.
5. Chou PY, Fasman GD. Prediction of the secondary structure of proteins from their amino acid sequence. Adv Enzymol Relat Areas Mol Biol. 1978;47:45–148.
6. Xu P, Shi M, Chen X-X. Antimicrobial peptide evolution in the Asiatic honey bee Apis cerana. PLoS ONE. 2009;4:e4239.
7. Romanos MA, Scorer CA, Clare JJ. Foreign gene expression in yeast: a review. Yeast. 1992;8:423–88.
8. Pichia Expression Kit, for expression of recombinant proteins in Pichia pastoris, Catalog No. K1710-01 [Internet]. 2014. p. 98. Available from: https://tools.thermofisher.com/content/sfs/manuals/pich_man.pdf. Accessed 5 Apr 2017.
9. Alves D, Olívia Pereira M. Mini-review: antimicrobial peptides and enzymes as promising candidates to functionalize biomaterial surfaces. Biofouling. 2014;30:483–99.
10. Sharp PM, Li WH. The codon adaptation index-a measure of directional synonymous codon usage bias, and its potential applications. Nucleic Acids Res. 1987;15:1281–95.
11. Clare JJ, Romanes MA, Rayment FB, Rowedder JE, Smith MA, Payne MM, et al. Production of mouse epidermal growth factor in yeast: high-level secretion using Pichia pastoris strains containing multiple gene copies. Gene. 1991;105:205–12.
12. Schagger H. Tricine-SDS-page. Nat Protoc. 2006;1:16–22.
13. Rohl CA, Strauss CEM, Misura KMS, Baker D. Protein structure prediction using Rosetta. Methods Enzymol. 2004;383:66–93.
14. Raman S, Vernon R, Thompson J, Tyka M, Sadreyev R, Pei J, et al. Structure prediction for CASP8 with all-atom refinement using Rosetta. Proteins Struct Funct Bioinforma. 2009;77:89–99.
15. Bradley P, Misura KMS, Baker DB. Toward high-resolution de novo structure prediction for small proteins. Science. 2005;309:1868–71.
16. Bonneau R, Strauss CEM, Rohl CA, Chivian D, Bradley P, Malmström L, et al. De novo prediction of three-dimensional structures for major protein families. J Mol Biol. 2002;322:65–78.
17. Bonneau R, Tsai J, Ruczinski I, Chivian D, Rohl C, Strauss CEM, et al. Rosetta in CASP4: progress in ab initio protein structure prediction. Proteins Struct Funct Genet. 2001;45:119–26.
18. Simons KT, Ruczinski I, Kooperberg C, Fox BA, Bystroff C, Baker D. Improved recognition of native-like protein structures using a combination of sequence-dependent and sequence-independent features of proteins. Proteins Struct Funct Genet. 1999;34:82–95.
19. Simons KT, Kooperberg C, Huang E, Baker D. Assembly of protein tertiary structures from fragments with similar local sequences using simulated annealing and Bayesian scoring functions. J Mol Biol. 1997;268:209–25.
20. Yang Y, Zhou Y. Specific interactions for ab initio folding of protein terminal regions with secondary structures. Proteins Struct Funct Genet. 2008;72:793–803.
21. Pettersen EF, Goddard TD, Huang CC, Couch GS, Greenblatt DM, Meng EC, et al. UCSF chimera—a visualization system for exploratory research and analysis. J Comput Chem. 2004;25:1605–12.
22. Rees JA, Moniatte M, Bulet P. Novel antibacterial peptides isolated from a European bumblebee, Bombus pascuorum (Hymenoptera, apoidea). Insect Biochem Mol Biol. 1997;27:413–22.
23. Shen X, Ye G, Cheng X, Yu C, Altosaar I, Hu C. Characterization of an abaecin-like antimicrobial peptide identified from a Pteromalus puparum cDNA clone. J Invertebr Pathol. 2010;105:24–9.
24. Reichhart JM, Meister M, Dimarcq JL, Zachary D, Hoffmann D, Ruiz C, et al. Insect immunity: developmental and inducible activity of the Drosophila diptericin promoter. EMBO J. 1992;11:1469–77.
25. Lee JH, Kim JH, Hwang SW, Lee WJ, Yoon HK, Lee HS, et al. High-level expression of antimicrobial peptide mediated by a fusion partner reinforcing formation of inclusion bodies. Biochem Biophys Res Commun. 2000;277:575–80.
26. Huang L, Leong SSJ, Jiang R. Soluble fusion expression and characterization of bioactive human beta-defensin 26 and 27. Appl Microbiol Biotechnol. 2009;84:301–8.
27. Buckholz RG, Gleeson MA. Yeast systems for the commercial production of heterologous proteins. Nat Biotechnol. 1991;9:1067–72.
28. Cregg JM, Vedvick TS, Raschke WC. Recent advances in the expression of foreign genes in Pichia pastoris. Nat Biotechnol. 1993;11:905–10.
29. Li L, Wang J-X, Zhao X-F, Kang C-J, Liu N, Xiang J-H, et al. High level expression, purification, and characterization of the shrimp antimicrobial peptide, Ch-penaeidin, in Pichia pastoris. Protein Expr Purif. 2005;39:144–51.
30. Jin F, Xu X, Wang L, Zhang W, Gu D. Expression of recombinant hybrid peptide cecropinA(1-8)-magainin2(1-12) in Pichia pastoris: purification and characterization. Protein Expr Purif. 2006;50:147–56.
31. Sang Y-X, Deng X-J, Yang W-Y, Wang W-X, Wen S-Y, Liu W-Q, et al. Secretive expression of insect antifungal peptide-encoded genes in Pichia pastoris and activity assay of the products. Agric Sci China. 2007;6:1209–16.

32. Chen Z, Wang D, Cong Y, Wang J, Zhu J, Yang J, et al. Recombinant anti-microbial peptide hPAB-β expressed in *Pichia pastoris*, a potential agent active against methicillin-resistant *Staphylococcus aureus*. Appl Microbiol Biotechnol. 2011;89:281–91.

33. Robinson TP, Aboaba OO, Kaloti A, Ocio MJ, Baranyi J, Mackey BM. The effect of inoculum size on the lag phase of *Listeria monocytogenes*. Int J Food Microbiol. 2001;70:163–73.

34. Blondelle SE, Houghten RA. Hemolytic and antimicrobial activities of the twenty-four individual omission analogues of melittin. Biochemistry. 1991;30:4671–8.

35. Blondelle SE, Simpkins LR, Pérez-Payá E, Houghten RA. Influence of tryptophan residues on melittin's hemolytic activity. Biochim Biophys Acta (BBA)/Protein Struct Mol. 1993;1202:331–6.

36. Saberwal G, Nagaraj R. Cell-lytic and antibacterial peptides that act by perturbing the barrier function of membranes: facets of their conformational features, structure-function correlations and membrane-perturbing abilities. Biochim Biophys Acta (BBA)-Rev Biomembr. 1994;1197:109–31.

37. Maloy WL, Kari UP. Structure-activity studies on magainins and other host-defense peptides. Biopolymers. 1995;37:105–22.

38. Otvos L. The short proline-rich antibacterial peptide family. Cell Mol Life Sci. 2002;59(7):1138–50.

39. Rahnamaeian M, Langen G, Imani J, Khalifa W, Altincicek B, Von Wettstein D, et al. Insect peptide metchnikowin confers on barley a selective capacity for resistance to fungal ascomycetes pathogens. J Exp Bot. 2009;60:4105–14.

40. Rahnamaeian M, Vilcinskas A. Defense gene expression is potentiated in transgenic barley expressing antifungal peptide metchnikowin throughout powdery mildew challenge. J Plant Res. 2012;125:115–24.

41. Rahnamaeian M, Cytryńska M, Zdybicka-Barabas A, Vilcinskas A. The functional interaction between abaecin and pore-forming peptides indicates a general mechanism of antibacterial potentiation. Peptides. 2016;78:17–23.

42. Rahnamaeian M, Cytryńska M, Zdybicka-Barabas A, Dobslaff K, Wiesner J, Twyman RM, et al. Insect antimicrobial peptides show potentiating functional interactions against gram-negative bacteria. Proc R Soc Lond B. 2015;282:20150293.

17

In vivo plug-and-play: a modular multi-enzyme single-cell catalyst for the asymmetric amination of ketoacids and ketones

Judith E. Farnberger[1], Elisabeth Lorenz[2], Nina Richter[1], Volker F. Wendisch[2]* (ID) and Wolfgang Kroutil[1,3]* (ID)

Abstract

Background: Transaminases have become a key tool in biocatalysis to introduce the amine functionality into a range of molecules like prochiral α-ketoacids and ketones. However, due to the necessity of shifting the equilibrium towards the product side (depending on the amine donor) an efficient amination system may require three enzymes. So far, this well-established transformation has mainly been performed in vitro by assembling all biocatalysts individually, which comes along with elaborate and costly preparation steps. We present the design and characterization of a flexible approach enabling a quick set-up of single-cell biocatalysts producing the desired enzymes. By choosing an appropriate co-expression strategy, a modular system was obtained, allowing for flexible plug-and-play combination of enzymes chosen from the toolbox of available transaminases and/or recycling enzymes tailored for the desired application.

Results: By using a two-plasmid strategy for the recycling enzyme and the transaminase together with chromosomal integration of an amino acid dehydrogenase, two enzyme modules could individually be selected and combined with specifically tailored E. coli strains. Various plug-and-play combinations of the enzymes led to the construction of a series of single-cell catalysts suitable for the amination of various types of substrates. On the one hand the fermentative amination of α-ketoacids coupled both with metabolic and non-metabolic cofactor regeneration was studied, giving access to the corresponding α-amino acids in up to 96% conversion. On the other hand, biocatalysts were employed in a non-metabolic, "in vitro-type" asymmetric reductive amination of the prochiral ketone 4-phenyl-2-butanone, yielding the amine in good conversion (77%) and excellent stereoselectivity (ee = 98%).

Conclusions: The described modularized concept enables the construction of tailored single-cell catalysts which provide all required enzymes for asymmetric reductive amination in a flexible fashion, representing a more efficient approach for the production of chiral amines and amino acids.

Keywords: Biocatalysis, Single-cell biotransformation, *Escherichia coli*, Asymmetric reductive amination, Transaminases, Chiral amines, α-Amino acids, Modular concept, Flexibility

Background

Enantiopure α-amino acids [1, 2] and amines [3, 4] represent classes of chiral chemicals with versatile application in chiral pharmaceutical and asymmetric synthesis. Biocatalysis [5] has contributed remarkably to the development of economically feasible and sustainable methods for the preparation of these compounds: examples of enzymes comprise imine reductases [6–8], monoamine oxidases [9, 10], amine dehydrogenases [11–14], hydrolytic enzymes [15, 16] or transaminases (TAs) [17–19] for asymmetric functionalization of keto groups. The latter are pyridoxal-phosphate dependent enzymes catalyzing simplified the amino group transfer from a donor

*Correspondence: volker.wendisch@uni-bielefeld.de; wolfgang.kroutil@uni-graz.at
[2] Genetics of Prokaryotes, Faculty of Biology & CeBiTec, Bielefeld University, 33501 Bielefeld, Germany
[3] Institute of Chemistry, University of Graz, NAWI Graz, BioTechMed Graz, Heinrichstrasse 28, 8010 Graz, Austria
Full list of author information is available at the end of the article

amine to a keto group to produce a chiral amine with a new stereocenter and thus a molecule with increased value. An efficient TA-catalyzed amination depends on the removal or recycling of the co-product to shift the reaction equilibrium towards the product side. A range of well-working techniques has been developed like the coupling of L-alanine dependent TAs with an alanine dehydrogenase (AlaDH) and an enzyme for nicotinamide cofactor regeneration, which can be for instance formate dehydrogenase or glucose dehydrogenase [20, 21]. This orthogonal three-enzyme cascade has successfully demonstrated its usefulness in a range of TA-catalyzed transformations and hence has occupied a permanent spot in cell-free biocatalysis involving TAs. Nevertheless, the assembly of multiple biocatalysts in vitro, which is generally termed "systems biocatalysis" [22–24], implies the individual production of each required enzyme, which is time-consuming and lacks elegance. Consequently, considerable effort has been dedicated to the development of microbial cell factories, which are designed to perform multistep biotransformations in vivo. However, instead of exploiting native or engineered metabolic pathways [25, 26], focus has lately been put on the introduction of whole artificial de novo pathways [27–30]. This approach leads to tailored single-cell biocatalysts, which excel related cell-free systems by offering a more inexpensive and easy catalyst preparation and a simplified overall process configuration. Required cofactors can be provided and/or regenerated by the cell using an internal cofactor recycling system coupled to the host's metabolism. Furthermore, the close proximity of biocatalysts within the confined space of one cell enables consecutive or even concurrent reaction steps in a highly efficient way [31]. Such in vivo cascades have been successfully developed for a variety of useful applications and are highlighted as well as opposed to in vitro approaches in several recent in-depth reviews [32–35]. While at the beginning mostly redox reactions including dehydrogenases, reductases and monooxygenases were considered for tailor-made designer cells [36–39], recently also TA-catalyzed reductive amination attracted attention to being included in artificial in vivo pathway construction [40, 41]. In this context especially the amino functionalization of alcohols by combining an alcohol dehydrogenase (ADH) with a TA in a redox self-sufficient fashion has been extensively studied [42, 43]. Moreover, transamination was investigated with recombinant yeast single-cells of *Saccharomyces cerevisiae*, exploiting cell metabolism for cofactor regeneration [44, 45]. To our knowledge however, there are no reports on transferring the classic TA-catalyzed reductive amination machinery into a single recombinant *E. coli* cell. Therefore, in the present study the orthogonal cascade was translated from an in vitro to an in vivo

mediated approach by generating all required enzymes within a bacterial single-cell system. In order to allow for a flexible interplay of multiple enzyme-catalyzed reactions within one cell, thorough catalyst design as well as a careful choice of co-expression strategy were crucial parameters. The microbial cell factory contained three basic enzyme modules, each of them catalyzing one of the reactions constituting the targeted orthogonal cascade: (1) asymmetric reductive amination, (2) amine donor regeneration and (3) nicotinamide cofactor regeneration. Each module offered various enzyme options, which were combined in a plug-and-play fashion for various demands, enabling broad applicability. The obtained toolbox of single-cell biocatalysts was successfully employed for the asymmetric reductive amination of α-ketoacids and prochiral ketones, investigating both fermentative transformations coupled to the host's metabolism and non-metabolic "in vitro-type" transformations.

Results

Design and construction of modular plug-and-play single-cell biocatalysts

The aim was the development of a recombinant *E. coli* single-cell system providing all required enzymes for the asymmetric reductive amination of prochiral α-ketoacids and ketones in a flexible plug-and-play fashion. The system consisted of three modules, each of them representing one essential reaction type of the amination system, namely (i) the amination of the target substrate (module I), (ii) the recycling of the amine donor molecule (module II) and (iii) the recycling of the reduced nicotinamide cofactor (module III) (Fig. 1). For each module a library of different constructs encoding specific enzymes was available. Upon combination of selected compatible constructs from each module a single bacterial cell was generated, capable of catalyzing the targeted orthogonal cascade reaction. For module I in our representative study one out of four alternative constructs encoding enzymes for the reductive amination of keto-functionalities can be chosen: a L-alanine–valine aminotransferase (encoded by *avtA*), a branched-chain amino acid aminotransferase either from *E. coli* MG1655 (encoded by $ilvE_{Ec}$) or *Streptococcus mutans* (encoded by $ilvE_{Sm}$) for the production of various α-amino acids or a (S)-selective transaminase from *Chromobacterium violaceum* [46] (encoded by ta_{Cv}), giving access to a broad range of optically pure amines. The required enzymes for regenerating the consumed amine donor are supplied by module II. While AvtA and Ta_{Cv} rely on L-alanine as amine donor, IlvE is consuming L-glutamate. Consequently, an L-alanine dehydrogenase (AlaDH) and a L-glutamate dehydrogenase (GluDH) from *Bacillus subtilis* were determined as the enzymes of module II. The third module finally

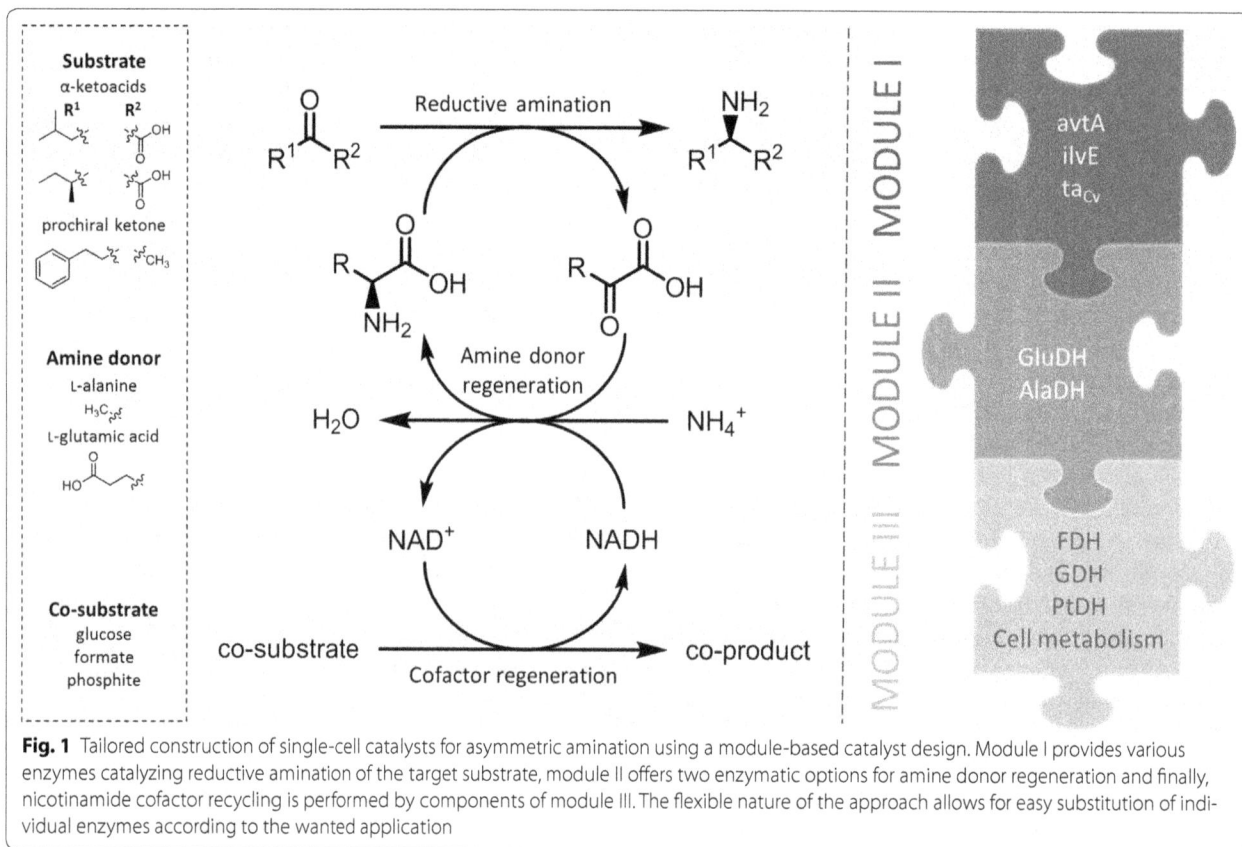

Fig. 1 Tailored construction of single-cell catalysts for asymmetric amination using a module-based catalyst design. Module I provides various enzymes catalyzing reductive amination of the target substrate, module II offers two enzymatic options for amine donor regeneration and finally, nicotinamide cofactor recycling is performed by components of module III. The flexible nature of the approach allows for easy substitution of individual enzymes according to the wanted application

provided various options for regenerating the nicotinamide cofactor NADH required for the oxidation reaction catalyzed by AlaDH and GluDH. NADH can either be regenerated in the traditional way by cellular metabolism, using a co-substrate like glucose, or by an enzyme-coupled approach. For the latter option, a toolbox was generated consisting of three constructs encoding either a formate dehydrogenase (FDH) from *Komagataella pastoris* GS115, a glucose dehydrogenase (GDH) from *Bacillus megaterium* and a phosphite dehydrogenase (PtDH) from *Pseudomonas stutzeri*. In order to enable a flexible catalyst design allowing for easy plug-and-play-like combination of desired enzymes, the corresponding genes had to be controlled separately from each other within one cell. Therefore, co-expression of the enzymes required for modules I–III was set-up by combining a two-plasmid strategy with the integration of one gene into the chromosome of the used *E. coli* host (Fig. 2). Genes encoding transaminases were individually cloned into the IPTG-inducible *E. coli* expression vector pTrc99A (plasmid 1) under the control of the P*trc* promoter and ribosome binding sites upstream of each gene (pTrc99A-*avtA*, pTrc99A-*ilvE*, pTrc99A-*ta*$_{Cv}$). Genes coding for AlaDH and GluDH were individually integrated into the genome of the used *E. coli* strain by homologous recombination

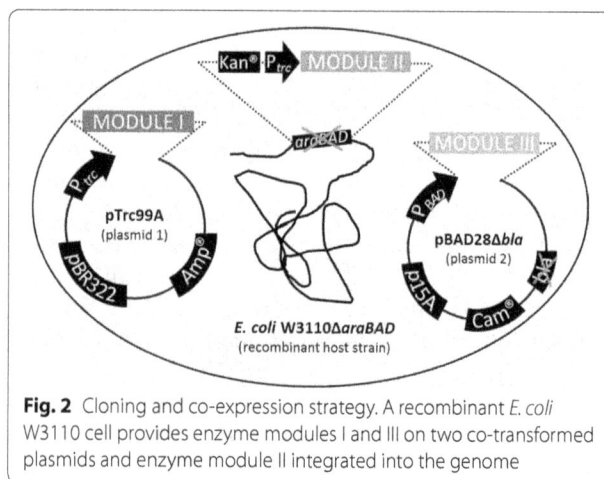

Fig. 2 Cloning and co-expression strategy. A recombinant *E. coli* W3110 cell provides enzyme modules I and III on two co-transformed plasmids and enzyme module II integrated into the genome

resulting in two hosts. For this purpose, the *araBAD* genes in the chromosome of *E. coli* W3110 were replaced with heterologous *ald* (AlaDH) or *rocG* (GluDH) gene, respectively under the control of IPTG-inducible strong P*trc* promoter [47] and ribosome binding sites upstream of each gene. The new strains were named *Ec*-AlaDH (*E. coli* W3110Δ*araBAD*::P*trc*-*ald*) and *Ec*-GluDH (*E. coli* W3110Δ*araBAD*::P*trc*-*rocG*) according to the amino acid dehydrogenase they provided. The genes encoding NADH-recycling enzymes were cloned into pBAD28 (plasmid 2) under control of the arabinose-inducible the P*BAD* promoter [48] and ribosome binding sites upstream of each gene. Due to a different origin of replication (p15Aori) and a chloramphenicol resistance, pBAD28 provides the appropriate features to be compatible with pTrc99A (pBR322ori, *bla*). However, since it also contains the same β-lactamase gene, we removed parts of it in order to inactivate the ampicillin resistance (pBAD28Δ*bla*). The newly obtained plasmids were designated pBAD28Δ*bla-fdh*, pBAD28Δ*bla-gdh*, pBAD28Δ*bla-ptdh*.

By co-transformation of the specific recombinant strains with the required compatible plasmid(s), the different enzyme modules were assembled flexibly within one cell and gave rise to a range of single-cell catalysts useful for various applications (Table 1). Due to the fact that transaminase IlvE depends on L-glutamate as amine donor, plasmid pTrc99A-*ilvE* was transformed into strain *Ec*-GluDH. In a first approach the NADH recycling depended on cell catabolism (Table 1, catalyst **1a** and **b**). The other transaminases AvtA and Ta$_{Cv}$ rely on L-alanine as amino donor and were thus transformed into strain *Ec*-AlaDH.

While AvtA was coupled both with glucose catabolism (catalyst 2) as well as with a regenerating enzyme for nicotinamide recycling (catalysts 3–5), the (*S*)-selective transaminase Ta$_{Cv}$ was just used in combination with the enzyme-coupled approach (catalysts 6–8). With the constructed single-cell biocatalysts in hand, optimization of co-expression of multiple enzymes within one cell was required. In order to validate the production of active proteins and an ideal ratio of enzyme activities, we performed photometric activity assays individually for each enzyme, ensuring that all the genes were functionally expressed. Depending on the expression conditions used, crude extracts displayed specific enzyme activities of approximately 0.1–0.3 U mg^{-1} for transaminase Ta$_{Cv}$ and IlvE, whereas a significantly higher transaminase activity of 0.7–1 U mg^{-1} was obtained for AvtA. While GluDH expressed from genome-encoded *rocG* indicated a specific activity of 0.4 U mg^{-1}, for AlaDH expressed from the genome-encoded *ald* widely differing specific activities in the range of 0.12–0.36 U mg^{-1} were obtained. In order to increase the amount of AlaDH present in the cell and consequently also its activity, a second *ald* gene copy was introduced to the system via plasmid pTrc99A-*avtA-ald*, which has already been constructed previously [49]. Additionally, another vector with a different gene order was constructed in this study, where AlaDH was encoded in the first position of the artificial operon with the P*trc* promoter upstream of *ald* (pTrc99A-*ald-avtA*). Indeed, activity assays showed that AlaDH activity was enhanced manifold to 2.1 U mg^{-1} with the latter plasmid. In comparison, crude extracts of the strain DH5α carrying pTrc99A-*avtA-ald* displayed only 0.3 U mg^{-1}. This result showed that not only the increased number of plasmid-encoded *ald* genes, but especially the distance between promoter and gene in the operon significantly affected expression efficiency (operon polarity). Nevertheless, both plasmid-based gene orders of the artificial *avtA-ald* and *ald-avtA* operons, respectively were used for further experiments and compared to the genome-based approach. With respect to the third module for nicotinamide regeneration all the required enzymes were functionally expressed, with FDH offering a specific activity of approximately 0.2 and 0.6 U mg^{-1} for PtDH. Concerning GDH, values ranged from 1.7 to 11.9 U mg^{-1}

Table 1 Constructed single-cell biocatalysts by flexible in vivo assembly of three modules

Catalyst	*E. coli* strain	Plasmid encoding Ta	Cofactor recycling	Model product
1a	*Ec*-GluDH	pTrc99A-*ilvE$_{Ec}$*	Cell metabolism	L-Leucine
1b	*Ec*-GluDH	pTrc99A-*ilvE$_{Sm}$*	Cell metabolism	L-Leucine
2a	*Ec*-AlaDH	pTrc99A-*avtA*	Cell metabolism	L-Isoleucine
2b	*Ec*-AlaDH	pTrc99A-*avtA-ald*	Cell metabolism	L-Isoleucine
2c	*Ec*-AlaDH	pTrc99A-*ald-avtA*	Cell metabolism	L-Isoleucine
3a	*Ec*-AlaDH	pTrc99A-*avtA*	pBAD28Δ*bla-fdh*	L-Isoleucine
3b	*Ec*-AlaDH	pTrc99A-*ald-avtA*	pBAD28Δ*bla-fdh*	L-Isoleucine
4a	*Ec*-AlaDH	pTrc99A-*avtA*	pBAD28Δ*bla-gdh*	L-Isoleucine
4b	*Ec*-AlaDH	pTrc99A-*ald-avtA*	pBAD28Δ*bla-gdh*	L-Isoleucine
5a	*Ec*-AlaDH	pTrc99A-*avtA*	pBAD28Δ*bla-ptdh*	L-Isoleucine
5b	*Ec*-AlaDH	pTrc99A-*ald-avtA*	pBAD28Δ*bla-ptdh*	L-Isoleucine
6	*Ec*-AlaDH	pTrc99A-*ta$_{Cv}$*	pBAD28Δ*bla-fdh*	4-Phenyl-2-butylamine
7	*Ec*-AlaDH	pTrc99A-*ta$_{Cv}$*	pBAD28Δ*bla-gdh*	4-Phenyl-2-butylamine
8	*Ec*-AlaDH	pTrc99A-*ta$_{Cv}$*	pBAD28Δ*bla-ptdh*	4-Phenyl-2-butylamine

depending on the expression conditions used. However, GDH turned out to be the most active enzyme in terms of cofactor regeneration.

Influence and toxicity of used substrates on the viability of *E. coli* W3110

Substrates used in single-cell biotransformations as well as the corresponding obtained products might have toxic effects on the living host cell and thus limit the efficiency and turnover of the biocatalyst. The potential toxicity of several α-keto acids which are common substrates for transaminases AvtA and IlvE has already been assessed in previous work [49]. Since in the present study the single-cell catalyzed transamination reaction has been further investigated with an enzyme-coupled approach for cofactor regeneration, some more substrates had to be tested. For instance, PtDH from *Pseudomonas stuzeri* accepts phosphite or hypophosphite as substrate for the regeneration of NADH, whereas FDH from *Komagataella pastoris* requires formate. Therefore, sodium hypophosphite monohydrate, sodium phosphite dibasic pentahydrate, sodium phosphate monobasic monohydrate and sodium formate were added in varying concentrations to W3110 SGA minimal medium cultures at inoculation and growth was followed by Biolector® cultivation system. Both phosphite substrates as well as phosphates had low inhibitory effects on *E. coli* W3110 cells in a similar range (Fig. 3). Even more than a half of maximum growth rate (μ) was observed at a concentration of 100 mM phosphite, hypophosphite or phosphate, respectively. In contrast, toxic effects based on addition of formate were more significant as the cell growth was reduced drastically by the presence of only 20 mM formate. Hence, the application of PtDH together with phosphite might be a promising way for regenerating consumed cofactors in single-cell catalyzed amination reactions.

Fermentative amination of α-ketoacids

The successful assembly and co-expression of either transaminase AvtA or IlvE of module I with AlaDH or GluDH, respectively from module II gave rise to single-cell biocatalysts for one-pot transformations of α-ketoacids to α-amino acids in a fermentative fashion. In the first part of the study, cofactor regeneration was performed via the cell metabolism-coupled approach, using glucose as carbon source. As a second strategy, the enzyme-coupled strategy for cofactor regeneration was investigated and compared with previous results.

Fermentative amination of α-ketoacids with metabolic NADH regeneration

The first biocatalyst to reductively aminate α-ketoacids combined an endogenous (catalyst 1a) or a heterologous

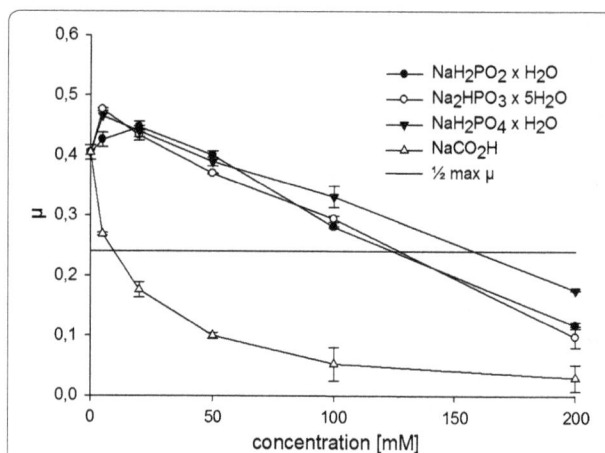

Fig. 3 Toxicity study. Growth of *E. coli* W3110 in SGA medium supplemented with varying concentrations of sodium hypophosphite monohydrate (*black circle*), sodium phosphite dibasic pentahydrate (*white circle*), sodium phosphate monobasic monohydrate (*Black inverted triangle*) and sodium formate (*white triangle*). The maximum growth rate (μ) is plotted against the concentration of substrates and products

L-glutamate dependent transaminase IlvE from *Streptococcus mutans* (catalyst 1b), respectively with NAD(P)H-dependent L-glutamate dehydrogenase (GluDH) from *B. subtilis*. Using L-glutamate as amine donor, IlvE transforms several 2-ketoacids to the corresponding amino acids, among them being the three branched-chain 2-keto-isocaproate (KIC), 2-keto-isovalerate (KIV) and 2-keto-3-methylvalerate (KMV), which are converted to L-leucine, L-valine and to L-isoleucine, respectively. Since preliminary studies (data not shown) revealed that the enzyme has its highest affinity for KIC, the biotransformation of KIC to L-leucine catalyzed by newly assembled *E. coli* single-cell catalysts **1a** (*Ec*-GluDH/pTrc99A-*ilvE*$_{Ec}$) and **1b** (*Ec*-GluDH/pTrc99A-*ilvE*$_{Sm}$) was investigated with and without the addition of amine donor (Fig. 4a). As a negative control *Ec*-GluDH strain transformed with empty pTrc99A vector was used. The cofactor NADH was regenerated by oxidative catabolism of the employed carbon source glucose. Fermentative biotransformations were performed with 50 mM KIC and L-leucine formation was monitored over a time period of 66 h. While catalyst **1a** showed only a low conversion of 16% after 43 h, catalyst **1b** with IlvE from *Streptococcus mutans* converted twice as much (32%) KIC to L-leucine within the same time. Remarkably, the negative control *Ec*-GluDH/pTrc99A without overexpressed transaminase showed 14% turnover after 43 h as well. However, addition of 250 mM L-glutamate improved the conversion of all strains overexpressing a transaminase to 90–92%, while 36% conversion was obtained for the control strain after

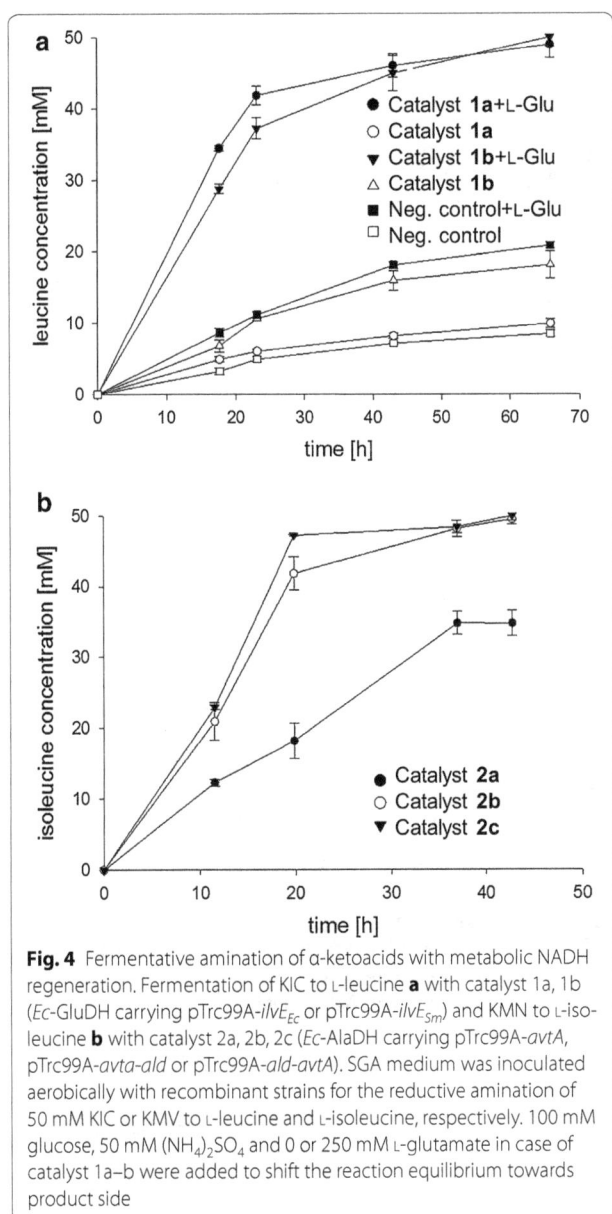

Fig. 4 Fermentative amination of α-ketoacids with metabolic NADH regeneration. Fermentation of KIC to L-leucine **a** with catalyst 1a, 1b (Ec-GluDH carrying pTrc99A-$ilvE_{Ec}$ or pTrc99A-$ilvE_{Sm}$) and KMN to L-isoleucine **b** with catalyst 2a, 2b, 2c (Ec-AlaDH carrying pTrc99A-$avtA$, pTrc99A-$avta$-ald or pTrc99A-ald-$avtA$). SGA medium was inoculated aerobically with recombinant strains for the reductive amination of 50 mM KIC or KMV to L-leucine and L-isoleucine, respectively. 100 mM glucose, 50 mM $(NH_4)_2SO_4$ and 0 or 250 mM L-glutamate in case of catalyst 1a–b were added to shift the reaction equilibrium towards product side

work, however, the concept was extended to a more flexible set-up with the *ald* gene being integrated into the host genome. Furthermore, the contribution of another *ald* gene copy being plasmid-encoded together with *avtA* as well as the effect of its cloning position was studied. It was assumed that the expression level of AlaDH played an important role in the turnover of the substrate. Biotransformations were again performed with 50 mM KMV in a fermentative way making use of the bacterial glucose catabolism for NADH recycling and the formation of L-isoleucine was monitored over 43 h (Fig. 4b). Employing catalyst **2a** (Ec-AlaDH/pTrc99A-$avtA$) for the fermentation, which offers only one genome-integrated copy of *ald*, moderate 68% of KMV were converted to L-isoleucine. However, when additional plasmid-born AlaDH was provided, conversion increased to 84% within 20 h using catalyst **2b** (Ec-AlaDH/pTrc99A-$avtA$-ald) and even 96% of product were formed with the help of catalyst **2c** (Ec-AlaDH/pTrc99A-ald-$avtA$) where a different gene order has been used. This strongly indicated that high expression of *ald* gene promoted the turnover of KMV to L-isoleucine, which coincided with the results obtained for catalyst **1**. The negative control Ec-AlaDH/pTrc99A did not produce any L-isoleucine. Due to the higher efficiency of the latter enzyme combination for reductive amination of α-ketoacids to α-amino acids, the single-cell catalyst composed of AvtA and AlaDH was chosen to be further investigated with the enzyme-coupled approach for cofactor regeneration based on GDH, FDH or PtDH.

43 h. Therefore, the presence of additional amine donor seems to play a key role under present reaction conditions, which might be explained by a too low expression level of GluDH and thus an insufficient regeneration of consumed L-glutamate.

The second single-cell biocatalyst for the fermentative production of amino acids relied on endogenous L-alanine dependent transaminase C (AvtA) co-expressed with NADH-dependent L-alanine dehydrogenase (AlaDH) from *B. subtilis*. These enzymes have already been investigated successfully for the single-cell catalyzed reductive amination of KMV to L-isoleucine using a two-plasmid based approach before [49]. In the present

Fermentative reductive amination of α-ketoacids with non-metabolic NADH regeneration

The tailor-made single-cell catalyst for reductive amination of KMV to L-isoleucine composed of AvtA and AlaDH was studied in more detail, using an enzyme-coupled approach for NADH cofactor regeneration instead of cellular oxidation of glucose. The gene for the required enzyme catalyzing the reoxidation of NAD^+ to NADH was provided by module III and expressed using plasmid pBAD28Δ*bla*, thus, it could be chosen independently from the other two enzymes. Three different strategies for cofactor regeneration were investigated: oxidation of formate to carbon dioxide by NAD^+-dependent formate dehydrogenase (FDH), oxidation of glucose to β-D-glucono-1,5-lactone by glucose dehydrogenase (GDH) from *Bacillus megaterium* and oxidation of phosphite to phosphate by NAD^+-dependent phosphite dehydrogenase (PtDH) from *Pseudomonas stutzeri*, respectively.

The reductive aminations were performed in a fermentative fashion (Fig. 5, Table 2), employing 50 mM KMV and 100 mM co-substrate. On the one hand, the three-enzyme-system was studied with the *ald* gene being

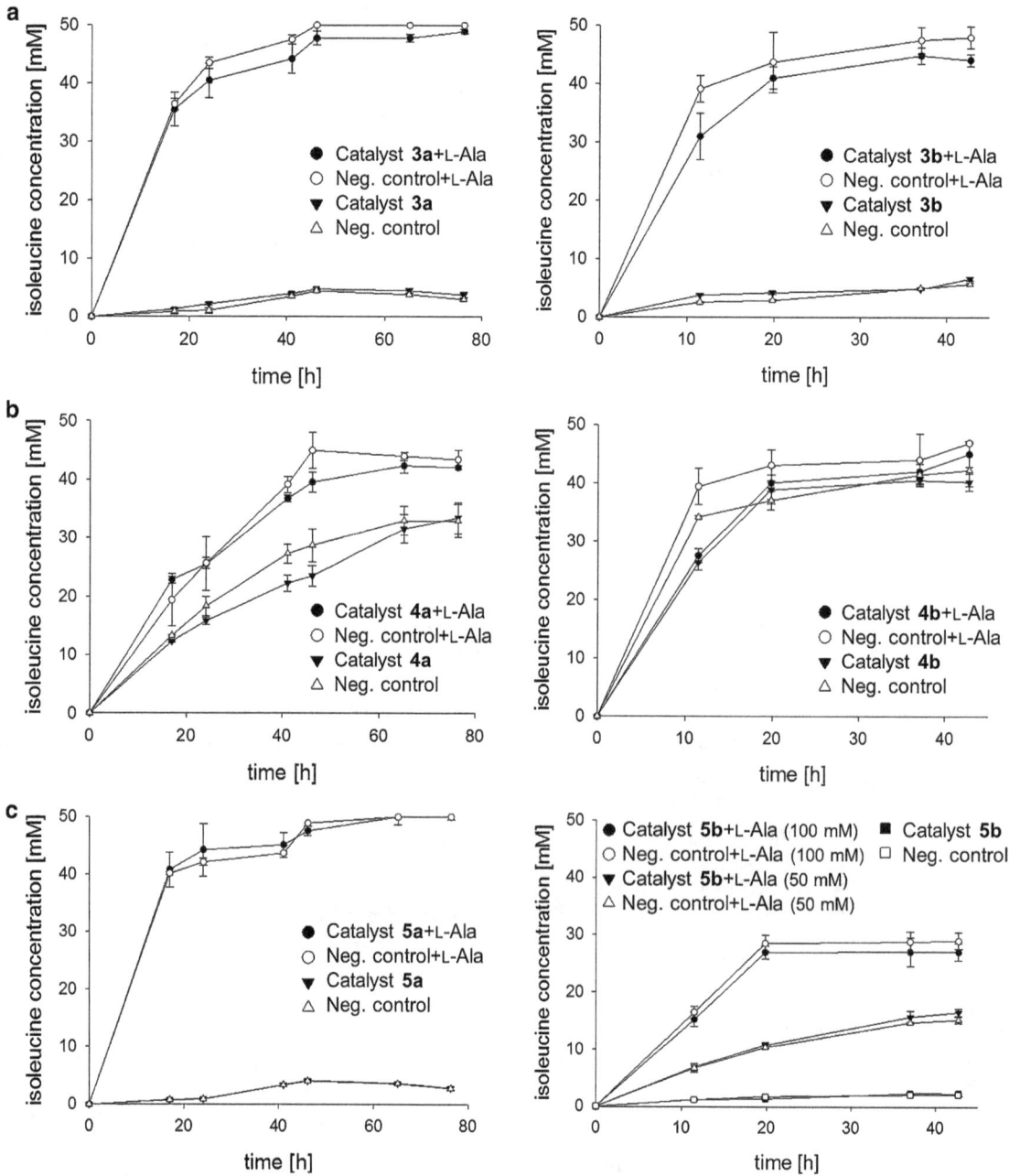

Fig. 5 Fermentative amination of α-ketoacids with enzyme-coupled NADH regeneration. Fermentative reductive amination of KMV to L-isoleucine with *Ec*-AlaDH carrying pTrc99A-*avtA* (catalyst **3a**, **4a**, **5a**) or pTrc99A-*ald-avtA* (catalyst **3b**, **4b**, **5b**) and using either FDH (**a**), GDH (**b**) or PtDH (**c**) for cofactor regeneration. Strains were inoculated in SGA medium with OD_{600} 4 and 100 mM co-substrate, 50 mM $(NH_4)_2SO_4$ and 0–250 mM L-alanine were added to push the reaction equilibrium towards product side

localized in the genome and thus present as a single copy (catalysts **3a**, **4a** and **5a**). In that case, the addition of high amounts of L-alanine (250 mM) was necessary. On the other hand, experiments were performed with another *ald* gene copy (catalysts **3b**, **4b** and **5b**) introduced via plasmid pTrc99A-*avtA*. As already observed before, the *ald* gene being positioned right behind the

P*trc* promoter (pTrc99A-*ald*-avtA) led to enhanced expression and activity, which is why this gene order was again used for additional AlaDH supply. In this case, the L-alanine concentration was reduced to 50–100 mM, as AlaDH expression and hence present amine donor amount was expected to be sufficient. For control reactions the respective strain was transformed with empty pBAD28Δ*bla* lacking the gene for cofactor regeneration, resulting in a fermentative amination which makes use of metabolic NADH regeneration and hence corresponds to catalysts **2a**, **2b** and **2c**, respectively. Coupling the L-alanine dependent reductive amination catalyzed by AvtA with an enzyme catalyzed NADH regeneration system did not improve the overall conversion of KMV to L-isoleucine in any case compared with the cellular regeneration approach (control strains). No significant turnover at all was achieved neither with the FDH-coupled catalyst **3a** (8% after 41 h), nor with the PtDH-coupled **5a** (7% after 41 h). However, 44% conversion was achieved in the same reaction time with catalyst **4a**, using GDH as recycling enzyme. This result coincided with the higher enzyme activities measured for GDH in contrast to the other recycling enzymes. Interestingly, comparable or even higher conversions were obtained in all three cases with corresponding control strains, indicating that the expression level of regeneration enzymes is not sufficient to outperform the cell-driven NADH recycling based on glucose (Table 2).

Moreover, again the genome-based *ald* expression was expected to be a limiting parameter for regeneration of consumed amine donor and thus for an efficient reductive amination. Indeed, the addition of 250 mM L-alanine increased overall turnover significantly both for all single-cell catalysts as well as for control strains, leading to 88–90% of product formation for catalyst **3a** and **5a** and 73% for **4a**. An additional cloning of *ald* on the plasmid pTrc99A-*ald-avtA* did not increase the catalytic efficiency of the *E. coli* cells considerably. While a turnover of 13% and comparable 11.5% were achieved with catalyst **3b** and its control strain after 42 h, catalyst **5b** led even to reduced product formation (5%). The only exception turned out to be catalyst **4b** making use of a GDH, where an additional *ald* gene copy seemed to be beneficial and increased turnover of KMV from 44 to 80%. Nevertheless, comparable turnover was obtained with the control strain too. Once more, a remarkable enhancement was possible when additional amine donor (50 mM) was present in the fermentation broth, leading to 88% product formation with catalyst **3b**. Similarly, turnover of catalyst **4b** could be slightly increased to 90%. However, still moderate conversions were achieved with catalyst **5b**, even when 50 or 100 mM L-alanine were added, resulting in 17 and 54% conversion, respectively.

"In vitro-type" single-cell catalyzed amination of prochiral ketones

Besides enabling the single-cell catalyzed reductive amination of α-ketoacids in a one-pot fermentation, the applicability of the modular plug-and-play concept was furthermore demonstrated for the synthesis of optically pure amines. For this purpose, a transaminase from *Chromobacterium violaceum* (Ta$_{Cv}$) from module I offering a broader substrate spectrum was coupled with AlaDH from module II for regenerating the consumed amine donor L-alanine as well as with each of the three available enzymes from module III for NADH cofactor recycling, resulting in three single-cell catalysts 6–8 (see Table 1). The asymmetric reductive amination of 4-phenyl-2-butanone was chosen as model reaction and the overall performance of each single-cell biocatalyst was compared by determining initial rates (Fig. 6) for the mentioned transformation. With respect to catalyst **6**, co-expressing Ta$_{Cv}$ together with AlaDH and FDH, an apparent overall activity of 0.30 ± 0.02 U g^{-1} was obtained, whereas catalyst **7** involving a GDH for cofactor regeneration showed an apparent activity of 0.22 ± 0.02 U g^{-1}. The best result, however, was obtained for catalyst **8**, combining the transamination machinery with a PtDH and reaching thereby an apparent activity of 0.99 ± 0.07 U g^{-1}. Then the reactions were monitored over a time period of 48 h, employing and comparing

Table 2 Obtained results for the reductive amination of KMV with and without addition of amine donor L-alanine

Catalyst	Conversion [%] without L-alanine		Conversion [%] with L-alanine	
Cofactor regeneration	Enzyme-coupled	Metabolic	Enzyme-coupled	Metabolic
3a (*Ec*-AlaDH/pTrc99a-*avtA*/pBADΔ*bla-fdh*)	8	8	88[a]	94[a]
3b (*Ec*-AlaDH/pTrc99a-*ald-avtA*/pBADΔ*bla-fdh*)	13	12	88[b]	96[b]
4a (*Ec*-AlaDH/pTrc99a-*avtA*/pBADΔ*bla-gdh*)	44	54	73[a]	78[a]
4b (*Ec*-AlaDH/pTrc99a-*ald-avtA*/pBADΔ*bla-gdh*)	80	84	90[b]	94[b]
5a (*Ec*-AlaDH/pTrc99a-*avtA*/pBADΔ*bla-ptdh*)	7	7	90[a]	87[a]
5b (*Ec*-AlaDH/pTrc99a-*ald-avtA*/pBADΔ*bla-ptdh*)	5	6	54[c]	58[c]

Conversion obtained for the enzyme-coupled system for cofactor regeneration was compared with the metabolic one

[a] Addition of 250 mM L-alanine

[b] Addition of 50 mM L-alanine

[c] Addition of 100 mM L-alanine

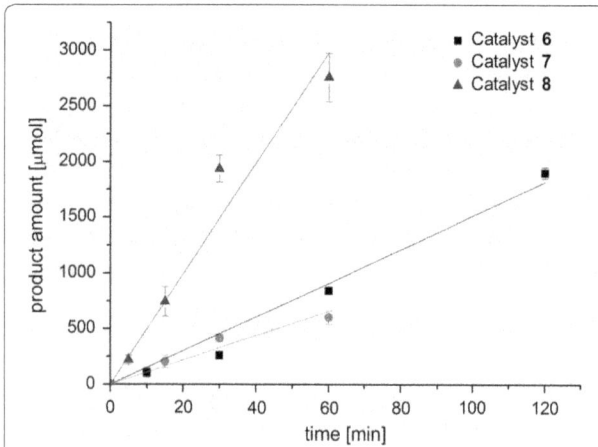

Fig. 6 Activities of singe-cell catalysts for "in vitro-type" amination of ketones. Initial rates of single-cell catalysts **6** (*Ec*-AlaDH/pTrc99A-*ta*$_{Cv}$/pBAD28Δ*bla-fdh*), **7** (*Ec*-AlaDH/pTrc99A-*ta*$_{Cv}$/pBAD28Δ*bla-gdh*) and **8** (*Ec*-AlaDH/pTrc99A-*ta*$_{Cv}$/pBAD28Δ*bla-ptdh*) in the reductive amination of 4-phenyl-2-butanone (25 mM), coupling Ta$_{Cv}$ with AlaDH and either FDH, GDH or PtDH for cofactor regeneration. Typical time curves are shown for the production of 4-phenyl-2-butylamine over a time period of 60 or 120 min, respectively using 50 mg resting cells per mL

different catalyst preparations (Fig. 7). Besides the use of resting cells, the biocatalyst was applied in the form of lyophilized single-cells, cell-free extract and lyophilized cell-free extract. For sake of better comparability the amount of used lyophilized cells, lysate and lyophilized lysate always correlated to 50 mg of wet cells. Employing resting cells of catalyst **6** in the model reaction (Fig. 7a) led to formation of 35% of the desired amine, which was slightly increased to 45% with a twofold catalyst loading (100 mg mL^{-1}). Similar results were obtained with catalyst **8**, yielding already 33% of amine with half the catalyst amount (20 mg mL^{-1}), which could be increased to 48–56% with one- to twofold of catalyst loading, respectively (Fig. 7c). Less efficient in the reductive amination of 4-phenyl-2-butanone seemed to be catalyst **7**, converting only 25–32% of the ketone depending on the amount of used resting cells (50–100 mg mL^{-1}). However, with catalyst **7** the occurrence of limitations during the reaction time became rather obvious (Fig. 7b), as within the first 4 h the reaction rate was accelerated proportional to the amount of employed catalyst, but with proceeding reaction this correlation decreased significantly and almost no amine was formed anymore after 24 h. With respect to GDH-catalyzed cofactor regeneration, especially local acidification within the cell caused by the co-product gluconic acid might be problematic, presumably leading to

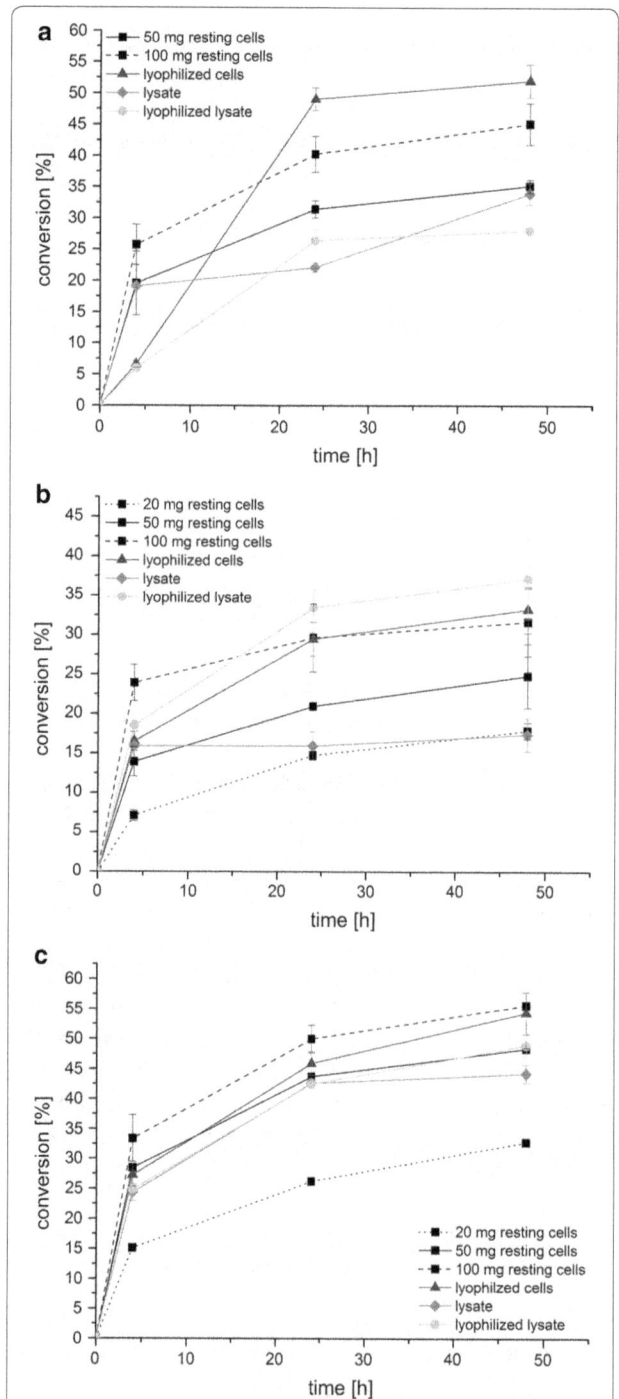

Fig. 7 "In vitro-type" amination of a prochiral ketone. Investigation of catalyst **6** (**a**), catalyst **7** (**b**) and catalyst **8** (**c**) in the reductive amination of 4-phenyl-2-butanone over 48 h. The biocatalyst was either applied as resting cells (*black square*), as lyophilized cells (*black triangle*), as cell-free extract (*grey diamond*) or as lyophilized cell-free extract (*grey circle*), respectively and the reaction was monitored over 48 h

enzyme inactivation. In general, applying the biocatalyst as lyophilized single-cells (corresponding to 50 mg of wet cells) significantly improved conversion levels in all cases, giving 52, 33 and 54% of amine with catalyst **6**, catalyst **7** and catalyst **8**, respectively. Most likely, the cell membrane was permeabilized due to lyophilization in a way that led to better access of reaction components. Surprisingly, performing the biotransformation either with cell-free or lyophilized cell-free catalyst preparations, respectively did not show further beneficial effect on the outcome of the reaction. This was at least the case for catalyst **6** and **8**. In contrast, the highest product formation (37%) with catalyst **7** was achieved when employed as lyophilized cell-free extract. The inability of *E. coli* cells to import glucose without permeabilization of the cell membrane is a major drawback of GDH-catalyzed cofactor regeneration [50], which is why the complete removal of the cell wall might be improving the overall catalyst performance. In order to further increase the amine production by the recombinant *E. coli* single-cell catalyst, the effect of a higher catalyst loading using catalyst **6** in form of lyophilized cells was investigated. With a tenfold amount of catalyst, conversion to the amine could be increased by a factor of 1.5 to maximum 77% within 48 h. This indicated once more the occurrence of limitations over time such as enzyme inactivation or inhibition. As already observed during one-pot fermentations of α-ketoacids before, also a too low activity level of AlaDH might lead to an inefficient reaction equilibrium and thus, to not quantitative conversions. Nevertheless, the concept of assembling the reaction cascade for reductive amination of prochiral ketones in vivo within one *E. coli* cell has been successfully proven, affording the (*S*)-amine in high yields with excellent to perfect optical purity (92–99%).

Discussion

"Tailor-made" single-cell biocatalysts co-expressing multiple enzymes enable efficient in vivo reaction cascades [28, 32]. Consequently, the design and application of such microbial cell factories has attracted attention in recent years, aiming for the production of a broad range of valuable chiral compounds. Since synthetic pathways commonly may involve cofactor-dependent redox reactions [51], various studies focused on the incorporation of reaction pathways in a host cell whereby the sequence itself enables a suitable cofactor regeneration resulting in redox self-sufficient single-cell catalysts [42, 52]. Alternatively, recycling of cofactors can be achieved by making use of the host's inherent metabolic pathways, e.g. the catabolism of carbon sources like glucose [53]. In this study, we established a modular platform to construct *E. coli* single-cell biocatalysts tailored for the in vivo amination

of ketoacids and prochiral ketones, exploiting both cell metabolism as well as additional enzyme-mediated strategies for cofactor regeneration. The fermentative transformation of KMV to L-isoleucine showed that the use of an enzyme-coupled approach for NADH regeneration led to similar conversion as obtained with the metabolism-coupled one. This strongly suggests that the expression level of regenerating enzymes FDH, GDH and PtDH, respectively, was insufficient, lowering the overall catalytic performance. As a matter of fact, impaired enzyme production is a well-known obstacle of whole-cell biocatalysis caused by an increased metabolic burden during cell growth due to the co-expression of multiple proteins within one cell [31]. In case of the amino acid dehydrogenases GluDH and AlaDH a low expression level became apparent too, since extra addition of amine donors L-glutamate and L-alanine, respectively, to the fermentation broth significantly improved the transformation of both KIC and KMV to the corresponding α-amino acids. The role of L-alanine in such in vivo amination cascades has already been investigated before concerning the redox self-sufficient amination of alcohols [43]. Since L-alanine is not only required as amine donor for the desired reaction but also as energy source for the cell to maintain viability and the protein biosynthesis machinery under stress conditions, its addition in certain amounts is to be recommended, or even necessary. However, it was possible to improve the amine donor regeneration system by including a second *ald* gene copy additionally to the one being genome-integrated. This change on the genetic level led to higher amounts of expressed AlaDH and consequently to increased product formation, confirming once more the assumption of reduced catalyst performance due to impaired enzyme synthesis. However, not only the additional plasmid-born *ald* gene but also its gene order contributed to enhanced protein expression. This emphasized the crucial role of a thoroughly reasoned co-expression strategy in order to fine-tune expression levels and to achieve a functional microbial cell as catalytic unit. A vast synthetic biology toolbox is available for that purpose, offering different promoter systems, plasmid types and strategies to compose the ideal co-expression cassette [28, 54, 55].

In this study, a recombinant two-plasmid-based gene expression was combined with a genome-integrated one, ideally enabling control and induction of protein production independent from the host's regulatory network. Furthermore, this modular system allowed for easy substitution of the individual enzymes according to the desired application of the single-cell catalyst. Exchanging for example the L-alanine–valine aminotransferase with the (*S*)-selective transaminase from *Chromobacterium violaceum* facilitated a broader substrate range and hence, a widened applicability of the approach. Thus, additionally to the amination of α-keto acids also the transformation of a prochiral ketone into an optically active amine was achieved. The amination of 4-phenyl-2-butanone was performed in a metabolism-independent fashion, making use of enzyme-coupled regeneration of redox cofactors. In such "in vitro-type" reactions decoupled from the host's metabolism parameters like activity, stability and concentration of enzymes co-expressed within the single-cell biocatalyst play a particularly important role [54]. Accordingly, limitations were observed during the amination of 4-phenyl-2-butanone, which might be attributed again to an insufficient and imbalanced enzyme expression. Next to a generally low enzymatic activity of AlaDH and NADH recycling enzymes also a loss thereof over time caused by inactivation or inhibition might be an issue. The adaption of the cellular metabolism of resting cells to the non-growth status results in a restricted self-regeneration and stress-handling capacity, leading to a decrease of active enzyme amounts and as a consequence to an intracellular NADH shortage [56]. This might have been the reason why it was not possible to perform the "in vitro-type" reactions satisfyingly without the external addition of cofactors. Conversions were much higher when NAD$^+$ was present in the reaction mixture. According to literature the NAD$^+$ level of E. coli lies in the range of 5.3 nmol mg^{-1} dry weight under standard growth conditions [57]. Thus, the application of whole-cell biocatalysts without additional coenzyme may be possible, but it has been reported before that by adding coenzymes the efficiency of the reaction is significantly enhanced [58]. Another crucial parameter in terms of non-fermentative transformations turned out to be the permeabilization of the cell wall. Mass transfer limitations were overcome by lyophilization or even complete removal of cell wall via disruption and thus, higher conversions were obtained for the amination of 4-phenyl-2-butanone.

Conclusions

The developed concept of a plug-and-play E. coli single-cell biocatalyst provides a library of enzymes for each of the different required reactions/modules of a reductive amination and suitable E. coli hosts, which can be combined depending on the desired target reaction. Consequently, it represents a quite promising and cost-efficient alternative to the combination of isolated enzymes. Especially, from an industrial point of view the described approach offers major advantages like the decrease of required fermentations and the reduction of costly and laborious isolation and purification steps. The flexible approach can be easily extended to other transaminases, allowing for a quick identification of the most suitable single-cell catalyst tailored for a specific substrate or a desired stereoselectivity. Depending on the required co-substrate the appropriate amino acid dehydrogenase as well as NADH recycling system can be chosen. A further improvement of the modular platform would be the addition of an alanine racemase, enabling the in situ racemization of L-alanine to D-alanine and thus the efficient use of (R)-selective ω-TAs. In summary, the presented approach offers a platform to construct a broad range of single-cell catalysts in a flexible plug-and-play fashion for reductive amination. In comparison with the in vitro transamination system using individual biocatalysts from separate preparations, the single-cell approach enables easy production of the required enzymes and cost-effective biotransformations.

Methods

Construction of bacterial strains and plasmids

Bacterial strains and plasmids used in this study are summarized in Table 3. Plasmid pRed/ET (tet^R) was obtained from the "Quick and Easy E. coli Gene Deletion Kit", Gene Bridges (Heidelberg, Germany) and plasmid pQE30 was received from Qiagen (Hilden, Germany). Plasmid isolation was performed with the QIAprep spin miniprep kit (Qiagen, Hilden, Germany). Standard DNA work like polymerase chain reaction (PCR), restriction and ligation were performed as described previously [59]. The oligonucleotides used in this study were obtained from Metabion GmbH (Martinsried, Germany) or Eurofins Genomics and are listed in Additional file 1: Table S1. For transformation CaCl$_2$-competent E. coli cells [60] were heat-shocked for uptake of respective DNA. All cloned DNA fragments were shown to be correct by sequencing performed by LGC Genomics GmbH (Berlin, Germany).

Construction of expression plasmids

Plasmids were constructed with fragments generated by PCR (KOD Hot Start Polymerase kit, Novagen, Darmstadt, Germany) using Bacillus subtilis as template for ald and rocG genes, E. coli MG1655 and Streptococcus mutans for ilvE genes, Chromobacterium violaceum for ta gene, Komagataella pastoris GS115 for fdh gene, Bacillus megaterium for gdh gene, Pseudomonas stuzeri for ptdh gene and E. coli W3110 for genes phnC, phnD, phnE. In this study two different cloning strategies based on the IPTG-inducible E. coli expression vector pTrc99A were used. In the first strategy the particular gene was inserted into the vector by using cut sites. In order to construct pTrc99A-ald-avtA, the oligonucleotides ald$_{Bs}$_RBS_fw/ald$_{Bs}$_rv were used and the PCR product was ligated into the vector pTrc99A-avtA via EcoRI restriction site. The construction of pTrc99A-ilvE$_{Ec}$ and pTrc99A-ilvE$_{Sm}$ was performed similarly with the primers

Table 3 List of bacterial strains and plasmids used in this work

	Relevant characteristics	Source/references
Strains		
E. coli TOP10	F⁻ *mcrA* Δ(*mrr-hsd*RMS-*mcr*BC) Φ80*lacZ*ΔM15 Δ *lacX74 recA1 araD139* Δ(*araleu*)7697 *galU galK rpsL* (StrR) *endA1 nupG*	Invitrogen
E. coli W3110	F⁻ λ⁻ INV(*rrnD-rrnE*)1 *rph-1*	[61]
E. coli AlaDH	W3110Δ*araBAD*::P*trc-ald*, gene for alanine dehydrogenase (*ald*) from *Bacillus subtilis* D196A/L197R integrated into *ara* locus in the choromosome and expressed from P*trc* promoter	This study
E. coli GluDH	W3110Δ*araBAD*::P*trc-rocG*, gene for glutamate dehydrogenase (*rocG*) from *Bacillus subtilis* D196A/L197R integrated into *ara* locus in the choromosome and expressed from P*trc* promoter	This study
Plasmids		
pTrc99A	P*trc*, pBR322ori, *rrn*B T1, *rrn*B T2, *lacIq*, *bla*, template for P*trc* Promoter	[47]
pTrc99A-*avtA*	pTrc99A carrying L-alanine–valine aminotransferase (*avtA*) from *E. coli* MG1655	[49]
pTrc99A-*avtA-ald*	pTrc99A carrying *avtA* from *E. coli* MG1655, *ald* from *Bacillus subtilis*	[49]
pTrc99A-*ald-avtA*	pTrc99A carrying *ald* from *Bacillus subtilis*, *avtA* from *E. coli* MG1655	This study
pTrc99A-*ilvE*_Ec_	pTrc99A carrying branched-chain amino-acid aminotransferase (*ilvE*) from *E. coli* MG1655	This study
pTrc99A-*ilvE*_Sm_	pTrc99A carrying *ilvE* from *Streptococcus mutans*	This study
pTrc99A-*ta*_Cv_	pTrc99A carrying (*S*)-selective ω-transaminase from *Chromobacterium violaceum*	This study
pBAD28	P*ara*, p15Aori, *rrn*B T1, *rrn*B T2, *bla*, Cam^R	[48]
pBAD28Δ*bla*	pBAD28 with deletion of β-lactamase gene	This study
pBAD28Δ*bla-fdh*	pBAD28Δ*bla* carrying formate dehydrogenase (*fdh*) from *Komagataella pastoris* GS115	This study
pBAD28Δ*bla-gdh*	pBAD28Δ*bla* carrying glucose dehydrogenase (*gdh*) from *Bacillus megaterium*	This study
pBAD28Δ*bla-ptdh*	pBAD28Δ*bla* carrying *ptdh* from *Pseudomonas stuzeri*	This study
pQE30	phage PT5 promoter, Col E1, 6xHis, *bla*, template for PT5	Qiagen
pKD13	*oriRY, bla*, Km^R, template for kanamycin cassette	[62]
pTrc99A-*kan*-P*trc-ald*	pTrc99A carrying FRT-flanked *kan* resistance gene of pKD13, P*trc* Promoter, *ald* from *Bacillus subtilis*	This study
pTrc99A-*kan*-P*trc-rocG*	pTrc99A carrying FRT-flanked *kan* resistance gene of pKD13, P*trc* Promoter, *rocG* from *Bacillus subtilis*	This study
pRed/ET	*Red* recombinase expression plasmid (ts), pSC101 based, Tc^R	Gene bridges
pCP20	*RepA101*(ts), *bla*, λ-red Flp recombinase for removal of resistance cassette	[62]

*ilvE*_Ec__RBS_fw/*ilvE*_Ec__rv, *ilvE*_Sm__RBS_fw/*ilvE*_Sm__rv and *Eco*RI restriction site. For pTrc99A-*ta*_Cv_ the PCR-amplified gene product using *ta*_Cv__fw/*ta*_Cv__rv was cut with *Eco*RI and *Sal*I and then ligated into pTrc99A digested with the same restriction enzymes. In the second strategy the Gibson assembly method [63] was performed as a second cloning strategy to construct pTrc99A-*ald-avtA-ptdh*. For this purpose the genes *ald* of *Bacillus subtilis*, *avtA* of *E. coli* MG1655 and *ptdh* of *Pseudomonas stuzeri* were amplified with the oligonucleotides GA_*ald*_RBS_*Eco*RI_fw/GA_*ald*_rv, GA_*avtA*_RBS_fw/GA_*avtA*_rv and GA_*ptdh*_RBS_fw/GA_*ptdh*_*Xba*I_rv, respectively and assembled with *Eco*RI/*Xba*I restricted pTrc99A using Gibson assembly method [63].

For expression of the genes encoding NAD(P)H-regenerating enyzmes the arabinose-inducible pBAD28 plasmid was used. For deletion of the *bla* gene, the plasmid was digested with the restriction enzymes *Alw44*I and *Nsb*I, treated with Klenow fragment and ligated. Subsequently, the obtained plasmid pBAD28Δ*bla* was used for generating plasmids pBAD28Δ*bla-fdh*

(*fdh*_RBS_fw/*fdh*_rv), pBAD28Δ*bla-gdh* (*gdh*_RBS_fw/*gdh*_rv) and pBAD28Δ*bla-ptdh* (*ptdh*_RBS_fw/*ptdh*_rv). Before ligation, the pBAD28Δ*bla* vector as well as the PCR-derived gene products were treated with *Sac*I and *Xba*I enzymes.

Construction of bacterial strains harbouring amino acid dehydrogenase genes

The open-reading coding region of the genes *araBAD* from *E. coli* W3110 genome were replaced with the gene *ald* or *rocG* under control of IPTG-inducible P*trc* promoter and a kanamycin cassette flanked by FLP recognition target sites by using modified one-step method for inactivation of genes [62]. For the construction of recombination plasmids, again the Gibson assembly method was applied. Linear DNA-fragments comprising FLP-*kan*-FLP cassette and P*trc* promoter as well as *ald* or *rocG* gene, respectively were obtained by PCR using primers GA_*kan*_*Eco*RI_fw/GA_*kan*_rv, GA_trc_fw/GA_trc_rv, GA_*ald*_RBS_fw/GA_*ald*_*Eco*RI_rv, GA_*rocG*_RBS_fw/GA_*rocG*_*Eco*RI_rv. For this purpose pTrc99A or pQE30 plasmid were used and genomic DNA from *B. subtilis* and pKD13, respectively as template. The linear pTrc99A vector (cut with *Eco*RI) and overlapping DNA fragments were assembled to pTrc99A-*kan*-P*trc*-*ald* and pTrc99A-*kan*-P*trc*-*rocG* plasmids. All gene cassettes were amplified by PCR using primers with homologous arms consisting of 50 nucleotides upstream (HS_*araCB*_fw) and downstream (HS_*araD*_rv) of the *araBAD* genes. The corresponding 3-kbp PCR products were purified, treated with *Dpn*I, and then transformed by electroporation into *E. coli* W3110(Red/ET) using "Quick and Easy *E. coli* Gene Deletion Kit" (Gene Bridges, Heidelberg) according to the manual provided by the supplier. Cells with homologous recombination were selected on an agar plate containing kanamycin and screened by colony PCR with *araC*_fw/Kt_rv primers. The antibiotic marker was removed by using a helper plasmid pCP20 encoding FLP-recombinase. The Km^R mutants were transformed with temperature-sensitive plasmid pCP20 and Amp^R transformants were selected at 30 °C. The elimination of antibiotic marker was verified by PCR (*araC*_fw/*polB*_rv). The helper plasmid was removed by strike out on non-selective plates at 43 °C. Newly obtained mutation strains were designated *Ec*-AlaDH (*E. coli* W3110Δ*araBAD*::P*trc*-*ald*) and *Ec*-GluDH (*E. coli* W3110Δ*araBAD*::P*trc*-*rocG*). For fermentative production experiments strain *Ec*-AlaDH was transformed with plasmids pTrc99A-*avtA*, pTrc99A-*avtA*-*ald* or pTrc99A-*ald*-*avtA* and pBAD28Δ*bla*, pBAD28Δ*bla*-*fdh*, pBAD28Δ*bla*-*gdh* or pBAD28Δ*bla*-*ptdh*, respectively. Alternatively, they were transformed with pTrc99A-*ald*-*avtA*-*ptdh* and pBAD28Δ*bla*-*phnCDE*. Strain *Ec*-GluDH was transformed with plasmids pTrc99A-*ilvE*$_{Ec}$ or pTrc99A-*ilvE*$_{Sm}$. For preparing the single-cell catalysts used for "*in vitro*-type" reactions, strain *Ec*-AlaDH was transformed with plasmids pTrc99A-*ta*$_{Cv}$ and pBAD28Δ*bla*-*fdh*, pBAD28Δ*bla*-*gdh* or pBAD28Δ*bla*-*ptdh*, respectively.

Media and cultivation conditions

Escherichia coli W3110 cells were grown in lysogeny broth (LB) complex medium (10 g L^{-1} of tryptone, 5 g L^{-1} of yeast extract, 10 g L^{-1} of sodium chloride) at 37 °C in baffled Erlenmeyer flasks (60–1000 mL) on a rotary shaker at 120 rpm. Ampicillin (100 mg L^{-1}), kanamycin (50 mg L^{-1}) and/or chloramphenicol (25 mg L^{-1}) were added when appropriate. The growth of *E. coli* was monitored by measuring the optical density at 600 nm (OD$_{600}$). Protein expression was induced at an OD$_{600}$ of approximately 0.5–0.7 by the addition of isopropyl-β-D-thiogalactopyranoside (IPTG, 0.25–1 mM) and/or L-arabinose (0.02–0.3 vol%). Afterwards shaking was continued over night at 120 rpm and 20, 25, 30 or 37 °C,

respectively. The cells were then harvested by centrifugation (2600–10,000×g for 10–30 min at 4 °C) and the resulting cell pellet was either applied directly in single-cell biotransformations or stored at −20 °C until further use. By lyophilizing the single-cell biocatalyst, its storage at 4 °C was possible for several weeks without any significant loss of enzyme activity. For this purpose the cell pellet was resuspended in a minimum amount of sodium phosphate buffer (NaP$_i$, 50 mM, pH 8, 0.5 mM PLP), frozen in liquid nitrogen and lyophilized over night.

Preparation of cell-free extracts

For preparation of cell-free extract the harvested cells were resuspended in NaP$_i$ buffer (50 mM, pH 8) yielding a 15 wt% cell solution. Cells were lysed by sonication treatment at 4 °C (15–60, 0.1 s sonicate, 0.4 s pause, 40% amplitude using a Branson Digital Sonifier®) and crude extract was centrifuged (13,000–16,000× for 10–30 min at 4 °C). The remaining cell debits were resuspended (15 vol%) in urea (6 M) for SDS-analysis. The supernatant was kept at 4 °C to be either applied directly in biotransformations or to be analyzed by activity assays and SDS-PAGE. Otherwise, the supernatant was frozen in liquid nitrogen and lyophilized overnight yielding in lyophilized cell-free extract.

Fermentative production of L-isoleucine and L-leucine

For the production of L-isoleucine and L-leucine strains derived from *E. coli* W3110 were grown in LB medium supplemented with 1 mM IPTG and/or 0.3% L-arabinose at 37 °C and 200 rpm. Then the cells were washed twice with the medium salts (49 mM KH$_2$PO$_4$; 76 mM K$_2$HPO$_4$ for FDH and GDH or 125 mM H$_3$O$_3$P for PtDH), resuspended and cultivated with an OD$_{600}$ of four in a chemically defined synthetic glycerol ammonium sulfate (SGA) medium as described previously [49]. As carbon source either 100 mM glucose for GDH, 100 mM sodium formate for FDH or 100 mM sodium hypophosphite for PtDH were used instead of glycerol. In the case of GDH the cells were cultured anaerobically for the production of L-isoleucine. When using PtDH, the phosphate components were replaced by phosphite components. For induction of expression 1 mM IPTG and/or 0.3% L-arabinose were added to the cultures. For production of L-isoleucine 50 mM MOPS, 100 mM (NH$_4$)$_2$SO$_4$, 50 mM 2-keto-3-methylvalerate (KMV) and 0–250 mM L-alanine were added to the medium. For production of L-leucine 50 mM MOPS, 100 mM (NH$_4$)$_2$SO$_4$, 50 mM 2-keto-isocaproate (KIC) and 0–250 mM L-glutamate were added to the medium. For quantification of extracellular amino acids, aliquots of the culture were taken, cells were removed by centrifugation at 13,000×g for 10 min, and the supernatants were frozen at −20 °C.

Toxicity tests

For the study of substrate toxicity batch cultivations with *E. coli* W3110 were performed in the Biolector® cultivation system (m2p Labs, Baesweiler) using 1 mL medium microtiter plates (Flower Plate®, m2p Labs, Baesweiler) at 1100 rpm at 37 °C. The cells were grown in SGA minimal medium with 1% Glucose and various concentrations (5, 20, 50, 100 and 200 mM) of sodium hypophosphite monohydrate ($NaH_2PO_2 \times H_2O$), sodium phosphite dibasic pentahydrate ($Na_2HPO_3 \times 5H_2O$), sodium phosphate monobasic monohydrate ($NaH_2PO_4 \times H_2O$) and sodium formate ($NaCO_2H$). The experiments were carried out in triplicates.

Activity assays

All assays were performed with cell-free extracts in triplicates and one unit of enzyme activity (U) was calculated as the amount of enzyme catalyzing the conversion of 1 µmol of substrate in 1 min. Protein concentration was determined by the Bio-Rad Protein Assay based on the method of Bradford [64] using bovine serum albumin as a reference standard.

Measurement of transaminase activity

The activities of AvtA and IlvE were measured as described before [65] using 10 mM keto acid (2-keto-3-methylvalerate (KMV) or 2-keto-isocaproate (KIC) and 50 mM amino donor (L-alanine for AvtA or L-glutamate for IlvE, respectively). The determination of Ta_{Cv} activity was performed via an indirect photometric assay based on the reductive amination of pyruvate to L-alanine using (S)-methylbenzylamine as amine donor. For this purpose a substrate solution (990 µL) containing PLP (0.1 mM), (S)-methylbenzylamine (10 mM) and sodium pyruvate (5 mM) in NaP$_i$ buffer (50 mM, pH 8) was mixed with cell-free extract (10 µL) in a semi-micro UV-cuvette. Immediately, the increase of acetophenone formation (initial rate $\Delta c/\Delta t$) was measured over time at 290 nm.

Measurement of activities of amino acid dehydrogenases

The activities of AlaDH and GluDH were assayed based on the conversion of pyruvate to L-alanine with concomitant NADH consumption, which was monitored spectrophotometrically at 340 nm as described before [65].

Measurement of activities of NADH-recycling enzymes

The increase of NADH absorption at 340 nm was used for determining activities of NADH-recycling enzymes. FDH-activity was calculated based on the oxidation of formate with concomitant reduction of NAD$^+$ to NADH. For measuring the activity of GDH β-D-glucose was converted to D-glucono-1,5-lactone and in order to assaying PtDH activity phosphite was oxidized to phosphate. In all cases cell-free extract (10 µL) was added to a substrate solution (970 µL) containing NAD$^+$ (1 mM) and either ammonium formate (100 mM), D-glucose (100 mM) or sodium phosphite (100 mM) in NaP$_i$ buffer (50 mM, pH 8) and the increase of NADH absorption at 340 nm over time (initial rate $\Delta c/\Delta t$) was measured immediately.

Biotransformations

Biotransformations were performed in NaP$_i$ buffer (50 mM, pH 8, 1 mM PLP) with L-alanine as amine donor and a substrate concentration of 25 mM. The particular co-substrate was applied according to the used cofactor-recycling enzyme. Together with the FDH catalyzed recycling system ammonium formate was used, while for the GDH system glucose and for the PtDH system sodium phosphite was required. For the GDH and the PtDH system additionally ammonium acetate was needed as nitrogen donor. The recombinant *E. coli* catalyst was employed either in form of resting cells (50 mg resuspended in 500 µL reaction buffer) or as cell-free extract (~350 µL, corresponding to 50 mg wet cells mixed with 150 µL reaction buffer) in an Eppendorf tube (1.5 mL). Alternatively, lyophilized single-cells (corresponding to 50 mg wet cells) or lyophilized cell-free extract (corresponding to 350 µL lysate) were rehydrated in reaction buffer for 15 min at 30 °C and 120 rpm prior to use. Then, reaction buffer (500 µL) containing L-alanine (250 mM, 5 eq), NAD$^+$ (2 mM), the particular co-substrate (300 mM, 6 eq) and ammonium acetate (150 mM, 3 eq) in case of GDH and PtDH recycling system was added. The reaction was started by the addition of the substrate 4-phenyl-2-butanone (3.8 µL, 25 mM) and the mixture was shaken at 30 °C and 800 rpm for up to 48 h using an orbital shaker. After 4, 24 and 48 h, respectively 200 µL of each sample were withdrawn and the reaction was quenched with aqueous NaOH solution (20 µL, 10 M). After extraction with EtOAc (2 × 400 µL) the combined organic layers were dried over Na$_2$SO$_4$, and conversion as well as *enantiomeric excess* (*ee*) were analyzed by GC.

Determination of initial rates of single-cell catalysts

The activity of the recombinant *E. coli* resting cells for the reductive amination of 4-phenyl-2-butanone to the corresponding amine was assayed over 1–2 h. For this purpose the biotransformation was performed according to the procedure described above. The reaction was stopped after 5, 10, 20, 40, 60 and 120 min, respectively by adding aqueous NaOH solution (20 µL, 10 M) and extracted with EtOAc (2 × 400 µL). The combined organic layers were dried over Na$_2$SO$_4$ and conversions were analyzed by GC. The enzyme activity (U) was defined as the amount of enzyme that catalyzes the conversion of 1 µmol of substrate per minute.

Analytical methods

Determination of conversion

Quantification of amino acid content in the fermentation reactions was performed using a high-pressure liquid chromatography system (HPLC, 1200 series, Agilent Techologies Deutschland GmbH, Böblingen, Germany) and an automatic precolumn derivatization with *ortho*-phthaldialdehyde. The amino acids were separated on a reversed phase column as described previously [66]. The conversion of 4-phenyl-2-butanone to 4-phenyl-2-butylamine was determined by GC-analysis (see chromatograms in Additional file 1: Figures S1, S2) on an Agilent 7890 A GC system equipped with a flame ionization detector (FID) using H_2 as carrier gas. Ketone and amine were separated on an achiral stationary phase using a 14% cyanopropylphenyl phase capillary column (J&W Scientific DB-1701; 30 m × 250 μm × 0.25 μm) with an injection and detection temperature of 250 °C (temperature program: 120 °C, 10 °C min^{-1} to 180 °C, 60 °C min^{-1} to 280 °C, hold 2 min).

Determination of enantiomeric excess

For the determination of the enantiomeric excess (*ee*) the regarding samples were derivatized by incubating them with pyridine (2 eq) and acetic anhydride (5 eq) for 2 h at 40 °C and 800 rpm. The reaction was quenched with aqueous saturated $NaHCO_3$ (250 μL) and extracted with EtOAc (2 × 250 μL). The combined organic layers were dried over Na_2SO_4 and subsequently analyzed on GC. The two amine enantiomers were separated on a chiral stationary phase (see chromatograms in Additional file 1: Figures S3, S4.) using a β-cyclodextrin capillary column (CP-ChiraSil-DEX CB; 30 m × 250 μm × 0.25 μm) with an injection and detection temperature of 250 °C (temperature program: 120 °C, 5 °C min^{-1} to 180 °C, hold 2 min).

Authors' contributions

VFW and WK designed the study. JEF, EL and NR performed the experiments. JEF, EL, NR, VFW and WK analyzed and interpreted the data. JEF, EL, VFW and WK wrote the manuscript. All authors read and approved the final manuscript.

Author details

¹ Austrian Centre of Industrial Biotechnology, ACIB GmbH, c/o University of Graz, Heinrichstrasse 28, 8010 Graz, Austria. ² Genetics of Prokaryotes, Faculty of Biology & CeBiTec, Bielefeld University, 33501 Bielefeld, Germany. ³ Institute of Chemistry, University of Graz, NAWI Graz, BioTechMed Graz, Heinrichstrasse 28, 8010 Graz, Austria.

Acknowledgements

COST Action CM1303 "Systems Biocatalysis" is acknowledged.

Competing interests

The authors declare that they have no competing interests.

Funding

This study was financed by the Austrian FFG, BMWFJ, BMVIT, SFG, Standortagentur Tirol and ZIT through the Austrian FFG-COMET-Funding Program.

References

1.   Soloshonok VA, Izawa K. Asymmetric synthesis and application of α-amino acids, vol. 1009. Washington: American Chemical Society; 2009.
2.   Wendisch VF. Microbial production of amino acids and derived chemicals: synthetic biology approaches to strain development. Curr Opin Biotechnol. 2014;30:51–8.
3.   Höhne M, Bornscheuer UT. Biocatalytic routes to optically active amines. ChemCatChem. 2009;1:42–51.
4.   Nugent TC, El-Shazly M. Chiral amine aynthesis—recent developments and trends for enamide reduction, reductive amination, and imine reduction. Adv Synth Catal. 2010;352:753–819.
5.   Bommarius AS. Biocatalysis: a status report. Annu Rev Chem Biomol Eng. 2015;6:319–45.
6.   Grogan G, Turner NJ. InspIRED by nature: NADPH-dependent imine reductases (IREDs) as catalysts for the preparation of chiral amines. Chem Eur J. 2016;22:1900–7.
7.   Schrittwieser JH, Velikogne S, Kroutil W. Biocatalytic imine reduction and reductive amination of ketones. Adv Synth Catal. 2015;357:1655–85.
8.   Gamenara D, de Maria PD. Enantioselective imine reduction catalyzed by imine reductases and artificial metalloenzymes. Org Biomol Chem. 2014;12:2989–92.
9.   Ghislieri D, Houghton D, Green AP, Willies SC, Turner NJ. Monoamine oxidase (MAO-N) catalyzed deracemization of tetrahydro-beta-carbolines: substrate dependent switch in enantioselectivity. ACS Catal. 2013;3:2869–72.
10. Li T, Liang J, Ambrogelly A, Brennan T, Gloor G, Huisman G, Lalonde J, Lekhal A, Mijts B, Muley S, Newman L, Tobin M, Wong G, Zaks A, Zhang XY. Efficient, chemoenzymatic process for manufacture of the boceprevir bicyclic [3.1.0]proline intermediate based on amine oxidase-catalyzed desymmetrization. J Am Chem Soc. 2012;134:6467–72.
11. Ahmed ST, Parmeggiani F, Weise NJ, Flitsch SL, Turner NJ. Chemoenzymatic synthesis of optically pure L- and D-Biarylalanines through biocatalytic asymmetric amination and palladium-catalyzed arylation. ACS Catal. 2015;5:5410–3.
12. Ye LJ, Toh HH, Yang Y, Adams JP, Snajdrova R, Li Z. Engineering of amine dehydrogenase for asymmetric reductive amination of ketone by evolving Rhodococcus phenylalanine dehydrogenase. ACS Catal. 2015;5:1119–22.
13. Zhang DL, Chen X, Zhang R, Yao PY, Wu QQ, Zhu DM. Development of beta-amino acid dehydrogenase for the synthesis of beta-amino acids via Reductive Amination of beta-keto acids. ACS Catal. 2015;5:2220–4.
14. Au SK, Bommarius BR, Bommarius AS. Biphasic reaction system allows for conversion of hydrophobic substrates by amine dehydrogenases. ACS Catal. 2014;4:4021–6.
15. Verho O, Bäckvall JE. Chemoenzymatic dynamic kinetic resolution: a powerful tool for the preparation of enantiomerically pure alcohols and amines. J Am Chem Soc. 2015;137:3996–4009.
16. Gotor-Fernández V, Busto E, Gotor V. Candida antarctica lipase B: an ideal biocatalyst for the preparation of nitrogenated organic compounds. Adv Synth Catal. 2006;348:797–812.
17. Nestl BM, Hammer SC, Nebel BA, Hauer B. New generation of biocatalysts for organic synthesis. Angew Chem Int Ed. 2014;53:3070–95.
18. Fuchs M, Farnberger JE, Kroutil W. The industrial age of biocatalytic transamination. Eur J Org Chem. 2015;15:6965–82.
19. Mathew S, Yun H. ω-Transaminases for the production of optically pure amines and unnatural amino acids. ACS Catal. 2012;2:993–1001.
20. Koszelewski D, Lavandera I, Clay D, Guebitz GM, Rozzell D, Kroutil W. Formal asymmetric biocatalytic reductive amination. Angew Chem Int Ed. 2008;47:9337–40.

21. Koszelewski D, Tauber K, Faber K, Kroutil W. Omega-transaminases for the synthesis of non-racemic alpha-chiral primary amines. Trends Biotechnol. 2010;28:324–32.

22. Fessner WD. Systems biocatalysis: development and engineering of cell-free "artificial metabolisms" for preparative multi-enzymatic synthesis. New Biotechnol. 2015;32:658–64.

23. Tessaro D, Pollegioni L, Piubelli L, D'Arrigo P, Servi S. Systems biocatalysis: an artificial metabolism for interconversion of functional groups. ACS Catal. 2015;5:1604–8.

24. Guo F, Berglund P. Transaminase biocatalysis: optimization and application. Green Chem. 2017;19:333–60.

25. Lee JW, Na D, Park JM, Lee J, Choi S, Lee SY. Systems metabolic engineering of microorganisms for natural and non-natural chemicals. Nat Chem Biol. 2012;8:536–46.

26. Keasling JD. Manufacturing molecules through metabolic engineering. Science. 2010;330:1355–8.

27. Quin MB, Schmidt-Dannert C. Designer microbes for biosynthesis. Curr Opin Biotechnol. 2014;29:55–61.

28. Bayer T, Milker S, Wiesinger T, Rudroff F, Mihovilovic MD. Designer microorganisms for optimized redox cascade reactions—challenges and future perspectives. Adv Synth Catal. 2015;357:1587–618.

29. Ladkau N, Schmid A, Buhler B. The microbial cell—functional unit for energy dependent multistep biocatalysis. Curr Opin Biotechnol. 2014;30:178–89.

30. Busto E, Gerstmann M, Tobol F, Dittmann E, Wiltschi B, Kroutil W. Systems biocatalysis: para-alkenylation of unprotected phenols. Catal Sci Technol. 2016;6:8098–103.

31. Wachtmeister J, Rother D. Recent advances in whole cell biocatalysis techniques bridging from investigative to industrial scale. Curr Opin Biotechnol. 2016;42:169–77.

32. France SP, Hepworth LJ, Turner NJ, Flitsch SL. Constructing biocatalytic cascades: in vitro and in vivo approaches to de novo multi-enzyme pathways. ACS Catal. 2017;7:710–24.

33. Schmidt-Dannert C, Lopez-Gallego F. A roadmap for biocatalysis—functional and spatial orchestration of enzyme cascades. Microb Biotechnol. 2016;9:601–9.

34. Muschiol J, Peters C, Oberleitner N, Mihovilovic MD, Bornscheuer UT, Rudroff F. Cascade catalysis—strategies and challenges en route to preparative synthetic biology. Chem Commun. 2015;51:5798–811.

35. Köhler V, Turner NJ. Artificial concurrent catalytic processes involving enzymes. Chem Commun. 2015;51:450–64.

36. Otte KB, Kittelberger J, Kirtz M, Nestl BM, Hauer B. Whole-cell one-pot biosynthesis of azelaic acid. ChemCatChem. 2014;6:1003–9.

37. Oberleitner N, Peters C, Muschiol J, Kadow M, Sass S, Bayer T, Schaaf P, Iqbal N, Rudroff F, Mihovilovic MD, Bornscheuer UT. An enzymatic toolbox for cascade reactions: a showcase for an in vivo redox sequence in asymmetric synthesis. ChemCatChem. 2013;5:3524–8.

38. Agudo R, Reetz MT. Designer cells for stereocomplementary de novo enzymatic cascade reactions based on laboratory evolution. Chem Commun. 2013;49:10914–6.

39. Gröger H, Chamouleau F, Orologas N, Rollmann C, Drauz K, Hummel W, Weckbecker A, May O. Enantioselective reduction of ketones with "designer cells" at high substrate concentrations: highly efficient access to functionalized optically active alcohols. Angew Chem Int Ed. 2006;45:5677–81.

40. Both P, Busch H, Kelly PP, Mutti FG, Turner NJ, Flitsch SL. Whole-cell biocatalysts for stereoselective C–H amination reactions. Angew Chem Int Ed. 2016;55:1511–3.

41. Wu SK, Zhou Y, Wang TW, Too HP, Wang DIC, Li Z. Highly regio- and enantioselective multiple oxy- and amino-functionalizations of alkenes by modular cascade biocatalysis. Nat Commun. 2016;7:11917.

42. Klatte S, Wendisch VF. Redox self-sufficient whole cell biotransformation for amination of alcohols. Bioorg Med Chem. 2014;22:5578–85.

43. Klatte S, Wendisch VF. Role of L-alanine for redox self-sufficient amination of alcohols. Microb Cell Fact. 2015;14:9.

44. Weber N, Gorwa-Grauslund M, Carlquist M. Exploiting cell metabolism for biocatalytic whole-cell transamination by recombinant Saccharomyces cerevisiae. Appl Microbiol Biotechnol. 2014;98:4615–24.

45. Weber N, Gorwa-Grauslund M, Carlquist M. Improvement of whole-cell transamination with Saccharomyces cerevisiae using metabolic engineering and cell pre-adaptation. Microb Cell Fact. 2017;16:3.

46. Kaulmann U, Smithies K, Smith MEB, HaileS HC, Ward JM. Substrate spectrum of omega-transaminase from Chromobacterium violaceum DSM30191 and its potential for biocatalysis. Enzyme Microb Technol. 2007;41:628–37.

47. Amann E, Ochs B, Abel KJ. Tightly regulated Tac promoter vectors useful for the expression of unfused and fused proteins in Escherichia coli. Gene. 1988;69:301–15.

48. Guzman LM, Belin D, Carson MJ, Beckwith J. Tight regulation, modulation, and high-level expression by vectors containing the arabinose P-Bad promoter. J Bacteriol. 1995;177:4121–30.

49. Lorenz E, Klatte S, Wendisch VF. Reductive amination by recombinant Escherichia coli: whole cell biotransformation of 2-keto-3-methylvalerate to L-isoleucine. J Biotechnol. 2013;168:289–94.

50. Ernst M, Kaup B, Müller M, Bringer-Meyer S, Sahm H. Enantioselective reduction of carbonyl compounds by whole-cell biotransformation, combining a formate dehydrogenase and a (R)-specific alcohol dehydrogenase. Appl Microbiol Biotechnol. 2005;66:629–34.

51. Kara S, Schrittwieser JH, Hollmann F, Ansorge-Schumacher MB. Recent trends and novel concepts in cofactor-dependent biotransformations. Appl Microbiol Biotechnol. 2014;98:1517–29.

52. Pazmiño DET, Riebel A, de Lange J, Rudroff F, Mihovilovic MD, Fraaije MW. Efficient biooxidations catalyzed by a new generation of self-sufficient baeyer–villiger monooxygenases. ChemBioChem. 2009;10:2595–8.

53. Blank LM, Ebert BE, Buehler K, Buhler B. Redox biocatalysis and metabolism: molecular mechanisms and metabolic network analysis. Antioxid Redox Signal. 2010;13:349–94.

54. Schrewe M, Julsing MK, Buhler B, Schmid A. Whole-cell biocatalysis for selective and productive C–O functional group introduction and modification. Chem Soc Rev. 2013;42:6346–77.

55. Chen H, Huang R, Zhang Y-HP. Systematic comparison of co-expression of multiple recombinant thermophilic enzymes in Escherichia coli BL21(DE3). Appl Microbiol Biotechnol. 2017;101:4481–93.

56. Walton AZ, Stewart JD. Understanding and improving NADPH-dependent reactions by nongrowing Escherichia coli cells. Biotechnol Prog. 2004;20:403–11.

57. Lilius EM, Multanen VM, Toivonen V. Quantitative extraction and estimation of intracellular nicotinamide nucleotides of Escherichia coli. Anal Biochem. 1979;99:22–7.

58. Richter N, Neumann M, Liese A, Wohlgemuth R, Weckbecker A, Eggert T, Hummel W. Characterization of a whole-cell catalyst co-expressing glycerol dehydrogenase and glucose dehydrogenase and its application in the synthesis of L-glyceraldehyde. Biotechnol Bioeng. 2010;106:541–52.

59. Sambrook J, Russell DW. Molecular cloning: a laboratory manual. 3rd ed. Cold Spring Harbor: Cold Spring Harbor Laboratoy Press; 2001.

60. Cohen SN, Chang AC, Hsu L. Nonchromosomal antibiotic resistance in bacteria—genetic transformation of Escherichia coli by R-factor DNA. Proc Natl Acad Sci USA. 1972;69:2110–4.

61. Hanahan D. Studies on transformation of Escherichia coli with plasmids. J Mol Biol. 1983;166:557–80.

62. Datsenko KA, Wanner BL. One-step inactivation of chromosomal genes in Escherichia coli K-12 using PCR products. Proc Natl Acad Sci USA. 2000;97:6640–5.

63. Gibson DG, Young L, Chuang RY, Venter JC, Hutchison CA, Smith HO. Enzymatic assembly of DNA molecules up to several hundred kilobases. Nat Methods. 2009;6:343–5.

64. Bradford MM. Rapid and sensitive method for quantitation of microgram quantities of protein utilizing principle of protein-dye binding. Anal Biochem. 1976;72:248–54.

65. Marienhagen J, Kennerknecht N, Sahm H, Eggeling L. Functional analysis of all aminotransferase proteins inferred from the genome sequence of Corynebacterium glutamicum. J Bacteriol. 2005;187:7639–46.

66. Schneider J, Eberhardt D, Wendisch VF. Improving putrescine production by Corynebacterium glutamicum by fine-tuning ornithine transcarbamoylase activity using a plasmid addiction system. Appl Microbiol Biotechnol. 2012;95:169–78.

18

Development of a silicon limitation inducible expression system for recombinant protein production in the centric diatoms *Thalassiosira pseudonana* and *Cyclotella cryptica*

Roshan P. Shrestha and Mark Hildebrand*

Abstract

Background: An inducible promoter for recombinant protein expression provides substantial benefits because under induction conditions cellular energy and metabolic capability can be directed into protein synthesis. The most widely used inducible promoter for diatoms is for nitrate reductase, however, nitrogen metabolism is tied into diverse aspects of cellular function, and the induction response is not necessarily robust. Silicon limitation offers a means to eliminate energy and metabolic flux into cell division processes, with little other detrimental effect on cellular function, and a protein expression system that works under those conditions could be advantageous.

Results: In this study, we evaluate a number of promoters for recombinant protein expression induced by silicon limitation and repressed by the presence of silicon in the diatoms *Thalassiosira pseudonana* and *Cyclotella cryptica*. In addition to silicon limitation, we describe additional strategies to elevate recombinant protein expression level, including inclusion of the 5′ fragment of the coding region of the native gene and reducing carbon flow into ancillary processes of pigment synthesis and formation of photosynthetic storage products. We achieved yields of eGFP to 1.8% of total soluble protein in *C. cryptica*, which is about 3.6-fold higher than that obtained with chloroplast expression and ninefold higher than nuclear expression in another well-established algal system.

Conclusions: Our studies demonstrate that the combination of inducible promoter and other strategies can result in robust expression of recombinant protein in a nuclear-based expression system in diatoms under silicon limited conditions, separating the protein expression regime from growth processes and improving overall recombinant protein yields.

Keywords: Diatom, *Thalassiosira pseudonana*, *Cyclotella cryptica*, Transformation, Inducible promoter, Silicon transporter, Silicon-limitation, Protein expression, Fatty acid synthesis inhibition

Background

A number of algal systems have been developed to produce nuclear-encoded recombinant proteins. In some algae, nuclear expression can be problematic due to epigenetic transgene silencing, in which an introduced gene loses its ability to be expressed [1]. In this case, several options are possible. The first relies on fusing the desired protein to a selectable marker, and maintaining selection conditions during expression [2, 3]. Incorporating a viral 2A amino acid sequence between the two fusion partners results in two separate protein products, so the desired recombinant protein can be generated. In one case in *Chlamydomonas reinhardtii*, this approach was used with a mutagenized cell line to reduce methylation that contributed to gene silencing [4]. Secretion of recombinant proteins into the growth media could be an alternative solution to problems in protein expression, including aggregation, incorrect folding or toxicity

*Correspondence: mhildebrand@ucsd.edu
Scripps Institution of Oceanography, University of California San Diego, La Jolla, CA, USA

[5]. A chloroplast expression system is another method [6]. Although there are many advantages to chloroplast expression [7–9], it is not suitable when recombinant proteins need post translational modifications, and in obligate phototrophic organisms that cannot grow on an external source of carbon, accumulation of a large quantity of a foreign protein could be detrimental to photosynthesis. The converse has been demonstrated; expression of recombinant GFP in the *C. reinhardtii* chloroplast was negatively affected by high light or low culture density [10], perhaps because of a drain on protein synthesis capability to replace photosynthetic protein turnover. Heterotrophic cultivation can alleviate this problem, but addition of extracellular carbon increases production cost and the risk of contamination. Nuclear-based expression can offer other advantages, for example, targeting of expressed protein to different intracellular compartments. In plants, protein expression yields are generally substantially improved when targeting is to the endoplasmic reticulum (ER), because of increased protein folding capability and stability in that compartment [11, 12]. There are algal species such as diatoms that do not suffer from epigenetic gene silencing [13], and developed algal nuclear expression systems include species of diatoms, green, and red algae, where a number of different proteins, including antigens and antibodies, and those capable of synthesizing bioplastics, were successfully expressed, as reviewed in [6, 14].

Transcriptional control elements such as the promoter exert a major control over gene transcript levels and ultimately protein production. Several constitutive promoters have been used to drive protein expression in algae [15]. Despite their advantages for production of proteins such as selectable markers or reporter proteins, they are not always desirable when expressing high levels of a protein for commercial purposes. For example, overproduced proteins can suppress host cell growth and metabolism by dominating the translation process and draining cellular biomolecules and energy from essential metabolic and growth processes. In fact, in the most highly developed recombinant protein expression systems, such as bacteria and yeasts, conditions in which cellular growth is blocked or severely retarded are desirable because they facilitate extra metabolic capability and energy flow into recombinant protein expression [16, 17]. Moreover, toxic proteins can kill the host organism or negatively affect growth [18]. In both scenarios, the use of inducible promoters in expression vectors is beneficial. These could be turned on or off via simple manipulations such as changing a nutrient concentration in the media or adding a chemical compound, and thus can control the timing of protein expression, for example,

during specific phase of the life or cell cycle, or under growth arrest conditions. Thus far, only a few inducible promoters are available for algal expression. The nitrate reductase (NR) promoter, which enables expression in the presence of nitrate or absence of nitrogen and repression by ammonium, is the best available inducible element. The NR promoter was developed for the pennate diatom *Cylindrotheca fusiformis* [19], and subsequently adapted for other diatoms including *T. pseudonana* [20] and *Phaeodactylum tricornutum* [21]. This promoter has been used to control expression of proteins making bioplastics [22] and IgG antibodies [23] in *P. tricornutum*. The NR promoter has been used in other algae, including *Chlorella ellipsoidea, C. vulgaris, Dunaliella salina* and *C. reinhardtii* [24–27]. In addition, a few other inducible promoter systems, which can be induced by chemicals, physical factors such as heat, and deficiency of certain elements and nutrients have been reported. The *Cpx1 and Cyc6* genes of *C. reinhardtii* are one of a few chemically regulated genes, the expression of which were elevated under copper deficiency or addition of nickel or cobalt in the medium [28]. The low CO_2-inducible *CAH1* promoter and heat inducible promoters of heat shock proteins such as *hsp70A* of *C. reinhardtii* were shown to be useful without adding any toxic heavy metals [29, 30]. The arylsulphatase promoter of *Volvox carteri* was shown to be a useful inducible promoter under sulfur starvation conditions [31, 32].

All of the published inducible promoters to date are inducible under conditions of cell growth, or involve conditions that are detrimental to the cell. Because the most highly developed recombinant protein expression systems in other organisms utilize conditions in which metabolic capacity and energy flow into recombinant protein expression is maximized [16, 17], we wanted to develop a similar capability in diatoms. Under silicon limited conditions, cell cycle progression and growth in diatoms is blocked, but other aspects of cellular metabolism are not negatively affected [33, 34]. We anticipate that these conditions will be amenable to recombinant protein expression. To this end, we report on the development of inducible expression systems for diatoms based on promoters driving expression of silicon transporters (SITs) and other genes concomitantly expressed with SITs in silicon (Si)-rich and Si-deficient media. The SITs are downregulated under sufficient silicic acid concentrations—a condition where silicic acid uptake occurs primarily by diffusion, and are highly upregulated during silicon starvation [35, 36]. This system allows separation of the cell growth phase and the recombinant protein production phase [37], which enables channeling of energy and metabolic capacity normally used for cell cycle progression.

Results

Identification of genes whose transcripts are upregulated by Si limitation

Thalassiosira pseudonana silicon transporter TpSIT1 (Thaps3_268895) and TpSIT2 (Thaps3_41392) mRNA levels are upregulated by Si limitation and downregulated under Si replete conditions, with TpSIT1 having a higher induction than TpSIT2 [35, 38]. Using whole genome microarray data from *T. pseudonana* [39], we identified other genes whose mRNA expression patterns clustered with TpSIT1 and 2 (Fig. 1a), and designated the expression elements promoting them silicon starvation inducible promoters (SSIPs). In a separate experiment, RNAseq transcriptome sequencing [39]

was done on a time course of Si limitation, which included a time point 2 h prior to placing cells in Si-free medium, corresponding to exponential growth in Si replete conditions. All clustered genes were induced during Si limitation, but TpSIT1 was induced far higher than the others (Fig. 1b). In fact, mRNA for TpSIT1 was the second most abundant transcript in the entire genome under Si limiting conditions. Plotting FPKM (Fragments Per Kilobase of transcript per Million mapped reads) for the −2 h time point (Fig. 1c) showed that TpSIT1 had the second lowest RNA abundance in the cluster under uninduced conditions, but Fig. 1d shows that transcript levels were induced over 500-fold by 8 h Si limitation.

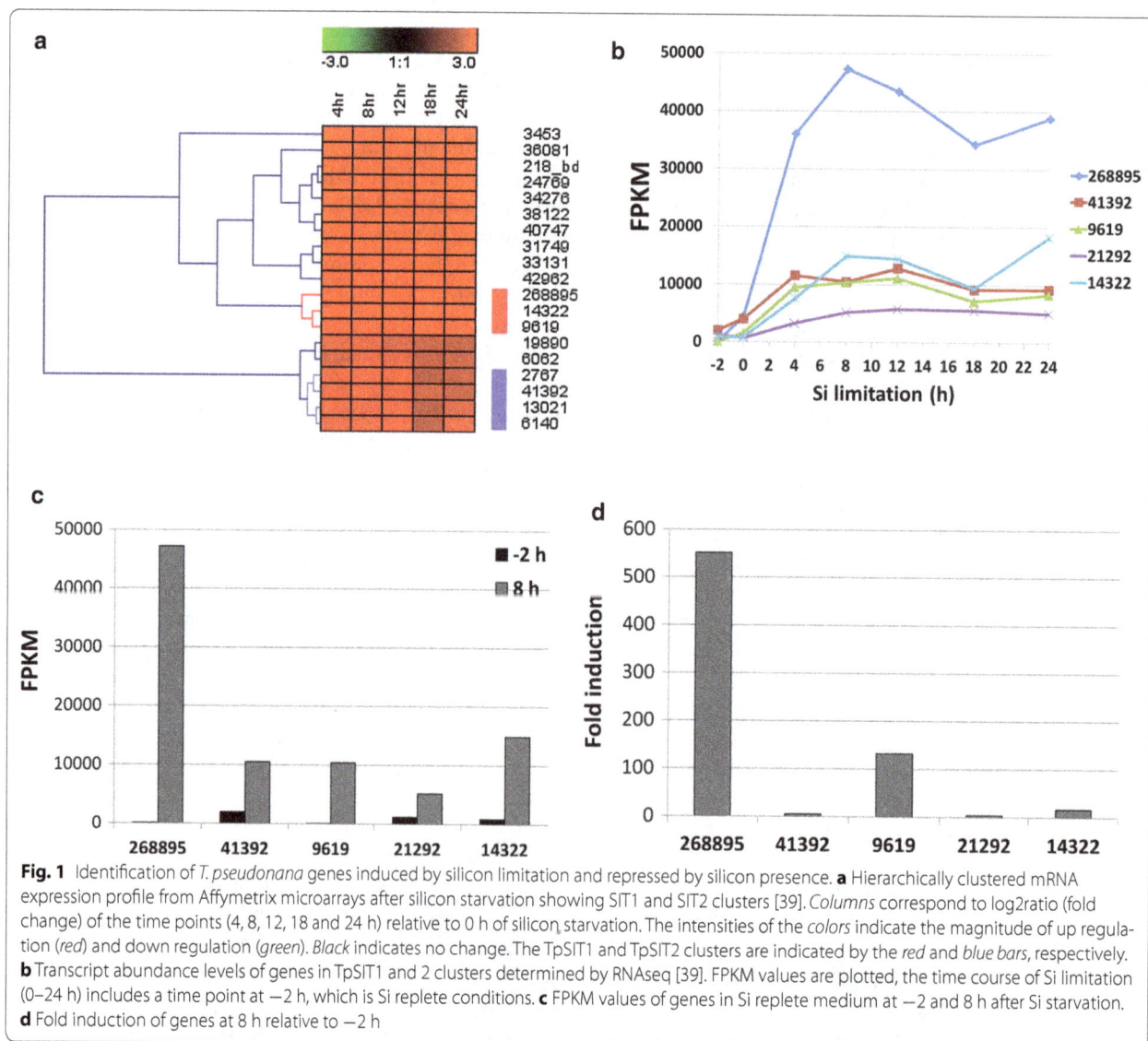

Fig. 1 Identification of *T. pseudonana* genes induced by silicon limitation and repressed by silicon presence. **a** Hierarchically clustered mRNA expression profile from Affymetrix microarrays after silicon starvation showing SIT1 and SIT2 clusters [39]. *Columns* correspond to log2ratio (fold change) of the time points (4, 8, 12, 18 and 24 h) relative to 0 h of silicon starvation. The intensities of the *colors* indicate the magnitude of up regulation (*red*) and down regulation (*green*). *Black* indicates no change. The TpSIT1 and TpSIT2 clusters are indicated by the *red* and *blue bars*, respectively. **b** Transcript abundance levels of genes in TpSIT1 and 2 clusters determined by RNAseq [39]. FPKM values are plotted, the time course of Si limitation (0–24 h) includes a time point at −2 h, which is Si replete conditions. **c** FPKM values of genes in Si replete medium at −2 and 8 h after Si starvation. **d** Fold induction of genes at 8 h relative to −2 h

SSIP promoters strongly induce protein expression during silicon deprivation

To initially evaluate the characteristics of the SSIPs with regard to protein expression, we constructed vectors expressing eGFP under control of their promoters for cytoplasmically targeted expression. Antibiotic-resistant transgenic clones were selected from agar plates, and we evaluated only the best GFP expressing lines by selecting clones with the brightest GFP fluorescence for further study. We also included a construct of a SIT gene from *C. cryptica*, CcSIT1_g10780.t1 to study expression in that species. Fluorescence microscopy revealed a high level of expression from all constructs under Si starvation, and eGFP occupied a large portion of the cytoplasm (Fig. 2).

Using imaging flow cytometry to evaluate large numbers of cells, we observed that not all cells in a clonal population expressed eGFP under Si-limiting conditions, and also detected a few cells expressing eGFP in Si-replete cultures. The lack of expression in all cells has been seen with other promoters under other culture conditions [40], so the phenomenon is common. These phenotypes are likely due to epigenetic effects. Because overall expressed protein yield depends on the level of

expression per cell and the percentage of cells in the population that are expressing, we evaluated three parameters related to expression, the mean eGFP fluorescence per cell, the percentage of cells expressing detectable levels of eGFP, and the net expression of eGFP, which was the product of the mean expression times the percentage of expressing cells. It should be noted that because on the order of 20,000 cells are being evaluated in these and subsequent measurements, the standard error is miniscule, and error bars are not typically visible.

Comparison of the SSIPs revealed different responses (Fig. 3). Variation was observed with the different promoters, but there was a consistent substantial increase in eGFP fluorescence and in the percentage of cells expressing under Si starvation (Fig. 3). Overall, at least 70% of the cells expressed eGFP after 24 h of Si starvation, and less than 6% of the cells expressed eGFP in Si-replete media. Amongst the *T. pseudonana* genes, SIT1 exhibited the highest gain in net expression, whereas Thaps3_9619 showed the highest absolute net expression. CcSIT1 also demonstrated a high degree of inducibility (Fig. 3).

Constitutive promoters expressing under both Si-replete and Si-deplete condition

To evaluate the effect of silicon limitation on protein expression in general, we also analyzed expression of eGFP under the control of other promoters commonly used in diatoms, namely the nitrate inducible NR promoter, and constitutive promoters of fucoxanthin chlorophyll a/c-binding protein (fcp) and ribosomal protein rpL41. In all cases, net eGFP expression was increased under Si limitation conditions, with improvements of 1.04-fold for rpL41, 1.2-fold for NR, and 1.5-fold for fcp (Fig. 4). Interestingly, transcript levels only for fcp were induced under these conditions, both NR and rpL41 were slightly reduced or there was no change (Additional file 1: Figure S1). This suggests that the improvement in yield is due to factors other than intrinsic capability to synthesize protein based on transcript abundance.

Strategies to increase expression level of recombinant proteins: inclusion of the N-terminal region of corresponding genes

Because TpSIT1 had the highest transcript level (Fig. 1) and most well-regulated eGFP expression response (Fig. 3) of all promoters, we focused on this promoter for subsequent evaluation and improvement of the protein expression system in *T. pseudonana*. An analysis of eGFP appearance under the control of the SIT1 promoter was performed using imaging flow cytometry to determine the timing of response of protein accumulation upon transfer of exponentially grown cells into Si-free medium. While most of the cell population was fluorescent by 24 h,

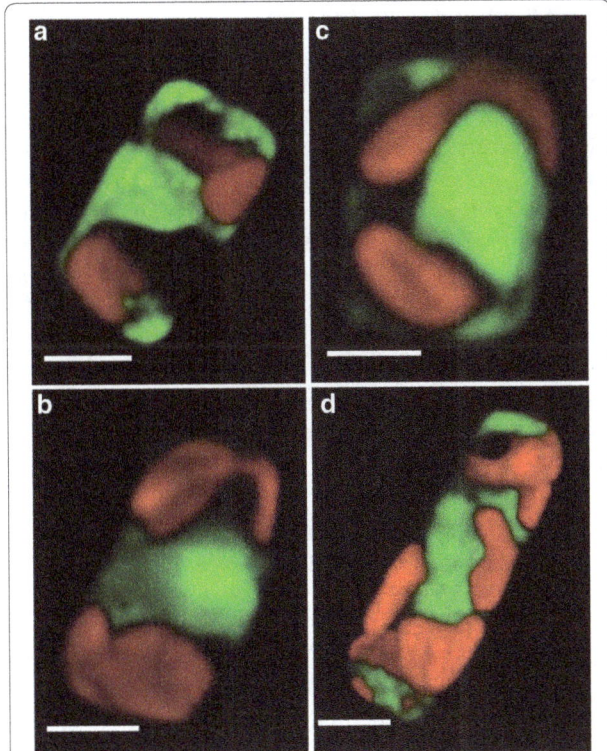

Fig. 2 Expression of eGFP under control of the Si-inducible promoters of *T. pseudonana* and *C. cryptica* after 24 h of Si starvation showing cytoplasmic localization. **a** TpSIT1 (Thaps3_268895); **b** TpSIT2 (Thaps3_41392); 9619, **c** (Thaps3_9619); and **d** CcSIT1, (g10780.t1). Scale bar 2 μM

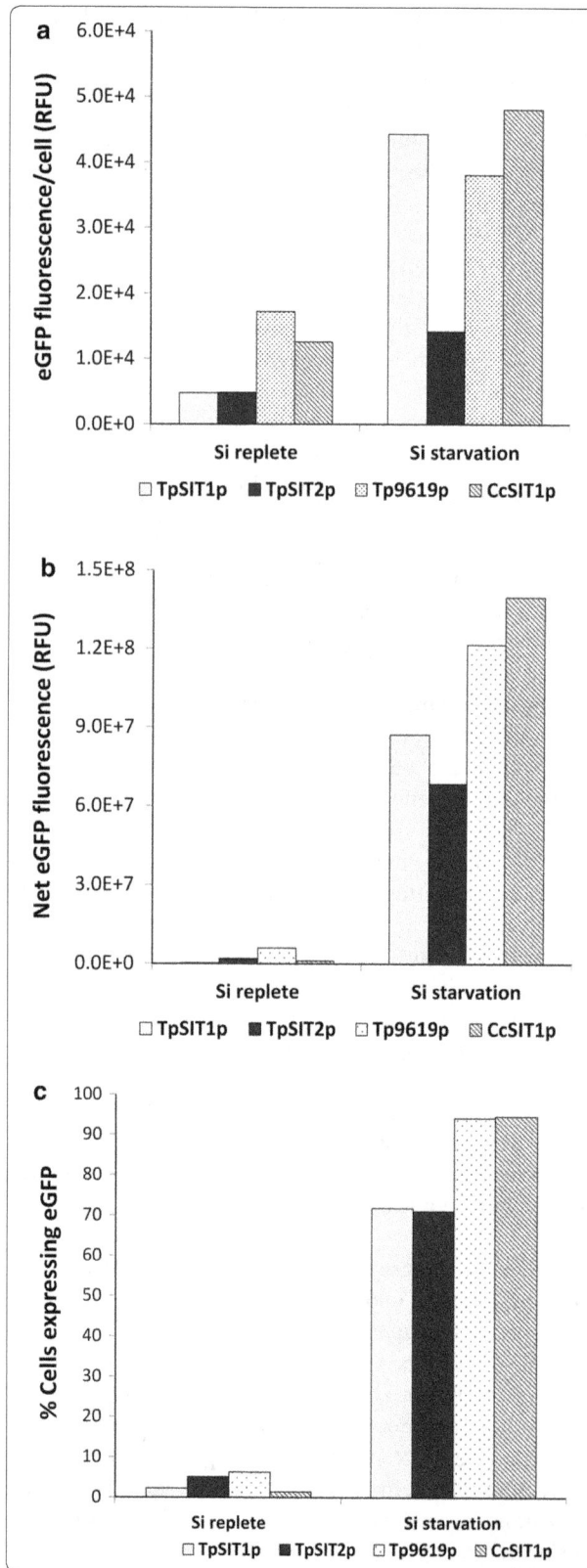

◀ **Fig. 3** Induction level of *T. pseudonana* and *C. cryptica* SSIPs in Si-replete media and after 24 h of Si starvation as determined by imaging flow cytometry. **a** Mean eGFP fluorescence per cell. **b** Net eGFP fluorescence calculated as product of eGFP intensity/cell and no. of cells expressing eGFP. **c** Percentage of cells expressing eGFP. eGFP was measured with a 488-nm laser with 50 mw power output without a blocking filter. *Error bars* represent standard errors of mean (SEM), n = 20,000. TpSIT1p, Thaps3_268895; TpSIT2p, Thaps3_41392; Tp9619p, Thaps3_9619; and CcSIT1p, g10780.t1

some cells expressed eGFP as early as 6 h (Fig. 5), showing cellular heterogeneity. Hence, we used the 24 h time point to compare overall eGFP expression level in this study.

We determined that eGFP had an unfavorable codon bias for *T. pseudonana* (Additional file 2: Figure S2), which could inhibit the translatability of the transcript. In attempt to enhance protein production with efficient translation initiation, we fused the 32-residues long N-terminal sequence of TpSIT1 gene (which has optimal codon usage—Additional file 2: Figure S2), to eGFP eliminating the amino-terminal codon bias of eGFP. We observed about twofold higher net expression of eGFP when theTpSIT1 N-terminus was present in comparison to eGFP alone (Fig. 6).

Quantification of eGFP yield

Although eGFP fluorescence is a simple way to estimate expression levels, it does not take into account possible expressed protein that may not be properly folded to enable fluorescence. We therefore quantified respective SIT1 promoter-driven eGFP production in *T. pseudonana* and *C. cryptica* after 24-h silicon-limitation by immunoblotting using different amounts of total soluble protein (TSP) extracted in PBS by sonication. The blot was probed with anti-GFP monoclonal antibodies, and the intensity compared against known amounts of purified recombinant *Aequorea victoria* GFP standard. The eGFP yield was about 1.8 µg per mg total soluble protein (TSP) (0.18% of TSP) in *T. pseudonana* (Fig. 7a) and 12.7 µg per mg TSP (1.3% of TSP) in *C. cryptica* (Fig. 7b).

Strategies to increase expression level of recombinant proteins: channeling metabolites towards recombinant protein production

In addition to selection and optimization of promoters and genes, we hypothesized that inhibition of the functions of chloroplasts, including synthesis of photosynthetic pigments and lipids, will channel metabolic capability and energy towards synthesis of recombinant protein. To test this hypothesis, various inhibitors were added to the media during Si starvation (Table 1). We

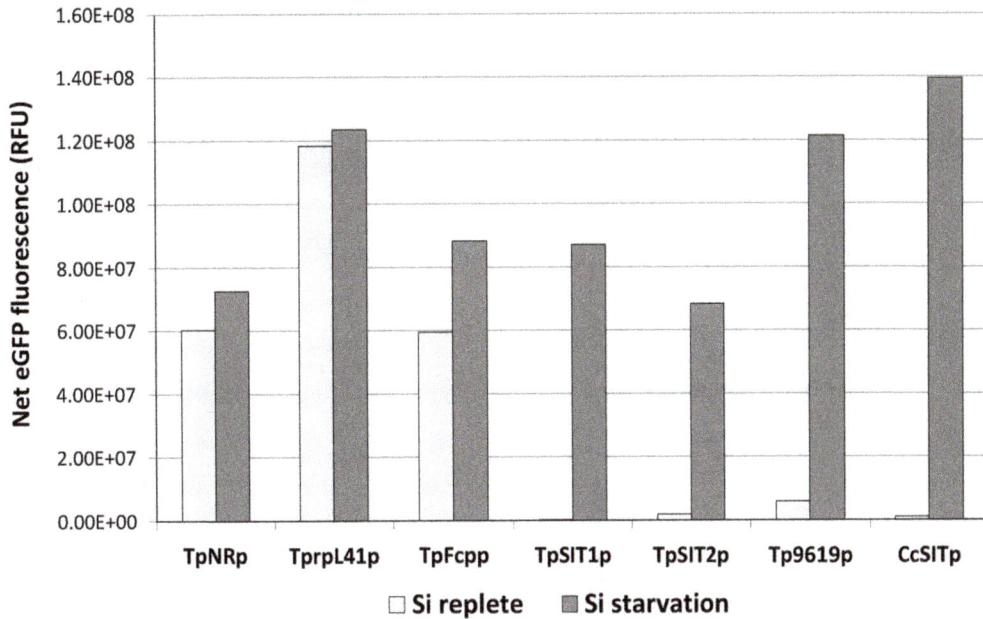

Fig. 4 Comparison of eGFP expression driven by Si inducible and non Si inducible promoters in Si-replete media and after 24 h of Si starvation. The net eGFP fluorescence was calculated as product of eGFP intensity/cell and no. of cells expressing eGFP derived from imaging flow cytometry. To compensate for different levels of expression, eGFP was measured with a 488-nm laser with 100 mW power output without filter (TpNRp, TprpL41p, TpFcpp) and 30 mW power output after blocking saturation with neutral density filter 1.0 (TpSIT1p, TPSIT2p, Thaps3_9619p, and CcSITp). *Error bars* represent standard errors of mean (SEM), n = 20,000. *NR* nitrate reductase promoter, *rpL41p* ribosomal protein L41 promoter, *Fcpp* fucoxanthin chlorophyll a/c binding protein promoter

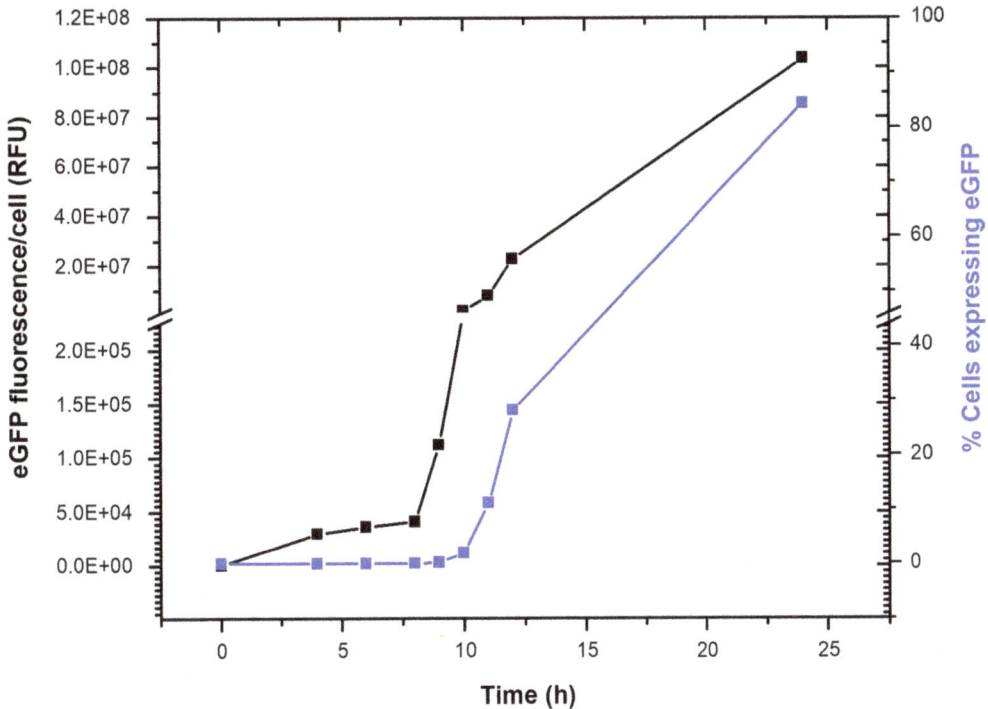

Fig. 5 A time-course expression analysis of eGFP expression under control of TpSIT1 promoter in *Thalassiosira pseudonana*. Exponentially growing cells were transferred to silicon-deprived media (0 h) and expression of eGFP was followed by measuring eGFP fluorescence using an imaging flow cytometer (a 488-nm laser with 100 mW power output without a blocking filter). *Error bars* represent standard errors of mean (SEM), n = 20,000

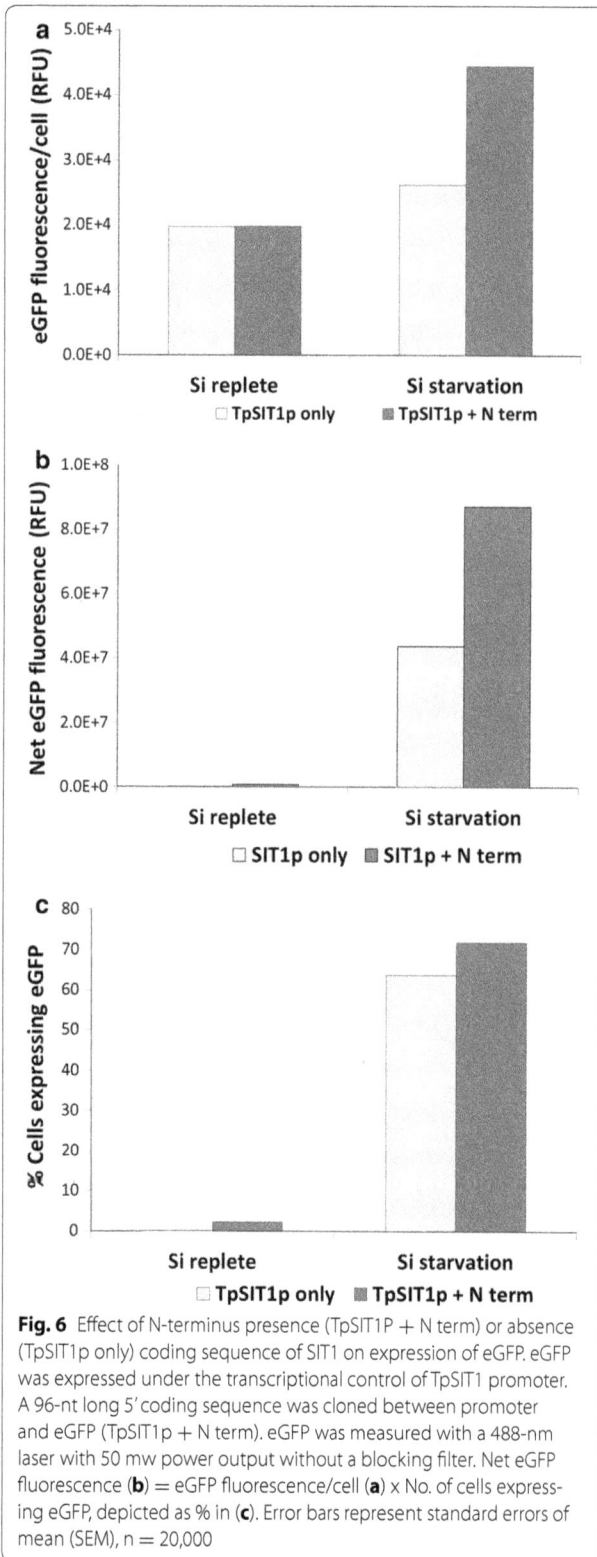

Fig. 6 Effect of N-terminus presence (TpSIT1P + N term) or absence (TpSIT1p only) coding sequence of SIT1 on expression of eGFP. eGFP was expressed under the transcriptional control of TpSIT1 promoter. A 96-nt long 5′ coding sequence was cloned between promoter and eGFP (TpSIT1p + N term). eGFP was measured with a 488-nm laser with 50 mw power output without a blocking filter. Net eGFP fluorescence (**b**) = eGFP fluorescence/cell (**a**) x No. of cells expressing eGFP, depicted as % in (**c**). Error bars represent standard errors of mean (SEM), n = 20,000

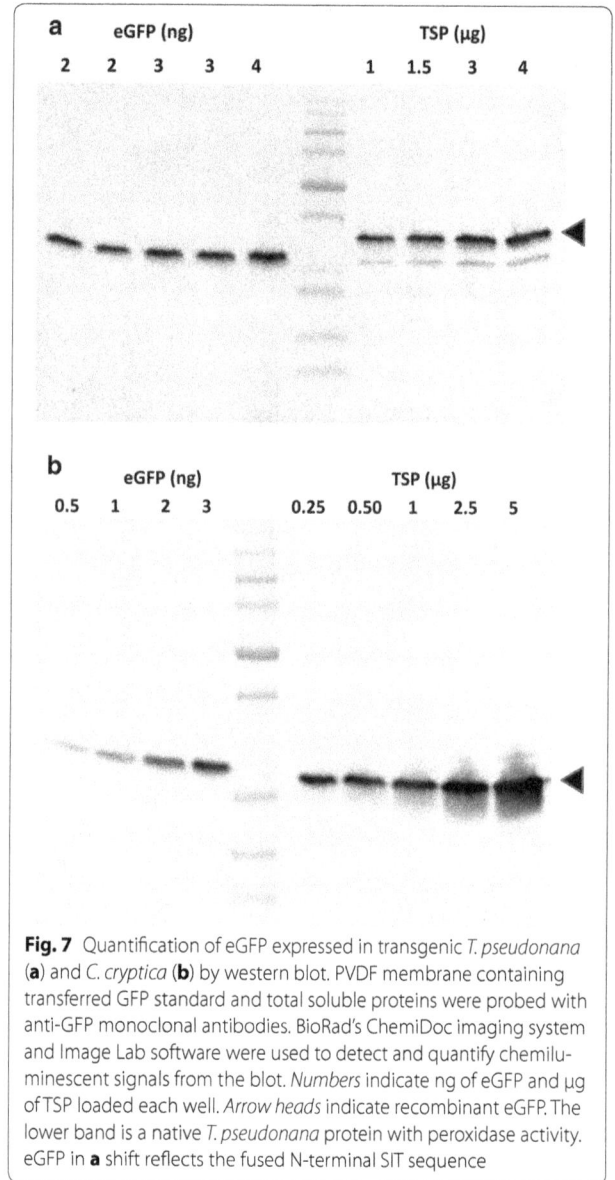

Fig. 7 Quantification of eGFP expressed in transgenic *T. pseudonana* (**a**) and *C. cryptica* (**b**) by western blot. PVDF membrane containing transferred GFP standard and total soluble proteins were probed with anti-GFP monoclonal antibodies. BioRad's ChemiDoc imaging system and Image Lab software were used to detect and quantify chemiluminescent signals from the blot. *Numbers* indicate ng of eGFP and μg of TSP loaded each well. *Arrow heads* indicate recombinant eGFP. The lower band is a native *T. pseudonana* protein with peroxidase activity. eGFP in **a** shift reflects the fused N-terminal SIT sequence

Table 1 List of inhibitors and their mode of action

Inhibitors	Mode of action
Carotenoid biosynthesis inhibitors	
Dithiothreitol (DTT)	Inhibits de-epoxidation violaxanthin to zeaxanthin [41]
Norflurazon	Inhibits phytoene desaturase [41]
Isoxaflutole	Inhibits 4-hydroxyphenylpyruvate dioxygenase (HPPD) [42]
Lipid biosynthesis inhibitors	
Sethoxydim	Inhibits acetyl CoA carboxylase (ACCase) [43]

prescreened the effect of various concentrations of these inhibitors on *T. pseudonana* cell viability, and selected only those concentrations at which cells were able to survive albeit with a slight decrease in cell growth. Figure 8 shows that dithiothreitol (DTT), isoxaflutole (IFT), norflurazon (NF) and sethoxydim increased recombinant

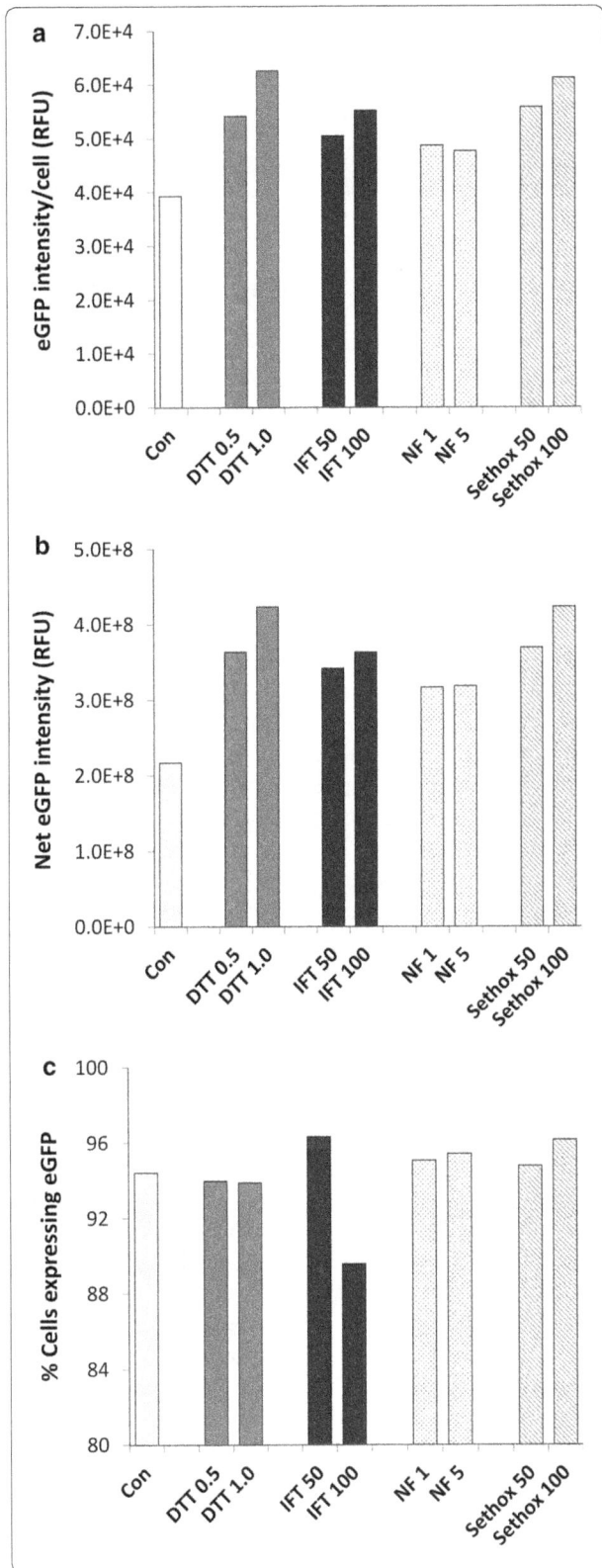

Fig. 8 Effect of carotenoid and lipid synthesis inhibitors on recombinant protein expression, under control of SSIP1 promoter, in *T. pseudonana* after 24 h of Si starvation. Expression of eGFP was measured using an imaging flow cytometer (488 nM laser/100 mW with neutral density filter 0.6). *Numbers under bars* indicate concentrations of the inhibitors dithiothreitol (DTT): mM, isoxaflutole (IFT): µg/mL, norflurazon (NF): µM, sethoxydim (Sethox): µM. **a** eGFP intensity/cell; **b** Net eGFP intensity; **c** percentage of cells expressing eGFP. *Error bars* represent standard errors of mean (SEM), n = 20,000. Control sample was without any addition of inhibitors

eGFP expression level in *T. pseudonana*, though all these treatments also affected normal cell metabolism as evident from the decrease in chlorophyll level (Additional file 3: Figure S3). Depending upon concentration, DTT and sethoxydim were the most efficient eGFP booster with up to twofold increase in net eGFP intensity. IFT and NF were also able to increase eGFP level, but less efficiently than DTT and sethoxydim. We also tested a combination of inhibitors to examine if they synergistically increased eGFP level. Cells were treated with combinations of DTT and sethoxydim. We observed that most of the combinations were too lethal for cells. Thus, the most useful inhibitors to increase recombinant protein were DTT and sethoxydim when used solely.

Time course of productivity

We evaluated the relation between initiation of induction by silicon starvation and eGFP yield to determine if there was an optimal time for maximum yield. Cells of *T. pseudonana* and *C. cryptica* expressing eGFP under respective SIT1 control elements were incubated in a 1-L flask under Si deprived conditions and eGFP fluorescence was monitored every 24 h. *T. pseudonana* cells in silicondeprived medium usually remain healthy only until 48 h, then cell intactness starts to deteriorate. We observed that, for *T. pseudonana*, the net eGFP intensity level increased up to the second day and then declined by day 3 (Fig. 9a). Net eGFP level on day 2 was about 40% higher than day 1. On the other hand, *C. cryptica* cells continued to express eGFP up to 4 days, with maximal productivity on the third day. Net eGFP intensity level on day 2 was twice that on day 1, which further increased up to 150% on day 3 (Fig. 9b).

In order to determine the maximum recombinant protein yield with strategies we developed, we further quantified eGFP expression in *C. cryptica* after treatment with DTT or sethoxydim and harvesting on the third day of silicon starvation. ELISA analysis showed that the productivity reached up to 18 µg eGFP per mg

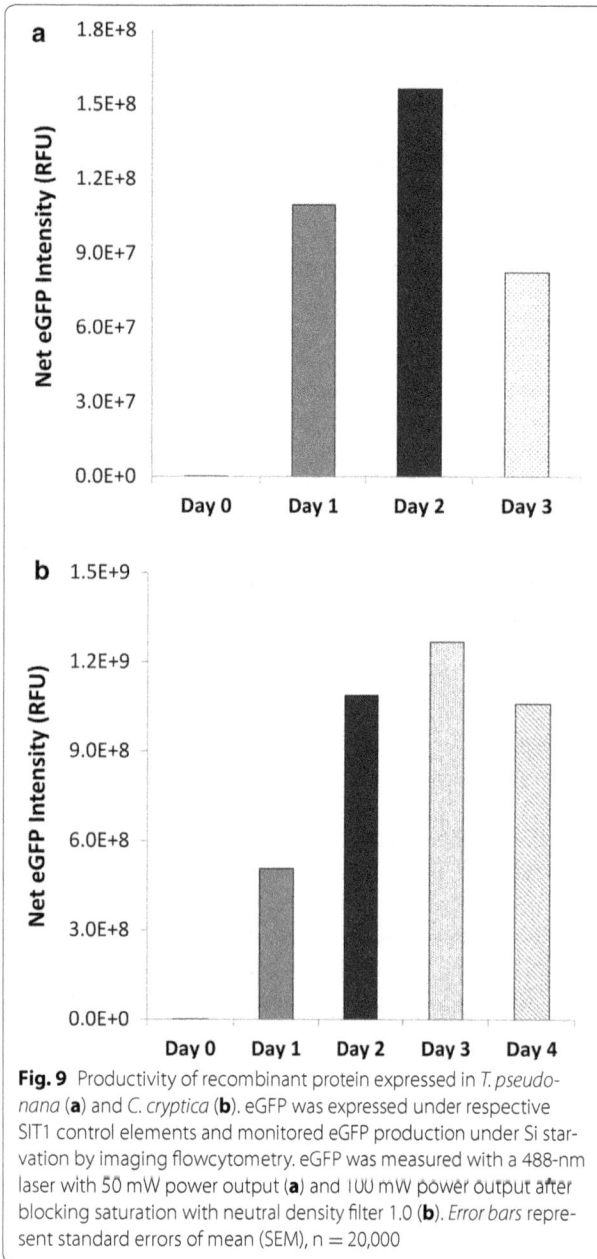

Fig. 9 Productivity of recombinant protein expressed in *T. pseudonana* (**a**) and *C. cryptica* (**b**). eGFP was expressed under respective SIT1 control elements and monitored eGFP production under Si starvation by imaging flowcytometry. eGFP was measured with a 488-nm laser with 50 mW power output (**a**) and 100 mW power output after blocking saturation with neutral density filter 1.0 (**b**). *Error bars* represent standard errors of mean (SEM), n = 20,000

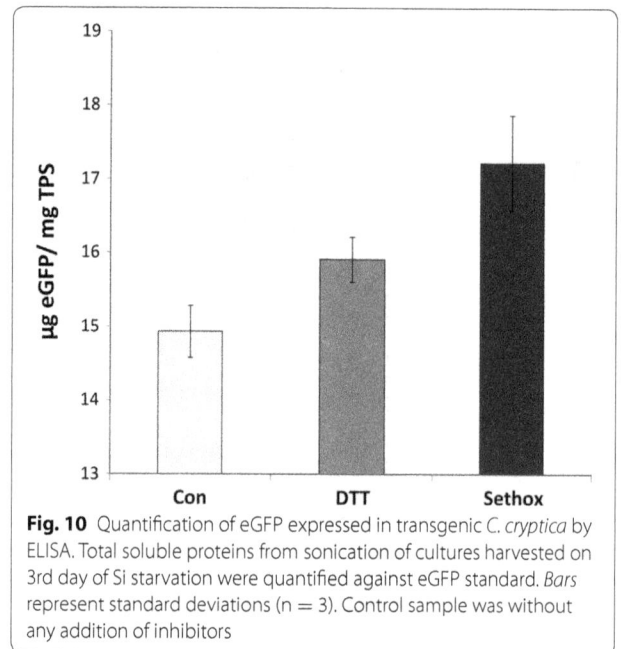

Fig. 10 Quantification of eGFP expressed in transgenic *C. cryptica* by ELISA. Total soluble proteins from sonication of cultures harvested on 3rd day of Si starvation were quantified against eGFP standard. *Bars* represent standard deviations (n = 3). Control sample was without any addition of inhibitors

TSP (1.8% of TSP), which corresponds to about a 19.8% increase when treated with sethoxydim relative to the control (Fig. 10).

Discussion

Inducible promoters for versatile and selective induction of recombinant proteins are one of the essential components for high level protein production. Such promoters can minimize detrimental effects on cell growth from overproduction of recombinant proteins or from toxic proteins. Moreover, it is highly desirable to enable expression under favorable metabolic conditions related to the availability of carbon and nitrogen, and during cessation of cell division, where cellular energy and metabolic capability can be channeled into protein synthesis [16, 44]. Only a few inducible algal promoters have been characterized thus far, and these typically rely on induction during growth (reviewed in [14]). Here, we present diatom inducible promoters that overexpress proteins under silicon-limitation conditions. A combination of high level expression coupled to lack of metabolic activities associated with cell division and ancillary processes synergistically increased protein production.

Silicon is a unique requirement for diatoms to make their silicified cell walls, and in nearly all species silicon is required for cell cycle progression [45]. In the absence of silicon, diatom cells cease to divide, but because silicon is not tightly tied into other aspects of cellular metabolism [33], other metabolic activities such as RNA and protein synthesis continue to occur [34]. Evidence for excess energy and carbon availability during silicon limitation is also seen by the ability of diatoms to accumulate large amounts of neutral lipids under these conditions [46–48].

The transport of silicon into diatom cells occurs by two mechanisms, (1) under lower extracellular Si concentrations, uptake is facilitated by silicon transporter (SIT) proteins [49], and (2) under higher extracellular Si concentrations unassisted transmembrane diffusion of the small uncharged silicic acid molecule occurs [50]. SIT transcripts are upregulated in diatoms under silicon limitation [35, 36], presumably as a scavenging response, but because of silicic acid diffusion, the transporters are not needed under silicon-replete conditions, and their transcripts are downregulated. Thus, robust growth occurs without the need for SIT-mediated transport. We have documented a correlation between SIT mRNA and protein levels and cell cycle processes, suggesting that the transporters may play a more dominant role in sensing the absence of silicon, rather than being the major means of import of silicon into the cell under Si-replete conditions [35]. The combination of potentially favorable metabolic and energetic conditions brought about by cell cycle cessation under silicon limiting conditions, and the upregulation of SITs during that time, led us to explore and develop the use of promoters from SITs and similarly regulated genes to drive recombinant protein expression in diatoms.

In *T. pseudonana* we examined a whole genome microarray dataset [39] for genes with expression characteristics similar to the SITs (Fig. 1), and identified several silicon starvation induced proteins. Comparison of RNAseq-based transcript profiles (Fig. 1) and fusion constructs to eGFP (Figs. 2, 3) enabled rapid semi-quantitative screening of the performance of different SSIPs. Transcript levels for TpSIT1 were the second highest in the entire transcriptome under Si limited conditions and were among the least abundant among the SSIP candidates during growth. From that standpoint, the TpSIT1 promoter had the most promising characteristics to drive recombinant protein expression under Si limitation, with minimal expression occurring during growth under Si replete conditions. We also identified a high performance SIT promoter from *C. cryptica* [51], which showed a higher level of eGFP accumulation than those from *T. pseudonana* (Fig. 3). Interestingly, three non-SSIP promoters tested also expressed GFP at higher levels during silicon starvation (Fig. 4). Although other factors may contribute, these results are consistent with an improved ability to synthesize protein due to greater availability of energy and metabolic capacity during silicon starvation division arrest than when cell division is occurring. In a different diatom species, over a 12 h period, total cellular protein levels increased 1.5-fold under Si deplete conditions, compared with 2.5-fold under Si-replete conditions [34]. Considering that protein related to cell division is not needed under Si deplete conditions, the data indicate little detrimental effect on the ability of Si limited cells to synthesize protein.

Our laboratory has pioneered the application of imaging flow cytometry to microalgae, and this approach provides unique insights into population-scale differences in a variety of cellular phenomena [52]. Among the benefits of this approach is the outcome of rigorous statistics by evaluating 10,000 cells in a population, standard error in these analyses is miniscule. The approach enables rapid assessment of detailed characteristics of thousands of cells in a population, and in our analysis, revealed one factor not commonly noted in protein expression. In addition to the mean eGFP fluorescence per cell, which is a proxy for expression level, we showed a substantial variation in the percentage of cells in the population that were expressing at all (Fig. 3). Since this occurs in a clonal population, the most likely explanation is epigenetic effects, perhaps related to differences in intracellular metabolic processes or silicon requirements. It is beyond the scope of the current investigation to explore this aspect of control over expression, however, the data suggest that significant gains in recombinant protein yield could result from overcoming epigenetic effects. Transgene localization is one of the main epigenetic factors [53], but current routine methods of gene transformation, including particle bombardment or electroporation, results in random integration of transgenes. Insertion of transgene at desired safe harbor sites in the genome [54, 55] or use of stable replicating episomes [56] that enable consistent high level expression without negative consequences to other essential genes' expression is highly desirable. The cloning at pre-determined chromosomal location will be only possible when an efficient homologous recombination method is developed for diatoms. No routine homologous recombination methods have been developed for eukaryotic algae yet except in the haploid eustigmatophyte *Nannochloropsis* sp. [57], but a recently developed CRISPR method for the model microalga *C. reinhardtii*, *T. pseudonana* and *P. tricornutum* is promising [58–60]. Until the availability of routine homologous recombination methods, flow cytometry can be used to select the highest eGFP expression lines, which could minimize both chromosomal location limitations, and epigenetic effects.

One of the strategies to improve recombinant protein yields is to reduce metabolic drain towards unwanted but major competing carbon sinks. Expression levels were consistently higher in the presence of various inhibitors (Fig. 8), consistent with this hypothesis. DTT and sethoxydim resulted in higher eGFP production after Si starvation for 24 h both in *T. pseudonana* and *C. cryptica* (Fig. 10). The resulting increase in recombinant protein could be due to inhibition of violaxanthin de-epoxidase [41], and thus decreased carbon flow towards violaxanthin and other carotenoids, including fucoxanthin, diadinoxanthin, and diatoxanthin [61]. We documented a reduction in chlorophyll fluorescence in the presence of the inhibitors, but it should be considered that was an indirect measure using 488-nm excitation, which is more optimal for excitation of accessory pigments which channel photons into chlorophyll. Similarly, sethoxydim inhibits AcetylCoA carboxylase [43] and should reduce neutral lipid formation, resulting in more carbon and energy available for production of recombinant proteins. Use of chemical inhibitors in a production system could be prohibitively expensive; however, similar phenotypic changes could be engineered into the cell by manipulation of genes involved in the carotenoid or lipid biosynthetic pathways. The knockdown approach using RNAi or antisense RNA [35] expression could be the best solution by inhibiting certain enzyme activities at certain stage of cell growth, for example by controlling expression with the SIT1 promoter. In addition, reducing fucoxanthin, the main accessory pigment in diatoms, could improve light utilization in a high-density culture and lead to an increased overall productivity [62, 63].

Other factors involved in enabling high level expression were examined. First, we found that incorporation of 32 N-terminal amino acids of TpSIT1 (upstream sequence of the first transmembrane domain) increased eGFP level by twofold in comparison to eGFP alone (Fig. 6). The positive effect seen here could be due to more efficient translation initiation related to codon usage (Additional file 2: Figure S2) [17, 64]. On the other hand, it has been shown in bacteria that mRNA structure but not the codon usage at the beginning of genes is the primary factor for efficient translation [65]. If this is true also in diatoms, fusion of native 5' coding sequence to a codon-optimized heterologous gene should be the method of choice to enhance protein yield. Secondly, we expressed eGFP in a larger diatom, C. cryptica, reasoning that larger cells should have a higher proportion of cytoplasmic volume to accumulate recombinant proteins and hence will result into a higher yield in terms per TSP. Expression levels were approximately eightfold higher in C. cryptica compared with T. pseudonana (Fig. 9), consistent with this. Moreover, productivity of C. cryptica continued to increase until 72 h, compared with maximal productivity at 48 h with T. pseudonana.

Quantitative evaluation of expression levels using ELISA indicated that we achieved 1.8% of total soluble protein expressing eGFP in C. cryptica (Fig. 10). It is important to appreciate that expressed protein yields are highly dependent on the individual protein, and that comparisons of yields between different proteins can be misleading. Moreover, caution should be taken when comparing "typical" protein yields in expression systems, the yield is very much dependent on the particular protein. Comparison of yield of the same protein between systems is valid. In C. reinhardtii, chloroplast-expressed yields of GFP were 0.5% TSP [10]. Improvement of nuclear-based expression of GFP in C. reinhardtii has been the focus of other work which included a mutagenesis and genetic screen to improve yields [66]. In spite of the sophistication of that approach, yields were only on the order of 0.2% TSP. Our results indicate a 3.6-fold

higher yield than that obtained with chloroplast expression and ninefold higher than nuclear expression in C. reinhardtii. Given their early stage of development, the ultimate capability of diatom-based expression systems is unknown, however expression of recombinant IgG under growth conditions in the non-silicified diatom P. tricornutum yielded 8.7% TSP [67].

A number of improvements could further increase expression efficiency. These include molecular optimization of the transgene such as codon optimization (eGFP described in this work was codon optimized) or interruption of coding sequences by heterologous introns [68, 69]. Protein subcellular localization is also an important factor. Unlike chloroplast-based expression, a potential advantage of nuclear-based expression is that recombinant proteins can be targeted to multiple intracellular compartments. We demonstrated that combined cytoplasmic and chloroplast targeting of a nuclear transgene increased protein yield by 17% in T. pseudonana over cytoplasmic targeting alone [40]. Endoplasmic reticulum targeting has proven extremely effective in increasing recombinant protein yields in plants due to improved protein folding and stability [11, 12].

In addition to the fundamental technology for nuclear-based recombinant protein expression in diatoms being established in this and other [22, 23, 67] studies, diatoms provide an attractive system for low-cost, large scale recombinant protein expression. Diatoms are among the most productive unicellular microalgae on the planet, and tend to outcompete other classes of algae for growth [70], indicating an innate efficiency for converting sunlight and nutrients into biomass. Large scale cultivation of diatoms and other algae in the context of biofuels and bioproducts will establish productive and large scale systems from which low cost recombinant protein could be one desirable product [71, 72].

Methods
Culture conditions

Thalassiosira pseudonana (CCMP1335) and C. cryptica (CCMP332) stock cultures were grown in ASW medium [34]. Cultures were grown in 125-mL Erlenmeyer flasks on an orbital shaker under continuous illumination of cool-white fluorescent lamps at 150 μE m^{-2} s^{-1} at 18 °C. For larger volume, cultures were grown in 2 or 8 L glass flasks on magnetic stirrers with air bubbling under continuous light as above.

Vector construction
Gateway expression vectors

Gateway frame B cassette was cloned into EcoRV site of pBluescript generating the destination vector pMHL_71. T. pseudonana Gateway™ expression vectors containing eGFP flanked by 5′ and 3′ flanking regions comprised of promoter and terminator of SIT of *T. pseudonana* and *C. cryptica* were constructed by LR clonase recombination of corresponding entry vectors and the destination vector pMHL_71 (Additional file 4: Figure S4). Genes of interest were PCR amplified using primers containing corresponding Gateway att sequences (Additional file 5: Table S1), cloned into the destination vectors using MultiSite Gateway® Pro kit (Life Technologies). Each transformation vector was cotransformed with pMHL_9 expressing *nat1* gene under control of ACCase promoter (received from N. Kroger), which confers resistance to antibiotics nourseothricin.

Diatom transformation

Expression vectors were introduced into *T. pseudonana* using microparticle bombardment using the Bio-Rad Biolistic PDS-1000/He particle delivery system [13, 19]. Briefly, exponentially grown cells were harvested and 1×10^8 cells were plated in a 5 cm diameter circle in the middle of ASW-agar plate lacking antibiotics. Transformation was then performed by bombarding plasmid DNA coated tungsten beads (M-17, 1.1 μm diameter, BioRad) onto cells plated on the agar plate (1.5% Bacto agar) under vacuum at a distance of 8 cm at 1350 psi. Plates were bombarded twice to obtain higher transformation efficiency. Immediately after bombardment, cells were overlaid with 10 mL of ASW, and placed under light for 24 h. The cells were then plated on NEPC agar containing 100 μg/mL nourseothricin (clonNAT, Werner BioAgents, Germany). Resistant colonies were then picked and transferred into each well of 24 well plates containing 2 mL of ASW with clonNat. Positive clones were then confirmed by PCR using clonNat specific primers as described below.

Genomic DNA extraction and PCR

Algal cultures grown in 24 well plates were centrifuged ($10,000g$, 2 min) after 1 week, washed once with PBS and resuspended in 25 μL of PBS. The cells were then subjected to three cycles of freeze/defreeze by placing in dry ice for 5 min and heating 2 min at 70 °C. The lysate was then heated for 5 min at 95 °C to denature DNAse. After centrifugation, the supernatants containing genomic DNA were stored at −20 °C until further use. Several antibiotic-resistant clones were screened by growing in 24 wells plates containing antibiotics, followed by confirmation by PCR.

Western blot

Cells were sonicated twice in ice for 30 s each using a microtip probe (50% duty cycle, Vibra Cell, Sonics & Materials, Inc. CT, USA). The lysate was centrifuged, and the concentration of the supernatant containing total protein extract was measured using Bio-Rad protein assay kit before loading equal amount of protein into each well of BioRad precast gel. PVDF membranes blotted with proteins were incubated with anti-GFP monoclonal antibody (*Living Colors® A.v.* JL-8, Clontech), followed by HRP-conjugated anti-mouse secondary antibodies (Pierce). The blots were then processed using Super signal West Pico kit (Pierce) and exposed on Bio-Rad ChemiDoc system to detect and quantify chemiluminescent signals using Bio-Rad Image Lab 4.1 software.

Imagestream flowcytometry

Cells transformed with eGFP-tag were analyzed with an ImageStream Imaging Flowcytometer X-100 (Amnis Corp.) using a 100-mW 488-nm laser excitation. At least 20,000 fluorescent as well as brightfield and side scatter cell images per sample were collected at 40× magnifications. Data were then analyzed using image analysis software (IDEAS v5, Amnis Corp).

Fluorescence microscopy

eGFP expressing cells were photographed with 63×/1.4 objective oil immersion plan APO using Axiovision 4.7.2 software in a Zeiss AxioObserver inverted microscope equipped with an Apotome (Carl Zeiss, NY). The GFP filter set used was Zeiss #38HE (Ex 470/40 nm, FT 495, Em 525/50 nm), and chlorophyll was imaged using filter set #05 (Ex 395–440 nm, FT 460 nm, Em 470 nm LP).

ELISA

Transgenic diatom cells were sonicated in ice. The total protein extracts were centrifuged, and collected supernatant containing soluble recombinant eGFP. GFP ELISA Kit (Cell Biolabs) was used to quantify eGFP according to manufacturer's instruction.

Additional files

Additional file 1: Figure S1. Transcript level change after silicon starvation as depicted by microarray analysis. A, FCP; B, rpL41 and NR.

Additional file 2: Figure S2. Codon bias of eGFP (upper panel) and 5′ end (96 bp) of TpSIT1 (lower panel) for *T. pseudonana*. The analysis was performed with Rare Codon Calculator (http://www.codons.org/).

Additional file 3: Figure S3. Effect of inhibitors dithiothreitol (DTT), isoxaflutole (IFT), norflurazon (NF) and sethoxydim (sethoxy) on cell no. (A) and chlorophyll level (B) measured using Imagestream imaging cytometer.

Additional file 4: Figure S4. Schematic diagrams of the Gateway-system based *T. pseudonana* transformation vectors expressing eGFP under the transcriptional control of silicon limitation inducible promoters (SSIPs). SSIP1, TpSIT1 Thaps3_268895; SSIP2, TpSIT2 Thaps3_41392; SSIP3, Thaps3_9619; and SSIP4, CcSIT1 g10780.t1. These vectors were cotransformed with a plasmid vector expressing *Nat1* conferring resistance to the antibiotic nourseothricin.

Additional file 5: Table S1. List of primers used.

Authors' contributions

RPS and MH conceived the ideas for this study. RPS constructed the plasmid vectors and performed experiments. RPS and MH wrote the manuscript. Both authors read and approved the final manuscript.

Acknowledgements

This study was supported by US Department of Energy Grant DEEE0003373, and a grant from the US Department of Agriculture (USDA AFRI # 2012-67015-30197).

Competing interests

The authors declare that they have no competing interests.

References

1. Specht E, Miyake-Stoner S, Mayfield S. Micro-algae come of age as a platform for recombinant protein production. Biotechnol Lett. 2010;32:1373–83.
2. Rasala BA, Lee PA, Shen Z, Briggs SP, Mendez M, Mayfield SP. Robust expression and secretion of xylanase1 in Chlamydomonas reinhardtii by fusion to a selection gene and processing with the FMDV 2A peptide. PLoS ONE. 2012;7:e43349.
3. Plucinak TM, Horken KM, Jiang W, Fostvedt J, Nguyen ST, Weeks DP. Improved and versatile viral 2A platforms for dependable and inducible high-level expression of dicistronic nuclear genes in Chlamydomonas reinhardtii. Plant J. 2015;82:717–29.
4. Kong F, Yamasaki T, Kurniasih SD, Hou L, Li X, Ivanova N, Okada S, Ohama T. Robust expression of heterologous genes by selection marker fusion system in improved Chlamydomonas strains. J Biosci Bioeng. 2015;120:239–45.
5. Sevastsyanovich YR, Leyton DL, Wells TJ, Wardius CA, Tveen-Jensen K, Morris FC, Knowles TJ, Cunningham AF, Cole JA, Henderson IR. A generalised module for the selective extracellular accumulation of recombinant proteins. Microb Cell Fact. 2012;11:69.
6. Rasala BA, Gimpel JA, Tran M, Hannon MJ, Miyake-Stoner SJ, Specht EA, Mayfield SP. Genetic engineering to improve algal biofuels production. In: Borowitzka MA, Moheimani NR, editors. Algae for biofuels and energy. Dordrecht: Springer; 2013. p. 99–113.
7. Fischer R, Stoger E, Schillberg S, Christou P, Twyman RM. Plant-based production of biopharmaceuticals. Curr Opin Plant Biol. 2004;7:152–8.
8. Heifetz PB. Genetic engineering of the chloroplast. Biochimie. 2000;82:655–66.
9. Michelet L, Lefebvre-Legendre L, Burr SE, Rochaix JD, Goldschmidt-Clermont M. Enhanced chloroplast transgene expression in a nuclear mutant of Chlamydomonas. Plant Biotechnol J. 2011;9:565–74.
10. Franklin S, Ngo B, Efuet E, Mayfield SP. Development of a GFP reporter gene for Chlamydomonas reinhardtii chloroplast. Plant J. 2002;30:733–44.
11. Conrad U, Fiedler U. Compartment-specific accumulation of recombinant immunoglobulins in plant cells: an essential tool for antibody production and immunomodulation of physiological functions and pathogen activity. Plant Mol Biol. 1998;38:101–9.
12. Faye L, Boulaflous A, Benchabane M, Gomord V, Michaud D. Protein modifications in the plant secretory pathway: current status and practical implications in molecular pharming. Vaccine. 2005;23:1770–8.
13. Dunahay TG, Jarvis EE, Roessler PG. Genetic transformation of the diatoms Cyclotella cryptica and Navicula saprophila. J Phycol. 1995;31:1004–12.
14. Shrestha RP, Haerizadeh F, Hildebrand M. Molecular genetic manipulation of microalgae: principles and applications. In: Richmond A, Hu Q, editors. Handbook of microalgal culture. 2nd ed. Oxford: Blackwell; 2013. p. 146–67.
15. Leon-Banares R, Gonzalez-Ballester D, Galvan A, Fernandez E. Transgenic microalgae as green cell-factories. Trends Biotechnol. 2004;22:45–52.
16. Ahmad M, Hirz M, Pichler H, Schwab H. Protein expression in Pichia pastoris: recent achievements and perspectives for heterologous protein production. Appl Microbiol Biotechnol. 2014;98:5301–17.
17. Rosano GL, Ceccarelli EA. Recombinant protein expression in Escherichia coli: advances and challenges. Front Microbiol. 2014;5:172.
18. Surzycki R, Greenham K, Kitayama K, Dibal F, Wagner R, Rochaix JD, Ajam T, Surzycki S. Factors effecting expression of vaccines in microalgae. Biologicals. 2009;37:133–8.
19. Poulsen N, Kröger N. A new molecular tool for transgenic diatoms. Control of mRNA and protein biosynthesis by an inducible promoter–terminator cassette. FEBS J. 2005;272:3413–23.
20. Poulsen N, Chesley PM, Kröger N. Molecular genetic manipulation of the diatom Thalassiosira pseudonana (Bacillariophyceae). J Phycol. 2006;42:1059–65.
21. Miyagawa A, Okami T, Kira N, Yamaguchi H, Ohnishi K, Adachi M. High efficiency transformation of the diatom Phaeodactylum tricornutum with a promoter from the diatom Cylindrotheca fusiformis. Phycol Res. 2009;57:142–6.
22. Hempel F, Bozarth AS, Lindenkamp N, Klingl A, Zauner S, Linne U, Steinbüchel A, Maier UG. Microalgae as bioreactors for bioplastic production. Microb Cell Fact. 2011;10:81.
23. Hempel F, Maier UG. An engineered diatom acting like a plasma cell secreting human IgG antibodies with high efficiency. Microb Cell Fact. 2012;11:126.
24. Niu Y, Zhang M, Xie W, Li J, Gao Y, Yang W, Liu J, Li H. A new inducible expression system in a transformed green alga, Chlorella vulgaris. Genet Mol Res. 2011;10:3427–34.
25. Wang P, Sun Y, Li X, Zhang L, Li W, Wang Y. Rapid isolation and functional analysis of promoter sequences of the nitrate reductase gene from Chlorella ellipsoidea. J Appl Phycol. 2004;16:11–6.
26. Xie H, Xu P, Jia Y, Li J, Lu Y, Xue L. Cloning and heterologous expression of nitrate reductase genes from Dunaliella salina. J Appl Phycol. 2007;19:497–504.
27. Llamas A, Igeño MI, Galván A, Fernández E. Nitrate signalling on the nitrate reductase gene promoter depends directly on the activity of the nitrate transport systems in Chlamydomonas. Plant J. 2002;30:261–71.
28. Quinn JM, Kropat J, Merchant S. Copper response element and Crr1-Dependent Ni2+-responsive promoter for induced, reversible gene expression in Chlamydomonas reinhardtii. Eukaryot Cell. 2003;2:995–1002.
29. Ferrante P, Diener D, Rosenbaum J, Giuliano G. Nickel and low CO$_2$-controlled motility in Chlamydomonas through complementation of a paralyzed flagella mutant with chemically regulated promoters. BMC Plant Biol. 2011;11:22.
30. Schroda M, Blöcker D, Beck CF. The HSP70A promoter as a tool for the improved expression of transgenes in Chlamydomonas. Plant J. 2000;21:121–31.
31. Hallmann A, Sumper M. Reporter genes and highly regulated promoters as tools for transformation experiments in Volvox carteri. Proc Natl Acad Sci USA. 1994;91:11562–6.
32. Hallmann A, Sumper M. An inducible arylsulfatase of Volvox carteri with properties suitable for a reporter-gene system. Eur J Biochem. 1994;221:143–50.
33. Claquin P, Martin-Jézéquel V, Kromkamp JC, Veldhuis MJ, Kraay GW. Uncoupling of silicon compared with carbon and nitrogen metabolisms and the role of the cell cycle in continuous cultures of Thalassiosira pseudonana (Bacillariophyceae) under light, nitrogen, and phosphorus control. J Phycol. 2002;38:922–30.
34. Darley W, Volcani B. Role of silicon in diatom metabolism. A silicon requirement for deoxyribonucleic acid synthesis in the diatom Cylindrotheca fusiformis Reimann and Lewin. Exp Cell Res. 1969;58:334–42.
35. Shrestha RP, Hildebrand M. Evidence for a regulatory role of diatom silicon transporters in cellular silicon responses. Eukaryot Cell. 2015;14:29–40.
36. Thamatrakoln K, Hildebrand M. Analysis of Thalassiosira pseudonana silicon transporters indicates distinct regulatory levels and transport activity through the cell cycle. Eukaryot Cell. 2007;6:271–9.
37. Werner S, Breus O, Symonenko Y, Marillonnet S, Gleba Y. High-level recombinant protein expression in transgenic plants by using a double-inducible viral vector. Proc Natl Acad Sci USA. 2011;108:14061–6.

38. Shrestha RP, Tesson B, Norden-Krichmar T, Federowicz S, Hildebrand M, Allen AE. Whole transcriptome analysis of the silicon response of the diatom *Thalassiosira pseudonana*. BMC Genom. 2012;13:499.

39. Smith SR, Glé C, Abbriano RM, Traller JC, Davis A, Trentacoste E, Vernet M, Allen AE, Hildebrand M. Transcript level coordination of carbon pathways during silicon starvation-induced lipid accumulation in the diatom *Thalassiosira pseudonana*. New Phytol. 2016;210:890–904.

40. Davis A, Crum LT, Corbeil LB, Hildebrand M. Expression of *Histophilus somni* IbpA DR2 protective antigen in the diatom *Thalassiosira pseudonana*. Appl Microbiol Biotechnol. 2017;101:5313–24.

41. Lohr M, Wilhelm C. Algae displaying the diadinoxanthin cycle also possess the violaxanthin cycle. Proc Natl Acad Sci USA. 1999;96:8784–9.

42. Dan Hess F. Light-dependent herbicides: an overview. Weed Sci. 2000;48:160–70.

43. Zhekisheva M, Zarka A, Khozin-Goldberg I, Cohen Z, Boussiba S. Inhibition of astaxanthin synthesis under high irradiance does not abolish triacylglycerol accumulation in the green alga *Haematococcus pluvialis* (Chlorophyceae). J Phycol. 2005;41:819–26.

44. Rosano GL, Ceccarelli EA. Recombinant protein expression in *Escherichia coli*: advances and challenges. Front Microbiol. 2014;5:172.

45. Martin-Jezequel V, Hildebrand M, Brzezinski MA. Silicon metabolism in diatoms: implications for growth. J Phycol. 2000;36:821–40.

46. Roessler PG. Effects of silicon deficiency on lipid composition and metabolism in the diatom *Cyclotella cryptica*. J Phycol. 1988;24:394–400.

47. Yu W-L, Ansari W, Schoepp NG, Hannon MJ, Mayfield SP, Burkart MD. Modifications of the metabolic pathways of lipid and triacylglycerol production in microalgae. Microb Cell Fact. 2011;10:91.

48. Traller JC, Hildebrand M. High throughput imaging to the diatom *Cyclotella cryptica* demonstrates substantial cell-to-cell variability in the rate and extent of triacylglycerol accumulation. Algal Res. 2013;2:244–52.

49. Hildebrand M, Volcani BE, Gassmann W, Schroeder JI. A gene family of silicon transporters. Nature. 1997;385:688–9.

50. Thamatrakoln K, Hildebrand M. Silicon uptake in diatoms revisited: a model for saturable and nonsaturable uptake kinetics and the role of silicon transporters. Plant Physiol. 2008;146:1397–407.

51. Traller JC, Cokus SJ, Lopez DA, Gaidarenko O, Smith SR, McCrow JP, Gallaher SD, Podell S, Thompson M, Cook O. Genome and methylome of the oleaginous diatom *Cyclotella cryptica* reveal genetic flexibility toward a high lipid phenotype. Biotechnol Biofuels. 2016;9:258.

52. Hildebrand M, Davis A, Abbriano R, Pugsley HR, Traller JC, Smith SR, Shrestha RP, Cook O, Sánchez-Alvarez EL, Manandhar-Shrestha K. Applications of imaging flow cytometry for microalgae. Methods Mol Biol. 2016;1389:47–67.

53. Dahodwala H, Sharfstein ST. Role of epigenetics in expression of recombinant proteins from mammalian cells. Pharm Bioprocess. 2014;2:403–19.

54. Ronda C, Maury J, Jakočiūnas T, Jacobsen SAB, Germann SM, Harrison SJ, Borodina I, Keasling JD, Jensen MK, Nielsen AT. CrEdit: CRISPR mediated multi-loci gene integration in *Saccharomyces cerevisiae*. Microb Cell Fact. 2015;14:97.

55. Lee JS, Kallehauge TB, Pedersen LE, Kildegaard HF. Site-specific integration in CHO cells mediated by CRISPR/Cas9 and homology-directed DNA repair pathway. Sci Rep. 2015;5:8572.

56. Karas BJ, Diner RE, Lefebvre SC, McQuaid J, Phillips AP, Noddings CM, Brunson JK, Valas RE, Deerinck TJ, Jablanovic J, Gillard JT, Beeri K, Ellisman MH, Glass JI, Hutchison CA, Smith HO, Venter JC, Allen AE, Dupont CL, Weyman PD. Designer diatom episomes delivered by bacterial conjugation. Nat Commun. 2015;6:6925.

57. Kilian O, Benemann CS, Niyogi KK, Vick B. High-efficiency homologous recombination in the oil-producing alga *Nannochloropsis* sp. Proc Natl Acad Sci USA. 2011;108:21265–9.

58. Jiang W, Brueggeman AJ, Horken KM, Plucinak TM, Weeks DP. Successful transient expression of Cas9 and single guide RNA genes in *Chlamydomonas reinhardtii*. Eukaryot Cell. 2014;13:1465–9.

59. Nymark M, Sharma AK, Sparstad T, Bones AM, Winge P. A CRISPR/Cas9 system adapted for gene editing in marine algae. Sci Rep 2016;6:24951.

60. Hopes A, Nekrasov V, Kamoun S, Mock T. Editing of the urease gene by CRISPR-Cas in the diatom *Thalassiosira pseudonana*. Plant Methods. 2016;12:49.

61. Lavaud J, Materna AC, Sturm S, Vugrinec S, Kroth PG. Silencing of the violaxanthin de-epoxidase gene in the diatom *Phaeodactylum tricornutum* reduces diatoxanthin synthesis and non-photochemical quenching. PLoS ONE. 2012;7:e36806.

62. de Mooij T, Janssen M, Cerezo-Chinarro O, Mussgnug JH, Kruse O, Ballottari M, Bassi R, Bujaldon S, Wollman F-A, Wijffels RH. Antenna size reduction as a strategy to increase biomass productivity: a great potential not yet realized. J Appl Phycol. 2015;27:1063–77.

63. Kirst H, Formighieri C, Melis A. Maximizing photosynthetic efficiency and culture productivity in cyanobacteria upon minimizing the phycobilisome light-harvesting antenna size. Biochim Biophys Acta Bioenerg. 2014;1837:1653–64.

64. Wang B, Wang Y, Kennedy C. 5'-coding sequence of the nasA gene of *Azotobacter vinelandii* is required for efficient expression. FEMS Microbiol Lett. 2014;359:201–8.

65. Bentele K, Saffert P, Rauscher R, Ignatova Z, Blüthgen N. Efficient translation initiation dictates codon usage at gene start. Mol Syst Biol. 2013;9:675.

66. Neupert J, Karcher D, Bock R. Generation of Chlamydomonas strains that efficiently express nuclear transgenes. Plant J. 2009;57:1140–50.

67. Hempel F, Lau J, Klingl A, Maier UG. Algae as protein factories: expression of a human antibody and the respective antigen in the diatom *Phaeodactylum tricornutum*. PLoS ONE. 2011;6:e28424.

68. Lacy-Hulbert A, Thomas R, Li X, Lilley C, Coffin R, Roes J. Interruption of coding sequences by heterologous introns can enhance the functional expression of recombinant genes. Gene Ther. 2001;8:649–53.

69. Hasannia S, Lotfi A, Mahboudi F, Rezaii A, Rahbarizadeh F, Mohsenifar A. Elevated expression of human alpha-1 antitrypsin mediated by yeast intron in *Pichia pastoris*. Biotechnol Lett. 2006;28:1545–50.

70. Carter C, Ross A, Schiel D, Howard-Williams C, Hayden B. In situ microcosm experiments on the influence of nitrate and light on phytoplankton community composition. J Exp Mar Biol Ecol. 2005;326:1–13.

71. Rasala BA, Mayfield SP. Photosynthetic biomanufacturing in green algae; production of recombinant proteins for industrial, nutritional, and medical uses. Photosynth Res. 2015;123:227–39.

72. Hildebrand M, Davis AK, Smith SR, Traller JC, Abbriano R. The place of diatoms in the biofuels industry. Biofuels. 2012;3:221–40.

PERMISSIONS

All chapters in this book were first published in MCF, by BioMed Central; hereby published with permission under the Creative Commons Attribution License or equivalent. Every chapter published in this book has been scrutinized by our experts. Their significance has been extensively debated. The topics covered herein carry significant findings which will fuel the growth of the discipline. They may even be implemented as practical applications or may be referred to as a beginning point for another development.

The contributors of this book come from diverse backgrounds, making this book a truly international effort. This book will bring forth new frontiers with its revolutionizing research information and detailed analysis of the nascent developments around the world.

We would like to thank all the contributing authors for lending their expertise to make the book truly unique. They have played a crucial role in the development of this book. Without their invaluable contributions this book wouldn't have been possible. They have made vital efforts to compile up to date information on the varied aspects of this subject to make this book a valuable addition to the collection of many professionals and students.

This book was conceptualized with the vision of imparting up-to-date information and advanced data in this field. To ensure the same, a matchless editorial board was set up. Every individual on the board went through rigorous rounds of assessment to prove their worth. After which they invested a large part of their time researching and compiling the most relevant data for our readers.

The editorial board has been involved in producing this book since its inception. They have spent rigorous hours researching and exploring the diverse topics which have resulted in the successful publishing of this book. They have passed on their knowledge of decades through this book. To expedite this challenging task, the publisher supported the team at every step. A small team of assistant editors was also appointed to further simplify the editing procedure and attain best results for the readers.

Apart from the editorial board, the designing team has also invested a significant amount of their time in understanding the subject and creating the most relevant covers. They scrutinized every image to scout for the most suitable representation of the subject and create an appropriate cover for the book.

The publishing team has been an ardent support to the editorial, designing and production team. Their endless efforts to recruit the best for this project, has resulted in the accomplishment of this book. They are a veteran in the field of academics and their pool of knowledge is as vast as their experience in printing. Their expertise and guidance has proved useful at every step. Their uncompromising quality standards have made this book an exceptional effort. Their encouragement from time to time has been an inspiration for everyone.

The publisher and the editorial board hope that this book will prove to be a valuable piece of knowledge for researchers, students, practitioners and scholars across the globe.

LIST OF CONTRIBUTORS

Tao Cheng
CAS Key Laboratory of Bio-based Materials, Qingdao Institute of Bioenergy and Bioprocess Technology, Chinese Academy of Sciences, No. 189 Songling Road, Laoshan District, Qingdao 266101, China
State Key Laboratory Base of Eco-Chemical Engineering, College of Chemistry and Molecular Engineering, Qingdao University of Science and Technology, Qingdao 266042, China

Hui Liu, Guang Zhao and Mo Xian
CAS Key Laboratory of Bio-based Materials, Qingdao Institute of Bioenergy and Bioprocess Technology, Chinese Academy of Sciences, No. 189 Songling Road, Laoshan District, Qingdao 266101, China

Huibin Zou
CAS Key Laboratory of Bio-based Materials, Qingdao Institute of Bioenergy and Bioprocess Technology, Chinese Academy of Sciences, No. 189 Songling Road, Laoshan District, Qingdao 266101, China
College of Chemical Engineering, Qingdao University of Science and Technology, Qingdao 266042, China

Ningning Chen and Mengxun Shi
College of Chemical Engineering, Qingdao University of Science and Technology, Qingdao 266042, China

Congxia Xie
State Key Laboratory Base of Eco-Chemical Engineering, College of Chemistry and Molecular Engineering, Qingdao University of Science and Technology, Qingdao 266042, China

Wentao Ding and Huifeng Jiang
Key Laboratory of Systems Microbial Biotechnology, Tianjin Institute of Industrial Biotechnology, Chinese Academy of Sciences, Tianjin, China

Xiaonan Liu
Key Laboratory of Systems Microbial Biotechnology, Tianjin Institute of Industrial Biotechnology, Chinese Academy of Sciences, Tianjin, China
University of Chinese Academy of Sciences, Beijing, China

Mario Carrasco, Jennifer Alcaíno, Víctor Cifuentes and Marcelo Baeza
Departamento de Ciencias Ecológicas, Facultad de Ciencias, Universidad de Chile, Las Palmeras 342, Casilla 653, Santiago, Chile

Xuhua Mo, Chunrong Shi, Yanjiao Zhang and Qingji Wang
Shandong Key Laboratory of Applied Mycology, School of Life Sciences, Qingdao Agricultural University, Qingdao 266109, China

Chun Gui and Jianhua Ju
CAS Key Laboratory of Tropical Marine Bio-resources and Ecology, Guangdong Key Laboratory of Marine Materia Medica, RNAM Center for Marine Microbiology, South China Sea Institute of Oceanology, Chinese Academy of Sciences, 164 West Xingang Rd., Guangzhou 510301, China

Qiuwei Zhao, Yin Li and Yanping Zhang
CAS Key Laboratory of Microbial Physiological and Metabolic Engineering, Institute of Microbiology, Chinese Academy of Sciences, No. 1 West Beichen Road Chaoyang District, Beijing 100101, China

Chunhua Zhao
CAS Key Laboratory of Microbial Physiological and Metabolic Engineering, Institute of Microbiology, Chinese Academy of Sciences, No. 1 West Beichen Road Chaoyang District, Beijing 100101, China
University of Chinese Academy of Sciences, Beijing 100049, China

Lijun Ye, Qingyan Li, Xueli Zhang and Changhao Bi
Tianjin Institute of Industrial Biotechnology, Chinese Academy of Sciences, Tianjin 300308, People's Republic of China
Key Laboratory of Systems Microbial Biotechnology, Chinese Academy of Sciences, Tianjin 300308, People's Republic of China

Ping He
Tianjin Institute of Industrial Biotechnology, Chinese Academy of Sciences, Tianjin 300308, People's Republic of China

Key Laboratory of Systems Microbial Biotechnology, Chinese Academy of Sciences, Tianjin 300308, People's Republic of China
School of Pharmacy, East China University of Science and Technology, Shanghai 200237, People's Republic of China

Toni Paasikallio, Anne Huuskonen and Marilyn G. Wiebe
VTT Technical Research Centre of Finland Ltd., P.O. Box 1000, 02044 Espoo, Finland

Shuting Xiong, Ying Wang, Mingdong Yao, Hong Liu, Xiao Zhou, Wenhai Xiao and Yingjin Yuan
Key Laboratory of Systems Bioengineering (Ministry of Education), Tianjin University, No. 92, Weijin Road, Nankai District, Tianjin 300072, People's Republic of China
SynBio Research Platform, Collaborative Innovation Center of Chemical Science and Engineering (Tianjin), School of Chemical Engineering and Technology, Tianjin University, Tianjin 300072, People's Republic of China

Shogo Yoshimoto, Yuki Ohara, Hajime Nakatani and Katsutoshi Hori
Department of Biomolecular Engineering, Graduate School of Engineering, Nagoya University, Furo-cho, Chikusa-ku, Nagoya 464-8603, Japan

Liang-Bin Xiong, Hao-Hao Liu, Li-Qin Xu, Wan-Ju Sun, Feng-Qing Wang and Dong-Zhi Wei
Liang-Bin Xiong and Hao-Hao Liu contributed equally to this work State Key Laboratory of Bioreactor Engineering, Newworld Institute of Biotechnology, East China University of Science and Technology, Shanghai 200237, China

Su Ma, Marita Preims, Daniel Kracher and Roland Ludwig
Department of Food Sciences and Technology, Vienna Institute of Biotechnology, BOKU-University of Natural Resources and Life Sciences, Vienna, Austria

François Piumi
UMR BDR, INRA, ENVA, Université Paris Saclay, 78350 Jouy en Josas, France

Lisa Kappel and Bernhard Seiboth
Research Area Biochemical Technology, Institute of Chemical Engineering, TU Wien, Gumpendorferstrasse 1a, Vienna, Austria

Eric Record
Aix Marseille Université, INRA, BBF (Biodiversité et Biotechnologie Fongiques), Marseille, France

Ida Lauritsen, Andreas Porse, Morten O. A. Sommer and Morten H. H. Nørholm
Ida Lauritsen and Andreas Porse contributed equally to this work Novo Nordisk Foundation Center for Biosustainability, Technical University of Denmark, 2800 Kongens Lyngby, Denmark

Xin-jing Yue, Xiao-wen Cui, Zheng Zhang, Ran Peng, Peng Zhang, Zhi-feng Li and Yue-zhong Li
State Key Laboratory of Microbial Technology, School of Life Science, Shandong University, Jinan 250100, China

Lei Yang, Jesper Vang, Mette Lübeck and Peter Stephensen Lübeck
Section for Sustainable Biotechnology, Department of Chemistry and Bioscience, Aalborg University Copenhagen, A. C. Meyers Vænge 15, 2450 Copenhagen SV, Denmark

Eleni Christakou
Section for Biotechnology, Aalborg University Copenhagen, Fredrik Bajers Vej 7H, 9220 Aalborg, Denmark

Yongping Xin, Tingting Guo, Yingli Mu and Jian Kong
State Key Laboratory of Microbial Technology, Shandong University, 27 Shanda Nanlu, Jinan 250100, People's Republic of China

Denis Prudencio Luiz, Nilson Nicolau-Junior and Carlos Ueira-Vieira
Genetics Laboratory, Institute of Genetics and Biochemistry, Federal University of Uberlândia, 1720 Pará, Uberlândia, MG 38400902, Brazil

Juliana Franco Almeida and Luiz Ricardo Goulart
Nanobiotechnology Laboratory, Institute of Genetics and Biochemistry, Federal University of Uberlândia, 1720 Pará, Uberlândia, MG 38400902, Brazil

Judith E. Farnberger and Nina Richter
Austrian Centre of Industrial Biotechnology, ACIB GmbH, c/o University of Graz, Heinrichstrasse 28, 8010 Graz, Austria

Elisabeth Lorenz and Volker F. Wendisch
Genetics of Prokaryotes, Faculty of Biology & CeBiTec, Bielefeld University, 33501 Bielefeld, Germany

Wolfgang Kroutil
Austrian Centre of Industrial Biotechnology, ACIB GmbH, c/o University of Graz, Heinrichstrasse 28, 8010 Graz, Austria

Institute of Chemistry, University of Graz, NAWI Graz, BioTechMed Graz, Heinrichstrasse 28, 8010 Graz, Austria

Roshan P. Shrestha and Mark Hildebrand
Scripps Institution of Oceanography, University of California San Diego, La Jolla, CA, USA

Index

www.ingramcontent.com/pod-product-compliance
Lightning Source LLC
Chambersburg PA
CBHW082024190326
41458CB00010B/3270